Rainer Burkard   Gerhard Woegi

T0250603

Algorithms – ESA '97

5th Annual European Symposium
Graz, Austria, September 15-17, 1997
Proceedings

Springer

Series Editors

Gerhard Goos, Karlsruhe University, Germany

Juris Hartmanis, Cornell University, NY, USA

Jan van Leeuwen, Utrecht University, The Netherlands

Volume Editors

Rainer Burkard
Gerhard Woeginger
Technical University of Graz, Institute of Mathematics
Steyrergasse 30, A-8010 Graz, Austria
E-mail: (burkard/gwoegi)@opt.math.tu-graz.ac.at

Cataloging-in-Publication data applied for

Die Deutsche Bibliothek - CIP-Einheitsaufnahme

Algorithms : 5th annual European symposium ; proceedings / ESA
'97, Graz, Austria, September 15 - 17, 1997. Rainer Burkhard ;
Gerhard Woeginger (ed.). - Berlin ; Heidelberg ; New York ;
Barcelona ; Budapest ; Hong Kong ; London ; Milan ; Paris ; Santa
Clara ; Singapore ; Tokyo : Springer, 1997
    (Lecture notes in computer science ; Vol. 1284)
    ISBN 3-540-63397-9

CR Subject Classification (1991): F.2, G.1-2, I.3.5, C.2, E.1, I.7.3

ISSN 0302-9743
ISBN 3-540-63397-9 Springer-Verlag Berlin Heidelberg New York

© Springer-Verlag Berlin Heidelberg 1997
Printed in Germany

Typesetting: Camera-ready by author
SPIN 10547703   06/3142 – 5 4 3 2 1 0   Printed on acid-free paper

Springer

Berlin
Heidelberg
New York
Barcelona
Budapest
Hong Kong
London
Milan
Paris
Santa Clara
Singapore
Tokyo

Lecture Notes in Computer Science   1284

Edited by G. Goos, J. Hartmanis and J. van Leeuwen

Advisory Board: W. Brauer   D. Gries   J. Stoer

Preface

The 5th European Symposium on Algorithms (ESA'97) was held September 15–17, 1997, at Graz University of Technology. This volume contains all contributed papers presented at the symposium.

ESA was established in 1993 as a major annual event for all researchers interested in algorithms, theoretical as well as applied. This international symposium covers research in the use, design, and analysis of algorithms as it is carried out in computer science, discrete applied mathematics and mathematical programming, with emphasis on questions of efficiency. Papers have been solicited describing results in all fields of algorithm research, such as approximation algorithms, combinatorial optimization, computational biology, computational geometry, data compression, distributed computing, evolutionary algorithms, graph and network algorithms, neural computing, number theory and computer algebra, on-line algorithms, parallel computing, pattern matching, symbolic computation, and so on. ESA'93 took place in Germany (Bad Honnef), ESA'94 was held in The Netherlands (Utrecht), ESA'95 in Greece (Corfu), and ESA'96 in Spain (Barcelona). The proceedings of these meetings appeared, respectively, as Volumes 726, 855, 979, and 1136 in the Springer series Lecture Notes in Computer Science.

In response to the call for papers for ESA'97, the program committee received 112 submissions, indicating a strong and growing interest in the symposium. The program committee for ESA'97 consisted of

Giorgio Ausiello (Roma)	Rolf Möhring (Berlin)
Rainer E. Burkard (Graz)	Pekka Orponen (Jyväskylä)
Josep Díaz (Barcelona)	Christos Papadimitriou (Berkeley)
Amos Fiat (Tel Aviv)	Bill Pulleyblank (Yorktown Heights)
Mark Jerrum (Edinburgh)	Zsolt Tuza (Budapest)
Jan Karel Lenstra (Eindhoven)	Gerhard J. Woeginger (Graz)
Kurt Mehlhorn (Saarbrücken)	

The program committee met on May 9 and May 10, 1997, in Graz and selected 38 contributed papers for inclusion in the scientific program of ESA'97. The selection was based on originality, quality, and relevance to the study of algorithms, and reflects many of the current directions in algorithm research. This was most probably the first meeting of a program committee in modern science that involved the rolling of a dice in combination with the usage of a compass, a voodoo doll, a pair of almost perfectly parallel mirrors, and a map of the southwestern part of the Kalahari. We wish to thank all those who submitted extended abstracts for consideration, and all subreferees and colleagues who helped in the extensive evaluation process. The list of the referees is as complete as we could make it, and we apologize for any possible omissions or errors.

Many thanks are due to the organizing committee of ESA'97 for making the symposium possible. The organizing committee consisted of Rainer Burkard, Karin Hirschbichler, Bettina Klinz, Hermann Maurer, Angela Kompöck-Poller, Franz Rendl, Günter Rote, Klaus Schmaranz, and Gerhard Woeginger.

ESA'97 was conducted in cooperation with the European Association for Theoretical Computer Science (EATCS) and with the Mathematical Programming Society (MPS), and it was sponsored by the Austrian Ministry of Science, by Graz University of Technology, by the Province of Styria, and by the City of Graz.

Graz, September 1997

Rainer E. Burkard
Gerhard J. Woeginger

List of Referees

K. Aardal
I. Adan
O. Aichholzer
S. Albers
P. Alimonti
N. Alon
H. Alt
N. Ascheuer
P. Auer
F. Aurenhammer
G. Ausiello
Y. Azar
L. Babel
G. Bacsó
R. Baeza-Yates
L. Becchetti
Th.C. Biedl
H. Bodlaender
T. Böhme
M.L. Bonet
F.J. Brandenburg
P. Brass
R.E. Burkard
T. Calamoneri
E. Cela
A.R. Conn
P.L. Crescenzi
F. d'Amore
J. de Jong
J. Díaz
M. Di Ianni
M. Dyer
A. Efrat
P.L. Erdős
U. Faigle
S. Fekete
S. Felsner
A. Fiat
P. Franciosa
B. Gerards
R. Giaccio
S. Gilmore
M.X. Goemans
J. Gustedt

E. Győri
J. Harant
S. Hartmann
M. Henzinger
E. Hexel
M. Hirvensalo
W. Hochstättler
F. Hoffmann
M. Hoffmann
H. Hoogeveen
M. Hujter
C. Hurkens
F. Hurtado
C. Iliopoulos
M. Jerrum
M. Joswig
S. Juvaste
B. Kalyanasundaram
D. Karger
J. Karhumäki
G.Y. Katona
M. Kaufmann
W. Kern
G. Kindervater
J. Kivinen
P. Kleinschmidt
B. Klinz
T. Kloks
E. Köhler
J. Komlós
U.H. Kortenkamp
E. Koutsoupias
D. Kratsch
K. Kriegel
J. Lagarias
A. Lenstra
J.K. Lenstra
S. Leonardi
V. Leppänen
J.M.Y. Leung
J. Levy
G. Liotta
M. Lübbecke
H. Mannila

A. Marchetti Spaccamela

C. Martinez

E.W. Mayr

M. Mendel

K. Mehlhorn

R. Möhring

H. Müller

M. Müller-Hannemann

U. Nanni

S. Naor

T. Nishizeki

O. Nurmi

P. Orponen

C. Papadimitriou

M. Penttonen

U. Pferschy

A. Piperno

K. Pruhs

B. Pulleyblank

J. Rambau

P. Ramos

F. Rendl

L. Rónyai

K. Roos

G. Rote

R. Rudolf

M. Ruszinkó

P. Scheffler

B. Schieber

I. Schiermeyer

A. Schrijver

R. Schultz

A.S. Schulz

A. Sebő

M.J. Serna

M. Sharir

D.B. Shmoys

M. Skutella

C. Stein

M. Stiebitz

F. Stork

L. Stougie

J.A. Telle

L. Trevisan

Zs. Tuza

E. Ukkonen

E. Upfal

J. van de Klundert

S. van Hoesel

A.P.A. Vestjens

M. Vicsek

L. Viennot

M. Voigt

D. Wagner

K. Weihe

R. Weismantel

G. Weiss

R. Wille

U. Wille

G.J. Woeginger

M. Zachariasen

Contents

Scheduling Independent Multiprocessor Tasks

A. K. Amoura[1], E. Bampis[2], C. Kenyon[3], and Y. Manoussakis[1]

[1] Université Paris Sud, LRI, Bâtiment 490,
91405 Orsay Cedex, France
[2] Université d'Evry, LaMI, Bvd des Coquibus,
91025 Evry Cedex, France
[3] ENS Lyon, LIP, URA CNRS 1398, 46 Allée d'Italie,
69364 Lyon Cedex 07, France

Abstract. We study the problem of scheduling a set of n independent multiprocessor tasks with prespecified processor allocations on a fixed number of processors. We propose a linear time algorithm that finds a schedule of minimum makespan in the preemptive model, and a linear time approximation algorithm that finds a schedule of length within a factor of $(1 + \epsilon)$ of optimal in the non-preemptive model.

1 Introduction

A scheduling problem is usually given by a set T of n tasks, with an associated partial order which captures data dependencies between tasks, and a set P_m of m target processors. The goal is to assign tasks to processors and time steps so as to minimize an optimality criterion, for instance the makespan, *i.e.* the maximum completion time $C_{\max}$ of any task. Depending on the model, tasks can be preempted or not. In the *non-preemptive* model, a task once started has to be processed (until completion) without interruption. In the *preemptive* model, each task can be at no cost interrupted at any time and completed later. If there are no precedence constraints, the tasks are said to be *independent*.

In the most classic model, each task is processed by only one processor at a time. However, as a consequence of the development of parallel computers, new theoretical scheduling problems have emerged to model scheduling on parallel architectures. The *multiprocessor task system* is one such model[1] [7]. In this model, a task may need to be processed simultaneously by several processors. In the *dedicated* variant of this model, each task requires the simultaneous use of a prespecified set of processors. Since each processor can process at most one task at a time, tasks that share at least one resource cannot be scheduled at the same time step and are said to be *incompatible*. Hence, tasks are subject to compatibility constraints.

The present paper is concerned with *scheduling independent multiprocessor tasks on dedicated processors*, in both the preemptive case and the non-preemptive case, and the goal is to minimize the makespan $C_{\max}$. Following the

---

[1] See [6] for some practical justifications for considering multiprocessor task systems, and a survey on the main results obtained in multiprocessor task scheduling.

notations of [11,20], we will denote the problem by $P_m|\text{fix}_j, \text{pmtn}|C_{\max}$ in the preemptive case, and $P_m|\text{fix}_j|C_{\max}$ in the non-preemptive case.

The preemptive problem. In [18], Kubale proved that the preemptive problem with an unbounded number of processors, $P_\infty|\text{fix}_j, \text{pmtn}|C_{\max}$, is strongly NP-hard even when $|\text{fix}_j| = 2$, (*i.e.* when each task requires the simultaneous use of at most 2 processors). He used a reduction from the edge multi-coloring problem. However, he also showed that the preemptive problem with a fixed number m of processors, $P_m|\text{fix}_j, \text{pmtn}|C_{\max}$, can be solved by linear programming under the restriction that $|\text{fix}_j| = 2$. On the other hand, Bianco et al. showed that if m is at most 4, then the preemptive problem $P_m|\text{fix}_j, \text{pmtn}|C_{\max}$ can be solved in linear time [1]. Recently, Krämer [16] showed that $P_m|\text{fix}_j, \text{pmtn}, r_j|L_{\max}$, which is a generalization of $P_m|\text{fix}_j, \text{pmtn}|C_{\max}$, can be solved by linear programming. In fact, he solves $O(\log n)$ linear programs each one with $O(n)$ constraints and $O(n^{m+1})$ variables.

In this paper, we also use a linear programming formulation to solve the preemptive problem with a fixed number m processors $P_m|\text{fix}_j, \text{pmtn}|C_{\max}$, in *linear time*, thereby improving the results of [16,18,1]. Our linear program is inspired from Kenyon and Remila's approximation scheme for strip-packing [14].

The non-preemptive problem. The non-preemptive three-processor problem, i.e. $P_3|\text{fix}_j|C_{\max}$, has been extensively studied. In [2,13], it is proved to be strongly NP-hard by a reduction from 3-partition. The best approximation algorithm is due to Goemans [9] who proposed a $\frac{7}{6}$-algorithm improving the previous best performance guarantee of $\frac{5}{4}$ [5]. For unit execution time tasks, *i.e.* $P_m|\text{fix}_j, p_j = 1|C_{max}$, the problem is solvable in polynomial time through an integer programming formulation with a fixed number of variables [13]. However, if the number of processors is part of the problem, *i.e.* $P|\text{fix}_j, p_j = 1|C_{\max}$, the problem becomes NP-hard. Furhermore, Hoogeveen et al. [13] showed that, for $P|\text{fix}_j, p_j = 1|C_{\max}$, there exists no polynomial approximation algorithm with performance ratio smaller than 4/3, unless $P = NP$.

In this paper, we use the linear programming formulation of the preemptive problem to get a polynomial time approximation scheme for the non-preemptive problem $P_m|\text{fix}_j|C_{\max}$ for fixed m. The rough idea is that it does not matter much whether short tasks are preemptable or not; hence the tasks are separated into long tasks and short tasks, and long tasks are dealt with separately by trying essentially every possible arrangement. This is similar in spirit to Hall and Shmoys' polynomial time approximation scheme for single-machine scheduling [12]. Note that the running time, although linear in terms of n, is super-exponential in terms of $1/\epsilon$.

2 The algorithm in the preemptive model

Given any schedule of the tasks, if we take a snapshot at instant t, we see a certain set of tasks, which are all being processed at t. Thus we are led to consider configurations, which correspond to collections of tasks which can be processed simultaneously. With a linear programming approach, our algorithm finds partial schedules associated to the configurations. A global schedule is then produced by simply merging the obtained partial schedules.

We first need a few definitions. The *type of a task* is its associated set of required processors, which is a non-empty subset of $\{1, 2, \ldots, m\}$. In general, a *type* τ is just a subset of $\{1, 2, \ldots, m\}$. We say that types τ and τ' are *compatible* if $\tau \cap \tau' = \emptyset$. We define a *configuration* to be a maximum set of compatible task types. *I.e.*, a configuration is a collection $\mathcal{C}$ of non-empty types $\tau_i \subseteq \{1, \ldots, m\}$ such that:

1. $\forall \tau_i, \tau_j \in \mathcal{C}, \ \tau_i \cap \tau_j = \emptyset$, *i.e.* types of $\mathcal{C}$ are pairwise compatible.
2. $\bigcup_{\tau_i \in \mathcal{C}} \tau_i = \{1, \ldots, m\}$, *i.e.* $\mathcal{C}$ is maximum.

Let $\#\mathcal{C}$ denote the total number of configurations, determined by the number m of processors.

Given: A fixed number m of processors.
Input: A set T of n multiprocessor dedicated preemptive tasks. Each task i is defined by its type and its processing time d_i.
Output: An optimal schedule.

The algorithm is in three stages:

1. [**Set up the Linear Program**]. We partition the tasks of T, according to their types, into subsets

$$T^\tau = \{i \in T \mid \text{task } i \text{ has type } \tau\},$$

one for each possible task type. (We have $T = \bigcup_{\tau \subseteq \{1, \ldots, m\}} T^\tau$.) We define D^τ as the total processing time of all the tasks of type τ, *i.e.*

$$D^\tau = \sum_{i \in T^\tau} d_i.$$

Let us assign a variable x_i to each configuration $\mathcal{C}_i$, whose interpretation will be the length of the time interval during which, if we take a snapshot of the schedule during that interval, the types that we see form configuration $\mathcal{C}_i$. The problem can now be formulated by the following linear program.

$$\text{minimize} \sum_{i=1}^{\#\mathcal{C}} x_i \tag{1}$$

$$\text{subject to: } \forall \tau \subseteq \{1, \ldots, m\} : \sum_{i \text{ s.t. } \tau \in \mathcal{C}_i} x_i \geq D^\tau \tag{2}$$

The optimization criterion (1) means that the makespan must have minimum length subject to constraints (2) which guarantee that all the tasks are completed.

2. **[Solve the Linear Program].** Let $(x_1^*, x_2^*, \ldots, x_{\#C}^*)$ denote the solution to the linear program.

3. **[Deducing a preemptive schedule].** A feasible optimal schedule can be constructed in a greedy way as follows. For each type τ, at any stage of the construction, some tasks of T^τ have already been completely scheduled, one task of T^τ may have been partially scheduled and preempted before completion, and the remaining tasks of T^τ have not yet been scheduled at all.

   Take the configurations C_i such that $x_i^* \neq 0$, in some arbitrary order. For each such configuration C_i, we construct a partial schedule of length x_i^* as follows. Consider all the types $\tau \in C_i$. For every such type τ, select first the partially scheduled task of T^τ (if there is one), then the remaining tasks of T^τ in arbitrary order, until the total processing time of the selected tasks becomes greater than or equal to x_i^* (or until all tasks of T^τ have been scheduled). If the total processing time is strictly greater than x_i^*, then the last selected task is preempted so that the total processing time is exactly x_i^*.

3 Analysis of the preemptive algorithm

Lemma 1. *The number of configurations is equal to $B(m)$, the mth Bell number.*

Proof. The number of configurations C having k elements, $1 \leq k \leq m$, is equal to the number of ways to partition a set of m elements into k nonempty disjoint subsets. But this is exactly the Stirling number of the second kind $S(m, k) = \{{k \atop m}\}$ [15]. Therefore, the number of all possible configurations is given by $B(m) = \sum_{k=1}^{m} S(m, k)$ (which corresponds to the mth Bell (or exponential) number [10]). $\square$

Remark 2. Using the fact that $S(m, k) = kS(m - 1, k) + S(m - 1, k - 1)$ (see [15] p.73) we get the crude upper bound $B(m) \leq m!$, hence the number of configurations is a constant (as the number of processors m is fixed) independent of the number n of tasks.

Theorem 3. $P_m|fix_j, pmtn|C_{\max}$ *can be solved in time $O(n)$.*

Proof. Correctness.

From the definition of a configuration, we see that within the same configuration tasks with different types are compatible and thus can be scheduled in parallel while tasks with the same type have to be scheduled consecutively. Hence, the obtained schedule is feasible. Furthermore, since tasks can be preempted, it is obvious that the above linear program gives the length $\sum_{i=1}^{\#C} x_i^*$ of the optimal schedule.

Running time.

Step 1: The list of tasks is scanned once, and so this step takes time $O(n)$.

Step 2: Since there are $(2^m - 1)$ types of tasks and $B(m)$ configurations, the linear program has $(2^m - 1)$ constraints and $\mathcal{B}(m)$ variables. Both values are constants and so the linear program can be solved in time $O(1)$.

Step 3: Each list T^τ is scanned once, and so this step takes time $O(n)$.

Overall the algorithm runs in time $O(n)$. □

4 The algorithm in the non-preemptive model

We say that a task is *preemptable* if it can be at no cost interrupted at any time and completed later. We will be led to consider mixed models where some tasks are preemptable and others not.

A simple approach to obtain a feasible non-preemptive schedule from a preemptive schedule S is to remove the preempted tasks from S and process them sequentially in a naïve way at the end of S. The produced schedule has a makespan equal to the makespan of S plus the total processing time of the delayed tasks. However, even if S is optimal preemptive, the schedule thus produced is almost certainly not close to optimal, since there can be a large number of possibly long preempted tasks.

Our algorithm does use a preemptive schedule to construct a non-preemptive schedule in the way just described. However, to ensure that the non preemptive solution is close to optimal, only "short" tasks are allowed to be preempted. The intuitive idea of the algorithm is the following: first, partition the set of tasks into "long" tasks and "short" tasks. For each possible schedule of the long tasks, complete it into a schedule of T, **assuming that the short tasks are preemptable**. From each such preemptive schedule construct a non-preemptive schedule. Finally, among all the schedules thus generated, choose the one that has the minimum makespan.

In what follows, for simplicity we give the algorithm for the three-processor case. Our approach can be directly generalized to deal with the m-processor case for any constant m.

We first need a few definitions. Given a set of tasks L, a *snapshot* of L is a set of compatible tasks of L. A *relative schedule* of L is a sequence of snapshots of L such that each task occurs in a subsequence of consecutive snapshots, and consecutive snapshots are different. Roughly speaking, each relative schedule corresponds to an order of processing of the long tasks. Note that to any given non-preemptive schedule of L, one can associate a relative schedule in a natural way by looking at the schedule at every instant where a task of L starts or ends, and writing the set of tasks of L being processed right after that transition. Also note that the last snapshot is then the empty set.

Recall that in section 2, we defined types and configurations. We now also need to define τ-configurations. Given a type τ, a τ-configuration is a collection of types $\tau_i \subseteq \tau$ such that:

1. $\forall \tau_i, \tau_j \in \mathcal{C}_\tau, \quad \tau_i \cap \tau_j = \emptyset$, *i.e.* types of $\mathcal{C}_\tau$ are pairwise compatible.
2. $\bigcup_{\tau_i \in \mathcal{C}} \tau_i = \tau$.

Let $\#\mathcal{C}_\tau$ denote the number of τ-configurations.

Let D denotes the total processing time of all the tasks, *i.e.* $D = \sum_{i=1}^{n} d_i$. The algorithm is in six stages.

Given: Three processors.

Input: A set T of n multiprocessor dedicated non-preemptive tasks, each task i being defined by its type and its processing time d_i; and a constant ϵ such that $0 < \epsilon < 1$.

Output: A schedule with makespan $C_{\max} \leq (1 + \epsilon) C_{\max}^{opt}$, where $C_{\max}^{opt}$ is the optimal makespan.

1. **[Partition tasks according to lengths].** Sort the $7^{1/\epsilon}$ longest tasks, of lengths $d_1 \geq \ldots \geq d_{7^{1/\epsilon}}$. Find $i \leq 1/\epsilon$ such that $d_{7^i} + \ldots + d_{7^{i+1}-1} < \epsilon$. Let $k = 7^i$. Perform the partition $T = T' \cup L$ where L is the set of tasks with the k longest execution times.

2. **[Construct the relative schedules].** List all possible relative schedules $M = (M(0), M(2), \ldots, M(f - 1))$ of L.

3. **[Set up linear programs.]** During this stage of the algorithm, we assume that the tasks of T' are preemptive. For each relative schedule M of L, we determine an optimal schedule for the tasks of T such that the associated relative schedule is M; this is done using the following linear program.

   Variables. There are three types of variables. First, there is one variable t_i for each snapshot of M; it represents the instant in the schedule corresponding to the snapshot $M(i)$. Thus t_{f-1} represents the time at which the last task of L terminates its execution. In addition, a variable t_f represents the time at which the last task of T terminates its execution. Second, there is one variable e_τ for each type τ, such that e_τ represents the total time during which the processors of $\{1, 2, 3\} \setminus \tau$ are processing tasks of L and the processors of τ are not processing tasks of L, *i.e.* processors of τ are either processing tasks of T' or idle. Third, for each type τ and each τ-configuration $C_i(\tau)$, we define a variable $x_i(\tau)$. These variables play the same role as the variables x_i of section 2, *i.e.* they help determine the schedule during the e_τ amount of time during which the processors of τ are available to process tasks of T'.

   Constraints. The constraints are the following:

   (a)

   $$\begin{cases} \text{for every } i \in \{1, \ldots, f\}: \ t_i > t_{i-1}, \\ \text{if } t_j = \text{finish time of task } r \text{ started at } t_i, \text{ then } t_j - t_i = d_r \end{cases} \tag{3}$$

   These constraints mean that the t_i's, together with M, define a feasible schedule of L.[2]

---

[2] Note that there are many more t_i's than equality constraints in general.

(b) The variables e_τ, $\tau \subseteq \{1,2,3\}$, are determined from the t_i's by linear equations[3].

$$\forall \tau \quad e_\tau = \sum_{i \text{ s.t. } \{\text{processors used by } M(i)\}=\{1,2,3\}\setminus\tau} t_{i+1} - t_i. \tag{4}$$

(c) The total lengths of the τ-configurations must fit in the corresponding amount of available time e_τ. For every variable e_τ, we must have one constraint.

$$\forall \tau \quad \sum_{C_i(\tau)} x_i(\tau) \leq e_\tau \tag{5}$$

(d) We must guarantee that we have enough time for the execution of all the tasks of T', i.e., for each $\tau' \subseteq \{1,2,3\}$, we put in the constraint

$$\forall \tau' \quad \sum_{\tau \supseteq \tau'} \sum_{i \text{ s.t. } \tau' \in C_i(\tau)} x_i(\tau) \geq D^{\tau'}, \tag{6}$$

where $D^{\tau'}$ is the total processing time of all the tasks of T' of type τ'. The linear program's objective is to minimize t_f.

4. [**Solve the linear programs**]. For each relative schedule M, we solve the associated linear program, hence a schedule Schedule(M).

5. [**Deducing a non-preemptive schedule**]. We now no longer assume that the tasks of T' are preemptive. To construct a non-preemptive schedule, we proceed as follows: for each relative schedule M, we consider the set of tasks of T' which are preempted in Schedule(M). We modify Schedule(M) by executing these tasks at the end of the schedule in a sequential manner.

6. [**Compare all schedules**]. Among all the schedules produced in stage 5, output the one that has the smallest makespan.

5 Example for the non-preemptive case

Consider the task system T given in Table 1. We have $D = 90$. Assume in addition that k (used in step 1 of the algorithm) is equal to 3. Hence, $L = \{1,2,3\}$ and $T' = \{4,5,6,7,8,9\}$. A possible relative schedule of the tasks of L is given by the sequence of snapshots $M = (\{3,1,2\},\{1,2\},\{1\},\emptyset)$ (see Fig. 1 (a)).

We shall now construct the linear program corresponding to M. There are five variables t_0, t_1, t_2, t_3 and t_4. Let $\tau = 1$, $\tau' = 13$ and $\tau'' = 123$ denote the various sets of processors available at the various times to execute tasks of T'.

Variable e_1 corresponds to the time steps where processors 2 and 3 are processing some tasks of L while processor 1 remains idle or processes some tasks of T'. We define, using similar arguments to those of e_1, variables e_{13} and e_{123}

---

[3] Thus the variables e_τ are not really necessary - they can be viewed as a convenient notation.

Table 1. The task system

Task name	1	2	3	4	5	6	7	8	9
Processing time	25	24	20	6	5	4	3	2	1
Required processors	2	3	1	2,3	1,3	1,2	3	2	1

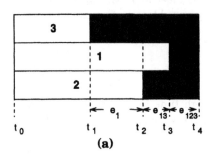

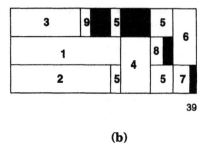

(a)   (b)

Fig. 1. *(a) A relative schedule of the long tasks (b) A preemptive schedule respecting the relative schedule of (a).*

(the other variables e_{τ_i} are all 0). Considering all possible partitions of τ, τ' and τ'', we have the variables $x_1(\tau)$, $x_{1,3}(\tau')$, $x_{13}(\tau')$, $x_{1,2,3}(\tau'')$, $x_{12,3}(\tau'')$, $x_{1,23}(\tau'')$, $x_{2,13}(\tau'')$, $x_{123}(\tau'')$.

The outline of the linear program is as follows.

minimize t_4 s.t.

$$t_0 = 0, \qquad t_1 - t_0 = 20, \qquad t_2 - t_0 = 24, \qquad t_3 - t_0 = 25$$

$$e_1 = t_2 - t_1 = 4$$
$$e_{13} = t_3 - t_2 = 1$$
$$e_{123} = t_4 - t_3$$

$$x_1(\tau) \le e_1$$
$$x_{1,3}(\tau') + x_{13}(\tau') \le e_{13}$$
$$x_{1,2,3}(\tau'') + x_{12,3}(\tau'') + x_{1,23}(\tau'') + x_{2,13}(\tau'') + x_{123}(\tau'') \le e_{123}$$

$$x_1(\tau) + x_{1,3}(\tau') + x_{1,2,3}(\tau'') + x_{1,23}(\tau'') \ge D^1 = d_9 = 1$$
$$x_{1,3}(\tau') + x_{1,2,3}(\tau'') + x_{12,3}(\tau'') \ge D^3 = d_7 = 3$$
$$x_{1,2,3}(\tau'') + x_{2,13}(\tau'') \ge D^2 = d_8 = 2$$
$$x_{13}(\tau') + x_{2,13}(\tau'') \ge D^{13} = d_5 = 5$$
$$x_{1,23}(\tau'') \ge D^{23} = d_4 = 6$$
$$x_{12,3}(\tau'') \ge D^{12} = d_6 = 4$$

Note that in this particular case the values of the variables t_1, t_2 and t_3 are fixed since tasks 1, 2 and 3 start at the same time as implied by the first snapshot $\{3, 1, 2\}$ of M (this is not true in general).

Figure 1 (b) represents the preemptive schedule given by the linear program in the stage 4 of the algorithm, and Fig.2 (a) gives the corresponding non preemptive schedule produced in stage 5. The optimal schedule is given in Fig.2 (b).

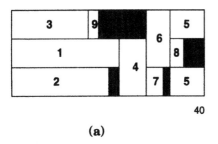
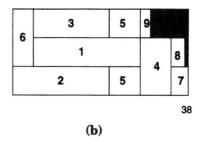

(a)   (b)

Fig. 2. *(a) A non-preemptive schedule given by the algorithm respecting the relative schedule of Fig.1.a (b) An optimal schedule.*

6   Analysis of the non-preemptive algorithm

6.1   Running time

Lemma 4. *The total number of relative schedules is at most $(k^3)^{2k}$.*

Proof. We create a new snapshot each time there is a new configuration with respect to L, *i.e.* when a task of L starts or ends its execution. Given that $|L| = k$, there are at most $2k$ such events and thus, the sequence contains at most $2k$ snapshots. Each snapshot consists of at most three tasks chosen among k possibilities, so there are at most k^3 possible snapshots. Consequently, there is a constant number, *i.e.* at most $(k^3)^{2k}$, of relative schedules for the tasks of L.   □

Lemma 5. *The number of variables of the linear program in stage 3 is $O(k)$.*

Proof. There are as many variables t_i as the number of snapshots of the considered relative schedule M, plus one. Hence, at most $2k + 1$ variables. In addition, there is one variable e_τ for each subset $\tau \subseteq \{1, 2, 3\}$, *i.e.* $O(1)$ variables. Finally, there is one variable of the third type, $x_i(\tau)$, for each τ-configuration, where $\tau \subseteq \{1, 2, 3\}$, hence $O(1)$ variables. The claim follows.   □

Lemma 6. *The number of constraints in the linear program of stage 3 is at most $O(k)$.*

Proof. There are at most as many equality constraints of type (3) as long tasks, *i.e. k*. Inequality constraints of type (3) are as many as variables t_i minus one, *i.e.* at most $2k$. There are only $O(1)$ other constraints. □

Proposition 7. *The algorithm presented in Section 4 has running time $O(n)$.*

Proof. A preprocessing phase is needed in order to determine the values of the total processing times of short tasks for each type τ, used in constraints (6). This phase must scan the list of tasks and thus takes $O(n)$ time. Then, it is sufficient to consider for each possible relative schedule the solution of the above linear program, and to choose the relative schedule that gives the schedule of minimum length. Given that the number of possible relative schedules is constant and the solution of the above linear program can be found in $O(1)$ time by the result of Lenstra [19], the result follows. □

6.2 Correctness

First of all notice that given a relative schedule M of the long tasks, the linear program of stage 3 of the algorithm gives the optimal schedule that respects M and where only short tasks are preemptable. Since at stage 3 of the algorithm all the possible relative schedules of the long tasks are considered, the following result is proved.

Lemma 8. *Under the assumption that only short tasks are preemptable, the schedule with the minimum length obtained in stage 4 of the algorithm is optimal.*
□

Thus the makespan of the best schedule obtained in stage 4 is a lower bound to the optimal solution of the non-preemptive problem. Now in order to show that the solution provided by the algorithm is close to the optimal schedule, let us assume that the sum D of the execution times of tasks equals 1. This assumption has no effect on the result since the problem is invariant by time scaling. In fact, given any problem instance T, we can multiply all the execution times by a factor of $1/\sum_{i \in T} d_i$ to create a new instance T_s, solve the problem on T_s, and then scale the schedule by $\sum_{i \in T} d_i$ to get a schedule of T.

The main delay is due to stage 5 of the algorithm. We assume that the execution times are sorted: $d_0 \geq d_2 \geq \ldots \geq d_{n-1}$. Recall that L is the set of the k longest tasks. The relative schedule (number of snapshots) has length at most $2k$, hence at most $m(2k) = 6k$ tasks of T' are preempted in the schedule of stage 4 and thus delayed to the end in stage 5. The delay thus incurred is at most $d_k + \ldots + d_{7k-1}$.

Choice of parameter k. we choose k so that $d_k + \ldots + d_{7k-1} < \epsilon$. This is possible while keeping $k = O(c^{1/\epsilon}) = O(1)$ as the following lemma shows.

Lemma 9. *Let $1 \geq d_0 \geq d_1 \geq \ldots \geq d_{n-1} \geq 0$ be a sequence of numbers such that $\sum_i d_i = 1$, and let $\epsilon > 0$. Then there exists $k \leq 7^{1/\epsilon}$ such that $d_k + d_{k+1} + \ldots + d_{7k-1} < \epsilon$.*

Proof. Decompose $d_0 + \ldots + d_{n-1}$ into blocks: $B_0 = d_0 + \ldots + d_6$, $B_1 = d_7 + \ldots + d_{7^2-1}, \ldots, B_i = d_{7^i} + \ldots + d_{7^{i+1}-1}$. Since the sum of all blocks equals 1, at most $1/\epsilon$ of them are greater than ϵ. Take the first block B_i which is less than ϵ. We have: $i \leq 1/\epsilon$ and for $k = 7^i$, $B_i = d_k + \ldots + d_{7k-1} < \epsilon$, hence the lemma. $\square$

Remark 10. If $n < 7^{1/\epsilon}$ then all the tasks may go into L and the algorithm finds the optimal solution in constant time.

All we have to do now is apply the algorithm setting $|L| = k \leq 7^{1/\epsilon}$, and we obtain a schedule of length at most $(1 + \epsilon)$ times the optimal length.

Theorem 11. *There is an algorithm A which given a set T of n independent multiprocessor tasks, a fixed number m of dedicated processors and a constant $\epsilon > 0$, produces, in time at most $O(n)$, a schedule of T whose makespan is at most $(1 + \epsilon)C_{\max}^{opt}$.*

7 Remarks

In this paper we proposed an optimal algorithm for scheduling n preemptable dedicated independent multiprocessor tasks on a fixed number of processors, and a polynomial time approximation scheme for the same problem when the tasks are non-preemptable. The approximation algorithm given is polynomial in the size of the problem but super-exponential in $1/\epsilon$. This result is the best performance guarantee one might hope for since no fully polynomial approximation scheme can be expected as the problem is strongly NP-hard [8]. Of course, it remains interesting to see whether there exists polynomial approximation schemes with lower time complexity in the number of processors m.

Note that we conjecture that our algorithm can be extended for the parallel variant of the model (non-dedicated case) as well (see [3] for related work).

Some natural questions arise. Do the linear programming ideas developed in this paper have applications to other variants of scheduling?

References

1. L. Bianco, J. Blazewicz, P. Dell'Olmo, and M. Drozdowski. *Scheduling Preemptive Multiprocessor Tasks on Dedicated Processors.* Performance Evaluation, **20**:361-371, (1994)
2. J. Blazewicz, P. Dell'Olmo, M. Drozdowski, and M.G. Speranza. *Scheduling Multiprocessor Tasks on Three Dedicated Processors.* Information Processing Letters, **41**:275-280, (1992)
3. J. Blazewicz, M. Drabowski, and J. Węglarz. *Scheduling multiprocessor tasks to minimize schedule length.* IEEE Transactions on Computing, **35**(5):389-393, (1986)
4. E. G. Coffman, M. R. Garey, D. S. Johnson, and A. S. Lapaugh. *Scheduling file transfers.* SIAM J. Compt., **14**(3):744-780, (1985)
5. P. Dell'Olmo, M. G. Speranza, and Z. Tuza. *Efficiency and Effectiveness of Normal Schedules on Three Dedicated Processors.* Submitted.

6. M. Drozdowski. *Scheduling Multiprocessor Tasks - An Overview.* Private communication, (1996)
7. J. Du, and J. Y-T. Leung. *Complexity of Scheduling Parallel Task Systems.* SIAM J. Discrete Math., 2(4):473-487, (1989)
8. M. Garey, D. Johnson. *Computers and Intractability, A Guide to the theory of NP-completeness.* W. H. Freemann and Company, San Francisco, CA, (1979)
9. M. X. Goemans. *An Approximation Algorithm for Scheduling on Three Dedicated Processors.* Disc. App. Math., 61:49-59, (1995)
10. R. L. Graham, D. E. Knuth, and O. Patashnik. *Concrete Mathematics.* Addison-Wesley, Reading, MA, (1990)
11. R. L. Graham, E. L. Lawler, J. K. Lenstra, and A. H. G. Rinnoy Kan. *Optimization and Approximation in Deterministic Sequencing and Scheduling: A Survey.* Ann. Disc. Math., 5:287-326, (1979)
12. L.A. Hall and D.B. Shmoys. *Jackson's rule for single-machine scheduling: making a good heuristic better.* Mathematics of Operations Research, 17:22-35, (1992)
13. J. Hoogeveen, S. L. van de Velde, and B. Veltman. *Complexity of Scheduling Multiprocessor Tasks with Prespecified Processors Allocation.* Disc. App. Math., 55:259-272, (1994)
14. C. Kenyon, E. Rémila. *Approximate Strip-Packing.* Proceedings of the 37th Symposium on Foundations of Computer Science (FOCS), 31-36, (1996)
15. D. E. Knuth. *The Art of Computer Programming,* Vol.1: *Fundamental Algorithms.* Addison-Wesley, Reading, MA, (1968)
16. A. Krämer. *Scheduling Multiprocessor Tasks on Dedicated Processors.* PhD Thesis, Fachbereich Mathematik/Informatik, Universität Osnabrück, (1995)
17. H. Krawczyk, and M. Kubale. *An approximation algorithm for diagnostic test scheduling in multicomputer systems.* IEEE Trans. Comput., C-34(9):869-872, (1985)
18. M. Kubale. *Preemptive Scheduling of Two-Processor Tasks on Dedicated Processors (in polish).* Zeszyty Naukowe Politechniki Śląskiej, Seria: Automatyka z.100, 1082:145-153, (1990)
19. H.W. Lenstra *Integer Programming with a Fixed Number of Variables.* Math. Oper. Res., 8:538-548, (1983)
20. B. Veltman, B. J. Lageweg, and J. K. Lenstra. *Multiprocessor Scheduling with Communication Delays.* Parallel Computing, 16:173-182, (1990)

On Local Search for Weighted k-Set Packing

Esther M. Arkin [*1] and Refael Hassin[2]

[1] Department of Applied Mathematics and Statistics
State University of New York
Stony Brook, NY 11794-3600, USA
E-mail: estie@ams.sunysb.edu
http://ams.sunysb.edu/~estie/estie.html
[2] Department of Statistics and Operations Research
School of Mathematical Sciences
Tel-Aviv University
Tel-Aviv 69978, Israel
E-mail: hassin@math.tau.ac.il
http://www.math.tau.ac.il/~hassin

Abstract. Given a collection of sets of cardinality at most k, with weights for each set, the maximum weighted packing problem is that of finding a collection of disjoint sets of maximum total weight. We study the worst case behavior of the t-local search heuristic for this problem proving a tight bound of $k - 1 + \frac{1}{t}$. This continues the work of Hurkens and Schrijver for unweighted packing problems.

1 Introduction

Maximum packing problems are among the most often studied in combinatorial optimization: Given a collection $X_1, \ldots, X_q$ of k-sets, find a largest collection of pairwise disjoint sets among them. One of the most fundamental packing problems is that of finding a maximum matching in a graph; this problem is polynomially solvable. However, many other packing problems are NP-hard, including maximum 3-dimensional matching, maximum triangle packing, maximum H matching, and maximum independent sets of axis parallel rectangles. See, for instance, [4] and [1].

One of the most natural heuristic algorithms for the maximum packing problem is that of local search. (For an analysis of this heuristic in the case of the related *minimum k-set cover* problem, see, [8].) Given a collection of disjoint sets constructed so far, check whether there are $p \le t$ sets that are not in the current collection that intersect at most $p - 1$ sets that are in it. If this happens, swap the sets to form a larger collection, and repeat; otherwise, terminate the search (the solution is *t-optimal*).

To analyze the performance of this heuristic for unweighted packing problems, Hurkens and Schrijver [9] presented the following remarkable theorem (see also, [7]):

---

* Partially supported by NSF Grant CCR-9504192

Theorem 1 *Let m, n, k and t be positive integers, $k \geq 3$. Let $E_1, \ldots, E_m$ be subsets of the set V of size n, such that we have the following:*

- *Each element of V is contained in at most k of the sets $E_1, \ldots, E_m$;*
- *For any $p \leq t$, the union of any p of the sets among $E_1, \ldots, E_m$ contains at least p elements of V.*

Then we have the following:

- $\frac{m}{n} \leq \frac{k(k-1)^r - k}{2(k-1)^r - k}$ *if $t=2r-1$;*
- $\frac{m}{n} \leq \frac{k(k-1)^r - 2}{2(k-1)^r - 2}$ *if $t=2r$.*

They also show that this bound is tight; namely, that for any fixed k and t, there exist m, n, and $E_1, \ldots, E_m$ subsets of V satisfying the two conditions, for which equality holds.

The theorem can be applied to bound the error ratio of local search algorithms for packing problems as follows. Let V be the index set of the sets in an optimal solution. Let E_j be the index set of the sets in a t-optimal solution which intersect X_j, $j \in V$. Then, the conditions of the theorem are satisfied and this proves that the right-hand side of the inequality bounds the error ratio $\frac{m}{n}$. The same tight bound was later obtained by Halldórsson, [7], who showed that it suffices to consider a restricted class of local improvements to obtain this ratio. This restricted class of local search may lead to more efficient algorithms to find an approximate solution. See also Khanna et al [10].

In this paper, we study the weighted version of the packing problem, in which each set is given a non-negative weight and the goal is to find a collection of disjoint sets of maximum total weight. We give the first analysis, with tight bounds, for the local search heuristic in weighted packing problems. We note that the stated bounds can be achieved (with an arbitrarily small deviation) in polynomial time [11].

We work within the following generalized setting. Consider a bipartite graph $G = (U, V, E_G)$ with color classes U and V and edge set E_G. For each node $i \in U \cup V$ let $E_i = \{j \in U \cup V | (i, j) \in E\}$ be the set of nodes adjacent to i and let $w_i \geq 0$ be a weight associated with it. For $X \subset U \cup V$ we denote $w(X) = \sum_{i \in X} w_i$. Suppose that

(i) $|E_v| \leq k$ for each $v \in V$;
(ii) Any subset $R \subseteq U$ of at most t nodes satisfies

$$w(R) \leq w(\cup_{u \in R} E_u). \tag{1}$$

The first assumption is analogous to the first condition of Theorem 1. We will later examine the effect of a similar restriction on the nodes in U. The second assumption refers to t-optimality in the weighted case and replaces the second condition of Theorem 1.

We are interested in bounding the ratio $\frac{w(U)}{w(V)}$. In Section 2 we show a trivial upper bound of k on this ratio. We then give a lower bound by presenting an example for which the ratio is $\frac{k}{t}(l - 1 + \frac{1}{t})$, where l is the maximum degree of a node in U. Adding the restriction that $k = l$ (which is the case in the application to local search algorithms for packing problems) yields a lower bound for this case of $k - 1 + \frac{1}{t}$. In Section 3 we present our main result, a matching upper bound for the $k = l$ case. Bafna, Narayanan and Ravi recently obtained the same bound for the case in which $t = k$, see [1]. Their proof is different and relies on the assumption that $t = k$. In Section 4 we conclude with a much simpler proof for the case of $k = l = 2$.

Feo, Goldschmidt and Khellaf considered the case in which a weighted graph is given and the weight of a subset of nodes is the sum of edge weights in the graph induced by it. Under this assumption they obtained a $\frac{1}{2}$ approximation for the cases of packing of node-disjoint 3-sets and 4-sets. For $k = 3$ this result matches ours while for $k = 4$ it is better. However, our framework is more general since the weights may be arbitrary. For example, our theorem also applies to packing of *edge-disjoint* sets of nodes (see Grable [6], for an analysis of algorithms for packing of edge-disjoint triangles).

A graph $G = (V, E)$ is *hypomatchable* if $|V|$ is odd and for every $i \in V$ there exists a matching that covers $V \setminus i$. Cornuéjols, Hartvigsen and Pulleyblank [3, 2] showed that the following special case is polynomially solvable: Each node has a weight, and the problem is to pack edges and subgraphs from a given family of hypomatchable subgraphs in order to maximize the weight of covered nodes.

We will use the following lemma from [9]:

Lemma 2 *For every k, l, and γ, there exists a bipartite graph of girth at least γ, with color classes U and V, say, such that each node in U has degree k, and each node in V has degree l.*

2 A bad example

We start with the following straightforward result:

Theorem 3 *For any given k such that $|E_v| \leq k$ $v \in V$ and for $t \geq 1$*

$$\frac{w(U)}{w(V)} \leq k.$$

Proof: By Inequality (1) for sets $R = \{u\}$, for each $u \in U$, $w_u \leq w(E_u)$. Summation over $u \in U$ gives that

$$w(U) \leq \sum_{u \in U} w(E_u) \leq kw(V).$$

∎

The next theorem shows that the bound is asymptotically achievable:

Theorem 4 *For any given k, l such that $|E_u| \leq l$ and $|E_v| \leq k$, and for $t \geq 1$, there is an instance in which*

$$\frac{w(U)}{w(V)} = \frac{k}{l}(l - 1 + \frac{1}{t}).$$

Proof: Consider a graph satisfying the conditions of Lemma 2 with $\gamma > 2t$. It follows that every p-set $R \subseteq U$ where $p \leq t$ has the following property:

$$| \cup_{u \in R} E_u | \geq lp - (p - 1). \tag{2}$$

Let

$$w_i = \begin{cases} l - 1 + \frac{1}{t} & i \in U \\ 1 & i \in V \end{cases}.$$

Condition (ii) is satisfied since by Inequality (2) for any $R \subseteq U$ such that $|R| = p \leq t$

$$w(R) = p(l - 1 + \frac{1}{t}) \leq | \cup_{u \in R} E_u | = w(\cup_{u \in R} E_u).$$

The theorem follows since

$$\frac{w(U)}{w(V)} = \frac{|U|(l - 1 + \frac{1}{t})}{|V|} = \frac{k}{l}(l - 1 + \frac{1}{t}).$$

$\blacksquare$

We note that in the applications to local search algorithms for packing problems, $k = l$. In the unweighted case analyzed by Hurkens and Schrijver, this property did not affect the bound and did not appear in the statement of Theorem 1. As we will see, in the more general case of unequal weights, this assumption leads to a better bound.

Corollary 5 *For any given k and t such that $|E_u| \leq k$ for each $u \in U$ and and $|E_v| \leq k$ for each $v \in V$, there exists an instance in which*

$$\frac{w(U)}{w(V)} = k - 1 + \frac{1}{t}.$$

Our goal in the next section is to show that there is a matching upper bound.

3   Main result

In this section we prove our main result:

Theorem 6 *For any given k and t and every instance $G = (U, V, E_G)$ satisfying*

(i) $|E_u| \leq k$ for each $u \in U$;
(ii) $|E_v| \leq k$ for each $v \in V$;
(iii) Any subset $R \subseteq U$ of at most t nodes satisfies $w(R) \leq w(\cup_{u \in R} E_u)$,

$$\frac{w(U)}{w(V)} \leq k - 1 + \frac{1}{t}.$$

3.1 Lemma on girth

Lemma 7 *Let γ be given, and suppose that there exists an instance satisfying the conditions of Theorem 6 such that $\frac{w(U)}{w(V)} = r$. Then, there exists, for the same k and t values, an instance satisfying the conditions of Theorem 6 such that $\frac{w(U)}{w(V)} = r$ and the girth of the graph is at least γ.*

Proof: Suppose an instance with ratio r consisting of a bipartite graph with node weights w is given. Suppose that the graph has girth g. We will show how to transform the instance to a new one with girth at least $g + 2$ and ratio r.

Given a graph $G = (U, V, E_G)$ with node weights w, we use the following type of transformation: Let $e \in E_G$ have $e = (u, v)$, $u \in U$, $v \in V$. We form a copy $G' = (U', V', E'_G)$ of G with the same weights as in G. We then perform a *crossing* operation by replacing the two copies (u, v) and (u', v') of e by new edges (u, v') and (u', v). Let $H = (U_H, V_H, E_H)$ be the resulting graph. It is obvious that the weights are feasible in $G \cup G'$ before the performance of the crossing operation. We claim that they remain feasible also afterwards. Conditions (i) and (ii) of the theorem are clearly satisfied in H, and we will show that also the third condition (iii) is satisfied. Consider a subset R of U_H.

- If R does not contain either of u and u' then the crossing is not relevant to (iii).
- If R contains both u and u' then again (iii) is not affected since the set of nodes adjacent to R is unchanged.
- Suppose that exactly one of u and u', say u' is in R. After the crossing operation, R has v' as an adjacent node (in addition to or instead of v), and since $w(v') = w(v)$, this may only increase the right hand side of (iii), so that the inequality still holds.

We call a cycle of length g a *small cycle*. Suppose that the transformation is performed on an edge e belonging to a small cycle C. The transformed graph does not contain copies of C. Instead it contains new cycles of length $2g$ at least, for example, the cycle consisting of the two copies of $C \setminus \{e\}$ and the two new edges. We say that C is *cancelled* by the transformation. We will perform a sequence of transformations of the above type until the resulting graph has no small cycles (i.e., has girth greater than g). At each step we cancel all the copies of a small cycle of the original graph. This is done as follows: Let $C_1, ..., C_q$ be the copies of a cycle C of G. Let $e_1, \ldots, e_q$ be copies of an edge $e \in C$. Form a copy G' of G with the same weights and where $e'_1, \ldots, e'_q$ are the respective copies of $e_1, \ldots, e_q$. Perform the crossing operation on each pair e_i and e'_i.

The transformed graph has no cycles of length $< g$ since such a cycle can be mapped into the original graph to produce also there a cycle with length $< g$, a contradiction. We will prove that small cycles are not generated, except for duplicate copies of cycles in the original graph. Therefore, the process will eventually construct a graph with no small cycles, and at this point the process terminates. Since the resulting weights are feasible for the associated graphs, this construction will prove the lemma.

We prove by induction that the new edges added through crossing are never contained in a small cycle. This will prove the above claim since a new cycle must use some of these edges. The claim clearly holds after the first transformation as we have shown above. Consider a general step of the process. Suppose, in contrast to the claim, that the transformation forms a new small cycle containing a newly added edge. Let G be the former graph and G' the new copy. The cycle has the general form

$$C = (a_1, b_1'), P_1', (b_2', a_2), P_2, (a_3, b_3'), P_3', \ldots, (b_m', a_m), P_m,$$

where P_i' is a path between b_i' and b_{i+1}' fully contained in G' for i odd, and P_i is a path between a_i and a_{i+1} fully contained in G for i even (with $a_{m+1} \equiv a_1$). The edges (a_i, b_i') were added by the current crossing operation (that replaced pairs $(a_i, b_i), (a_i', b_i') \in G$ by $(a_i, b_i'), (b_i, a_i')$). We use the copies, $P_1, P_3, \ldots, P_{m-1}$ of $P_1', P_3', \ldots, P_{m-1}'$, respectively, in G and edges $(a_1, b_1), (a_2, b_2), \ldots, (a_m, b_m)$ (that are eliminated by the crossing operation) to obtain a cycle

$$\bar{C} = (a_1, b_1), P_1, (b_2, a_2), P_2, (a_3, b_3), P_3, \ldots, (b_m, a_m), P_m$$

in G with the same length as C. The edges (a_i, b_i) are copies of the same original edge. Since C was assumed to be a small cycle, so is $\bar{C}$. Therefore, by the inductive assumption, $\bar{C}$ does not contain any edges added through crossings. The only way for this condition to hold is that $m = 2$. But then, $C = (a_1, b_1'), P_1', (a_1', b_1), P_1$ has twice as many edges as the former cycle $(a_1, b_1), P_1$, so that C cannot be a small cycle. This is a contradiction to the conclusion that $|\bar{C}| = |C|$. ∎

3.2 Proof of the theorem

The following linear program computes for a given graph, a set of weights on $U \cup V$ such that the worst possible ratio, $\frac{w(U)}{w(V)}$, is obtained:

$$\max \sum_{u \in U} w_u$$

$$\sum_{v \in V} w_v = 1 \tag{3}$$

$$\sum_{u \in R} w_u \leq \sum_{v \in U_{u \in R} E_u} w_v \quad R \subseteq U, \quad |R| \leq t \tag{4}$$

$$w_i \geq 0 \quad i \in U \cup V.$$

The dual problem is:

$$\min z$$

$$\sum_{R \ni u} y_R \geq 1 \quad u \in U \tag{5}$$

$$z \geq \sum_{R:R\cap E_v \neq \emptyset} y_R \quad v \in V \tag{6}$$

$$y_R \geq 0 \ R \subseteq U.$$

We apply the above programs to a given instance. If the solution contains indices $u \in U$ for which $w_u = 0$ we delete these nodes from the graph to obtain another instance with the same ratio and with $w_u > 0$ for all $u \in U$. The fact that in the optimal solution $w_u > 0$ for all $u \in U$ implies, by the complementary conditions, that an optimal dual solution satisfies the corresponding dual constraints (5) with equality. Consequently, (6) can be written in the following way:

$$z \geq \sum_{R:R\cap E_v \neq \emptyset} y_R$$

$$= \sum_{u \in E_v} \sum_{R \ni u} y_R - \sum_{R:|R\cap E_v|\geq 2} (|R \cap E_v| - 1)y_R$$

$$= |E_v| - \sum_{R:|R\cap E_v|\geq 2} (|R \cap E_v| - 1)y_R \quad v \in V.$$

Thus, the optimal value z satisfies

$$z = \max\{ |E_v| - \sum_{R:|R\cap E_v|\geq 2} (|R \cap E_v| - 1)y_R \ : \ v \in V\}. \tag{7}$$

Let $V_1 = \{v \in V : |E_v| = k\}$, $V_2 = V \setminus V_1$. Our goal is to show that $z \leq k - 1 + \frac{1}{t}$. By (7), since the right hand side is smaller than $|E_v|$, it is sufficient to show that a dual solution exists such that for $v \in V_1$:

$$\sum_{R:|R\cap E_v|\geq 2} (|R \cap E_v| - 1)y_R \geq \frac{t-1}{t}. \tag{8}$$

By Linear Programming duality, this implies the required bound for the primal problem.

For the rest of the proof, it is convenient to assume that the given instance has node degrees of *exactly* k in both U and V. To achieve this property, we extend U and V by adding slack nodes with zero weights and slack edges between original nodes and slack nodes and possibly between two slack nodes. Now, let $U = U_1 \cup U_2$ where U_1 is the original set of nodes and U_2 are the slack nodes added to U. We will assign values y_R to subsets $R \subseteq U$ of size t. We will project these values into U_1. The actual weight given to a subset $R' \subset U_1$ will then be $\sum_{R \subseteq U:R'=R\cap U_1} y_R$. For every $u \in U_1$ the sum $\sum_{R \subseteq U_1:R \ni u} y_R$ is preserved by this projection, while since for $v \in V_1$ $E_v \subset U_1$, for such a v also the left hand side of Inequality (8) is preserved. Consequently, the projected solution is dual feasible and proves our claim.

Define the integer c by $t = c(k - 1) + 1 + p$, or $c = \frac{t-(1+p)}{k-1}$, where p is an integer such that $0 \leq p \leq k - 2$.

We define a *chain* as a set $R \subseteq U$, $|R| = t$, with the following properties: There are nodes $v_1, \ldots, v_{c+1} \in V$, such that $v_i \neq v_j$ for $i \neq j$, and for $i = 1, \ldots, c$ we have $E_{v_i} \cap E_{v_{i+1}} \neq \emptyset$ and

$$R = E_{v_1} \cup \cdots \cup E_{v_c} \cup D,$$

$$D \subseteq E_{v_{c+1}} \setminus E_{v_c}, \text{ and } |D| = p.$$

We call the nodes $v_1, \ldots, v_{c+1}$ the *links* of the chain. If $p = 0$, then $D = \emptyset$ and we consider only nodes $v_1, \ldots v_c$. Applying Lemma 2, we may assume that the girth of the graph is greater than $2t$. Therefore the number of nodes in a chain is exactly t. Informally, a chain is given by the entire neighborhoods of c nodes in V, in addition to a partial neighborhood of another node, such that every two consecutive such neighborhoods intersect in exactly one node.

We let A be the number of chains that contain a node $u \in U$. Note that it is easy to see that each node in U is contained in the same number of chains. Similarly, we let B denote the number of chains that have v as a link for $v \in V$ (i.e., the chain contains at least two nodes of E_v). By the fact that the girth of the graph is large, this implies that either the chain contains all of E_v or it contains exactly $p + 1$ nodes of E_v, in the case $p > 0$.

For each chain R, we let $y_R = \beta$, where $\beta = \frac{c}{tB} = \frac{t-1}{t} \frac{1}{B(k-1)}$ if $p = 0$, and $\beta = \frac{c+1}{tB}$ if $p > 0$.

For convenience, we will sometimes consider a chain as an ordered set of nodes of U. Each unordered chain gives rise to a constant number of ordered chains depending only on k and t, which we denote ν. Thus, each $u \in U$ is contained in $A' = \nu A$ ordered chains, and each $v \in V$ is a link in $B' = \nu B$ such chains.

Consider first the case in which $p = 0$. In this case, for each v such that $|R \cap E_v| > 1$ we have $E_v \subseteq R$ and therefore $|R \cap E_v| = k$. By the definition of B and β, we get:

$$\sum_{R: |R \cap E_v| \geq 2} (|R \cap E_v| - 1) y_R = B(k-1)\beta = \frac{t-1}{t}.$$

Next, we wish to compute for each $u \in U$ the sum $\sum_{R \ni u} y_R$. To do so, denote by $v_1^u, \ldots v_k^u$ the neighbors of u. Of all ordered chains that have v_1^u as a link, a fraction of $\frac{1}{c}$ contain the neighborhood of that link first (i.e., $v_1 = v_1^u$). Of these chains, a fraction of $\frac{1}{k}$ contain u as the first node. Therefore, the number of ordered chains that contain $E_{v_1^u}$, such that u is the first node is $\frac{B'}{ck}$. Summing this over all neighbors of u, $v_1^u, \ldots v_k^u$, we get that the number of ordered chains in which node u appears first is $\frac{B'}{c}$. By symmetry, this implies that A', the number of chains in which node u appears in any one of the t possible positions, is $\frac{tB'}{c}$, so $A' = \frac{tB'}{c}$, and thus $A = \frac{tB}{c}$. Finally,

$$\sum_{R \ni u} y_R = \beta A = \frac{c}{tB} \frac{tB}{c} = 1.$$

Our second case is that in which $p > 0$. In this case, for each v such that $|R \cap E_v| > 1$ we have either that $E_v \subseteq R$ and therefore $|R \cap E_v| = k$, or exactly $p + 1$ nodes of E_v are in the chain, and therefore, $|R \cap E_v| = p + 1$.

Of all (unordered) chains that have v as a link for some node $v \in V$ (i.e., those containing at least two nodes of E_v), a fraction of $\frac{1}{c+1}$ have $v = v_{c+1}$ (i.e., only $p + 1$ of the k neighbors of v are in the chain). Thus, the number of chains that contain exactly $p + 1$ nodes of E_v is $\frac{B}{c+1}$. Similarly, the number of chains that contain E_v for $v = v_r$, $r < c + 1$ is $B\frac{c}{c+1}$. We get:

$$\sum_{R:|R\cap E_v|\geq 2} (|R \cap E_v| - 1)y_R = B(k-1)\beta\frac{c}{c+1} + Bp\beta\frac{1}{c+1}$$

$$= \frac{B\beta}{c+1}((k-1)c + p)$$

$$= \frac{B\beta(t-1)}{c+1} = \frac{t-1}{t}.$$

Next, again, we wish to compute for each $u \in U$ the sum $\sum_{R\ni i} y_R$. This is done similarly to the previous case: Denote by $v_1^u, \ldots v_k^u$ the neighbors of u. The number of ordered chains that contain $E_{v_1^u}$, such that $v_1^u = v_{c+1}$ is $\frac{B'}{c+1}$. A fraction of $\frac{1}{k}$ of these chains contain u in the last (t-th) place. Therefore, summing this over all neighbors of u, $v_1^u, \ldots v_k^u$, we get that the number of ordered chains in which node u appears last is $\frac{B'}{c+1}$. By symmetry, this implies that the number of ordered chains in which node u appears in any one of the t possible positions, is $\frac{tB'}{c+1}$, which by our definition is equal to A', $\frac{tB'}{c+1} = A'$, implying that $\frac{tB}{c+1} = A$. Finally,

$$\sum_{R\ni u} y_R = \beta A = \frac{c+1}{tB}\frac{tB}{c+1} = 1.$$

4   Concluding remarks

Hurkens and Schrijver's theorem assumes $k \geq 3$. The case $k = 2$ can be solved polynomially since this is a matching problem. However, it may be of interest to note the performance of local search also for this case. Indeed, the bound for this case is $1 + \frac{1}{t}$ is obtained from their formula if we apply L'Hôpital's rule. For the weighted case, we insert $k = 2$ in Theorem 6 and get the following corollary:

Corollary 8 *Let OPT be a maximum matching with weight opt. Let APX be t-optimal with weight apx. Then, $\frac{opt}{apx} \leq 1 + \frac{1}{t}$.*

With Corollary 5, this bound is also tight. It is interesting to note that in this case the bounds for the weighted and unweighted cases coincide, in contrast to the $k \geq 3$ case in which the weighted case gives rise to a worse bound that is only marginally better than the obvious bound of k.

Corollary 8 has a simple direct proof that we now present. To simplify the presentation we consider the maximum perfect matching problem on a complete graph, possibly with zero weight edges.

Proof: The assumption of t-optimality implies that no $p \leq t$ edges not in APX have strictly larger weight then the edges of APX that are incident to at least one of them. Then, $OPT \cup APX$ is a collection of edges (of $OPT \cap APX$) and cycles with edges alternating between OPT and APX. The cost of the OPT edges in each cycle is at least that of the corresponding APX edges. Consider a cycle $A_1, O_1, A_2, O_2, \ldots, A_m, O_m$ in which this relation is strict, where $O_i \in OPT$ and $A_i \in APX$ for $i = 1, \ldots, m$.

By t-optimality, $m \geq t + 1$, and for each i, $w(A_i) + \cdots w(A_{i+t+1}) \geq w(O_i) + \cdots w(O_{i+t})$, where the indices are taken modulo m. Summing over $i = 1, \ldots, m$ we use each edge A_i $t + 1$ times and each edge O_i t times. We conclude that $(t + 1)apx \geq t \cdot opt$. ∎

References

1. V. Bafna, B. Narayanan, and R. Ravi, "Nonoverlapping local alignments (weighted independent sets of axis parallel rectangles)", *Proceedings of the Workshop for Algorithms and Data Structures (WADS)*, 1995.
2. G. Cornuéjols and D. Hartvigsen, "An extension of matching theory", *Journal of Combinatorial Theory, Series B* **40** 1986, 285-296.
3. G. Cornuéjols, D. Hartvigsen and W. Pulleyblank, "Packing subgraphs in a graph", *Operations Research Letters* 1 1982, 139-143.
4. P. Crescenzi and V. Kann, *A compendium of NP optimization problems*, http://www.nada.kth.se/nada/theory/problemlist.html.
5. T. Feo, O. Goldschmidt and M. Khellaf, "One-half approximation for the k-partition problem", *Operations Rsearch* **40** 1992, S170-S172.
6. D. A. Grable, "On random greedy triangle packing", Technical Report, Institute of Computer Science, Humboldt University, Berlin, 1996.
7. M. M. Halldórsson, "Approximating discrete collections via local improvements", *Proceedings of the 6th annual ACM-SIAM symposium on discrete algorithms (SODA)* 160-169, 1995.
8. M. M. Halldórsson, "Approximating k-set cover and complementary graph coloring", *Proceedings of the 5th conference on Integer programming and Combinatorial Optimization (IPCO)*, 1996.
9. C. A. J. Hurkens and A. Schrijver, "On the size of systems of sets every t of which have an SDR, with an application to the worst-case ratio of heuristics for packing problems", *SIAM J. on Discrete Mathematics* **2**, 68-72, 1989.
10. S. Khanna, R. Motwani, M. Sudan, and U. Vazirani, "On syntactic versus Computational views of approximability", *Proceedings of the 35th annual IEEE symposium on Foundations of Computer Science (FOCS)*, 819-830, 1994.
11. Shlomi Rubinstein, private communication.

On-Line Machine Covering

Yossi Azar[1], Leah Epstein[2]

[1] Dept. of Computer Science, Tel-Aviv University. ***
[2] Dept. of Computer Science, Tel-Aviv University. †

Abstract. We consider the problem of scheduling a sequence of jobs to m parallel machines as to maximize the minimum load over the machines. This situation corresponds to a case that a system which consists of the m machines is alive (i.e. productive) only when all the machines are alive, and the system should be maintained alive as long as possible. It is well known that any on-line deterministic algorithm for identical machines has a competitive ratio of at least m and that greedy is an m competitive algorithm. In contrast we design an on-line randomized algorithm which is $\tilde{O}(\sqrt{m})$ competitive and a matching lower bound of $\Omega(\sqrt{m})$ for any on-line randomized algorithm. In the case where the jobs are polynomially related we design an optimal $O(\log m)$ competitive randomized algorithm and a matching tight lower bound for any on-line randomized algorithm. In fact, if F is the ratio between the largest job and the smallest job then our randomized algorithm is $O(\log F)$ competitive.

A sub-problem that we solve which is interesting by its own is the problem where the value of the optimal algorithm is known in advance. Here we show a deterministic (constant) $2 - \frac{1}{m}$ competitive algorithm. We also show that our algorithm is optimal for two, three and four machines and that no on-line deterministic algorithm can achieve a better competitive ratio than 1.75 for $m \geq 4$ machines.

For related machines we show that there is no on-line algorithm, whose competitive ratio is a function of the number of machines. However, for the case where the value of the optimal assignment is known in advance, and for the case where jobs arrive in non increasing order, we show that the exact competitive ratio is m. We show a constant 2 competitive algorithm for the intersection of the above two cases, i.e. the value of the optimal assignment is known in advance and the jobs arrive in non increasing order.

1 Introduction

We consider the problem of scheduling a sequence of jobs to m parallel machines as to maximize the minimum load over the machines. This situation is motivated by the following scenario. A system consists of m (identical or related)

*** E-Mail: azar@math.tau.ac.il. Research supported in part by Allon Fellowship and by the Israel Science Foundation administered by the Israel Academy of Sciences.
† E-Mail: lea@math.tau.ac.il.

machines. The system is alive (i.e. productive) only when all the machines are alive. In order to keep a machine alive it requires resources (e.g. tanks of fuel). The various size resources arrive one after the other, and each resource should be assigned immediately upon its arrival to one of the machines. The goal is clearly to keep the system alive as long as possible. The above problem has applications also in the sequencing of maintenance actions for modular gas turbine aircraft engines [15]. To conform with the standard scheduling terminology we view the resources as jobs. Thus, jobs are assigned to machines as to maximize the minimum load. If all the machines are identical then the problem corresponds to the identical machines scheduling/load-balancing problem and if each machine has its own size (of the engine that operates on the fuel) then it corresponds to the related machines problem.

We give a formal definition of the problem discussed above. Consider a set of m identical machines and a set of jobs that arrive on-line. Each job j has a weight w_j. The load of a machine i is the sum of the weights of the jobs assigned to it. That is, $l_i = \sum_{j \in J_i} w_j$, where J_i is the of jobs assigned to machine i. If machine i has a speed v_i (related machines case) then $l_i = \sum_{j \in J_i} \frac{w_j}{v_i}$. The goal is to assign the jobs to the machines as to maximize the minimum load over the machines. The problems are on-line versions of classical covering problems and are called the machine covering problems. Note that these problems are different from the bin covering problems [2, 4, 5, 13] where the goal is to maximize the number of covered bins, i.e. bins of load of at least 1.

We use the standard definition of the competitive ratio. Denote by $V_{on}(\sigma)$ (or just V_{on}) the value of the on-line algorithm for a sequence σ which is the load incurred on the least loaded machine. The value of the optimal assignment for this sequence would be denoted by $V_{opt}(\sigma)$ (or just V_{opt}). The algorithm is c competitive (the competitive ratio is c) if for every sequence σ, $V_{opt}(\sigma) \leq c \cdot V_{on}(\sigma)$.

Known results: The off-line problem of maximizing the load of the least loaded machine, is known to be NP-complete in the strong sense [17]. Woeginger [22] designed a polynomial time approximation scheme for the identical machines case. He also showed that the greedy algorithm is m competitive. (It is well known that no deterministic algorithm can achieve a better competitive ratio.) Deuermeyer, Friesen and Langston [14] studied a semi on-line problem on identical machines. They examined the LPT-heuristic which order the jobs by non increasing weights and assigns each job to the least loaded machine at the moment. It is shown in [14] that the competitive ratio of this heuristic is at most $\frac{4}{3}$. The tight ratio $\frac{4m-2}{3m-1}$ is given by Csirik, Kellerer and Woeginger [12].

Our Results: We consider the on-line version of the machine covering problems. We show the following results for the identical machines case.

- There is a randomized $O(\sqrt{m} \log m)$ competitive algorithm and any randomized algorithm is at least $\Omega(\sqrt{m})$ competitive. This is in contrast to the competitive ratio of the best possible deterministic algorithm which is m.
- There is a randomized $O(\log F)$ competitive algorithm for jobs of weights which may vary up to a factor of F. In particular, there is an $O(\log m)$

competitive randomized algorithm for polynomially related weight jobs. (The deterministic lower bound of m holds already for jobs of weights that are polynomially related.) Also, any randomized algorithm is at least $\Omega(\log m)$ competitive for polynomially related weight jobs.

A sub-problem that we solve which is interesting by its own is the problem where the value of the optimal algorithm is known in advance. Here we show the following results:

- There is a deterministic $2 - \frac{1}{m}$ competitive algorithm.
- The algorithm is optimal for two, three and four machines and no on-line deterministic algorithm can achieve a better competitive ratio than 1.75 for $m \geq 4$ machines.

For related machines we show the following results

- There is no algorithm whose competitive ratio is a function of the number of machines.
- For the case where the value of the optimal assignment is known in advance the exact competitive ratio is m.
- For the case where jobs arrive in non increasing order the exact competitive ratio is also m.
- For the case where both the value of the optimal assignment is known in advance and jobs arrive in non increasing order there is 2 competitive algorithm.

Other related work: The original scheduling problem of minimizing the maximum load over all machines has been widely studied. The problem was introduced by Graham [18, 19] who gave a greedy algorithm "List Scheduling" which is $2 - \frac{1}{m}$ competitive. It turns out that the algorithm of Graham is not optimal for minimizing the maximum load (for all $m \geq 4$) [16, 11]. Bartal et al. [8] were the first to show an algorithm whose competitive ratio is strictly below $c < 2$ (for all m). More precisely, their algorithm achieves a competitive ratio of about $2 - \frac{1}{70}$. Later, the algorithm was generalized by Karger, Phillips and Torng [20] to yield an upper bound of 1.945. Very recently, Albers [1] designed 1.923 competitive algorithm and improved the lower bound to 1.852 (the previous lower bound was 1.8370 [7]). Clearly randomized algorithms for the problem may improve the bounds only by a constant factor. Unlike the scheduling problem, there are no constant competitive algorithms for the covering problem on identical machines. Moreover, as we mentioned our results show that the competitive ratio of randomized algorithms are significantly better than the competitive ratio of the deterministic ones.

For the related machines case it was proved in [3] that there is a constant 8 competitive algorithm for minimizing the maximum load. (the constant was recently improved by [9].) If the value of the optimal assignment is known in advance then the competitive ratio is 2. This is in contrast to the "more difficult" covering problem where in order to get comparable results one need to assume that the jobs arrive in non-increasing order and the optimal value is known in advance.

It is interesting to note that a measure which is different than the maximum load has been already used for scheduling problems. Specifically, Awerbuch et al.[6] studied the case of minimizing the sum of the squares of the load (and general L_p norm). It is shown that there is a constant competitive algorithm for a general class of scheduling problems.

Structure of the paper: In section 2 we consider the sub-problem where the value of the optimal assignment is known in advance. In section 3 we use an algorithm from section 2 as a procedure, and discuss randomized algorithms for identical machines. In section 4 we consider algorithms for related machines.

2   Known optimal value

In this section we show a simple algorithm for identical machines for the case where the value of the optimal assignment is known in advance. We assume without loss of generality that $V_{opt} = 1$. The competitive ratio of our algorithm is $2 - \frac{1}{m}$, which is the same ratio as for the algorithm of Graham for scheduling. In contrast to the algorithm of Graham which tries to fill all the machines evenly, our algorithm fills the machines one by one. We define $\beta = \frac{m}{2m-1}$ and call a machine of load at least β a full machine. A non empty machine which is not full is called active. Our algorithm maintains at most one active machine.

Algorithm FILL:

If there are no empty machines, assign a new job to the least loaded machine. Otherwise, if the weight of the new job is at least β assign it to an empty machine (the machine becomes full). If the weight of the job is less than β assign it to the active machine if exists, and otherwise to an empty machine (which becomes active).

Theorem 1. *The algorithm Fill has a competitive ratio of $\frac{1}{\beta} = 2 - \frac{1}{m}$.*

Proof. We show that when the algorithm terminates, all machines are full. This implies that the competitive ratio is at most $\frac{1}{\beta}$. Assume that there is a non full machine. First we may replace all jobs of weight more than 1 by jobs of weight 1, which would not influence the optimal algorithm or our algorithm. We show that all jobs of weight at least β were assigned to empty machines. Otherwise the first such job that was assigned to a non empty machine was assigned to the only active machine, and after it was assigned all machines becomes full which contradicts the assumption. We estimate the load of each machine. Denote the load of the least loaded machine by h. The load of machines that have only one job is clearly at most 1. If there are at least two jobs on a machine, its load is bounded by 2β, since just before the last job was placed the load was less than β and the weight of the last job is also less than β. Thus the load of each machine is bounded by 2β. Since the optimal value is 1 then

$$h \geq m - (m-1) \cdot 2\beta \geq m - (m-1)\frac{2m}{2m-1} = \frac{m}{2m-1} = \beta$$

which completes the proof.

Notice that later we will use algorithm Fill with $\beta' \leq \beta$ which is clearly $\frac{1}{\beta'}$ competitive. The algorithm Fill with the parameter β turns out to be optimal for two, three and four machines.

Theorem 2. *The competitive ratio of any algorithm for two machines is at least* 1.5 .

Proof. We consider the following sequence of jobs. First two jobs of weight 2/3 arrive. There are two cases:

- If both jobs were placed on one machine, then two jobs of weight 1/3 arrive. The optimal algorithm assigns one small job and one big job on each machine. The best value the on-line is 2/3 by assigning both small jobs to the empty machine. Thus the competitive ratio in this case is at least 1.5 .
- If each job was placed on a different machine, then one job of weight 1 arrives. The optimal algorithm assignment the two jobs of weight 2/3 on one machine, and the unit job on the other machine. The on-line places the unit job on one machine, and the load of the other machine is still 2/3. The competitive ratio in this case is also at least 1.5 .

Theorem 3. *The competitive ratio of any algorithm for three machines is at least* 5/3 .

Proof. We construct a sequence such that $V_{on} \leq 0.6$ and thus the competitive ratio is at least 5/3. We consider the following sequence of jobs. First two jobs of weight 0.6 arrive. If the on-line algorithm assigns the jobs on different machines, two unit jobs arrive. The minimum load of the on-line is 0.6 while the optimal algorithm assigns the jobs of weight 0.6 on one machine, and one unit job on each other machine.

If the on-line algorithm assigns them to the same machine, we continue the sequence by two jobs of weight 0.3 . Clearly, the on-line will not use the machine with the first two jobs any more, because it already has load of at least 1. Thus we are left with two machines. There are two cases:

- If the two jobs of weight 0.3 are placed on different machines, then two jobs of weight 0.1 and one unit job arrive. Clearly, the best value the on-line can obtain is 0.5 (by assigning the unit job to one of the two machines, and both jobs of weight 0.1 to the other). The optimal algorithm assigns the unit job on one machine, and three jobs of weights $0.6, 0.3, 0.1$ on each of the other two machines and achieves a minimum load of 1.
- If the two jobs of weight 0.3 are placed on the same machine, then a job of weight 0.6 arrives. If it is placed on the only empty machine, a unit job arrives, and since there are two on-line machines with load 0.6, $V_{on} = 0.6$. The optimal algorithm assigns two jobs of weight 0.6 on one machine, the unit job on another machine and all the other jobs on the third machine and thus achieves a minimum load of 1. If the job of weight 0.6 is placed on the

machine with load 0.6, three jobs with the weights $0.4, 0.1, 0.1$ arrive. Even if they are all placed on the third machine, $V_{on} = 0.6$. The optimal algorithm assigns a job of weight 0.6 with a job of weight 0.4 on one machine and a job of weight 0.6 with two jobs of weights 0.1 and 0.3 on each of the other two machines. Thus the competitive ratio is at least 5/3 for all the cases.

The proof of the following theorem is omitted.

Theorem 4. *The competitive ratio of any algorithm for four machines is at least* 1.75 .

We give a general lower bound for $m > 4$ machines by noticing that the competitive ratio of the problem is non-increasing as a function of the number of machines.

Lemma 5. *Let r be a lower bound for k machines, then r is also a lower bound for $m > k$ machines.*

Proof. We assume that r is a lower bound for k machines. We convert it into a lower bound for m machines, where $m > k$. We begin the lower bound with $m - k$ unit jobs. After those jobs arrive, we continue with the original lower bound for k machines. The on-line and the off-line algorithms both assign the first $m - k$ jobs to $m - k$ different machines and do not use those machines later. Thus, the competitive ratio of any algorithm for $m > k$ machines is at least r.

Corollary 6. *The competitive ratio of any algorithm for $m \geq 4$ machines is at least* 1.75 .

3 The identical machines case

We recall that no deterministic algorithm can achieve a better competitive ratio than m (consider a sequence of m jobs of weight 1 which may be followed with $m - 1$ jobs of weight m) and that greedy is m competitive. Here we show that randomization really helps in reducing the competitive ratio. We first introduce an algorithm whose competitive ratio depends on the maximum ratio between the largest job and the smallest job. We denote this ratio by F. We show an $O(\log F)$ competitive algorithm for the case that F is known in advance. If F is not known in advance we can apply the standard technique of [21] and get an $O(\log^{1+\varepsilon} F)$ competitive algorithm.

Lemma 7. *Assume that it is known in advance that $\lambda \leq V_{opt} < 2\lambda$. Then algorithm Fill with parameter λ as the known optimal value and the parameter $\beta' = \frac{1}{2}$ is 4 competitive.*

Proof. The results of running the algorithm Fill with these parameters, is the same as running the algorithm with the correct value of V_{opt} and $\beta'' = \frac{\lambda}{2V_{opt}}$. Since $\frac{1}{4} < \beta'' \leq \frac{1}{2}$ then by Theorem 1 the algorithm succeeds and is 4 competitive.

Our randomized algorithm normalizes the weight of the first job to 1. Thus the weights of the jobs are bounded between F and $\frac{1}{F}$. We first define two procedures that the algorithm uses.

Procedure RFILL:
The procedure chooses an integer i^* uniformly at random, where 2^{i^*} is in the range $\frac{1}{2F} \leq 2^{i^*} \leq 2F$ (Assume that F is known in advance). Then the procedure runs the algorithm Fill with the value 2^{i^*} as the known optimal value and $\beta' = 1/2$.

Lemma 8. *The procedure RFill is $O(\log F)$ competitive if $F \geq V_{opt}/2$.*

Proof. Let us bound the value of V_{opt}. If $V_{opt} = 0$, then the lemma is trivially correct. Thus we may assume that the optimal algorithm has at least one job on each machine, and since the weight of a job is at least $\frac{1}{F}$ then $\frac{1}{F} \leq V_{opt} \leq 2F$. For each possible value of V_{opt}, a suitable value of i^* such that $2^{i^*} \leq V_{opt} < 2^{i^*+1}$ was chosen with probability $O(\frac{1}{\log F})$. Hence the procedure is $O(\log F)$ competitive. If F is not known in advance we choose an integer value i, $-\infty < i < \infty$ with probability $\frac{1}{(|i|)^{1+\epsilon}+1}$ and achieve competitive ratio of $O(\log^{1+\epsilon} F)$.

Procedure GREEDY:
Assign each new job to a machine that has the current minimum load.

Lemma 9. *The procedure is 2 competitive if $F \leq V_{opt}/2$*

Proof. Denote by h the minimum on-line load, by H the maximum on-line load and by x a machine with load H. Clearly $V_{opt} \leq H$ and $V_{on} = h$. Let Y be the last job assigned to x. By the definition of F, $w(Y) \leq F$. Clearly at the time that Y was assigned, x had the minimum load and hence $H - w(Y) \leq h$. Thus

$$H - h \leq w(Y) \leq F \leq V_{opt}/2 \leq H/2$$

which implies that $h \geq H/2$ and $\frac{V_{opt}}{V_{on}} \leq \frac{H}{h} \leq 2$.

Algorithm RFILL-GREEDY: Run RFill with probability 1/2 and Greedy with probability 1/2.

Theorem 10. *The algorithm RFill-Greedy is $O(\log F)$ competitive.*

Proof. The proof follows immediately from the proofs of Lemmas 8 and 9.

The above theorem immediately implies the following:

Corollary 11. *The algorithm RFill-Greedy is $O(\log m)$ competitive for polynomially related weight jobs.*

In fact, we cannot achieve a better bound up to a constant factor.

Theorem 12. *Any randomized algorithm for a sequence for which the ratio between the largest and the smallest jobs is at least m is $\Omega(\log m)$ competitive.*

Proof. We use an adaptation of Yao's theorem for randomized on-line algorithms which states that a lower bound on the competitive ratio of deterministic algorithms on any distribution on the input is also a lower bound for randomized algorithms, and is given by $\max\{1/E\left(V_{on}/V_{opt}\right), E(V_{opt})/E(V_{on})\}$ [10]. Assume for simplicity that $m = 2^k$. Consider the following sequence. At first m jobs of weight 1 arrive. After those unit jobs are placed then $m - 2^{k-i^*}$ big jobs of weight 2^{i^*} arrive, where $0 \leq i^* \leq k$ is an integer chosen uniformly at random. The optimal off-line assignment would be to put one big job on each machine, and 2^{i^*} unit jobs on the remaining 2^{k-i^*} machines, yielding the value 2^{i^*}. It is easy to see that the best strategy for the on-line algorithm is to assign the big jobs to the least loaded machines. After the first phase, we sort the machines in non decreasing order, according to the number of jobs assigned to each machine: $M_1, M_2, ..., M_m$, where M_1 is the most loaded machine. Let $a_i, 0 \leq i \leq k$ be

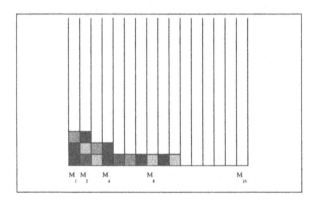

Fig. 1. The machines sorted by non increasing load

the number of jobs on the machine M_{2^i}. Since there are m unit jobs, and the machines are sorted, $a_0 + \sum_{1 \leq i \leq k} a_i \cdot (2^i - 2^{i-1}) \leq m$. Now we can estimate the value of the on-line algorithm for each one of the $k + 1$ cases. For a fixed value of i^*, there are $m - 2^{k-i^*}$ big jobs, that are placed on the least loaded machines, which implies a load of at least 2^{i^*} on those machines. After that, the least loaded machine is $M_{2^{k-i^*}}$, which has load a_{k-i^*}. The competitive ratio in this case is $\frac{2^{i^*}}{a_{k-i^*}}$. Hence,

$$E\left(\frac{V_{on}}{V_{opt}}\right) \leq \frac{1}{k+1} \sum_{0 \leq i \leq k} \frac{a_{k-i}}{2^i} = \frac{1}{k+1} \sum_{0 \leq i \leq k} \frac{a_i}{2^{k-i}}$$

$$= \frac{1}{2^{k-1}(k+1)} \sum_{0 \leq i \leq k} a_i 2^{i-1} \leq \frac{1}{2^{k-1}(k+1)} \left(a_0 + \sum_{1 \leq i \leq k} a_i \cdot 2^{i-1}\right)$$

$$\leq \frac{1}{2^{k-1}(k+1)} \cdot m = \frac{2}{k+1}$$

Thus, by Yao's theorem the competitive ratio is at least $\frac{k+1}{2} = \Omega(\log m)$.

Next we provide an algorithm whose performance is as a function only of the number of machines. We introduce a randomized algorithm Partition which is $O(\sqrt{m}\log m)$ competitive. Assume without loss of generality that m is a power of 4. We round down the weights of jobs in the sequence to powers of 2. Clearly the competitive ratio of a general weight sequences is at most twice the competitive ratio of power of 2 sequences. Next with probability $1/2$ we apply the procedure Greedy and with probability $1/2$ we choose $0 \leq i \leq \frac{1}{2}\log m$ uniformly at random and apply the following partitioning procedure with $k = 2^i$. The procedure partitions the machines into two parts. The left part consists of k machines and the right part consists of the remaining machines. The idea of the procedure is to classify each new job to one of the sets, and assign it to a machine in this set. The procedure keeps a guess λ for the value of $\frac{V_{opt}}{\sqrt{m}}$, which is initialized to 0. Whenever a job j arrives, if $\lambda \geq w(j)$, the job is considered below threshold, and would be assigned to the left k machines. Otherwise, with probability $\frac{1}{2m}$ the procedure decides to increase λ to be $w(j)$, and the job which becomes below threshold is assigned to the left machines. If λ was not changed, the job is assigned to the right machines. Jobs on the right machines are assigned greedily, i.e. on a least loaded right machine. Jobs on the left machines are placed in a round Robin manner for each weight separately.

Theorem 13. *The algorithm Partition is $O(\sqrt{m}\log m)$ competitive.*

Proof. We will prove the following claim: For every sequence, either greedy is $O(\sqrt{m})$ competitive or there exists a choice of k, for which the partitioning procedure is $O(\sqrt{m})$ competitive. Since there are $O(\log m)$ choices of k, all with equal probability, the algorithm is $O(\sqrt{m}\log m)$ competitive.

For a fixed sequence we define a job j as big if $w(j) \geq \frac{V_{opt}}{8\sqrt{m}}$ and otherwise small. Let r be the number of big jobs in the sequence and $k_1 = \max(0, m-r)$.

We first show that if $k_1 = 0$, the simple greedy algorithm yields the desired result. We show that the load of every machine is at least $\frac{V_{opt}}{8\sqrt{m}}$. The definition of k_1 implies that there are at least m big jobs. If at least one big job was placed on each machine, the claim on the load holds. Otherwise, there a machine with at least two big jobs. At the moment the second job was placed, this machine had the minimum load among all machines, which is at least the weight of one big job, and thus all machines had load of at least $\frac{V_{opt}}{8\sqrt{m}}$.

Next we show that if $k_1 > \sqrt{m}$, the simple greedy algorithm also yields the desired result. Here we show that the small jobs give enough load for all machines. Consider the greedy assignment. Assume that machine z has minimum load h. Since the algorithm is greedy, if we remove the latest job from all other machines, they all would have loads that do not exceed h. We bound the total load achieved from small jobs. It is h for machine z and h plus one small job for all other machines. The total is bounded by $hm + (m-1) \cdot V_{opt}/(8\sqrt{m})$. Since the total load of small jobs is at least $k_1 V_{opt} \geq \sqrt{m}V_{opt}$, we conclude that $h \geq \frac{V_{opt}}{2\sqrt{m}}$.

Finally we show that there is an appropriate choice of k for all cases $0 < k_1 \leq \sqrt{m}$. Assume $2^{i_1} < k_1 \leq 2^{i_1+1}$. We show that the choice $k = 2^{i_1+1}$ yields the load $\Omega\left(\frac{V_{opt}}{\sqrt{m}}\right)$ with constant probability. Notice that for this choice $2k_1 > k \geq k_1$. We first show that with constant probability all the big jobs, were placed on the right machines. In order for some of the big jobs to be placed on the left machines, the value λ should have been changed for at least one of them. The probability that λ was not changed for a single job is $1 - 1/(2m)$. The probability that for $r \leq m$ big jobs λ was not changed is $(1 - \frac{1}{2m})^r \geq (1 - \frac{1}{2m})^m \geq \frac{1}{2}$. Thus, with probability at least $1/2$, r big jobs are placed on the $m - k$ right machines. Since $m - k \leq m - k_1 = r$ the load of each right machine is at least $\frac{V_{opt}}{8\sqrt{m}}$. Next we estimate the expected minimum load on the left machines. We first prove the following lemma:

Lemma 14. *The expected minimum load on the left machines induced by the small jobs, is at least $\frac{V_{opt}}{32\sqrt{m}}$ for the appropriate choice of k.*

Proof. There are at most $2k_1$ left machines. We denote the total weight of the small jobs by W. We define a success for a set of $2\sqrt{m}$ jobs of equal weight w by the event that after the arrival of the first $\sqrt{m}$ jobs, the value of λ is at least w (and thus the last $\sqrt{m}$ jobs were below threshold, and were assigned to the left machines).

The failure probability for a set of $2\sqrt{m}$ jobs is most $(1 - \frac{1}{2m})^{\sqrt{m}} \leq e^{-\frac{1}{2\sqrt{m}}}$. Since for $0 \leq x \leq 1$, $e^{-x} \leq 1 - \frac{x}{2}$, the failure probability is at most $1 - \frac{1}{4\sqrt{m}}$. Hence the success probability in a set of $2\sqrt{m}$ jobs is at least $\frac{1}{4\sqrt{m}}$. If there was a success in a subsequence of $2\sqrt{m}$ jobs of equal weight, then at least $\sqrt{m}$ jobs were assigned to all left machines, thus at least $\frac{\sqrt{m}}{k}$ jobs to each one of the left machines. We partition all small jobs into sets of size $2\sqrt{m}$. The last set, which may contain less than $2\sqrt{m}$ jobs, is ignored. The weight that is lost by ignoring those sets is bounded by

$$2\sqrt{m}\sum_{j \geq 0}\frac{w_{max}}{2^j} \leq 4\sqrt{m}w_{max} \leq 4\sqrt{m} \cdot \frac{V_{opt}}{8\sqrt{m}} \leq \frac{V_{opt}}{2}$$

where w_{max} is the weight of the largest small job. Denote the set of sets that are not ignored by A, and the total weight of A by W_A. Since the total weight of small jobs is at least V_{opt} then $W_A \geq \frac{W}{2}$. The expected of the minimum load over the left machines is at least

$$\frac{1}{4\sqrt{m}}\sum_{a \in A}\frac{\sqrt{m}w_a}{k} = \frac{1}{8k\sqrt{m}}\sum_{a \in A}2\sqrt{m}w_a \geq \frac{1}{8k\sqrt{m}}W_A \geq \frac{1}{8k\sqrt{m}}\frac{W}{2}$$

Clearly $W \geq k_1 V_{opt}$ and hence the expected minimum load on the left machines is at least

$$\frac{1}{16k\sqrt{m}}k_1 V_{opt} \geq \frac{k}{32k\sqrt{m}}V_{opt} \geq \frac{V_{opt}}{32\sqrt{m}}$$

We conclude that with probability of at least $1/2$ the expected minimum load over all the machines is $\Omega\left(\frac{V_{opt}}{\sqrt{m}}\right)$ and thus the partitioning procedure is $O(\sqrt{m})$ competitive, for the appropriate choice of k.

Surprisingly, no algorithm can achieve a significantly better competitive ratio.

Theorem 15. *Any randomized algorithm is $\Omega(\sqrt{m})$ competitive.*

Proof. By Yao's theorem it is enough to show a lower bound for some distribution on the input on $\max\{1/E\left(V_{on}/V_{opt}\right), E(V_{opt})/E(V_{on})\}$.

Assume for simplicity that $m = k^2 \geq 4$. The first phase of the sequence contains k parts, each of k jobs. All the jobs of part i have equal weight of $s_i = k^{2i}$ (for $1 \leq i \leq k$). In the second phase, an integer $1 \leq i^* \leq k$ is chosen uniformly at random, and additional $ki^* - 1$ jobs of weight k^{2i^*+1} arrive. Clearly the sequence can be assigned by the optimal algorithm so that the load of each machine is at least k^{2i^*+1}. This can be done since there are $k^2 - 1$ jobs of weight at least k^{2i^*+1} and k jobs of weight k^{2i^*}.

We consider the on-line assignment after the first phase. We sort the on-line machines in non decreasing order according to the load on each machine: $M_1, M_2, ..., M_m$, where M_1 is the most loaded machine. Notice that for each i, a job of weight s_i is larger than the sum of all smaller jobs, since

$$k^{2i} > 2k \cdot k^{2(i-1)} \geq k(k^{2(i-1)} + k^{2(i-2)} + ... + 1) .$$

Thus the largest ki jobs are assigned to machines $M_1, ..., M_{ki}$, for all $1 \leq i \leq k$ (not necessary all of them). All the jobs of weight s_{k-i+1} jobs are assigned to machines $M_1, ..., M_{ki}$. The largest weight of a job on machine M_{1+ki} is at most s_{k-i}. Denote by b_i the number of jobs of weight s_{k-i+1} jobs on the machine $M_{1+k(i-1)}$, and by l_i the load of this machine. Notice that $l_i \leq (b_i + 1)s_{k-i+1}$.

Next we show that $B = b_1 + b_2 + ... + b_k \leq 3k$. Let j be the minimum index (if exists) such that $\sum_{i<j}(b_i - 1) \geq k$. Clearly, all the kj largest jobs were assigned to $M_1, .., M_{k(j-1)}$. Thus $b_j = 0$ and similarly $b_i = 0$ for all $i \geq j$. Since by the definition of j $\sum_{i<j-1}(b_j-1) < k$ and since $b_{j-1} \leq k$ then $\sum_{i<j}(b_j-1) < 2k$. If j is not defined, then $\sum_{1 \leq i \leq k}(b_j - 1) < k < 2k$ as well. Thus $\sum_{1 \leq i \leq k}(b_j - 1) < 2k$ and $B < 3k$.

We now evaluate $E(V_{on}/V_{opt})$. Consider the case $V_{opt} = k^{2i^*+1}$. It is easy to see that the best for the on-line algorithm is to assign the jobs of the second phase, to the least loaded machines. Thus $ki^* - 1$ jobs are assigned to machines $M_{k(k-i^*)+2}, ..., M_m$, and the ratio is

$$\frac{ks_{k-i-1}}{l_{k(k-i^*)+1}} \leq \frac{ks_{k-i-1}}{(b_{k-i^*+1} + 1)s_{k-i-1}} = \frac{k}{b_{k-i^*+1} + 1} .$$

Thus

$$E\left(\frac{V_{on}}{V_{opt}}\right) \leq \frac{1}{k} \sum_{1 \leq i \leq k} \frac{b_{k-i+1}+1}{k} \leq \frac{1}{k} \sum_{1 \leq i \leq k} \frac{b_i+1}{k} \leq \frac{1}{k^2}(B+k) \leq \frac{4k}{k^2} = \frac{4}{\sqrt{m}}$$

which completes the proof of the theorem.

4   The related machines case

In this section we consider the case where the machines are related. Machine j has a speed v_j, and if job i is assigned to machine j, the load of the machine is increased by w_i/v_j. Most of the proofs in this section are omitted. First we show that it is impossible to design an algorithm whose competitive ratio is a function of the number of the machines.

Lemma 16. *There is no on-line algorithm for related machines whose competitive ratio is a function of the number of the machines (already when the number of machines is 2).*

We consider the case where the value of the optimal off-line assignment is known in advance. We show that the exact competitive ratio in this case is m.

Theorem 17. *The competitive ratio of any deterministic algorithm for related machines, even when the value of the optimal off-line assignment is known in advance, is at least m.*

Now, we show an m competitive algorithm for related machines case in which the value of the optimal off-line algorithm is known in advance. We use a constant $\alpha = \frac{1}{m}$. Assume without loss of generality that the value of the optimal off-line algorithm is 1.
Algorithm SLOW-FAST: Assign each new job on the fastest machine whose current load is less than α and its load with the job would be at least α. If no such machine exists, put the job on the slowest machine that has load less than α. If all machines have load of at least α, put the job on an arbitrary machine.

Theorem 18. *The algorithm Slow-Fast is m competitive for the related machines case in which the optimal value is known in advance.*

We call an algorithm "semi on-line" if jobs arrive in non increasing weight order.

Theorem 19. *Any semi on-line algorithm for related machines has competitive ratio of at least m.*

We show a matching m competitive algorithm.
Algorithm BIASED-GREEDY: Assign each new job to the machine with the current minimum load. In case of ties assign the job to the fastest machine among those with the minimum load.

Theorem 20. *The semi on-line algorithm Biased-Greedy is m competitive for related machines.*

If, in addition, the value of the optimal off-line algorithm is to be known in advance, it is possible to reduce the competitive ratio. We show a constant competitive semi on-line algorithm for related machines, for which the value of the optimal off-line algorithm is known in advance. We assume without loss of generality that this value is 1.

Algorithm NEXT-COVER: Assign a new job to the fastest machine, whose current load does not exceed $\frac{1}{2}$. If the load of all machines is at least $\frac{1}{2}$, assign it to an arbitrary machine.

Theorem 21. *The semi on-line algorithm Next-Cover is 2 competitive for the related machines case in which the optimal value is known in advance.*

Proof. From the definition of the algorithm it is clear that if the algorithm fails, the slowest machine has load less than $\frac{1}{2}$. We assume by contradiction that the slowest machine has load less than $1/2$. We sort the machines by non increasing speed, i.e. machine 1 is the fastest machine, and machine m is the slowest machine.

We show the following invariant. Let W_i be the total weight of the jobs assigned to machines 1 through i by the on-line algorithm and let W_i^* be the respective value for the optimal algorithm. We show that there exists an optimal assignment such that $W_i \leq W_i^*$ for all i. In particular $W_{m-1} \leq W_{m-1}^*$, and the slowest machine of the on-line is loaded by at least the same load of the optimal algorithm which is at least 1 which is a contradiction.

For $i = 0$ the invariant is trivially true. Assume it for $i - 1$ and prove it for i. If the on-line load on machine i does not exceed 1, the claim is trivially true. (Since in the optimal assignment, the load on this machine is at least 1). Thus, we may assume that the on-line load on the machine exceeds 1. We consider two cases according to the number of jobs on machine i.

Assume first that there are at least two jobs on the machine. The last job is larger than half the speed of the machine, since before it was placed, the load of the machine was less than $1/2$, and after it was placed, the load is larger than 1. Since the jobs arrive in non increasing order, then the first job on the machine is also larger than half the speed which contradicts the fact that the load was less than $1/2$.

Thus, we may assume that only one job placed on the machine i. Consider the set S of all jobs that arrived before the only job on machine i (including the job). Clearly the on-line assigned those jobs and only those jobs on machines 1 through i, thus $W_i = W(S)$. If all jobs in S are assigned by the optimal algorithm to machines 1 through i, then $W_i = W(S) \leq W_i^*$. Otherwise change the off-line packing in the following way: take one job of S that is assigned to a machine x, $x > i$, and put it on machine i; this gives load of at least 1 to machine i, since even the smallest job of S is at least of weight v_i. Put on x all jobs that were on i. Since x is a slower machine, its load is also at least 1. Hence $W_i \leq W_i^*$.

References

1. S. Albers. Better bounds for on-line scheduling. In *Proc. 29th ACM Symp. on Theory of Computing*, 1997. To appear.
2. N. Alon, J. Csirik, S. V. Sevastianov, A. P. A. Vestjens, and G. J. Woeginger. On-line and off-line approximation algorithms for vector covering problems. In *Proc. 4th European Symposium on Algorithms*, LNCS. Springer, 1996.

3. J. Aspnes, Y. Azar, A. Fiat, S. Plotkin, and O. Waarts. On-line load balancing with applications to machine scheduling and virtual circuit routing. In *Proc. 25th ACM Symposium on the Theory of Computing*, pages 623–631, 1993.

4. S.F. Assmann. Problems in discrete applied mathematics. Technical report, Doctoral Dissertation, Mathematics Department, Massachusetts Institute of Technology, Cambridge, Massachusetts, 1983.

5. S.F. Assmann, D.S. Johnson, D.J. Kleitman, and J.Y.-T. Leung. On a dual version of the one-dimensional bin packing problem. *J. Algorithms*, 5:502–525, 1984.

6. B. Awerbuch, Y. Azar, E. Grove, M. Kao, P. Krishnan, and J. Vitter. Load balancing in the l_p norm. In *Proc. 36th IEEE Symp. on Found. of Comp. Science*, pages 383–391, 1995.

7. Y. Bartal, H. Karloff, and Y. Rabani. A better lower bound for on-line scheduling. *Information Processing Letters*, 50:113–116, 1994.

8. Yair Bartal, Amos Fiat, Howard Karloff, and R. Vorha. New algorithms for an ancient scheduling problem. In *Proc. 24th ACM Symp. on Theory of Computing*, 1992.

9. P. Berman and M. Karpinski. A note on on-line load balancing for related machines. Unpublished notes.

10. A. Borodin and R. El-Yaniv. On randomization in online computations. In *Computational Complexity*, 1997.

11. B. Chen, A. van Vliet, and G. Woeginger. New lower and upper bounds for on-line scheduling. *Operations Research Letters*, 16:221–230, 1994.

12. J. Csirik, H. Kellerer, and G. Woeginger. The exact lpt-bound for maximizing the minimum completion time. *Operations Research Letters*, 11:281–287, 1992.

13. J. Csirik and V. Totik. On-line algorithms for a dual version of bin packing. *Discr. Appl. Math.*, 21:163–167, 1988.

14. B. Deuermeyer, D. Friesen, and M. Langston. Scheduling to maximize the minimum processor finish time in a multiprocessor system. *SIAM J. Discrete Methods*, 3:190–196, 1982.

15. D. Friesen and B. Deuermeyer. Analysis of greedy solutions for a replacement part sequencing problem. *Math. Oper. Res.*, 6:74–87, 1981.

16. G. Galambos and G. Woeginger. An on-line scheduling heuristic with better worst case ratio than graham's list scheduling. *Siam Journal on Computing*, 22(2):349–355, 1993.

17. M.R. Garey and D.S. Johnson. *Computers and Intractability*. W.H. Freeman and Company, San Francisco, 1979.

18. R.L. Graham. Bounds for certain multiprocessor anomalies. *Bell System Technical Journal*, 45:1563–1581, 1966.

19. R.L. Graham. Bounds on multiprocessing timing anomalies. *SIAM J. Appl. Math*, 17:263–269, 1969.

20. D. Karger, S. Phillips, and E. Torng. A better algorithm for an ancient scheduling problem. In *Proc. 5th ACM-SIAM Symp. on Discrete Algorithms*, 1994.

21. R. J. Lipton and A. Tomkins. Online interval scheduling. In *Proc. of the 5th ACM-SIAM Symposium on Discrete Algorithms*, pages 302–311, 1994.

22. G. Woeginger. A polynomial time approximation scheme for maximizing the minimum machine completion time. Technical Report, 1995.

Area-Efficient Static and Incremental Graph Drawings *

Therese C. Biedl[1]   Michael Kaufmann[2]

[1] RUTCOR, Rutgers Univ., P.O. Box 5062, New Brunswick, NJ 08903, USA
e-mail: therese@rutcor.rutgers.edu.
[2] Wilhelm-Schickard-Institut, Universität Tübingen, Sand 13,
72076 Tübingen, Germany
e-mail: mk@informatik.uni-tuebingen.de.

Abstract. In this paper, we present algorithms to produce orthogonal drawings of arbitrary graphs. As opposed to most known algorithms, we do not restrict ourselves to graphs with maximum degree 4. The best previous result gave an $(m-1) \times (\frac{m}{2}+1)$-grid for graphs with n nodes and m edges.

We present algorithms for two scenarios. In the static scenario, the graph is given completely in advance. We produce a drawing on a grid of size at most $\frac{m+n}{2} \times \frac{m+n}{2}$, or on a larger grid where the aspect ratio of the nodes is bounded. Furthermore, we give upper and lower bounds for drawings of the complete graph K_n in the underlying model. In the incremental scenario, the graph is given one node at a time, and the placement of previous nodes can not be changed for later nodes. We then come close to the bounds achieved in the static case and get at most an $(\frac{m}{2}+n) \times (\frac{2}{3}m+n)$-grid. In both algorithms, every edge gets at most one bend, thus, the total number of bends is at most m.

Then we focus on planar graphs and outer-planar graphs. We obtain planar drawings in an $(m-n+1) \times \min\{\frac{m}{2}, m-n+1\}$-grid with $m-n$ bends for planar triconnected graphs. The best previous result here was an $m \times m$-grid and m bends, if the boxes of the nodes are constrained to be small.

All algorithms work in linear time.

1   Background

In recent years, the subject of graph drawings has created intense interest, due to numerous applications. Different drawing styles have been investigated (see [3] for an overview). One possible drawing technique is to produce orthogonal graph drawings, where only horizontal and vertical lines are employed. For example, in networking and data base applications, graph drawings serve as a tool to help display large diagrams efficiently. Specific uses of orthogonal graph drawings include Data Flow Diagrams and Entity Relationship Diagrams. The goal is to obtain an aesthetically pleasing drawing, and common objectives are small area, few bends, and few crossings.

---

* The research was partly funded by the NIST Advanced Technology Program Award No. 70NANB5H1162" and by the German Research Society, Grant DFG Ka/4-2.

For graphs with maximum degree 4, the usual definition of an orthogonal drawing is an embedding in the plane with nodes drawn as points, and edges drawn as sequences of horizontal and vertical line segments. For graphs with higher maximum degree, it is not possible to drawn the nodes as points, since no overlap among edges is allowed. Several attempts to generalize the known results for graphs with maximal degree 4 have been made. In Giotto [15], the high-degree nodes are split into several 'small' nodes and the previous techniques could be applied. Unfortunately, no theoretical bounds have been achieved and even worse, the final boxes of the nodes might be stretched unrelated to the degree.

In this paper, we forbid that nodes may be stretched far, which can be described in one sentence as "the nodes are not bigger than they need to be in order to accommodate all incident edges", see also [1]. For our presentation, we use the simpler constraint that the half-perimeter of the box of each node is at most $deg(v)$. In the same paper, a generic scheme was presented how to create orthogonal drawings of graphs by placing first nodes, then bends, and then ports. We will describe our algorithms using this scheme to simplify our presentation.

Most of our drawings will be in the so-called *Kandinsky-model* introduced by Fößmeier and Kaufmann [7]. In such a drawing, there are two different types of grid-lines. The grid-lines of a coarse grid are used to place the nodes. The grid-lines of a finer grid are added to allow more than one edge to attach on each side of a node. A set of neighboring fine grid-lines, called a *slot*, is assigned to each coarse grid-line. Any two slots are disjoint. Additionally, the Kandinsky-model imposes the *bend-or-end property*. If e is an edge attaching at the top of v, and if w is another node placed in the same column as v, and above v, then either e must have the other endpoint w, or e must be drawn with a bend *below* the lowest row of w. The same holds for the other directions analogously.

Fößmeier and Kaufmann presented an algorithm that computes an orthogonal drawing in the Kandinsky-model with the minimum number of bends [7]. However, this algorithm works only for planar graphs, the nodes all have the same sizes and the running time is $\mathcal{O}(n^2 \log n)$. The required area has not been analyzed. In a subsequent paper, Fößmeier, Kant and Kaufmann gave a linear-time heuristic to achieve, in the same model, an $m \times m$-grid and one bend per edge for planar triconnected graphs [6]. Another recent result by Papakostas and Tollis draws biconnected graphs with width $m - 1$ and height $\frac{m}{2} + 2$ using a less restrictive model [13].

We develop a new algorithm, which yields an $\frac{m+n}{2} \times \frac{m+n}{2}$-grid, and one bend per edge, and improves on the previous algorithms in various ways. It works for any simple graph, without demands on the connectivity. It takes only linear time. It improves the grid-size, apart from differences in the chosen model, by a factor of close to 2, under the reasonable assumption that m is significantly larger than n. Furthermore, the size of the box of each node v is bounded by $\frac{deg(v)}{2}$. However, the aspect ratio of each node can be unbounded, since the box of the node may appear to have size $1 \times \frac{deg(v)}{2}$. A variation of our algorithm achieves an aspect ratio of at most 1:2 for each node, while the grid-size is at most $(\frac{3}{4}m + \frac{n}{2}) \times (\frac{3}{4}m + \frac{n}{2})$.

To explore the Kandinsky-model more thoroughly, we study the special case of the complete graph. We give a construction in a grid of width and height $\frac{m}{2} + \frac{3}{8}n$ with $m - \frac{n}{2}$ bends. Furthermore, we show that any drawing of the complete graph in this model must have $m - n$ bends, thus we are close to optimality.

In incremental scenarios the graph is given one node at a time, and the next node has to be inserted into a fixed previous drawing. This incremental scenario is a first important step towards full interactive scheme and has been considered for 4-graphs in [12]. The critical point for the interactivity is that insertion of new nodes should not change the previous drawing, or at least we can modify it only in a very restricted way. In [15] as well as in [1] interactive schemes have been presented. The second paper yields an $(m + n) \times (m + n)$-grid and m bends. We modify our static algorithm and get area bounds which are only about a factor of 4/3 away from our results for the static scenario.

In the third part of the paper, we consider static algorithms for drawing planar graphs. For triconnected planar graphs we produce drawings in a grid of size of at most $(m - n + 1) \times \min\{\frac{m}{2}, m - n + 1\}$, where the number of bends is $m - n$. The height and width of each box is at most $deg(v)$, which distinguishes these drawings from k-visibility representations (e.g. in [16]), where we have less bends and a smaller area, but in exchange the nodes are bigger.

2 The static scenario

Assume that the full graph G is given in advance. We present one generic algorithm, which works for any node order and edge orientation. Using special implementations, we obtain two different results on high-degree drawings.

2.1 A generic algorithm

Assume some arbitrary node order $\{v_1, \ldots, v_n\}$ and some arbitrary edge orientation of G is given. An edge directed from v_i to v_j is called *good* if $i < j$ and *bad* otherwise. A *predecessor (successor)* of v_j is a neighbor v_i where the edge (v_i, v_j) is incoming (outgoing) at v_j. A predecessor is *good* if the according edge is good. We denote the number of incoming edges of v_j as $indeg(v_j)$, and the number of good and bad incoming edges of v_j as $indeg^{good}(v_j)$, and $indeg^{bad}(v_j)$, respectively. Similarly we define $outdeg(v)$, $outdeg^{good}(v)$ and $outdeg^{bad}(v)$.

We create the drawing by first computing the coarse grid-lines for the nodes, this corresponds to the "node placement" phase introduced in [1]. We assign one row for each node, and one column for each node. No two nodes will be placed in the same row or the same column. These rows and columns will later be expanded into horizontal and vertical slots.

We compute rows for the nodes by processing them in forward order. Consider v_i, $i = 1$ to n. If it has no good predecessor, then we create a new row at an arbitrary place. Otherwise, we add a row close to the median of the rows of the good predecessors. Precisely, let $r_1, \ldots, r_s$ be the rows of the good predecessors of v_i. If s is odd, add a row before or after $r_{\frac{s+1}{2}}$. If s is even, add a row somewhere between $r_{\frac{s}{2}}$ and $r_{\frac{s}{2}+1}$. Place v_i in this row.

We compute a column for each node similarly, but this time in backward order. For v_i, $i = n$ down to 1, place v_i in a new column. This new column is created near the median of the columns of the good successors of v_i, if v_i has good successors, and at an arbitrary place otherwise.

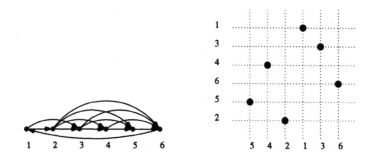

Fig. 1. An example of the node placement.

Next, we assign an approximate place for each bend, which corresponds to the "edge routing" phase in [1]. If $e = (v_i, v_j)$ is an edge directed from v_i to v_j, then we place a temporary bend in the row of v_i and the column of v_j. This edge routing does not yield a feasible drawing, but it gives a first sketch, enough to analyze the drawing. For each node v, let $b(v)$ be the number of edges that attach at the bottom of v. Similarly, define $l(v)$, $r(v)$ and $t(v)$ as the number of incident edges at the left, right, and top side of v.

Lemma 1. *For each node,* $\lfloor \text{outdeg}(v)/2 \rfloor \leq r(v), l(v) \leq \lceil \text{outdeg}^{good}(v)/2 \rceil + \text{outdeg}^{bad}(v)$, *and* $\lfloor \text{indeg}(v)/2 \rfloor \leq t(v), b(v) \leq \lceil \text{indeg}^{good}(v)/2 \rceil + \text{indeg}^{bad}(v)$.

Proof. Consider $b(v)$. By the bend-placement, any bend at the bottom of v belongs to an incoming edge of v. By the node placement, at most half (rounded up) and at least half (rounded down) of the good predecessors are below v. The bad predecessors can be, but need not be, below v. The result follows for $b(v)$, and is similar for the other three sides.

As described in [1], we can get a feasible drawing from this sketch easily. Consider a row r. In this row, there is one node v, and some number of bends. We add $max\{r(v), l(v), 1\} - 1$ rows above the row of v. Then, we distribute the bends among these rows such that no two edges on one side cross, as demonstrated in Fig. 2.

This algorithm can be implementing in $\mathcal{O}(m + n)$ time, using the data structure by Dietz and Sleator [4], the linear-time median-finding algorithm and bucket sort (see for example [2]).

Fig. 2. We assign edges to newly added rows and columns in such a way that there are no crossings between edges from the same side.

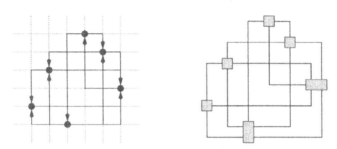

Fig. 3. Continuing the example of Fig. 1, we show the edge routing and how the edges are distributed in the final drawing.

2.2 A small grid-size

In this section, we show how to achieve a small grid-size by choosing a special node order and orientation.

Definition 2. A node order together with an edge orientation is called *polar-free almost-acyclic*, if $indeg(v) \geq 1$ and $outdeg(v) \geq 1$ for all $v \in V$. Furthermore,
(a) $indeg^{bad}(v) \leq 1$, if $indeg^{good}(v) > 0$ then $indeg^{bad}(v) = 0$, and
(b) $outdeg^{bad}(v) \leq 1$, if $outdeg^{good}(v) > 0$ then $outdeg^{bad}(v) = 0$.

Lemma 3. *Let G be a simple graph without nodes with degree ≤ 1. Then G has a polar-free almost-acyclic order and orientation. It can be found in $\mathcal{O}(m)$ time.*

Proof (Sketch). If G is biconnected, compute an *st*-order [11] of it, such that the nodes v_1 and v_n are adjacent. Direct the edges according to it, and reverse edge (v_1, v_n). This can be done in $\mathcal{O}(m)$ time [5].

If G is not biconnected, compute an *st*-order for every biconnected component B of G. If B contains at least two cut-nodes, choose two cut-nodes as first and last node. Otherwise, choose two adjacent nodes in B that are not cut-nodes, and reverse the edge between them. Merge all these orderings such that the order in each component stays unchanged.

Applying the generic algorithm with a polar-free almost-acyclic order and orientation leads to good worst-case bounds on the grid-size.

Theorem 4. *Let G be a simple graph without nodes of degree ≤ 1. Then G has an orthogonal drawing in an $\frac{m+n}{2} \times \frac{m+n}{2}$-grid with one bend per edge. The box size of each node v is at most $\frac{\deg(v)}{2} \times \frac{\deg(v)}{2}$. It can be found in $\mathcal{O}(m)$ time.*

Proof. We will only prove the claim on the height, the claim on the width is similar. After the node placement, we had n rows. For each node v, we add $max\{r(v), l(v), 1\} - 1$ rows. Thus, the height is $\sum_{v \in V} max\{1, r(v), l(v)\}$. By Lemma 1 and the conditions of the polar-free almost-acyclic ordering, we have $r(v), l(v) \leq \lceil \frac{outdeg(v)}{2} \rceil$. Furthermore, since $outdeg(v) \geq 1$ for all nodes, we also have $1 \leq \lceil \frac{outdeg(v)}{2} \rceil$. Therefore, the height is at most $\sum_{v \in V} \lceil \frac{outdeg(v)}{2} \rceil \leq \sum_{v \in V} \frac{outdeg(v)+1}{2} = \frac{m+n}{2}$. The height of the box of node v is $max\{1, r(v), l(v)\} \leq \lceil \frac{outdeg(v)}{2} \rceil \leq \lceil \frac{deg(v)-1}{2} \rceil \leq \frac{deg(v)}{2}$, since $indeg(v) \geq 1$.

A remark here on the condition of "no nodes of degree ≤ 1". Such nodes should be pre-processed and removed from the graph. They can later be re-inserted, by adding only one grid-line and no bend per node. We skip the details here, and only mention that we can achieve a width and height of $\lceil \frac{m+n}{2} \rceil$ for the grid and $\lceil \frac{deg(v)+1}{2} \rceil$ for each node.

2.3 Nodes with bounded aspect ratio

In the previous algorithm, the aspect ratio of a node may be unbounded, since the box of node v may appear as a $1 \times deg(v)/2$ box. In this section, we add the requirement to the model that the nodes should have a bounded aspect ratio. This can be ensured by an orientation via eulerian circuits. Therefore, we make the graph first eulerian by adding new edges between pairs of nodes with odd degree. Then we compute the eulerian circuits which determine the orientation of the edges. For the resulting orientation, we have $indeg(v), outdeg(v) \leq \lceil deg(v)/2 \rceil$.

Now we want a node order $\{v_1, \ldots, v_n\}$ that minimizes the number of bad edges. This problem is $\mathcal{NP}$-complete, since it is the feedback arc problem [8]. But we can always find a node order such that there are at most $m/2$ bad edges.

Applying the generic algorithm with this node order and edge orientation, we obtain good bounds on the aspect ratio of each node. Precisely, one can see from Lemma 1 that the height of v is at most $indeg(v) \leq outdeg(v) + 1$, and the width is at least $(outdeg(v) + 1)/2$, therefore the aspect ratio of v is at most 1:2.

The area of the resulting drawing is determined by the bad edges. If m_g and m_b is the number of good and bad edges, respectively, then the width of the grid is at most $\frac{m_g}{2} + \frac{n}{2} + m_b \leq \frac{3}{4}m + \frac{n}{2}$.

Theorem 5. *Let G be a simple graph without nodes of degree ≤ 1. Then G has an orthogonal drawing in an $(\frac{3}{4}m + \frac{n}{2}) \times (\frac{3}{4}m + \frac{n}{2})$-grid with one bend per edge where each node has aspect ratio at most 1:2. It can be found in $\mathcal{O}(m)$ time.*

We expect that with a suitable choice of a heuristic to determine the node order, the expected area of the drawing can be improved tremendously.

3 The complete graph in the Kandinsky-model

In this section, we study the behavior of the Kandinsky-model for graphs with many edges. We derive upper and lower bounds for the drawing of K_n, the complete graph with n nodes.

Theorem 6. *If n is divisible by 8, then K_n can be embedded in the Kandinsky-model in a grid of width and height $\frac{m}{2} + \frac{3}{8}n$ with $m - \frac{n}{2}$ bends.*

Proof. We divide the nodes into four groups of equal size. Enumerate the nodes in each group as $\{v_1^{(i)}, \dots, v_{n/4}^{(i)}\}$. We place the nodes "as a diamond," i.e., on a coarse $\frac{n}{2} \times \frac{n}{2}$-grid such that node $v_k^{(i)}$ is placed in the ith quadrant and such that the absolute value of the coordinates is $(k, \frac{n}{4} + 1 - k)$.

For each $1 \leq k \leq \frac{n}{4}$, we define the *k-square* to be the four nodes $v_k^{(i)}$, $i = 1, \dots 4$. These nodes induce a K_4 with 6 edges. Four of these edges are drawn straight, the other two edges are drawn with two bends each. We use two columns for these bends if $k \leq \frac{n}{8}$, and two rows otherwise.

For any $1 \leq i \leq \frac{n}{4}$, any $j > i$, and $k = 1, \dots, 4$, we define the *kth i,j-cross* as the group of nodes $\{v_i^{(k)}, v_j^{(k)}, v_i^{(k-1)}, v_j^{(k+1)}\}$, where all additions are modulo 4. In an i,j-cross there are four edges that were not in a k-square. We draw these four edges with one bend each, using two rows and two columns. This is possible since by $j > i$ the ith vertical and the jth horizontal coarse grid-line cross inside the diamond. See also Fig. 4.

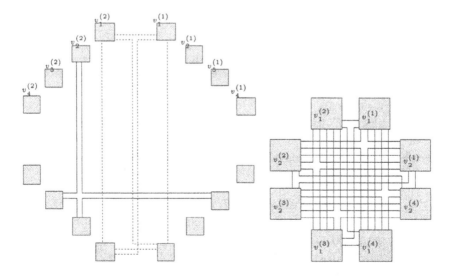

Fig. 4. The construction in the Kandinsky-model. We show the 1-square (dashed), and the third 2,3-cross (solid), and the completed drawing of K_8.

We have $4\sum_{i=1}^{n/4} i(\frac{n}{4} - i - 1) = \frac{n^2}{8} - \frac{n}{2}$ many crosses, each accounts for 4 edges and 4 bends and uses two rows and columns. We have $\frac{n}{4}$ squares, each accounts for 6 edges and 4 bends. Half of the squares use 4 columns and 5 rows, the other half uses 5 columns and 4 rows. Since $4(\frac{n^2}{8} - \frac{n}{2}) + 6\frac{n}{4} = \frac{n^2}{2} - \frac{n}{2} = m$, all edges are either in a square or in a cross. Thus, the total number of bends is $4(\frac{n^2}{8} - \frac{n}{4}) + 4\frac{n}{4} = m - \frac{n}{2}$. The width and height each is $2(\frac{n^2}{8} - \frac{n}{2}) + 4\frac{n}{8} + 5\frac{n}{8} = \frac{n^2}{4} + \frac{n}{8} = \frac{m}{2} + \frac{3}{8}n$.

Variants of this technique lead to other drawings in other models. If we drop the bend-or-end property, but still keep the dimensions of the nodes limited, then we can improve the bounds to a grid of width and height $\frac{m}{2} + \frac{n}{4} - 1$ with $m - 2n + 2$ bends. If we also drop the constraint on the size of the nodes but let them grow arbitrarily (similar as in 2-visibility representations), we can even prove that the grid has width and height $\frac{m}{2} - \frac{3}{4}n + 3$ with $m - 6n + 20$ bends. This is optimal in the number of bends, since any orthogonal drawing has at most $6n - 20$ edges drawn as straight lines [9].

Now we prove a lower bound on the number of bends.

Theorem 7. *Any drawing of the K_n in the Kandinsky-model has at least $m - n$ bends.*

Proof. Assume we have a drawing Γ of K_n. Let A be the number of vertical slots that contain nodes. Let B be the number of horizontal slots that contain nodes. The number of nodes in the ith vertical slot is denoted a_i, while the number of nodes in the jth vertical slot is denoted b_i, so $\sum_{i=1}^{A} a_i = \sum_{i=1}^{B} b_i = n$.

The edges split into three groups. E_a are the edges where the two endpoints are in the same vertical slot, E_b are the edges where the endpoints are in the same horizontal slot. We have $E_a \cap E_b = \emptyset$, by property of the Kandinsky-model. E_c are the remaining edges.

We have $|E_a| = \sum_{i=1}^{A} \binom{a_i}{2}$. Of these edges, $\sum_{i=1}^{A}(a_i - 1)$ can be drawn without bends. On the other hand, all remaining edges in E_a must have at least two bends by the bend-or-end property. Therefore, the number of bends in E_a is at least $2\sum_{i=1}^{A}[\binom{a_i}{2} - (a_i - 1)]$. Similarly, the number of bends in E_b is at least $2\sum_{i=1}^{B}[\binom{b_i}{2} - (b_i - 1)]$.

Every edge in E_c has at least one bend. So the total number of bends is

$$m - \sum_{i=1}^{A}\binom{a_i}{2} - \sum_{i=1}^{B}\binom{b_i}{2} + 2\sum_{i=1}^{A}\left[\binom{a_i}{2} - (a_i - 1)\right] + 2\sum_{i=1}^{B}\left[\binom{b_i}{2} - (b_i - 1)\right]$$

$$= m + \sum_{i=1}^{A} a_i^2/2 + \sum_{i=1}^{B} b_i^2/2 - 5n + 2A + 2B.$$

Given fixed values of A, B, the minimum of this expression is achieved if all a_i's, respectively all b_i's, are the same; so $a_i = \frac{n}{A}$ and $b_i = \frac{n}{B}$. The number of bends then is $m + n^2/2A + n^2/2B - 5n + 2A + 2B$. Minimizing this for A and B, we arrive at $A, B = \frac{n}{2}$, therefore the number of bends is at least $m + 2n - 5n + 2n = m - n$.

4 Incremental drawing of graphs with high degrees

We now study the *incremental scenario* where the nodes are given one by one, but we have to fix one placement of a node before the next is given. An insertion of a new row or column is allowed only at those positions where such an operation does not stretch any node box unnecessarily large. We naturally generalize the relative-coordinates scenario introduced in [12].

Assume the order of the nodes given is $\{v_1, \ldots, v_n\}$. Since the static algorithm also worked with a node order, it is an obvious idea to try to use the same algorithm with the user-defined node order and the induced edge orientation. Two main differences occur: We have no information about the node order, and in particular it is possible that the in-degree or out-degree of a node is 0. Secondly, we have no information about the successors of node v_i when placing v_i, and therefore have to find a different column-choice strategy.

We need to maintain a valid drawing, i.e., a drawing where nodes are boxes and no two edges overlap. Thus, we will represent the nodes as boxes throughout the algorithm and keep the invariant that for any two boxes, the x-intervals of the boxes as well as the y-intervals are disjoint.

So assume nodes $v_1, \ldots, v_{i-1}$ are drawn in this way. Compute all predecessors of node v_i, and find their median as in the static scenario. Add a new row at this median place, such that it does not intersect any existing node (this is possible since the y-intervals are disjoint).

Add $w_i = max\{1, \lceil \frac{indeg(v_i)}{2} \rceil\}$ new columns for v_i, and add these either at the extreme left or at the extreme right. Clearly, in a practical implementation, one would allow to place a new node at any new column in the middle of the drawing, but here we analyze only the situation where placements to the right or left hand side of the drawing is allowed. Thus, these columns do not intersect any existing node. v_i will be drawn as $w_i \times 1$-box in the beginning, and will increase in height later, when we add more outgoing edges.

To route an edge (u, v_i), we may have space left at the correct side of u, or we may have to increase the height of u to make space for the edge. We can increase u by adding a new row, this will not intersect any other node. We make sure that this new row is between the upward-bending and the downward-bending edges on either side. The edge (u, v_i) will leave u on the right or left (depending on where we placed v_i), and enter v_i at the top or bottom side of the box of v_i. It bends exactly once.

Let n_s be the number of nodes with in-degree 0. Since for each node v_i the width of the box is $max\{1, \lceil \frac{indeg(v_i)}{2} \rceil\}$, the total width now is $n_s + (m + n - n_s)/2 = (m + n + n_s)/2 \leq \frac{m}{2} + n$.

Since we choose a column without knowledge about the successors, any number of outgoing edges may attach on the right side or on the left side of the box. Thus, we can estimate the height of node v_i only by $h_i = max\{1, outdeg(v_i)\}$. So if n_t is the number of nodes with out-degree 0, then the total height may be as much as $n_t + m$.

We reduce this bound by choosing the column for v_i wisely. At a fixed time, if a node v has $r(v)$ incident edges on the right side and $l(v)$ edges on its left side,

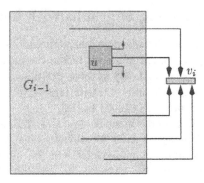

Fig. 5. v_i is inserted at the extreme left or right end, and at the median of the rows of its predecessor.

and if $r(v) > l(v)$ then we say $l(v)$ edges on each side are *mated*, and $r(v) - l(v)$ edges are *(right) free*. Obviously, the height of v is $r(v)$. We call v *right-weighted* since $r(v) > l(v)$.

If f is the number of left or right free edges at the end, then $m - f$ is the number of mated edges. The total height of the drawing is n_t plus half of the mated edges plus the number of free edges, which amounts to $n_t + \frac{m-f}{2} + f = n_t + \frac{m}{2} + \frac{f}{2}$. So we want to minimize the number of free edges.

Our approach is greedy and chooses that side for the placement of node v_i which appears better with respect to the number of free edges. If the number of right-weighted predecessors of v_i is smaller than the number of left-weighted predecessors, then we place v_i on the right hand side. Therefore, the left-weighted predecessors loose one free edge each. Otherwise, we place v_i on the left side, and the right-weighted predecessors loose one free edge each.

We estimate the number f of free edges. There are only $n - n_t$ nodes with outgoing edges, and only those may have free edges. So the number of first free edges is at most $n - n_t$. Consider a right free edge $e = (v, w)$ which is not the first on its node v. Edge e was inserted when we placed w on the right side. Hence the number a of left-weighted predecessors of w was at least as big as the number b of right-weighted predecessors. We created $2a$ mated edges (the incoming edges of w, and the edges that caused the predecessors to be left-weighted), and only b right-free edges. So we can assign two mated edges to every free edge that was not the first on its node. This gives $f \leq n - n_t + \frac{1}{2}(m - f)$, or $f \leq \frac{2}{3}(n - n_t) + \frac{1}{3}m$. Therefore, the total height is at most $\frac{1}{3}n + \frac{2}{3}n_t + \frac{2}{3}m \leq \frac{2}{3}m + n$.

Theorem 8. *Assume G is given incrementally. Then we can achieve an $(\frac{m}{2} + n) \times (\frac{2}{3}m + n)$-grid and 1 bend per edge.*

Next, we show that the analysis above is tight (at least for this algorithm). We define a graph with $n + 1$ nodes, where n is divisible by 3, as follows:

Insert nodes v_1, v_2, v_3 without any edges.

For $i = 1$ to $n/3 - 1$ **do**

    let $j = 3 \cdot i$;

    insert v_{j+1} adjacent to $v_{j/3+1}, ..., v_j$; (* inserted to the left hand side *)

    insert v_{j+2} adjacent to $v_1, ..., v_{2j/3}$; (* inserted to the right *)

    insert v_{j+3} adjacent to $v_1, ..., v_{j/3}, v_{2j/3+1}, ..., v_{3j}$; (* inserted to the right *)

od;

insert v_{n+1} adjacent to $v_{n/3}, ..., v_{n-1}$; (* inserted to the right *)

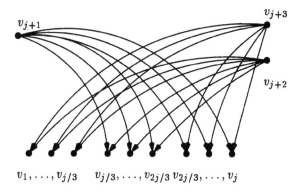

Fig. 6. A graph where incremental drawing performs badly.

We can prove for this graph with $n + 1$ nodes, and a total number of edges of $m = n^2/3 - n/3$, a bound for the height of the drawing of $1 + \frac{2}{3}m + \frac{2}{9}n$. This almost closes the gap between the guaranteed behavior of the greedy algorithm and its behavior on a specific example.

5 Planar graphs

We now present a new linear-time heuristic that works for triconnected planar graphs, and that gives an $(m-n+1) \times \min\{\frac{m}{2}, m-n+1\}$-grid and $m-n$ bends. Every edge has at most one bend. Thus, we improve the grid-size by a factor of 2, and we decrease the number of bends by n, compared to the best previous bounds of an $m \times m$-grid and m bends [6].

5.1 The canonical ordering

Assume from now on that G is a triconnected, simple, planar graph. For such graphs, we can use the *canonical ordering* as introduced by Kant [10]. For a node ordering $\{v_1, \ldots, v_n\}$, let $G(i)$ be the graph induced by $v_1, \ldots, v_i$, in the planar embedding as induced by G.

Lemma 9. *[10] Let G be a planar simple triconnected graph with a fixed planar embedding. Then G has a node ordering* $V = \{v_1, \ldots, v_n\}$, *called a* canonical ordering, *such that the following holds:*

- (v_1, v_2) *is an edge and belongs to the outer-face, with v_1 clockwise after v_2 on the outer-face.*
- v_n *belongs to the outer-face and has at least three neighbors.*
- *For $3 \leq j \leq n-1$, v_j is in the outer-face of $G(j-1)$, and one of the following holds:*
  - *Either "v_j is a new single node", i.e., v_j has at least three neighbors in $G(j-1)$ and at least one neighbor in $G - G(j)$. $G(j)$ is biconnected.*
  - *Or "v_j is part of a new chain", i.e., there exists i, k, $i \leq j \leq k$, such that for all $i < l < k$ v_l is adjacent to v_{l-1} and v_{l+1}, has no other neighbor in $G(k)$, and at least one neighbor in $G - G(k)$. Furthermore, v_i and v_k have each exactly one neighbor in $G(i-1)$ and at least one neighbor in $G - G(k)$. $G(k)$ is biconnected.*

Fig. 7. The two different possibilities for v_j.

Given an element other than the first one of the ordering, let the *left foot-point* be the clockwise first node after v_1 on the outer-face connected to a node in the new element. We define the *right foot-point* similarly.

We call a node on the outer-face of G_i *open* if it still has neighbors in $G - G_i$, otherwise we call it *closed*. We distinguish the chain-elements further. A chain is *left-free* if its left foot-point is closed after adding the chain. Or, in other words, the highest outgoing edge of the left foot-point, which is the edge leading to the node with the highest number in the ordering, is the one that is incoming to the chain. Similarly a chain is *right-free* if its right foot-point is closed after adding the chain. A chain is *free* if it is either right-free or left-free, and *non-free* otherwise.

5.2 The placement

We add the nodes following the elements of the canonical ordering. Throughout the algorithm, we maintain the invariant that the open nodes have disjoint intervals, and they are sorted from left to right when going around the outer-face from v_1 to v_2 in clockwise direction.

We start with the **placement of** v_1 **and** v_2. We add one row and two columns for these two nodes, and draw the edge between them as a straight line.

Assume we want to **add a new non-free chain** $v_k, \ldots, v_l$ with left and right foot-point c_α respectively c_β. Let r_α be the highest row used by any incident edge of c_α on the right side of c_α. Similarly, let r_β be the highest row used by any incident edge of c_β on the left side. We will embed the chain in the row above $max\{r_\alpha, r_\beta\}$ (we add a new row on top if there was no such row yet). Call this row r_c. Add $l - k + 1$ new columns between the columns of c_α and c_β. Place $v_k, \ldots, v_l$ in these columns and in r_c. For lack of space, we skip the proof that this placement does not create any overlap.

If $r_c = r_\alpha + 1$, then, if necessary, we increase the height of the node c_α by 1 unit, so that it now overlaps r_c as well. In this case, the edge (c_α, v_k) is routed as horizontal line. Otherwise, we increase the width of c_α by one, and add a new column between the left-continuing and right-continuing outgoing edges of c_α. We route the edge (c_α, v_k) using this column with a bend above c_α. Similarly we proceed for the edge (c_β, v_l). All edges (v_i, v_{i+1}), $i = k, \ldots, l - 1$ are routed horizontally along r_c.

Assume we next want to **place a left-free chain** $v_k, \ldots, v_l$ (the placement of a right-free chain is symmetric). Let the foot-points be again c_α and c_β. Add a new row r_c on top of the drawing. Add $k - l$ columns between the columns of c_α and c_β. All nodes of the chain will be placed in r_c. Place v_k in the column of c_α (this does not violate the invariant, since c_α is closed after adding the chain). Place $v_{k+1}, \ldots, v_l$ in the newly created columns. The edge (c_α, v_k) is routed vertically. All edges (v_i, v_{i+1}), $i = k, \ldots, l - 1$ are routed horizontally. The edge (v_l, c_β) is routed with a bend above c_β, this adds a new column to c_β.

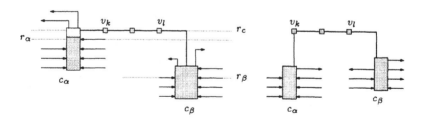

Fig. 8. Placement of a non-free chain and left-free chain, respectively.

Finally, assume we next want to **place a new node** v_i. Let $w_1, \ldots, w_d$ be the predecessors of v_i, sorted from left to right. Nodes with in-degree 2 are chain-elements, so $d \geq 3$. Add $\lceil \frac{d-1}{2} \rceil$ rows on top of the existing drawings. v_i will overlap all these rows. Add a new column to each predecessor of v_i. Place v_i in the new column of $w^* = w_{\lceil \frac{d}{2} \rceil}$, which is closed after adding v_i since $d \geq 3$.

Route the edge (w^*, v_i) vertically, while all other edges (w_j, v_i) are routed with a bend above w_j. This adds a new column to c_α and c_β. Assign rows to the incoming edges of v_i such that there is no crossing among them.

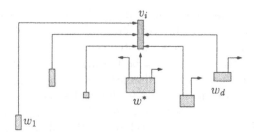

Fig. 9. Placement of new node.

5.3 Bounds

Let us now consider the height of the obtained drawing. Placing v_1 and v_2 requires one row. Placing a chain $v_k, \ldots, v_l$ (free or not) requires $1 = \frac{indeg(v_l)}{2} \leq \sum_{i=k}^{l} \frac{indeg(v_i)}{2}$ rows. Placing a node requires $\lceil \frac{indeg(v_i)-1}{2} \rceil$ rows. Since $indeg(v_2) = 1$, the total height is therefore at most $\frac{1}{2} + \sum_{v \in V} \frac{indeg(v)}{2} = \frac{m+1}{2}$. With a clever choice of v_n, we can shave off this $\frac{1}{2}$-term, and get a bound of $\frac{m}{2}$. Another estimation on the height can be obtained as follows: For v_1 and v_2 we use $1 = 2 + \sum_{i=1}^{2} (indeg(v_i) - 1)$ rows. For every chain we use $1 = \sum_{i=k}^{l} (indeg(v_i) - 1)$ rows. For placing a node we use $\lceil \frac{indeg(v)-1}{2} \rceil \leq indeg(v) - 2$ rows, since the in-degree is at most 3. Since we have at least one node-element in v_n, the total number of rows therefore is at most $1 + \sum_{v \in V} (indeg(v) - 1) = m - n + 1$.

Now let us consider the width. We will count added columns when we route incoming edges. Thus, to place v_1 and v_2 we do not count any of the used columns – they will be accounted for when placing the highest outgoing edges of v_1 and v_2. Similarly, we do not count any columns when adding a non-free chain, and only one column when adding a free chain.

Finally, when adding a node v with in-degree d, we count the $d-1$ columns of the predecessors that are not the median predecessor w^*. The column of w^* will be accounted for when placing the last outgoing edge of v. The only exception to this is the case $v = v_n$, in which case we must count d columns. Since the second element of the ordering is always a non-free chain, the total number of columns is at most $\sum_{v \in V} indeg(v) - 1 + 1 = m - n + 1$. Similarly one can estimate the number of bends as $m - n$. We increase the height or width of a node only if we add a new incident edge to it. Every node has an incident horizontally attaching edge (one of the incoming edges), and an incident vertically attaching edge (the last outgoing edge). Therefore, the half-perimeter of each node is at most $deg(v)$.

It is quite straightforward to show that the algorithm can be implemented in linear time.

Theorem 10. *There exists a linear-time heuristic to draw a planar triconnected graph orthogonally in an $(m-n+1) \times \min\{m-n+1, \frac{m}{2}\}$-grid with $m-n$ bends. Every edge has at most one bend. The half-perimeter of the box of each node is at most $\deg(v)$.*

5.4 Variations of the algorithm

We now show how to achieve related results with slight variations of the algorithms. For lack of space, we have to skip all proofs.

A graph is called *outer-planar* if we can add a dummy-node v^* connected to all nodes and the graph stays planar. If we choose this dummy-nodes as last node, then we can show that we get an $(n-1) \times n$-drawing where all edges are routed horizontally. So the produced drawing is a 1-visibility representation. As opposed to all previous algorithms (e.g. [16,14,9,6]), in our drawings there are known bounds on the height of a node.

Theorem 11. *Let G be an outer-planar graph. Then G has a 1-visibility representation in an $(n-1) \times n$-grid. Every node v has height at most $\deg(v)$.*

In our drawings of planar graphs, the half-perimeter of each node is at most $deg(v)$, but the drawing is not necessarily in the Kandinsky-model, since we may violate the bend-or-end property when adding a non-free chain. By changing the placement of non-free chains, we can get a drawing in the Kandinsky-model at the cost of introducing more bends.

Theorem 12. *There exists a linear-time heuristic to draw a planar simple tri-connected graph orthogonally planar in the Kandinsky-model in an $(m-1) \times \min\{m - n + 1, \frac{m}{2}\}$-grid with $m-2$ bends. Every edge has at most one bend.*

A slight modification of our algorithm produces 2-visibility drawings of small area, where the best previous bound was a half-perimeter of $2n$ [6].

Theorem 13. *There exists a linear-time heuristic to draw a planar simple graph without bends or crossings as a 2-visibility drawing in an $(n-1) \times (n-1)$-grid.*

6 Conclusion

In this paper, we presented an algorithm to compute an orthogonal drawing of a simple graph with arbitrary degrees. We achieved a grid-size of $\frac{m+n}{2} \times \frac{m+n}{2}$. Furthermore, every edge has exactly one bend, thus the number of bends is m. This result improves previous results by a factor that approaches 2 as m gets large relative to n. We also, for the first time, managed to show a non-trivial bound on the box size of at most $\frac{deg(v)}{2}$ for node v. For the incremental scenario, where the drawing has to be produced as the nodes are given in a sequence, we achieve a grid-size of $(\frac{m}{2} + n) \times (\frac{2}{3}m + n)$, which is surprisingly close to the results for the static case.

Many open problems remain:

- In the static scenario, we would like to remove the "$\frac{n}{2}$" terms. It arises if the in-degree or out-degree of a node is odd, since we need to add a $\frac{1}{2}$ as a correction term for rounding. At least one of the two terms cannot be avoided for nodes with odd degrees. But can we reduce it if there are many nodes with even degree?

– It is not clear how much improvement can be achieved in other models. If we drop the restrictions on the size of boxes, does the grid-size get smaller?

– Our algorithm for the static case does not consider planarity of the graph, and in fact, may create $\mathcal{O}(n^2)$ crossings for a planar graph. Can a grid-size of (roughly) $\frac{m}{2}$ in both directions be achieved for planar drawings as well?

References

1. T. Biedl, B. Madden, and I. Tollis. The Three-Phase Method: A Unified Approach to Orthogonal Graph Drawing. Tech. Report, University of Texas at Dallas, UT-DCS 03-97, 1997.

2. T.H. Cormen, C.E. Leiserson, and R.L. Rivest. *Introduction to Algorithms.* MIT Press, McGraw-Hill Book Company, 1990.

3. G. Di Battista, P. Eades, R. Tamassia, and I. Tollis. Algorithms for drawing graphs: an annotated bibliography. *Computational Geometry: Theory and Applications,* 4(5):235–282, 1994.

4. P.F. Dietz and D.D. Sleator. Two algorithms for maintaining order in a list. In *Proc. 19th Annual ACM Symp. Theory of Computing,* pp. 365–372, 1987.

5. S. Even and R.E. Tarjan. Computing an *st*-numbering. *Theoretical Computer Science,* 2:436–441, 1976.

6. U. Fößmeier, G. Kant, and M. Kaufmann. 2-visibility drawings of planar graphs. In S. North, editor, *Symp. on Graph Drawing, GD 96,* volume 1190 of *Lecture Notes in Computer Science,* pp. 155-168. Springer Verlag, 1997.

7. U. Fößmeier and M. Kaufmann. Drawing high degree graphs with low bend numbers. In F. Brandenburg, editor, *Symp. on Graph Drawing, GD 95,* volume 1027 of *Lecture Notes in Computer Science,* pp. 254–266. Springer Verlag, 1996.

8. M. Garey and D. Johnson. *Computers and Intractability: A Guide to the Theory of NP-Completeness.* Freeman, 1979.

9. J. Hutchinson, T. Shermer, and A. Vince. On representation of some thickness-two graphs. In F. Brandenburg, editor, *Symp. on Graph Drawing, GD 95,* volume 1027 of *Lecture Notes in Computer Science,* pp. 324–332. Springer Verlag, 1996.

10. G. Kant. *Algorithms for Drawing Planar Graphs.* PhD thesis, Univ. Utrecht, 1993.

11. A. Lempel, S. Even, and I. Cederbaum. An algorithm for planarity testing of graphs. In *Proc. of Theory of Graphs, Int. Symp. Rome,* pp. 215–232. 1966.

12. A. Papakostas and I. Tollis. Issues in interactive orthogonal graph drawing. In F. Brandenburg, editor, *Symp. on Graph Drawing, GD 95,* volume 1027 of *Lecture Notes in Computer Science,* pp. 419–430. Springer Verlag, 1996.

13. A. Papakostas and I. Tollis. High-degree orthogonal drawings with small grid-size and few bends. *Workshop on Algorithms and Data Structures, WADS 97,* Lecture Notes in Computer Science. To appear.

14. P. Rosenstiehl and E. Tarjan. Rectilinear Planar Layouts and Bipolar Orientations of Planar Graphs. *Discrete Computational Geometry,* 1:343–353, 1986.

15. R. Tamassia, G. Di Battista, and C. Batini. Automatic graph drawing and readability of diagrams. *IEEE Trans. on Systems, Man and Cybernetics,* 18(1), 1988.

16. R. Tamassia and I. Tollis. A unified approach to visibility representations of planar graphs. *Discrete Computational Geometry,* 1:321–341, 1986.

Denesting by Bounded Degree Radicals

Johannes Blömer

Institut für Theoretische Informatik, Eidgenössische Technische Hochschule Zürich, ETH Zentrum, CH-8092 Zurich, Switzerland

Abstract. Given a nested radical α involving only d^{th} roots we show how to compute an optimal or near optimal depth denesting of α by a nested radical that only involves D^{th} roots, where D is an arbitrary multiple of d. As a special case the algorithm computes denestings as in [9]. The running times of the algorithms are polynomial in the description size of the splitting field for α.

1 Introduction

Simplification or denesting of radical expressions is a natural simplification problem that algebraic and symbolic manipulation systems face. Denestings are useful for manipulating large formulas as well as understanding the final result. Accordingly, starting in the mid-70's the problem has been studied intensively in Computer Algebra or Algorithmic Algebra (see for example [5],[14],[4],[9],[10],[6],[2], [3], and in particular the survey article by S. Landau [11]). Without doubt, many researches were also attracted by the following seemingly mysterious equations, which can be found in Ramanujan's notebook and which nicely illustrate and explain the general problem.

$$\sqrt[3]{\sqrt[3]{2}-1} = \sqrt[3]{1/9} - \sqrt[3]{2/9} + \sqrt[3]{4/9}$$

$$\sqrt[6]{7\sqrt[3]{20}-19} = \sqrt[3]{5/3} - \sqrt[3]{2/3}.$$

In each of these equations the depth 2 formula on the left is denested by a depth 1 formula on the right.

In denesting radicals an important question is which field to consider as the ground field, the field of constants, so to speak. In the examples given above the ground field is the field of rational numbers. In general, it is not feasible to consider arbitrary ground fields. Dealing symbolically with radicals, nested or just plain roots, usually requires the presence of appropriate roots of unity, that is, roots of the polynomials $X^d - 1$, $d \in \mathbf{N}$.

Horng/Huang [6] denest a nested radical expression using arbitrary roots. To do so they consider a ground field containing all possible roots of unity. The nesting depth they achieve is the minimal possible one over this extension of the original field. Computing in a field containing all roots of unity is computationally infeasible and Horng/Huang show that a finite number of roots of unity suffice. The bound on the degree of the roots of unity that one needs to consider is, in the worst case, double-exponential in the degrees of the roots appearing in the original expression.

Landau [9] sticks closer to the field over which the original expression was defined. She adjoins to this field roots of unity whose degree is related to the Galois group of the nested radical. Over this extension field she computes a denesting such that the depth of this denesting differs at most by 1 from the depth of the optimal denesting over the original field. The worst case bound on the degree of the roots of unity that are adjoined to the original field is single-exponential in the degrees of the roots involved in the original expression.

In this paper, we follow a different strategy than Landau and Horng/Huang. We believe that adjoining large degree roots of unity may actually hide interesting information about the original expression. As an example, consider $\sqrt{5 + 2\sqrt{6}}$. It denests over the rational numbers to $\sqrt{2} + \sqrt{3}$. However, $\sqrt{2}$ and $\sqrt{3}$ are contained in the extension of $\mathbf{Q}$ containing all roots of unity. Hence, instead of finding the denesting $\sqrt{2} + \sqrt{3}$ the algorithms in [6], for example, will return some expression involving roots of unity. The most innocent one being $\sqrt{-1}(\zeta_3 + \zeta_3^2) + (\zeta_8 + \zeta_8^7)$, where ζ_3 and ζ_8 are primitive 3rd and primitive 8th roots of unity.

Therefore, in this paper we do not try to obtain denestings using arbitrary roots. Instead, in our algorithms the user can specify an integer D, with the understanding that only roots whose degree divides D are allowed in the denesting. Quite naturally, we require that D is a multiple of the least common multiple of the degrees in the original expression. Accordingly, the only root of unity we have to adjoin to the ground field is a primitive D^{th} root of unity. The algorithm then computes a denesting that is optimal for the class of nested radicals that use D^{th} roots over this extension field. The depth of the denesting is at most the depth of the optimal denesting that is defined over the original field and that uses only D^{th} roots.

In particular, for nested radicals defined over an arbitrary field involving only square roots the algorithm finds the optimal denesting using square roots without changing the underlying field. Thus for $\sqrt{5 + 2\sqrt{6}}$ and setting $D = 2$ the algorithm produces the denesting $\sqrt{2} + \sqrt{3}$ rather than some expression in roots of unity. With respect to nested radicals of square roots our result generalizes results in [4], where a restricted class of nested radicals involving square roots was considered.

The algorithm is similar to the algorithm in [9] and for a special choice of D it is the same. Also the running time of our algorithms is comparable to the running time of Landau's algorithm and the analysis is similar, too. The running time is polynomial in size of the splitting field of the input expression and can therefore be in the worst case exponential in the input size. Depending on the choice of the parameter D, compared to Landau's algorithm we need to work in field extensions of smaller degree. In these cases the algorithm will be more efficient. Both, the algorithms in this paper and in [9] are always more efficient than the algorithms in [6].

The restriction that D is a multiple of the least common multiple of the degrees in the original expression can be avoided. Without this restriction the original expression may not be expressible at all using only D^{th} roots. Therefore we show that for arbitrary D and an arbitrary algebraic number α it can be decided in time polynomial in the size of the minimal polynomial of α whether α can be expressed as a nested radical involving only D^{th} roots. Here we have to assume that the ground field contains a primitive D^{th} root of unity. This result is a consequence of the breakthrough result on solvability by radicals obtained in [12] and the techniques presented in our paper.

The techniques of this paper provide alternative proofs to the structure theorems in [9]. Our proofs are simpler and more direct than the original proofs in [9]. In many respects we reverse the order of arguments in [9]. Instead of deriving field theoretic consequences from group theoretic arguments, we obtain the necessary field theoretic facts directly. Then we use these results to derive group theoretic consequences, which in turn lead to the algorithms. For lack of space, in this extended abstract we omit a detailed discussions of these issues.

The paper is organized as follows. Section 2 recalls the basic definitions and facts about (nested) radicals. Section 3 contains the basic structure theorems, while Section 4 shows how these structure theorems yield denesting algorithms. Section 5 briefly deals with issues of solvability by radicals.

2 Definitions and Basic Facts

Except for some minor changes we follow the notation used in [6]. Let $K \supseteq \mathbf{Q}$ be an algebraic number field. In the following op $\in \{+, -, \times, \div\}$. Nested radicals and their nesting depth are defined inductively. An element $a \in K$ is a *nested radical of depth* 0 over K, denoted $\text{depth}_K(a) = 0$. For any two nested radicals α, β over K, α op β is a nested radical over K. The depth of α op β is defined as $\text{depth}_K(\alpha \text{ op } \beta) = \max\{\text{depth}_K(\alpha), \text{depth}_K(\beta)\}$. If α is a nested radical over K and $d \in \mathbf{N}, d > 1$, then $\sqrt[d]{\alpha}$ is a nested radical over K. The depth of $\sqrt[d]{\alpha}$ is $\text{depth}_K(\alpha) + 1$.

Let α be a nested radical over K. The *set of associated simple radicals* of α, denoted $S(\alpha)$, is defined as follows. If $\alpha = a \in K$ then $S(\alpha) = \{a\}$. If $\alpha = \beta$ op γ then $S(\alpha) = S(\beta) \cup S(\gamma)$. Finally, if $\alpha = \sqrt[d]{\beta}$ then $S(\alpha) = S(\beta) \cup \{\sqrt[d]{\beta}\}$.

Due to the ambiguity of the symbol $\sqrt[d]{\ }$ a nested radical α over K does not refer to a unique algebraic number over K. A *radical valuation* resolves this ambiguity. Let R be the set of all nested radicals over K. A radical valuation v is a function $v : R \to \mathbf{C}$ such that $v(a) = a$ for all $a \in K$, $v(\alpha \text{ op } \beta) = v(\alpha) \text{ op } v(\beta)$, for all $\alpha, \beta \in R$, and $(v(\sqrt[d]{\alpha}))^d = v(\alpha)$, for all $\alpha \in R$ and all $d \in \mathbf{N}$.

If $a \in K$, then a radical valuation v relates the symbol $\sqrt[d]{a}$ to some specific d^{th} root of a. All the other d^{th} roots of a are given by the values of $\sqrt[d]{1}^k \sqrt[d]{a}$, where $k \in \{0, \ldots, k-1\}$ and the value of $\sqrt[d]{1}$ is some primitive d^{th} root of unity. Similar remarks apply to the values of $\sqrt[d]{\alpha}$, where $\alpha \in R$. To simplify the notation, we will not mention the radical valuation v explicitly, writing $\sqrt[d]{\alpha}$, instead of $v(\sqrt[d]{\alpha})$. But it is always understood that $\sqrt[d]{\alpha}$ refers to some fixed algebraic number over K.

Let α, β be nested radicals over k. β is called a *denesting of* α iff the values of α and β are the same and $\mathrm{depth}_K(\beta) \leq \mathrm{depth}_K(\alpha)$. The nested radical β is called a *denested form of* α iff in addition for every denesting $\gamma \in R$ of α we have $\mathrm{depth}_K(\beta) \leq \mathrm{depth}_K(\gamma)$.

A radical α of depth 1 over K is called an *order d radical* iff $\alpha^d \in K$. In this paper we will mostly deal with radicals whose order is a divisor of some fixed but arbitrary positive integer. An extension $K(\alpha_1, \ldots, \alpha_n)$ is a *radical extension of K* iff α_i is a depth 1 radical for $i = 1, \ldots, n$. The extension $K(\alpha_1, \ldots, \alpha_n)$ is called an *order d radical extension* iff α_i is an order d radical for $i = 1, \ldots, n$. By this definition an order d radical (an order d radical extension) is also an order D radical (an order D radical extension) for all D that are divisible by d.

As was done in [9] and [6] we will relate the nesting depth of nested radicals to the length of certain field towers. A field tower $K = K_0 \subseteq K_1 \subseteq \ldots \subseteq K_n$ is called a *radical tower over K* iff K_i is a radical extension of K_{i-1} for all $i = 1, \ldots, n$. The integer n is called the *length* of the radical tower.

If $S(\alpha)$ is the set of associated radicals of some nested radical α, then we denote by $S_i \subseteq S(\alpha)$ the set of elements in $S(\alpha)$ of depth at most i. The *radical tower associated to* α is given by the fields $K_i = K(S_i)$. Using these notions we can generalize the notion of order d radicals to that of order d nested radicals. A nested radical α is called an *order d nested radical* iff its associated radical tower is an order d radical tower. As before, an order d nested radical is also an order D nested radical for all D divisible by d.

From these definitions we immediately get

Lemma 1. *A nested radical α has depth n over K iff the associated radical tower has length n. The nesting depth of a denested form of α is the length of a shortest radical tower $K \subset K_1 \subset \ldots \subset K_n$, such that $\alpha \in K_n$.*

In this paper we show how to compute denestings for order d radicals by order d radicals. Hence we make the following, final definition of this section.

Definition 2. *Let α be an order d nested radical over K. An order d nested radical β is called an order d denesting of α iff β has the same value as α and $\mathrm{depth}_K(\beta) \leq \mathrm{depth}_K(\alpha)$. β is called an order d denested form of α if in addition for all order d nested radicals γ having the same value as α, $\mathrm{depth}_K(\beta) \leq \mathrm{depth}_K(\gamma)$.*

Frequently, the following basic facts about order d radicals and order d radical extensions will be used. Proofs for these facts can be found in [7]

Theorem 3. *Let $K \supseteq \mathbf{Q}$ be a field that contains a primitive d^{th} root of unity. If α is an order d radical over K, then its minimal polynomial over K is of the form $X^t - a$, where t divides d and $a \in K$. The extension $K(\alpha)$ is a Galois extension of K with cyclic Galois group of order t.*

Theorem 3 fails if K does not contain a primitive d^{th} root of unity. A simple counterexample is given by the 3rd roots of unity. The 3rd roots of unity are the roots of $X^3 - 1$ which factors as $(X - 1)(X^2 + X + 1)$. Only the minimal polynomial of 1 has the form stated in the theorem. For the primitive 3rd roots of unity the minimal polynomial is $X^2 + X + 1$. Hence only if K contains a primitive d^{th} root of unity can we assume that the degree of the field extension generated by an order d radical over K is a divisor of d. Theorem 3 is the reason that almost any algorithm dealing symbolically with radicals assumes that the ground field contains appropriate roots of unity. The exception being the algorithms described in [2],[3]. Those algorithms deal with real radicals over real fields, in which case a theorem similar to Theorem 3 holds.

If the assumptions of Theorem 3 hold, the structure of order d radical extensions is well understood.

Theorem 4. *Assume $K \supseteq \mathbf{Q}$ contains a primitive d^{th} root of unity.*

(i) *If $L = K(\alpha_1, \ldots, \alpha_n)$ is an order d radical extension of K, then L is a Galois extension of K. The Galois group $G(L/K)$ of L over K is an abelian group of exponent d.*
Conversely, if L is a Galois extension of K with abelian Galois group of exponent d, then L is an order d radical extension $K(\alpha_1, \ldots, \alpha_n)$ of K.
(ii) *Every subfield of an order d radical extension $K(\alpha_1, \ldots, \alpha_n)$ is an order d radical extension of K.*

3 Algebraic Structure of Denestings

Since any d^{th} root can also be considered a Dth root, provided d divides D, denesting a nested radical involving only d^{th} roots by a nested radical involving only D^{th} roots is a special case of denesting a nested radical with d^{th} roots by a nested radical also containing only d^{th} roots. Therefore we will state all our results in this simpler form. However, the reader should bear in mind the more general formulation.

Let $K \supseteq \mathbf{Q}$ be an algebraic number field. Let α be an order d nested radical over K. By ζ_d we denote a primitive d^{th} root of unity. The field $K(\zeta_d)$ is denoted by F. The Galois closure of $F(\alpha)$ over F is the smallest Galois extension of F containing $F(\alpha)$. It is the field generated by α and all its conjugates over F. The Galois closure is denoted by L.

The goal of this section is to prove the following theorem.

Theorem 5. *If β is an order d denesting of α over K with depth $\mathrm{depth}_K(\beta) = n$, then there is an order d radical tower $F = F_0 \subseteq F_1 \subseteq \ldots \subseteq F_n = L$ such that F_i is a Galois extension of F for all $i = 1, \ldots, n$.*

As will be seen later, the interesting part in computing a denesting for α is to determine the radical tower $F \subseteq F_1 \subseteq \ldots \subseteq F_n$. Once the tower has been computed standard techniques can be used to express α as a nested radical of depth at most n.

The theorem guarantees a near optimal denesting of α. It is not optimal because the ground field for the radical tower is F rather than K. As will be seen in the proof, this is because Theorem 3 fails for fields not containing appropriate roots of unity. If we consider a primitive d^{th} root of unity an order d radical the theorem shows how to find a denesting of α over K whose depth differs from the optimal depth by 1.

To prove Theorem 5, first it is shown that the basic properties of radical towers are preserved by taking Galois closures.

Lemma 6. *Let* $K(\zeta_d) = F = F_0 \subseteq F_1 \subseteq \ldots \subseteq F_n$ *be an order d radical tower over F. By L_i denote the Galois closure of F_i over F. Then* $F = L_0 \subseteq L_1 \subseteq \ldots \subseteq L_n$ *is an order d radical tower over F.*

Proof. The proof is by induction on the position i, $i = 0, \ldots, n$, in the radical tower $F = L_0 \subseteq L_1 \subseteq \ldots \subseteq L_n$. For $i = 0$ there is nothing to prove.

For the induction step let $F_i = F_{i-1}(\sqrt[d_1]{\gamma_1}, \ldots, \sqrt[d_k]{\gamma_k})$, where $\gamma_j \in F_{i-1}$ and $d_j | d$ for all $j = 1, \ldots, k$. By Theorem 3 the degree of $\sqrt[d_j]{\gamma_j}$ is a divisor of d_j. Without loss of generality we assume it is exactly d_j. Let $\gamma_j^{(1)}, \ldots, \gamma_j^{(n_j)}$ be the conjugates of γ_j over F. By induction each of these conjugates is contained in L_{i-1}. Since L_{i-1} contains a primitive d^{th} root of unity and therefore a primitive d_j^{th} root of unity, Theorem 3 implies that the conjugates of $\sqrt[d_j]{\gamma_j}$ are the d_j^{th} roots of $\gamma_j^{(1)}, \ldots, \gamma_j^{(n_j)}$. Hence

$$L_i = L_{i-1}\left(\sqrt[d_1]{\gamma_1^{(1)}}, \ldots, \sqrt[d_1]{\gamma_1^{(n_1)}}, \ldots, \sqrt[d_k]{\gamma_k^{(1)}}, \ldots, \sqrt[d_k]{\gamma_k^{(n_k)}} \right).$$

This proves the lemma. □

In the proof of Theorem 5 we will use the following well-known theorem from Galois theory (see [7]).

Theorem 7. *Let k be an arbitrary field. Let M be a Galois extension of k and E an arbitrary extension of k. Then the field ME is a Galois extension of E and M is a Galois extension of $M \cap E$. The Galois group $G(ME/E)$ of ME over E is isomorphic to the Galois group $G(M/M \cap E)$ of M over $M \cap E$.*

The theorem is best visualized by Figure 1, in which we indicated the Galois groups that are isomorphic.

Proof. (Theorem 5) Let $K \subseteq K_1 \subseteq \ldots \subseteq K_n$ be the radical tower associated to the denesting β. Hence $n = \text{depth}_K(\beta)$. Let L be the Galois closure of $F(\alpha)$ over F.

Let L_i be the Galois closure of $K_i F$ over F. Since the fields $K_i F$ form an order d radical tower over F, Lemma 6 shows that the fields L_i also form an order radical tower over F. Since $\alpha \in K_n$, we have $L \subseteq L_n$.

Define $F_i = L_i \cap L$. Note that as an intersection of two Galois extensions $F_i = L_i \cap L$ is a Galois extension of F. We claim that the fields $F_i, i = 1, \ldots, n$, form an order d radical tower over F. Since $F_n = L$, this will prove the theorem.

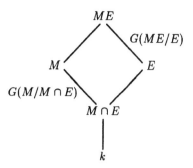

Fig. 1. Illustration of Theorem 7

To prove the claim we have to show that F_i is an order d radical extension of $F_{i-1}, i = 1, \ldots, n$. We apply Theorem 7 with $k = F, F_i = M$, and $E = L_{i-1}$ (recall Figure 1).

The extension $F_i L_{i-1}$ of L_{i-1} is a subfield of the order d radical extension L_i of L_{i-1}. By Theorem 4 it is itself an order d radical extension of L_{i-1}. Since F, and hence L_{i-1}, contains a primitive d^{th} root of unity, Theorem 4 also shows that $F_i L_{i-1}$ is Galois over L_{i-1} with abelian Galois group of exponent d. By Theorem 7 the extension F_i over $L_{i-1} \cap F_i = L_{i-1} \cap L = F_{i-1}$ is Galois with abelian Galois group of exponent d. Applying Theorem 4 shows that F_i is an order d radical extension of F_{i-1}. □

We remark that similar proofs can be used to prove the two main structure theorems in [9]. The proofs we obtain are simpler and more direct than the original proofs in [9]. In this extended abstract we have to omit a more detailed discussion of these issues.

4 Computing an Order d Denesting

As in the previous sections, $K \supseteq \mathbf{Q}$ is an algebraic number field. $F = K(\zeta_d)$, where ζ_d is a primitive d^{th} root of unity. α is an order d nested radical over K and L is the Galois closure of $F(\alpha)$ over F. By $G = G(L/F)$ denote the Galois group of L over F.

By Theorem 5 the depth of a denested form β of α is related to the length of radical towers between F and L. Hence we make the following definition.

Definition 8. *A radical tower* $F = F_0 \subset F_1 \subset \ldots \subset F_n = L$ *is called the shortest radical tower for L iff for any other radical tower* $F = L_0 \subseteq L_1 \subseteq \ldots \subseteq L_m = L$ *between F and L*

(i) $n \leq m$
(ii) $L_i \subseteq F_i$ *for all* $i = 1, \ldots, m$, *setting* $F_j = F_n$ *for* $j > n$.

The order d radical tower $F = F_0 \subset F_1 \subset \ldots \subset F_n = L$ is the shortest radical tower of order d for L if these conditions are met by any order d radical tower between F and L.

Due to condition (ii) the shortest radical tower is unique. Lemma 6 shows that in the shortest order d radical tower $F \subset F_1 \subset \ldots \subset F_{n-1} \subset F_n = L$ each intermediate field F_i is a Galois extension of F.

By Galois theory such an order d radical tower corresponds to a chain of normal subgroups $G = G_0 \supseteq G_1 \supseteq \ldots \supseteq G_{n-1} \supset G_n = \{\text{id}\}$, where the factor groups $G_i/G_{i+1}, i = 0, \ldots, n-1$, are abelian of exponent d. We will call such a chain of groups an *abelian tower of exponent d for G*. The *shortest abelian tower of exponent d* is a chain of groups $G = G_0 \supset G_1 \supset \ldots \supset G_{n-1} \supset G_n = \{\text{id}\}$ such that for any other tower $G = H_0 \supseteq H_1 \supseteq \ldots \supseteq H_{n-1} \supset H_m = \{\text{id}\}$, that is abelian and of exponent d, we have $n \leq m$ and $H_i \supseteq G_i$ for all $i = 1, \ldots, m$. Here $G_j = \{\text{id}\}$ for $j > n$. Due to the condition $H_i \supseteq G_i$, the shortest abelian tower of exponent d is unique.

We say that the multiplicatively written group G is given by a group table, if we have a table that contains for every pair of elements $x, y \in G$ its product $z = xy$.

Theorem 9. *Let G be given by a group table. The shortest abelian tower of exponent d for G can be computed in time polynomial in $|G|$.*

Proof. We need some terminology from group theory. Let H be an arbitrary group. For any subset S of H the *normal closure* of S is the smallest normal subgroup of H containing S. The *commutator subgroup* of H is the smallest normal subgroup $C = C(H)$ of H whose factor group H/C is abelian. C is uniquely defined, that is, any normal subgroup N of H with abelian factor group H/N contains the commutator C.

To prove the theorem we define a specific abelian tower of exponent d, prove that it is the shortest abelian tower of exponent d, and finally show that it can be computed in polynomial time.

Let $G_0 = G$. Assuming G_i has already been defined and is not the trivial group, G_{i+1} is defined as follows. By G_i^d denote the set of all d^{th} powers of elements in G_i. By $C(G_i)$ denote the commutator subgroup of G_i. Then G_{i+1} is defined as the normal closure of $G_i^d \cup C(G_i)$ in G_i.

Let $G = H_0 \supseteq H_1 \supseteq \ldots \supseteq H_{m-1} \supseteq H_m = \{\text{id}\}$ be an arbitrary abelian tower of order d. We show by induction that $G_i \subseteq H_i$. Obviously, $G_0 \subseteq H_0$. So assume that $G_j \subseteq H_j$ has already been proven for all $j \leq i$. We need to show $G_{i+1} \subseteq H_{i+1}$.

By definition H_i/H_{i+1} is abelian of exponent d, hence $H_i^d \subseteq H_{i+1}$. Since $G_i \subseteq H_i$ we also have $G_i^d \subseteq H_{i+1} \cap G_i$. By the isomorphism theorems for groups (see [7]) $H_{i+1} \cap G_i$ is a normal subgroup of G_i and $G_i/(H_{i+1} \cap G_i)$ is isomorphic to $G_i H_{i+1}/H_{i+1}$. As a subgroup of H_i/H_{i+1} the group $G_i H_{i+1}/H_{i+1}$ is abelian. Hence $H_{i+1} \cap G_i$ contains the commutator of G_i.

So far it has been shown, that the normal subgroup $H_{i+1} \cap G_i$ contains G_i^d and the commutator $C(G_i)$, but then it contains the normal closure of $G_i^d \cup C(G_i)$ in G_i, proving that $G_{i+1} \subseteq H_{i+1}$.

It remains to prove that the G_i's can be computed in polynomial time. Given G_i the set G_i^d can be computed in polynomial time. The commutator $C(G_i)$ can also be computed in polynomial time. An efficient algorithms for this problem can be found in [1]. Finally, given G_i^d and $C(G_i)$ the normal closure of their union can again be computed by using algorithms described in [1]. □

Before we can state and prove the main theorem of this paper some terminology from algorithmic algebra is needed.

Every algebraic number field has the form $\mathbf{Q}(\gamma)$, where we can assume that γ is not only algebraic but is an algebraic integer. The length $|p|_2$ of a polynomial $p = \sum_{i=0}^{n} p_i X^i \in \mathbf{Q}[X]$ is defined by $\left(\sum_{i=0}^{n} |p_i|^2 \right)^{\frac{1}{2}}$.

If $\beta = \frac{1}{b} \sum_{i=0}^{n-1} b_i \gamma^i$ is an element of the algebraic number field $\mathbf{Q}(\gamma)$, the representation size $[\beta]$ of β is defined as $\left(|b|^2 + \sum_{i=0}^{n-1} |b_i|^2 \right)^{\frac{1}{2}}$. For a polynomial f in $\mathbf{Q}(\gamma)[X]$ its representation size is defined as the sum of the representation sizes of its coefficients.

Theorem 10. *Let $K = \mathbf{Q}(\gamma)$ be an algebraic number field, where the field generator is an algebraic integer with minimal polynomial $p(X) \in \mathbf{Z}[X]$. Let α be an order d nested radical over K. If an order d denested form of α over K has depth n, then an order d denesting over $F = K(\zeta_d)$ of α of depth at most n can be found time polynomial in $|p|_2$, the representation size of the minimal polynomial of α over K and the degree of the Galois closure of $F(\alpha)$ over F.*

In the theorem we assume that the minimal polynomial of α over K is given. If that is not the case, that is, if we only have a nested radical expression for α, the minimal polynomial of α can be computed in time polynomial in the representation size of the minimal polynomial (see [9]).

Proof. If α has a denested form of depth n over K, then Theorem 5 shows that there is a radical tower $F = L_0 \subseteq L_1 \subseteq \ldots \subseteq L_{n-1} \subset L_n = L$ of length n between F and the Galois closure L of $F(\alpha)$. The fields in this tower are Galois over F. Hence this radical tower corresponds to an abelian tower of exponent d for the Galois group G of L over F. Therefore a shortest abelian tower of exponent d has length at most n.

The group G can be computed within the time bounds stated in the theorem [8]. By Theorem 9 within this time bound we can compute the shortest abelian tower of exponent d for G. This tower corresponds to an order d radical tower $F = F_0 \subset F_1 \subset \ldots \subset F_\ell = L$ between F and L, with $\ell \leq n$.

To obtain the desired denesting for α it suffices to compute for each extension $F_i : F_{i-1}$ a set of order d radicals that generate the extension and to express α as an element of F_ℓ. For these problems we can use the algorithms described in [9]. Also, the analysis given there can be applied to the present case. □

For $d = 2$ a primitive root is given by -1. Hence the theorem implies that over arbitrary number fields for a nested radical α involving only square roots an order 2 denested form can be computed within the time bounds stated in the theorem. This result generalizes the algorithms in [4].

5 Solvability by Order d Radicals

The algorithm leading to Theorem 10 can be used to determine whether the root α of a polynomial f over an algebraic number field $K = \mathbf{Q}(\gamma)$, which contains a primitive d^{th} root of unity, can be expressed as an order d nested radical. First compute the Galois group of $K(\alpha)$ over K, then try to compute an abelian tower of exponent d for this group. If this fails, α cannot be expressed as an order d nested radical. Otherwise, the algorithm will find such an expression for α.

However, if we are only interested in deciding whether α can be expressed as an order d nested radical, without necessarily computing a denesting of minimal depth, we can do much better. In fact, the decision problem can be solved in polynomial time.

A polynomial f over a field K is called solvable by radicals over K iff there is a nested radical α over K that solves f, that is, $f(\alpha) = 0$. In [12] Landau and Miller showed that, given a polynomial f over some algebraic number field K, the question whether the polynomial is solvable by radicals can be decided in time polynomial in the representation size of f, as defined in the previous section.

We say that a polynomial $f \in K[X]$ is solvable by order d radicals iff there is an order d nested radical α over K with $f(\alpha) = 0$. Combining the methods of [12] with the results presented in the previous sections the following theorem can be shown. The proof is omitted in this extended abstract.

Theorem 11. *Let $K = \mathbf{Q}(\gamma)$ be an algebraic number field, where the field generator γ is an algebraic integer with minimal polynomial $p(X) \in \mathbf{Z}[X]$. Assume that K contains a primitive d^{th} root of unity. Let f be a polynomial over K. It can be decided in time polynomial in $|p|_2$ and the representation size of f whether f is solvable by order d radicals over K.*

As a special case of the result, we obtain a polynomial time algorithm that decides for polynomials over arbitrary number fields whether they can be solved by a nested radical involving only square roots.

The algorithms that decides solvability by order d nested radicals can be extended, so that it also computes a nested radical that solves the polynomial. The running time is still polynomial. The depth of the expression will not be optimal, nor will the expression solve the polynomial exactly, that is, not all possible values of the expression are roots of the polynomial.

References

1. L. Babai, E. Luks, Á. Seress: Fast management of permutation groups. Proc. 29th Symposium on Foundations of Computer Science (1988) pp. 272-282.

2. J. Blömer: Computing Sums of Radicals in Polynomial Time. Proc. 32nd Symposium on Foundations of Computer Science (1991) pp. 670-677.
3. J. Blömer: Denesting Ramanujan's Nested Radicals. Proc. 33nd Symposium on Foundations of Computer Science (1992) pp. 447-456.
4. A. Borodin, R. Fagin, J. E. Hopcroft, M. Tompa: Decreasing the Nesting Depth of Expressions Involving Square Roots. Journal of Symbolic Computation 1 (1985) pp. 169-188.
5. B. Caviness, R. Fateman: Simplification of Radical Expressions. Proc. 1976 ACM Symposium on Symbolic and Algebraic Computation (1976).
6. G. Horng, M. -D. Huang: Simplifying Nested Radicals and Solving Polynomials by Radicals in Minimum Depth. Proc. 31st Symposium on Foundations of Computer Science (1990) pp. 847-854.
7. S. Lang: Algebra, 3rd edition. (1993) Addison-Wesley.
8. S. Landau: Factoring polynomials over algebraic number fields. SIAM Journal on Computing 14(1) (1985) pp. 184-195.
9. S. Landau: Simplification of Nested Radicals. SIAM Journal on Computing 21(1) (1992) pp 85-110.
10. S. Landau: A Note on Zippel-Denesting. Journal of Symbolic Computation 13 (1992) pp. 41-46.
11. S. Landau: How to Tangle with a Nested Radical. Mathematical Intelligencer 16(2) (1994) pp. 49-55.
12. S. Landau, G. L. Miller: Solvability by Radicals is in Polynomial Time. Journal of Computer and System Sciences 30 (1985) pp. 179-208.
13. S. Ramanujan: Problems and Solutions, Collected Works of S. Ramanujan (1927) Cambridge University Press.
14. R. Zippel: Simplification of Expressions Involving Radicals. Journal of Symbolic Computation 1 (1985) pp. 189-210.

A Linear Time Algorithm for the Arc Disjoint Menger Problem in Planar Directed Graphs
(Extended Abstract)

Ulrik Brandes and Dorothea Wagner

Universität Konstanz, Fakultät für Mathematik und Informatik
Fach D 188, D-78457 Konstanz, Germany
{Ulrik.Brandes , Dorothea.Wagner}@uni-konstanz.de

Abstract. Given a graph $G = (V, E)$ and two vertices $s, t \in V$, $s \neq t$, the Menger problem is to find a maximum number of disjoint paths connecting s and t. Depending on whether the input graph is directed or not, and what kind of disjointness criterion is demanded, this general formulation is specialized to the directed or undirected vertex, and the edge or arc disjoint Menger problem, respectively.

For planar graphs the edge disjoint Menger problem has been solved to optimality [Wei97], while the fastest algorithm for the arc disjoint version is Weihe's general maximum flow algorithm for planar networks [Wei94], which has running time $\mathcal{O}(|V| \log |V|)$. Here we present a linear time, i.e. asymptotically optimal, algorithm for the arc disjoint version in planar directed graphs.

1 Introduction

Due to their importance – in their own right as well as in bottleneck routines of other algorithms – disjoint path problems have been studied extensively. The famous Menger Theorems [Men27] are structural in nature. However, they have not only been generalized to capacitated versions like the max-flow/min-cut theorem, but also extended to algorithms actually constructing disjoint paths, separators, or cuts.

A generic formulation of Menger's problem is the following: Given a graph $G = (V, E)$ and two distinct vertices $s, t \in V$, find a maximum cardinality set of disjoint paths connecting s and t. This leads to four concrete versions of the problem. The instances are either directed or undirected, and the (s, t)–paths have to be vertex or edge (arc) disjoint.

For planar undirected graphs, linear time algorithms exist for both the vertex [RLWW97] and edge disjoint case [Wei97]. However, there is no such algorithm for either case when the planar input graphs are directed. In any graph the arc disjoint Menger problem obviously corresponds to a maximum flow problem with unit capacities [AMO93]. The first algorithm tailored to solve the maximum flow problem with arbitrary capacities especially in planar graphs was presented in [IS79]. Faster algorithms have subsequently been developed, e.g. [JV82] and

[KRRHS94]. By now, the fastest algorithm is that of [Wei94] yielding a running time of $\mathcal{O}(n \log n)$, where $n = |V|$. Here we concentrate on the more special Menger problem and present a linear time solution. Our algorithm is not only faster than the max-flow algorithm, but also considerably simpler.

Our approach is based on right-first-search, which appears to be extremely suitable for path problems in planar graphs (cf. [RLWW95]). In particular, the optimal algorithms for the Menger problem in undirected graphs are based on this variant of depth-first-search [RLWW97,Wei97]. In a right-first-search, arcs are chosen according to a *right-hand-rule*, i.e. the continuation arc is the counterclockwise next in the adjacency list of the vertex entered by the current arc. One of the main difficulties encountered by this strategy is the treatment of right cycles. Similar to [Wei97], we therefore use an observation of [KNK93] to restrict the set of input instances to graphs without right cycles.

Roughly speaking, the algorithm successively occupies arcs in order to build a set of (s, t)–paths. The paths in this set are frequently reorganized, such that the determination of consecutive arcs on the same path becomes intricate. Another problem is the efficient choice of an arc to backtrack with when the path that is currently built can no longer be extended. Together, these problems make a linear time implementation rather difficult. The obstacles are overcome by a careful analysis of partial solutions which leads to local characterizations resolving both problems.

In Sect. 2, we introduce our basic terminology and show how to restrict the problem to certain input instances. Section 3 gives a description of the algorithm on an abstract level, providing a better understanding of the underlying ideas. Its correctness is proved in Sect. 4. In Sect. 5, properties of partial solutions are examined. Based on these properties, a linear time implementation of the algorithm is described in Sect. 6.

2 Preliminaries

Let us first introduce our basic assumptions and terminology. We are given an embedded planar graph $G = (V, A)$ with distinct vertices $s \neq t$. The *adjacency list* of a vertex $v \in V$ is a cyclic list of all arcs incident to v, arranged in the order in which they appear around v in the embedding. We will often make use of this ordering, and say that an arc a is the *first arc after* b in (counter)clockwise order around v, if b is an immediate successor of a when the adjacency list of v is traversed in a (counter)clockwise fashion.

With the assumption of a fixed embedding, we can make heavy use of spatially descriptive terms, e.g. left and right, inside and outside, etc. For example, the right side of a directed path is its right-hand-side when following its arcs' directions. A directed cycle divides the plane into two disjoint regions, its left-hand-region and its right-hand-region. The region containing the outer face is called its *exterior*, the other is called *interior*. A cycle is called a *left (right) cycle*, if its interior equals its left-hand-region (right-hand-region). Cycles with s in their interior are called *orbits*.

Note that a maximum collection of arc disjoint directed (s, t)–paths in G corresponds to a maximum flow from s to t, if all arcs have unit capacity. Conversely, it is easy to construct a maximum collection of (s, t)–paths from an integral maximum flow. Also, given a maximum integral flow, a partition of the vertices inducing a minimum cut can always be found in linear time by a simple labelling algorithm.

Moreover, the maximum flow value does not change when a set of right cycles is replaced by left cycles which are obtained by simply altering arc orientations. Therefore, let C be a set of right cycles. Then $G_C = (V, A_C)$ is called the *residual graph*, where A_C is the set of all arcs (v, w), with $(v, w) \in A$ and (v, w) does not belong to a cycle in C, or $(w, v) \in A$ and (w, v) does belong to a cycle in C. Note that reversion of arcs may introduce multiple arcs, which makes A_C a multiset. If $f_C : A_C \to \{0, 1\}$ is a maximum integral flow in the residual graph G_C, then a maximum integral flow $f : A \to \{0, 1\}$ in G is obtained by setting $f(v, w) = f_C(v, w)$, if $(v, w) \in A$ and $(v, w) \in A_C$, and $f(v, w) = 1 - f_C(w, v)$, if (w, v) is the replacement of (v, w) in A_C. From [KNK93] it can be seen that there always exists a set C of right cycles such that the residual graph G_C contains left cycles only[1]. Moreover, this special set C can always be found in linear time using a breadth-first-search in the planar dual of G, which can also be obtained in linear time. Thus, a linear time algorithm solving the arc disjoint Menger problem in G_C is sufficient to provide a linear time solution for the problem in G. This technique was also used in [Wei97], where further details can be found.

In the remainder we assume that we are given a planar directed graph $G = (V, A)$ that is embedded in the plane, such that t is on the boundary of the outer (i.e. the infinite) face[2] and contains no right cycle. We may further assume that there are no arcs entering s, and no arcs leaving t, since these obviously do not affect the maximum number of arc disjoint directed (s, t)–paths.

3 The Algorithm

In this section, we present an algorithm that determines a maximum set of possibly non-simple) arc disjoint (s, t)–paths in an embedded planar directed graph with t on the outer face and no right cycle (cf. the previous section). For convenience, we here describe the algorithm on an abstract level, which both facilitates understanding and displays the basic simplicity of our approach. Nonetheless, it is not at all obvious how a linear worst case complexity can be obtained.

The algorithm applies a special variant of depth-first-search, namely right-first-search, which is suitable for many problems involving paths in planar graphs [RLWW95]. As a by-product, the resulting solution is rightmost in the sense that no path can be routed further to the right without changing others.

---

[1] This set C corresponds to the *bottom element* of the distributive lattice formed by all circulations of G with appropriately defined operations for *meet* and *join* [KNK93].

[2] Such an embedding can always be obtained in linear time [HT74].

All paths and cycles in this section are directed. After each step, the partial solution consists of a *search path*, which starts at s and ends at some vertex $v \neq t$, and a set of (s,t)–paths and left cycles, such that every arc belongs to at most one path or cycle. Given such a set of arc disjoint directed paths and cycles, each path (cycle) induces a straightforward (cyclic) *traversal order* on its arcs. For a directed (sub-)path, its *first* and *last* arc are well defined, then. We say that two arcs are *consecutive*,

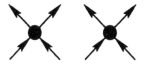

Fig. 1. Crossing (left) and non-crossing (right) pairs of consecutive arcs

if they are immediate successors in the traversal order of a path or cycle, respectively. The last arc of the search path is called the *leading arc*, and its head is called the *leading vertex*. We say that two pairs of consecutive arcs form a *crossing*, if they share their middle vertex v, and their arcs are encountered alternatingly when traversing the cyclic order of arcs incident to v. See Fig. 1. A set of arc disjoint paths and cycles is said to be *non-crossing*, if no two pairs of consecutive arcs form a crossing. All partial solutions will be non-crossing.

The algorithm uses only three basic operations: *search steps*, *backtracking steps*, and *realignments*.

Search Step. An unsearched arc leaving the leading vertex is added to the search path. Among all unsearched arcs, the counterclockwise first after the current leading arc in the adjacency list of the leading vertex is chosen (right-hand-rule).

Backtracking Step. Some arc of the search path entering the leading vertex is removed from the graph. Which arc exactly need not be specified in this general version of the algorithm. In the implementation presented in Sect. 6, this choice is subject to certain configurations and the stage of the algorithm. If a non-simple search path has more than one arc entering the leading vertex, the removal may split the search path into the new search path starting at s and ending at the removed arc's tail, and a left-over subpath starting and ending at the leading vertex, say v. We then modify the traversal order with respect to the arcs

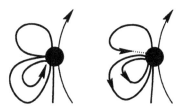

Fig. 2. After backtracking with the second arc of the search path entering the leading vertex, there are two resulting left cycles (containing four irrelevant arcs incident to the formerly leading vertex). The removed arc is indicated by the dotted line.

of the cut-off end of the search path that are incident to v, such that the subpath is transformed into a set of left cycles that do not cross at v. Each of these cycles is constrained to have exactly two arcs incident to v. See Fig. 2 and note that there is a unique reassignment of consecutive arcs satisfying these conditions.

We refer to the reassignment of consecutive arcs during a backtracking step as *closing left cycles*, and to the arcs that are reassigned as *irrelevant*. Then, every arc that belongs to a path or cycle and is not irrelevant is called *relevant*[3].

In order to introduce the third operation, some more terminology is needed. For a vertex $v \in V \setminus \{s, t\}$, we define a *passage through* v, or v–*passage* for short, of a path or cycle to be an (inclusion-)maximal subpath with the following properties: Its first arc is a relevant arc entering v, and its last arc is a relevant arc leaving v. Moreover, if it is non-simple, then s is in the exterior of every cycle formed by the subpath. Fig. 3 (a) gives an example. If existent, the last v–passage of the search path is called *leading* v–*passage*. An arc (u, v) is said to *hit* some v–passage p from the right (left), if it is on the v–passage's right (left) side, and p and (u, v) are on the same sides of every other v–passage. An arc (v, w) is said to *leave* a v–passage to the right (left) in the analogous situation. Two v–passages *touch* at v, if they do not cross at v, and each of them contains an arc hitting or leaving the other. Paths and cycles are hit, left or touched at a vertex v, if one of their v–passages is. Fig. 3 (b) summarizes these definitions.

Realignment. Let v be the leading vertex, and let some v–passage be hit from the right by the leading arc. Since there are no right cycles, there must be consecutive arcs (u, v) and (v, w) of the v–passage hit, such that the leading arc appears between (u, v) and (v, w) in the counterclockwise cyclic order of arcs incident to v. The search path is said to be realigned with the corresponding path or cycle, if the leading arc is made consecutive with (v, w), such that (u, v) becomes the new leading arc.

We are now ready to state our algorithm in simple terms. The bottom line is that we always try to go as far to the right as possible. The contribution of realignments is two-fold: on one hand, they prevent crossings, and on the other hand, they ensure that the search path is, in a sense, leftmost at the leading vertex when backtracking has to be performed.

(a)

(b)

Fig. 3. (a) Two paths making up for three v–passages. (b) Three passages p, q, r, and two arcs a, b incident to the same vertex. Passage p touches passage q on the right, r touches q on the left, while p and r do not touch at all. Arc a hits q from the right, and p from the left. However, it does not hit r. Arc b leaves p to the right, but neither q nor r. a and b are not consecutive.

---

[3] Actually, arcs ought to be called relevant or irrelevant *with respect to* one of their incident vertices. We omit this distinction, because from context it should always be clear, with respect to which vertex an arc is relevant or irrelevant, respectively.

Algorithm 1 Menger Algorithm

---

for each *outgoing arc a of s* **do**
  let the search path consist of arc a
  while *the leading vertex is neither s nor t* **do**
    if *the leading arc hits some passage from the right* **then**
      realign the search path with the corresponding path or cycle
    else
      if *there is an unsearched arc leaving the leading vertex* **then**
        perform a search step
      else
        perform a backtracking step

---

4 Correctness

In this section, we outline the steps to prove that after termination of Algorithm 1 the set of (s, t)–paths generated is maximum. The proofs themselves are given in the full version of this paper [BW97]. A set of arcs whose removal disconnects s and t is called a (directed) (s, t)–*cut*. By Menger's theorem, a set of (s, t)–paths is maximum, if and only if its cardinality equals the cardinality of a minimum (s, t)–cut. Such (s, t)–cuts are called *saturated*. Three observations are trivial:

Lemma 1. *During the execution of Algorithm 1, the following statements hold:*

  a) Just before a search or backtracking step, the leading arc does not hit any passage through the leading vertex from the right.
  b) Just before a search step, there is no irrelevant arc incident to the leading vertex and no removed arc entering the leading vertex.
  c) Just before a backtracking step, every arc leaving the leading vertex has already been searched.

There are two easy cases, for which the following lemma yields correctness of Algorithm 1.

Lemma 2. *If either all or none of the arcs leaving s have been removed by Algorithm 1, the arcs leaving s that have not been removed form a saturated directed (s, t)–cut.*

Correctness is based on the fact that the solution produced by Algorithm 1 is maximal and rightmost, i.e. no (s, t)–path can be routed further to the right without changing others, too. This notion of rightmost is characterized by the following four invariants:

Lemma 3. *During the execution of Algorithm 1, the following properties remain invariant:*

(P1) All paths and left cycles are arc disjoint and non-crossing.
(P2) No unsearched arc leaves a passage to the right.
(P3) No removed arc hits a passage from the right.
(P4) No two passages mutually touch their right sides.

Based on the above observations, Algorithm 2 determines a saturated directed (s, t)–cut in the output of Algorithm 1. This cut is induced by the exterior of a cycle enclosing s. This obviously suffies to prove correctness. From the discussion in Sect. 6, it is easy to see that Algorithm 2 can also be implemented with linear running time. It is based on the analogous variant of depth-first-search, left-first-search. Arcs removed or not searched by Algorithm 1 are searched in forward direction, while arcs belonging to paths or cycles are searched in backward direction.[4] When Algorithm 2 backtracks, the backtrack arc is said to be *discarded* from the search path. Algorithm 2 terminates, when the search path hits itself from the left, such that the resulting cycle is a right cycle[5] surrounding s. The arcs on (s, t)–paths having their tail on this cycle and their head in the exterior then form the desired cut.

Algorithm 2 Saturated Cut Algorithm

---

if *all or none of the arcs leaving s are removed* **then**
  let the search path consist of s only
else
  let the search path consist of a removed arc leaving s
  repeat
    if *there is an unsearched candidate arc* **then**
      search the clockwise next candidate arc
    else
      discard the leading arc from the search path
  until *the search path consists of s only*
        and the clockwise next candidate arc has been searched
    or *the clockwise next candidate arc is part of the search path*
        and s does not lie in the exterior of the resulting cycle

---

To avoid confusion, an arc is denoted *ignored*, if it was not searched by Algorithm 1. An arc is then called a *candidate arc*, if it leaves the leading vertex and was either removed or ignored, or if it enters the leading vertex and is part of an (s, t)–path or left cycle produced by Algorithm 1.

Lemma 4. *Algorithm 2 never discards an arc from the search path that belongs to an (s, t)–path produced by Algorithm 1.*

---

[4] Basically, the algorithm tries to find an augmenting path.
[5] Since some arcs are searched in backward direction, right cycles are possible.

Lemma 5. *If an ignored arc is searched by Algorithm 2, it lies in the interior of one of the left cycles produced by Algorithm 1.*

Lemma 6. *The leading arc of Algorithm 2 never enters an (s,t)–path produced by Algorithm 1 from the right.*

After termination of Algorithm 2, the search path either consists of s alone, or it hits itself from the left while surrounding s. In the second case, let $C = (v_1, a_1, v_2, a_2, \ldots, v_k = v_1)$ be the sequence of vertices and arcs of the search path beginning with the leading vertex and the clockwise next arc on the search path, and ending with the leading arc and the leading vertex. In the first case, let C be the trivial cycle s.

Lemma 7. *If C does not equal s, it is a right cycle with s in its interior or on its boundary.*

Lemma 8. *The set of arcs on (s,t)–paths produced by Algorithm 1 having their tail on C and their head in its exterior form a saturated directed (s,t)–cut.*

Corollary 9. *The set of arc disjoint (s,t)–paths determined by Algorithm 1 is maximum.*

5   Properties of Partial Solutions

In Sect. 3, an algorithm solving the arc disjoint Menger problem in planar directed graphs was described. In this section now, we state a number of invariants that are used to efficiently implement this algorithm. Again, details are given in the full version of this paper [BW97].

Since our goal is to achieve linear running time, the possibly more than linear number of realignments cannot actually be performed. While the arc to be searched next is still computed easily, it can be difficult to identify an arc on the search path that may be used for backtracking, when it is not known which arcs are consecutive. We subsequently analyze the structure of partial solutions. This will permit an implementation that does not need to explicitly represent which arcs are consecutive.

A first structural insight is the relative orientation of passages. Two passages p, q through the same vertex are said to be *oriented likewise*, if p is completely to the left of q, while q is completely to the right of p. In Fig. 3 (b), passages p, q are oriented likewise, while passages p, r and q, r are oriented differently. The following lemma shows that at most the last v–passage of the search path can be oriented different than other v–passages.

Lemma 10. *During the execution of Algorithm 1, the following property remains invariant:*

(P5) For all $v \in V \setminus \{s,t\}$, all v–passages are oriented likewise, possibly except for the leading v–passage.

By the above property, all but at most one specific v–passage (which then is the leading v–passage) of a vertex $v \in V \setminus \{s,t\}$ are oriented likewise. We define the *leftmost* v–passage to be the unique leftmost of these, and *lastLeft*(v) to be its last arc. As an immediate, yet crucial, consequence of (P5) the following corollary states that knowledge of *lastLeft*(v) is sufficient to identify an arc that may be used in a backtracking step (i.e. an arbitrary arc of the search path entering the leading vertex).

Corollary 11. *During the execution of Algorithm 1, the following property remains invariant:*

(P6) If $v \in V \setminus \{s,t\}$ is the leading vertex and the search path does not hit a v–passage from the right, then the counterclockwise next relevant arc after lastLeft(v) is an incoming arc of the search path.

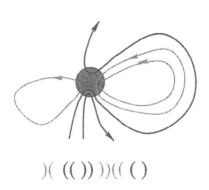

Fig. 4. A vertex v and its associated parenthesis expressions. $M(v)$ is depicted darker than $m(v)$, *lastLeft*(v) is the arc leaving v upwards.

We can gain even further knowledge from arc *lastLeft*(v): If the relevant arcs incident to v are traversed in counterclockwise order such that *lastLeft*(v) is encountered last, define $M(v)$ to be a string of parentheses, one opening for each incoming relevant arc, and one closing for each outgoing relevant arc of v. If v is the leading vertex, the leading arc does not contribute to $M(v)$. Analogously, let $m(v)$ be the string of parentheses for the irrelevant arcs. We say that a string of parentheses has *parenthesis structure*, if the number of closing parentheses does not exceed the number of opening parentheses in any prefix of the string. Likewise, it has *inverse parenthesis structure*, if the number of closing parentheses in a prefix is never less than the number of opening parentheses. Two strings of (possibly inverse) parenthesis structure are called *interleaved* (in a common superstring), if no parenthesis of one string is positioned between a pair of matching parentheses in the other.

The following property is not only useful to recompute pairs of consecutive arcs from *lastLeft*(v), but interesting in its own right.

Lemma 12. *During the execution of Algorithm 1, the following property remains invariant:*

(P7) For $v \in V \setminus \{s,t\}$, $M(v)$ has parenthesis structure, $m(v)$ has inverse parenthesis structure, $M(v)$ and $m(v)$ are interleaved, and every pair of consecutive relevant or irrelevant arcs incident to v corresponds to a pair of matching parentheses in $M(v)$ or $m(v)$, respectively.

Even though *lastLeft(v)* provides all the information needed to implement Algorithm 1 efficiently, we are not yet done, because it is not always possible to update *lastLeft(v)* correctly based on local knowledge only (an example is given in Fig. 5).

Define *lastLeading(v)* to be the last arc of the search path leaving *v*. We now give a simple rule to mark an arc *last(v)*, which equals at least one of *lastLeft(v)* and *lastLeading(v)*. Consider, for a vertex $v \in V \setminus \{s, t\}$ that is not the leading vertex, the arc that has most recently been searched or removed. If it has been searched, it is outgoing and we let *last(v)* be just this arc. If it is removed, it is incoming, and we let *last(v)* be the clockwise next relevant outgoing arc. Now, if *v* is the leading vertex, it will be clear from context, whether *last(v)* refers to the current or next arc with the above properties. Furthermore we define *first(v)* to be the incoming relevant arc corresponding to the opening parenthesis immediately after *last(v)* in a cyclic traversal of *M(v)*.

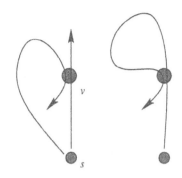

Fig. 5. In the left situation there is an (s, t)-path touched on the left by the search path. Here, *lastLeft(v)* is the arc leaving *v* on the (s, t)-path, and hence *lastLeft(v)* $\neq$ *lastLeading(v)* = *last(v)*. In the right situation, the search path returns to *v*, and we have *last(v)* = *lastLeading(v)* = *lastLeft(v)*. Note that these situations cannot be distinguished solely based on the arcs incident to *v* and the order in which they were searched.

The next lemma states that *last(v)* is the last arc of the *v*–passage that is to the left of all likewise oriented *v*–passages. Recall from (P5) that at most the leading *v*–passage is oriented differently, and observe that *lastLeft(v)* equals *lastLeading(v)*, if and only if *v* has a leading *v*–passage that is to the left of every other *v*–passage and oriented likewise.

Lemma 13. *During the execution of Algorithm 1, the following property remains invariant:*

> (P8) *For every* $v \in V \setminus \{s, t\}$, *last(v) equals lastLeft(v) or lastLeading(v). If last(v) equals lastLeading(v), the leading v–passage is to the left of every other v–passage.*

Our final lemma gives a sufficient condition for *last(v)* to equal *lastLeft(v)*.

Lemma 14. *Let* $v \in V \setminus \{s, t\}$ *be the leading vertex. If the leading arc is positioned between first(v) and last(v) in a cyclic traversal of M(v), then last(v) = lastLeft(v).*

6 Linear Time Implementation

In this section, we show that Algorithm 1 can be realized with linear running time. Since the number of changes caused by realignments can be more than linear in the number of arcs, it is crucial to avoid an explicit maintainance of consecutive arcs, at least in general. Fortunately, due to the highly structured partial solutions generated by the algorithm, realignments need not be performed explicitly. We show that the arc to be searched next can be determined from the current leading arc, while an arc to backtrack with can be determined from $first(v)$ and $last(v)$. Sometimes arcs need to be paired according to a substring of $M(v)$ and $m(v)$. Therefore, procedures match_all, match_left, and match_right compute the preceding arc a.pred for outgoing arcs a of a particular vertex, where the first one matches all pairs of arcs, while the other match maximal substrings of $m(v)$ at the left and right end of $M(v)$, respectively.

In each iteration of the main loop, the search path is initialized with the particular arc leaving s, and the inner loop is executed. The implementation of the inner loop consists of search and backtracking steps only. A variable global_mode takes values SEARCH or BACKTRACK, and is used to determine which step is to be performed next. For a search step, consider the next arc that is searched. By (P2), it is the counterclockwise next unsearched arc after the leading arc, even if no realignment is performed. Algorithm 3 is a more formal description of possible cases. For each vertex $v \in V \setminus \{s, t\}$, a field v.mode is used to store a *local mode*, which is either FORWARD (v still has outgoing unsearched arcs), TRANSITION (all outgoing arcs are searched, but $lastLeft(v)$ is not yet identified), SKIP (all outgoing arcs are searched and $lastLeft(v) = last(v)$), or DONE (all outgoing arcs of the vertex are removed).

For a backtracking step, consider the step immediately afterwards. Let (u, v) be the arc that is removed, then the next step is either a search or a backtracking step with leading vertex u. If it is a search step, we again have that the counterclockwise next unsearched arc after (u, v) is the next unsearched arc chosen by Algorithm 1 (possibly after a number of realignments). If it is a backtracking step, an arc of the search path must be determined that is removed next. When it is known that $last(u)$ equals $lastLeft(v)$, this is done according to Corollary 11, otherwise the preceding arc of (u, v) is retrieved from a temporary assignment of consecutive arcs. According to (P7), this can easily be computed the first time that there is no outgoing arc at u. By Lemma 14 it is only needed for a certain transition phase, in which there is no (not even implicit) realignment at u. Algorithm 4 gives a more formal description.

Theorem 15. *A maximum set of arc disjoint (non-crossing) (s, t)-paths of G can be determined in linear time.*

Proof. Above, we have described an implementation of Algorithm 1, which by Corollary 9 computes a maximum solution. For its linear time realization, observe that every time an arc is used, its state is altered from unsearched (flow 0) to searched (flow 1), or from searched to removed (no longer present). The

---

Algorithm 3 Search Step

```
v := head(leading_arc)
if there is an outgoing arc of v with flow 0 then
    leading_arc := first arc after leading_arc
                   in counterclockwise order around v
                   that is outgoing and has flow 0
    leading_arc.flow := 1
    v.last := leading_arc
else
    case v.mode of
        FORWARD:
            match_all(v)
            v.first := first arc after v.last
                       in counterclockwise order around v
                       that is incoming with flow 1
                       and does not equal leading_arc
            v.mode := TRANSITION
        TRANSITION:
            if leading_arc is to the right of {v.first,v.last} then
                v.mode := SKIP
            else
                global_mode := BACKTRACK
        SKIP:
            if leading_arc is to the right of {v.first,v.last} then
                leading_arc := v.first
                inverse_match_left(v)
            global_mode := BACKTRACK
        DONE:
            global_mode := BACKTRACK
```

---

predecessor of an arc is computed at most twice, and because of (P7) a simple stack algorithm matches consecutive arcs in linear time. Thus it is sufficient to show that – computation of consecutive arcs not accounted for – a single search or backtracking step can be implemented with constant (amortized) running time. The only critical operation of a search step is the determination of the counterclockwise next arc after the current leading arc. It was shown in [WW95] how Gabow and Tarjan's technique for the efficient implementation of certain *union-find*-structures [GT85] can be adapted to determine this arc in constant (amortized) time. The corresponding operation needed during a backtracking step can be performed on the same data structure. One easily verifies that during the update of fields first and last in modes FORWARD and TRANSITION every incident arc of a vertex v needs to be traversed at most once. $\square$

Acknowledgments. The authors would like to thank Annegret Liebers and Karsten Weihe for their helpful comments and suggestions.

Algorithm 4 Backtrack Step

```
u := tail(leading_arc)
dummy := leading_arc
if leading_arc is the only outgoing arc of u then
    leading_arc := the incoming arc with flow 1
    u.mode := DONE
else if there is an outgoing arc of u with flow 0 then
    leading_arc := first arc after leading_arc
                   in counterclockwise order around u
                   that is outgoing and has flow 0
    leading_arc.flow := 1
    u.last := leading_arc
    global_mode := SEARCH
else
    case u.mode of
        FORWARD:
            match_all(u)
            u.first := first arc after u.last
                       in counterclockwise order around u
                       that is incoming with flow 1
                       and does not equal leading_arc
            u.mode := TRANSITION
        TRANSITION:
            if leading_arc = u.last then
                leading_arc := leading_arc.pred
                u.last := first arc after leading_arc
                          in clockwise order around u
                          that is outgoing and has flow 1
                u.first := first arc after leading_arc
                           in counterclockwise order around u
                           that is incoming and has flow 1
            else
                u.mode := SKIP
        SKIP:
            if leading_arc is to the left of {u.first,u.last} then
                leading_arc := leading_arc.pred
            else
                if leading_arc = u.last then
                    inverse_match_right(u)
                leading_arc := u.first
                inverse_match_left(u)
remove dummy from graph
```

References

[AMO93]   Ravindra K. Ahuja, Thomas L. Magnanti, and James B. Orlin. *Network flows*. Prentice Hall, 1993.

[BW97]   Ulrik Brandes and Dorothea Wagner. A linear time algorithm for the arc disjoint Menger problem in planar directed graphs. Konstanzer Schriften in Mathematik und Informatik 29, Universität Konstanz, 1997.

[GT85]   Harold N. Gabow and Robert E. Tarjan. A linear-time algorithm for a special case of disjoint set union. *J. of Computer and System Sciences*, 30:209–221, 1985.

[HT74]   John E. Hopcroft and Robert E. Tarjan. Efficient planarity testing. *J. of the Association for Computing Machinery*, 21:549–568, 1974.

[IS79]   Alon Itai and Yossi Shiloach. Maximum flows in planar networks. *SIAM J. Comput.*, 8:135–150, 1979.

[JV82]   David S. Johnson and S.M. Venkatesan. Using divide and conquer to find flows in directed planar networks in $\mathcal{O}(n^{3/2} \log n)$ time. In *Proceedings 20th Ann. Allerton Conf. Comm., Control, and Comp.*, pages 898–905, 1982.

[KNK93]   Samir Khuller, Joseph (Seffi) Naor, and Philip Klein. The lattice structure of flow in planar graphs. *SIAM J. Discrete Math.*, 6(3):477–490, 1993.

[KRRHS94] Philip Klein, Satish B. Rao, Monika Rauch-Henzinger, and S. Subramanian. Faster shortest–path algorithms for planar graphs. In *Proceedings of the 26th Annual ACM Symposium on Theory of Computing, STOC'94*, pages 27–37, 1994.

[Men27]   Karl Menger. Zur allgemeinen Kurventheorie. *Fund. Math.*, 10:95–115, 1927.

[RLWW95]   Heike Ripphausen-Lipa, Dorothea Wagner, and Karsten Weihe. Efficient algorithms for disjoint paths in planar graphs. In William Cook, Laszlo Lovász, and Paul Seymour, editors, *DIMACS Series in Discrete Mathematics and Computer Science*, volume 20, pages 295–354. American Mathematical Society, 1995.

[RLWW97]   Heike Ripphausen-Lipa, Dorothea Wagner, and Karsten Weihe. The vertex-disjoint Menger problem in planar graphs. *SIAM J. Comput.*, 1997. to appear.

[Wei94]   Karsten Weihe. Maximum (s,t)–flows in planar network in $O(n \log n)$ time. In *Proceedings of the 35th Annual Symposium on Foundations of Computer Science, FOCS'94*, pages 178–189, 1994.

[Wei97]   Karsten Weihe. Edge–disjoint (s,t)–paths in undirected planar graphs in linear time. *J. of Algorithms*, 23:121–138, 1997.

[WW95]   Dorothea Wagner and Karsten Weihe. A linear time algorithm for edge-disjoint paths in planar graphs. *Combinatorica*, 15:135–150, 1995.

Distance Approximating Trees for Chordal and Dually Chordal Graphs*

(Extended Abstract)

Andreas Brandstädt[1], Victor Chepoi[2] ** and Feodor Dragan[1] **

[1] Universität Rostock, Fachbereich Informatik, Lehrstuhl für Theoretische Informatik, Albert-Einstein-Str. 21, D–18051 Rostock, Germany
{ab,dragan}@informatik.uni-rostock.de
[2] Laboratoire de Biomathématiques, Université d'Aix Marseille II, 27 Bd Jean Moulin, F–13385 Marseille Cedex 5, France aria@pacwan.mm-soft.fr

Abstract. In this note we show that, for each chordal graph G, there is a tree T such that T is a spanning tree of the square G^2 of G and, for every two vertices, the distance between them in T is not larger than the distance in G plus two. Moreover, we prove that, if G is a strongly chordal graph or even a dually chordal graph, then there exists a spanning tree T of G which is an additive 3–spanner as well as a multiplicative 4–spanner of G. In all cases the tree T can be computed in linear time.

1 Introduction

Many combinatorial and algorithmic problems concern the distance d_G on the vertices of a possibly weighted graph $G = (V, E)$. Approximating d_G by a simpler distance (in particular, by tree–distance) is useful in many areas such as communication networks, data analysis, motion planning, image processing, network design, and phylogenetic analysis (see [1, 2, 4, 8, 10, 20, 25, 26, 28, 30]). The goal is, for a given graph G, to find a sparse graph $H = (V, E')$ with the same vertex–set, such that the distance $d_H(u, v)$ in H between two vertices $u, v \in V$ is reasonably close to the corresponding distance $d_G(u, v)$ in the original graph G. There are several ways to measure the quality of this approximation, two of them leading to the notion of a spanner. For $t \geq 1$ a spanning subgraph H of G is called a *multiplicative t–spanner* of G [26, 10, 25] if $d_H(u, v) \leq t \cdot d_G(u, v)$ for all $u, v \in V$. If $r \geq 0$ and $d_H(u, v) \leq d_G(u, v) + r$ for all $u, v \in V$, then H is called an *additive r–spanner* [20].

For many applications (e.g. in numerical taxonomy or in phylogeny reconstruction) the condition that H must be a spanning subgraph of G can be dropped (see [2, 28, 30]). In this case there is a striking way to measure how sharp d_H approximates d_G, based on the notion of a pseudoisometry between two metric spaces. This idea is borrowed from the geometry of hyperbolic groups

---

* Second and third author supported by VW, Project No. I/69041, third author supported by DFG.
** On leave from the Universitatea de stat din Moldova, Chişinău

[13, 16]. For graphs and finite metric spaces a related notion of a near–isometry has been already used by Linial et al [21]. For our purposes we present a simplified version of this notion (the interested reader can consult [13, pp.71–72] and [16] for the general definition and related material).

Let $t \geq 1$ and $r \geq 0$ be real numbers. Two graphs $G = (V, E)$ and $H = (V, E')$ are called (t, r)–*pseudoisometric* if

$$d_H(u, v) \leq t \cdot d_G(u, v) + r \quad \text{and} \quad d_G(u, v) \leq t \cdot d_H(u, v) + r$$

for all $u, v \in V$. In this case we will say that H is a *distance (t, r)–approximating graph* for G (and conversely, G will be a distance (t, r)–approximating graph for H). The graphs G and H are $(t, 0)$–pseudoisometric iff

$$\frac{1}{t} \cdot d_G(u, v) \leq d_H(u, v) \leq t \cdot d_G(u, v)$$

for $u, v \in V$. If, in addition, H is a spanning subgraph of G, then we obtain the notion of the multiplicative t–spanner. Clearly, G and H are $(1, r)$–pseudoisometric iff $|d_G(u, v) - d_H(u, v)| \leq r$ for $u, v \in V$. Again, if H is a spanning subgraph of G, this is the usual notion of the additive r–spanner.

Recently Cai and Corneil [8] have considered multiplicative tree spanners in graphs. They showed that for a given graph G and integer t, the problem to decide whether G has a tree t–spanner is NP–complete for $t \geq 4$ and is linearly solvable for $t = 1, 2$. The status of the case $t = 3$ is still open. Tree 3–spanners exist for interval and permutation graphs and they can be found in linear time [22]. Similar results are known for the additive tree r–spanner problem. [27] proposes a simple approach to construct additive tree 2–spanners in interval and distance–hereditary graphs and such 4–spanners in cocomparability graphs. Both papers [8, 27] ask which important graph classes have tree t–spanners and r–spanners with small t and r. As it is mentioned in [27], McKee showed that for every fixed integer t there is a chordal graph without tree t–spanners (additive as well as multiplicative). Nevertheless, from the metric point of view chordal graphs look like trees. In this note we prove that for every chordal graph G there exists a tree T (actually, T is a spanning tree of the square G^2) such that

$$d_T(u, v) \leq 3 \cdot d_G(u, v) \quad \text{and} \quad d_T(u, v) \leq d_G(u, v) + 2$$

for all vertices u, v of G. In other words, T is a $(3, 0)-$ and $(1, 2)-$approximating tree for G. Moreover, if G is a strongly chordal graph then there exists a spanning tree T of G which is an additive 3–spanner and a multiplicative 4–spanner. Thus, this answers the question whether strongly chordal graphs have tree t–spanners with small t, posed in [27]. Furthermore we show that the method elaborated for strongly chordal graphs works for a more general graph class, for the dually chordal graphs. In all cases the tree T can be computed in linear time.

2 Preliminaries

All graphs occurring in this note are connected, finite, undirected, loopless and without multiple edges. A graph $G = (V, E)$ is *chordal* [7, 11] if it does not contain any induced (chordless) cycle of length at least four. In a graph G the *length* of a path from a vertex v to a vertex u is the number of edges in the path. The *distance* $d_G(u, v)$ between the vertices u and v is the length of a shortest (u, v)–path and the *interval* $I(u, v)$ between these vertices is the set $I(u, v) = \{w \in V : d_G(u, v) = d_G(u, w) + d_G(w, v)\}$. Let k be a positive integer. The *kth power* G^k of G has the same vertices as G and two vertices are joined by an edge in G^k if and only if the distance in G is at most k. The *disk* of radius k centered at v is the set of all vertices with distance at most k to v, $D_k(v) = \{w \in V : d_G(v, w) \leq k\}$, and the *kth neighbourhood* $N_k(v)$ of v is defined as the set of all vertices at distance k to v, that is $N_k(v) = \{w \in V : d_G(v, w) = k\}$. By $N(v)$ we denote the *neighbourhood* of v, i.e., $N(v) := N_1(v)$. More generally, for a subset $S \subseteq V$ let $N(S) = \cup_{v \in S} N(v)$ denote the *neighbourhood* of S.

A subset $S \subseteq V$ of a graph G is called *m–convex* [15] if for any pair of vertices $u, v \in S$ each induced path connecting u and v is contained in S. If two vertices x and y of a m–convex set S can be joined outside S by a path, then by definition of m–convexity x and y must be adjacent.

Proposition 1 [15]. *Any disk $D_k(v)$ of a chordal graph G is m–convex.*

This basic metric property of chordal graphs has some immediate but important consequences (some of them being already used by different authors). In the subsequent results the graph G is assumed to be chordal and u is an arbitrary but fixed vertex of G.

Lemma 1. *For every vertex v of G and every k, $0 < k < d_G(u, v)$, the set $N_k(u) \cap I(u, v)$ induces a complete subgraph of G.*

Lemma 2. *If two vertices $v, w \in N_k(u)$ are adjacent, then they have a common neighbour in the set $N_{k-1}(u)$.*

Lemma 3. *For any connected component S of the subgraph of G induced by $N_k(u)$, the set $N(S) \cap N_{k-1}(u)$ induces a complete subgraph. Moreover, there exists a vertex $w \in S$ such that $N(S) \cap N_{k-1}(u) = N(w) \cap N_{k-1}(u)$.*

Proof. Let $x, y \in N(S) \cap N_{k-1}(u)$. Then we can find two vertices $v, w \in S$ such that x is adjacent to v and y is adjacent to w. Consider a path P of S connecting the vertices v and w. Then P together with the edges xv and yw will give a path which intersects the disk $D_{k-1}(u)$ only in the vertices x and y. Since $D_{k-1}(u)$ is m–convex, we deduce that the vertices x and y are adjacent.

Now consider a vertex $w \in S$ with maximal number of neighbours in $N_{k-1}(u)$ and assume that $wy \notin E$ for some $y \in N(S) \cap N_{k-1}(u)$. Among neighbours of y in S we choose a vertex v for which the distance $d_S(w, v)$ is minimal. Let P be a shortest path in S connecting the vertices w and v. For every neighbour

x of w in $N_{k-1}(u)$ consider the cycle formed by path P and the edges wx, xy and yv. Since G is chordal and $d_S(w,v)$ is minimal, the vertices x and v must be adjacent. So, every neighbour of w in $N_{k-1}(u)$ is a neighbour of v as well. But then from $wy \notin E$ and $vy \in E$ we get a contradiction with the choice of the vertex w.   □

Let $q = \max \{d_G(u,v) : v \in V\}$. For a given k, $0 \le k \le q$, let $S_1^k, \ldots, S_{p_k}^k$ be the connected components of the subgraph of G induced by the kth neighbourhood of u. Define a new graph Γ whose vertices are the connected components S_i^k, $k = 0, \ldots, q$ and $i = 1, \ldots, p_k$. Two vertices S_i^k and S_j^{k-1} are adjacent if and only if there is an edge of G with one end in S_i^k and another end in S_j^{k-1}. Lemma 3 implies that every S_i^k, $k > 0$ is adjacent in Γ to exactly one connected component of $N_{k-1}(u)$. This shows that the following holds.

Lemma 4. Γ *is a tree.*

In what follows we will consider Γ rooted at the vertex $u = S_1^0$. As usually, the *nearest common ancestor* $nca(S_i^k, S_j^l)$ of two vertices S_i^k and S_j^l of Γ is the root of the smallest subtree of Γ that contains both the vertices S_i^k and S_j^l.

3 Distance Approximating Trees for Chordal Graphs

In this section for a given chordal graph $G = (V, E)$ we construct a tree $T = (V, E')$ which is a distance $(3,0)-$ and $(1,2)-$approximating tree for G. As in the previous section we fix an arbitrary vertex u of G. To construct T, for each connected component S_j^k $(k \ge 1)$ we select a vertex $v \in N(S_j^k) \cap N_{k-1}(u)$ and make v adjacent in T to all vertices of S_j^k (a formal description is given below).

Procedure 1.

> $E' := \emptyset$;
> **for** $k := q$ **downto** 1 **do**
> > **for** $j := 1$ **to** p_k **do**
> > > pick an arbitrary vertex v in $N(S_j^k) \cap N_{k-1}(u)$;
> > > **for all** $x \in S_j^k$ add xv to E'.

By Lemma 4 it follows easily that T is a tree. For any edge xv of T which is not an edge of G we have $d_G(x,v) = 2$. Indeed, by Lemma 3 every neighbour of x in $N_{k-1}(u)$ is adjacent to v. Therefore, T is a spanning tree of G^2. Since for constructing T we need only to find the connected components of the kth neighbourhoods of u, the complexity of this procedure is $O(|V| + |E|)$.

Theorem 1. T *is a distance $(3,0)-$ and $(1,2)-$approximating tree of G.*

Proof. First we will show that for any edge xy of G we have $d_T(x, y) \leq 3$. If $d_G(u, x) = d_G(u, y) = k$, then x and y belong to a common connected component of $N_k(u)$. Therefore, in T both x and y are adjacent to the same vertex v. In this case $d_T(x, y) = 2$. Now suppose that $d_G(u, x) - 1 = d_G(u, y) = k$ and xy is not an edge of T. Let v be the neighbour of x on the path connecting u and x in the tree T. By Lemma 3 the vertices v and y are adjacent. Since $d_G(u, v) = k$, from the previous case we obtain that $d_T(v, y) = 2$. Hence $d_T(x, y) = 3$. On the other hand, as we mentioned above, for every edge xy of T, $d_G(x, y) \leq 2$.

Now consider two arbitrary vertices x and y of G and a shortest (x, y)–path. Applying to every edge of this path the obtained inequalities, we will get

$$\frac{1}{2} \cdot d_G(x, y) \leq d_T(x, y) \leq 3 \cdot d_G(x, y).$$

To prove that $|d_G(x, y) - d_T(x, y)| \leq 2$ we proceed as follows. Suppose that $x \in S_i^k$ and $y \in S_j^l$. Let $S := nca(S_i^k, S_j^l)$ and assume that S is a connected component of the sth neighbourhood of u. Denote by S' and S'' the neighbours of S in Γ on the paths between S, S_i^k, and S, S_j^l, respectively. From the definition of the trees Γ and T we obtain that in T and G the distances between u and any other vertex of V are the same. Any shortest path in G between u and each of the vertices x and y shares a common vertex with S. Therefore, one can select two vertices $x', y' \in S$ such that $d_G(x, x') = k - s$ and $d_G(y, y') = l - s$ (note that x' and y' are vertices of S closest to x and y, respectively). Since Γ is a tree, one can easily show that in G any shortest (x, y)–path $P(x, y)$ between x and y passes through the set S. Let x'', y'' denote the first and the last vertices of $P(x, y)$ in S. Since $x', x'' \in N(S')$ and $y', y'' \in N(S'')$, by Lemma 3, $d_G(x, x'') = k - s$ and $d_G(y, y'') = l - s$. Therefore, $d_G(x, y) = k + l - 2 \cdot s + d_G(x'', y'')$. From Lemma 3 we have $d_G(x'', y'') \leq 3$. On the other hand, from the algorithm we obtain that $d_T(x, y) = k + l - 2 \cdot s + \alpha$, where $\alpha = 0$ if in T the nearest common ancestor of x and y is a vertex of S, and $\alpha = 2$ if this ancestor belongs to the father of S in Γ. In the first case necessarily $N(S')$ and $N(S'')$ share a vertex in S. Since by Lemma 3 $N(S') \cap S$ and $N(S'') \cap S$ are complete subgraphs, one can easily show that $d_G(x'', y'') \leq 2$. Therefore, in this case $|d_G(x, y) - d_T(x, y)| \leq 2$. The same inequality is evidently true if $\alpha = 2$. This completes the proof of the theorem. $\square$

It is an open problem whether the distance matrix D of a chordal graph $G = (V, E)$ can be computed in $O(|V|^2)$ time. From the second assertion of Theorem 1 we obtain that within these time bounds we can compute the elements of D with an error at most 2. Even more, given a pair of vertices $x, y \in V$ by the algorithm of Harel and Tarjan [18] the nearest common ancestor $nca(x, y)$ of x and y in T can be computed in $O(1)$ time after a linear time preprocessing of T. Since $d_T(x, y) = d_T(x, u) + d_T(y, u) - 2 \cdot d_T(u, nca(x, y))$, the distance $d_T(x, y)$ can be found using a constant number of operations. Therefore, after a linear time preprocessing, in only $O(1)$ time we can compute $d_G(x, y)$ with an error at most 2.

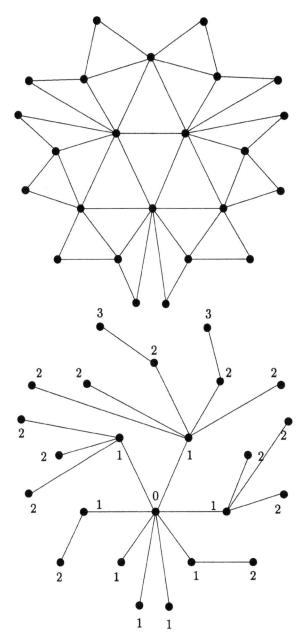

FIGURE 1. A chordal graph and a distance (1,2)–approximating tree of it.

A chordal graph together with some distance $(1, 2)$–approximating tree produced by Procedure 1 is given in Figure 1. Note that this chordal graph has no
additive tree r–spanner for $r \leq 3$ (see [27]). It remains an open question whether
every chordal graph admits an $(1, 1)$–approximating tree.

4 Spanners of Strongly Chordal Graphs

As we already mentioned in the introduction, chordal graphs do not have multiplicative or additive t–spanners with a fixed t. However, we will show that strongly chordal graphs have multiplicative 4–spanners and additive 3–spanners. We present an $O(|V| + |E|)$ time algorithm for computing such spanners.

A k–sun S_k $(k \geq 3)$ is a graph consisting of a complete subgraph $U = \{u_1, \ldots, u_k\}$ and an independent set $W = \{w_1, \ldots, w_k\}$, such that every w_i is adjacent only to u_i and $u_{i+1}(\bmod k)$ (S_3 and S_4 are presented in Figure 2).

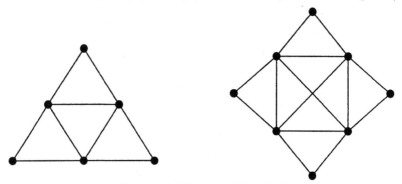

FIGURE 2. The suns S_3 and S_4.

A chordal graph G is called *strongly chordal* if it does not contain any sun S_k as an induced subgraph [9, 14]. For equivalent definitions and properties of strongly chordal graphs see [14] and for the recognition problem see [24, 29].

By a *two–set* of a graph G we will mean a subset $M \subseteq V$ such that $d_G(x, y) \leq 2$ for any $x, y \in M$.

Lemma 5. *Let G be a chordal graph which does not contain S_3 as an induced subgraph. For any two–set M of the subgraph of G induced by $N_k(u)$ there exists a vertex $v \in N_{k-1}(u)$ which is adjacent to all vertices of M.*

PROOF IS OMITTED.

Pick an arbitrary vertex u of G, and let $S_1^{k+1}, \ldots, S_{p_{k+1}}^{k+1}$ be the connected components of the subgraph of G induced by $N_{k+1}(u)$, $k \geq 1$. Let $C_j = N(S_j^{k+1}) \cap N_k(u)$ $(j = 1, \ldots, p_{k+1})$ (due to Lemma 3 C_j is a clique) and denote by H_k the graph with $\cup_{j=1}^{p_{k+1}} C_j$ as the vertex set and two vertices x, y are adjacent in H_k if and only if they belong to a common clique C_j.

Lemma 6. *For a strongly chordal graph G, every connected component of the graph H_k $(k \geq 1)$ is a two–set of G.*

PROOF IS OMITTED.

Now we have all prerequisites to formulate the main result of this section. Let G be a strongly chordal graph. Consider an arbitrary fixed vertex u and a tree Γ

rooted at u which is defined as in Section 2. Let $q := max\{d(u,v) : v \in V\}$ and $S_1^k, \ldots, S_{p_k}^k$ be the connected components of the kth neighbourhood $N_k(u)$ of u. We construct a distance approximating spanning tree $T = (V, E')$ of G step by step, starting from the leaves of Γ, i.e. from $N_q(u)$. Initially E' is empty. At first, for each vertex $v \in N_q(u)$, we add to E' one edge of the form vw, where w is a neighbour of v in $N_{q-1}(u)$. Now consider an arbitrary k which runs from $q-1$ to 1 and the cliques $C_j = N(S_j^{k+1}) \cap N_k(u)$ $(j = 1, \ldots, p_{k+1})$. By Lemma 6 every connected component F of the graph H_k is a two–set. Therefore all vertices of F have a common neighbour v_F in $N_{k-1}(u)$ (see Lemma 5). Now if x is a vertex of H_k, say $x \in F$ for some connected component F of H_k, then we add to the current E' the edge xv_F. For every vertex y of $N_k(u)$ which does not belong to H_k, we add to E' an edge connecting y with an arbitrary neighbour of it in $N_{k-1}(u)$.

Procedure 2.

> $E' := \emptyset$;
> **for** every $v \in N_q(u)$ pick a neighbour w in $N_{q-1}(u)$ and
>   add the edge vw to E';
> **for** $k := q - 1$ **downto** 1 **do**
>   compute the connected components $S_1^{k+1}, \ldots, S_{p_{k+1}}^{k+1}$ of $N_{k+1}(u)$;
>   determine the graph H_k and compute its connected components;
>   **for** every $y \in N_k(u) \setminus H_k$ pick a neighbour w in $N_{k-1}(u)$ and
>   add the edge yw to E';
>   **for** every connected component F of the graph H_k **do**
>   choose in $N_{k-1}(u)$ a common neighbour v_F of all vertices of F;
>   **for** every $x \in F$ add the edge xv_F to E'.

One can easily show that the graph $T = (V, E')$ constructed by this procedure is a spanning tree of G. Next we will show that the procedure can be implemented in linear time. In the preprocessing step we apply the breadth–first search to compute the kth neighbourhoods of the vertex u in $O(|V| + |E|)$ time. Denote by $deg(v)$ the degree of a vertex v in G. The second line of the procedure requires at most $\sum_{v \in N_q(u)} deg(v)$ operations. We spend $O(|N_{k+1}(u)| + \sum_{v \in N_{k+1}(u)} deg(v))$ time for computing the connected components in $N_{k+1}(u)$.

To determine the graph H_k we need to find the cliques $C_j = N(S_j^{k+1}) \cap N_k(u)$, $(j = 1, \ldots, p_{k+1})$. To do this we proceed in the following way. By Lemma 3 a vertex w_j of the connected component S_j^{k+1} which has the maximum number of neighbours in $N_k(u)$ obeys the condition $N(w_j) \cap N_k(u) = N(S_j^{k+1}) \cap N_k(u)$. Therefore, we have to find the vertices $\{w_1, \ldots, w_{p_{k+1}}\}$ and put $C_j := N(w_j) \cap N_k(u)$. This can be done in $\sum_{v \in N_{k+1}(u)} deg(v)$ time. Having the cliques $C_1, \ldots, C_{p_{k+1}}$, the connected components of the graph H_k can be computed by constructing a special bipartite graph $B_k = (W, K; U)$. In this graph $W = \{w_1, \ldots, w_{p_{k+1}}\}$, $K = \cup_{i=1}^{p_{k+1}} C_i$, and a vertex of W and a vertex of K are adjacent in B_k if and only if they are adjacent in G. This graph B_k can be constructed in $O(\sum_{j=1,\ldots,p_{k+1}} |C_j|) = O(\sum_{v \in N_{k+1}(u)} deg(v))$ time. One can easily see that the connected components of H_k are exactly the intersections of the connected components of B_k with the set K, and thus can be found within the same time bounds.

Finally, to decide which connected components of the graph H_k are contained in the neighbourhood $N(v)$ of some vertex $v \in N_{k-1}(u)$, we do the following. In total $O(deg(v))$ time, for each connected component F_i $(i = 1, \ldots, r)$ with $N(v) \cap F_i \neq \emptyset$, we compute the value n_i which is the number of vertices of F_i from $N(v)$. If $n_i = |F_i|$ then we put $v_{F_i} := v$. Thus, this line of the procedure can be implemented in $O(\sum_{v \in N_{k-1}(u)} deg(v))$ time, too.

Summarizing, the whole procedure requires only

$$O(\sum_{k=1}^{q}(\sum_{v \in N_k(u)} deg(v) + |N_k(u)|)) = O(|V| + |E|)$$

time.

A strongly chordal graph together with some additive tree 3–spanner produced by our procedure is given in Figure 3. Note that this strongly chordal graph has no additive tree 2–spanner [19]. The following theorem shows that not only this graph but every strongly chordal graph has an additive tree 3–spanner. So this result is best possible.

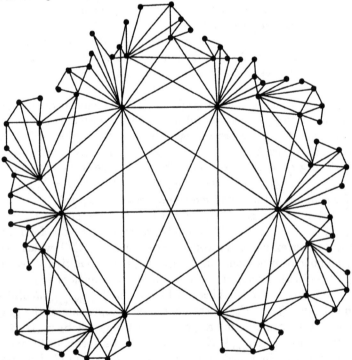

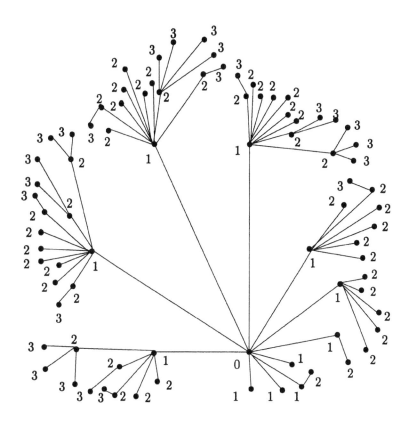

FIGURE 3. A strongly chordal graph and an additive tree 3-spanner of it.

Theorem 2. *Let G be a strongly chordal graph. The tree $T = (V, E')$ constructed by the Procedure 2 is a multiplicative 4–spanner as well as an additive 3–spanner of G.*

Proof. First we will show that $d_T(x, y) \leq 4$ holds for any edge xy of G. Suppose that $d_G(u, x) = d_G(u, y) = k$, say $x, y \in S_j^k$. If x and y belong to a common connected component F of the graph H_k, then in T they have a common father v_F, and thus $d_T(x, y) = 2$. Otherwise, let x' and y' be the fathers of x and y, respectively, in T. Since $x', y' \in C_j$, in H_{k-1} the vertices x' and y' lie in a common connected component. As we already showed $d_T(x', y') = 2$. Therefore, in this case $d_T(x, y) = 4$.

Now suppose that $x \in N_k(u)$ and $y \in N_{k-1}(u)$. Let x' be the father of x in T. If $x' = y$ we are done. Assume $x' \neq y$. Both x' and y have some neighbor in the same component of $N_k(u)$, and thus they are adjacent in H_{k-1}. According to the algorithm, for an edge of this type, $d_T(x', y) = 2$ holds. Hence $d_T(x, y) = 3$.

Now consider two arbitrary vertices v and w of G and a shortest (v, w)–path. Applying to every edge xy of this path the inequality $d_T(x, y) \leq 4$, we will get

$d_T(v, w) \leq 4 \cdot d_G(v, w)$, i.e., T is a multiplicative 4–spanner for G.

That T is an additive 3–spanner of G follows already from the previous part of our proof and from [27, Lemma 1]. We present here another proof. For this suppose that T is rooted at the vertex u. From the algorithm it follows that the distances in G and T between a vertex and any of its ancestors are the same. Pick two vertices $v, w \in V$ and proceed by induction on $d_G(v, w)$. If v and w are adjacent, then we are done, because then $d_T(v, w) \leq 4$. Now suppose that $d_G(v, w) = s \geq 2$ and let z be a neighbour of v on a shortest path between v and w. From the induction assumption we have $d_T(z, w) \leq s - 1 + 3 = s + 2$ and $d_T(v, z) \leq 4$. Let $a = nca(v, z)$ be the nearest common ancestor of v and z in the tree T. Since $d_T(a, v) = d_G(a, v)$, $d_T(a, z) = d_G(a, z)$ and $vz \in E$ we obtain that $|d_T(v, a) - d_T(z, a)| \leq 1$.

We can additionally assume that $d_T(z, w) < d_T(v, w) - 1$, otherwise we immediately conclude that $d_T(v, w) \leq d_G(v, w) + 3$. From this and the previous inequality we deduce that the vertex $nca(w, z)$ lies on the path of T between the vertices a and z. Therefore, a is an ancestor of w, and thus $d_T(a, w) = d_G(a, w)$. Notice that the distance sums $d_T(v, w) + d_T(a, z)$ and $d_T(v, z) + d_T(a, w)$ are equal. Hence

$$d_T(v, w) = d_T(a, w) - d_T(a, z) + d_T(v, z) =$$

$$d_G(a, w) - d_G(a, z) + d_T(v, z) \leq d_G(w, z) + 4 \leq d_G(v, w) + 3,$$

concluding the proof. □

5 Spanners of Dually Chordal Graphs

Below we show that the Procedure 2 from the previous section can be applied to produce multiplicative 4–spanners and additive 3–spanners in dually chordal graphs. These graphs were introduced in [12] as a generalization of strongly chordal graphs (which are the hereditary dually chordal graphs) where the Steiner tree problem and many domination-like problems still have efficient solutions. It turns out that the dually chordal graphs are exactly the intersection graphs of maximal cliques of chordal graphs (see [31, 5]).

To define dually chordal graphs, we need some notions from the theory of hypergraphs [3]. Let $\mathcal{E}$ be a hypergraph with underlying set V, i.e. $\mathcal{E}$ is a collection of subsets of V. The *dual hypergraph* $\mathcal{E}^*$ has $\mathcal{E}$ as its vertex set and for every $v \in V$ a hyperedge $\{e \in \mathcal{E} : v \in e\}$. The *line graph* $L(\mathcal{E}) = (\mathcal{E}, E)$ of $\mathcal{E}$ is the intersection graph of $\mathcal{E}$, i.e. $ee' \in E$ if and only if $e \cap e' \neq \emptyset$. A *Helly hypergraph* is one whose edges satisfy the Helly property, that is, any subfamily $\mathcal{E}' \subseteq \mathcal{E}$ of pairwise intersecting edges has a nonempty intersection. A hypergraph $\mathcal{E}$ is a *hypertree* if there is a tree T with vertex set V such that every edge $e \in \mathcal{E}$ induces a subtree in T. Equivalently, $\mathcal{E}$ is a hypertree if and only if the line graph $L(\mathcal{E})$ is chordal and $\mathcal{E}$ is a Helly hypergraph. A hypergraph $\mathcal{E}$ is a *dual hypertree* (α–*acyclic hypergraph*) if there is a tree T with vertex set $\mathcal{E}$ such that, for every vertex $v \in V$, $T_v = \{e \in \mathcal{E} : v \in e\}$ induces a subtree of T. Observe that $\mathcal{E}$ is a hypertree if and only if $\mathcal{E}^*$ is a dual hypertree.

For a graph G by $\mathcal{C}(G) = \{C : C$ is a maximal clique in $G\}$ we denote the *clique hypergraph*. Let also $\mathcal{D}(G) = \{D_k(v) : v \in V, k$ a nonnegative integer$\}$ be the *disk hypergraph* of G. A graph G is called *dually chordal* if the clique hypergraph $\mathcal{C}(G)$ is a hypertree [5, 12, 31]. In [5, 12] it is shown that dually chordal graphs are the graphs G whose disk hypergraphs $\mathcal{D}(G)$ are hypertrees (see [5, 12] for other characterizations, in particular in terms of certain elimination schemes, and [6] for their algorithmic use). From the definition of hypertrees we deduce that for dually chordal graphs the line graphs of the clique and disk hypergraphs are chordal. Conversely, if G is a chordal graph then $\mathcal{C}(G)$ is a dual hypertree, and, therefore, the line graph $L(\mathcal{C}(G))$ is a dually chordal graph, justifying the term "dually chordal graphs". Finally note that graphs being both chordal and dually chordal were dubbed *doubly chordal* and investigated in [5, 12, 23].

Henceforward, we will suppose that G is a dually chordal graph. To prove that the Procedure 2 indeed constructs an additive 3–spanner of G, we need some properties of 2–sets and graphs H_k constructed from G.

Lemma 7. *For any two–set $M \subseteq N_k(u)$ of G there exists a vertex $v \in N_{k-1}(u)$ which is adjacent to all vertices of M.*

PROOF IS OMITTED.

Lemma 8. *Let S be a connected component of the subgraph of G induced by $N_k(u)$. Then $M := N(S) \cap N_{k-1}(u)$ is a 2–set, and, moreover, any two nonadjacent vertices of M have a common neighbour in M. In particular, the graph Γ, defined in the Section 2, is a tree.*

PROOF IS OMITTED.

As in Procedure 2, let $S_1^{k+1}, \ldots, S_{p_{k+1}}^{k+1}$ denote the connected components of the subgraph of G induced by $N_{k+1}(u)$, $k \geq 1$. Set $M_j = N(S_j^{k+1}) \cap N_k(u)$, $j = 1, \ldots, p_{k+1}$ (the 2–sets M_j play the role of the cliques C_j in the case of strongly chordal graphs). Denote by H_k the graph with $\cup_{j=1}^{p_{k+1}} M_j$ as the vertex set and two vertices x, y are adjacent in H_k if and only if they belong to a common set M_j. We continue with the following result analogous to Lemma 6.

Lemma 9. *Every connected component F of the graph H_k ($k \geq 1$) is a two–set of G.*

Proof. Let x, y be two nonadjacent vertices of a connected component F of the graph H_k. Then we can find a collection of 2–sets $M_{i_1}, M_{i_2}, \ldots, M_{i_h}$ such that $x \in M_{i_1}, y \in M_{i_h}$ and $M_{i_j} \cap M_{i_{j+1}} \neq \emptyset$ for all $j = 1, \ldots, h-1$. Pick $z_j \in M_{i_j} \cap M_{i_{j+1}}$, $j = 1, \ldots, h-1$ and let $z_0 := x$ and $z_h := y$. Since $z_{j-1}, z_j \in M_{i_j}$, $j = 1, \ldots, h$, we can find two vertices $v_j', v_j'' \in S_{i_j}^{k+1}$ adjacent to z_{j-1} and z_j, respectively. Let P_j be a path of $S_{i_j}^{k+1}$ connecting the vertices v_j' and v_j''. The disk $D_{k-1}(u)$ together with $D_1(x), D_1(y)$ and the disks of the family $\{D_1(z) : z \in \cup_{j=1}^h P_j\}$ forms a cycle in the line graph $L(\mathcal{D}(G))$. From chordality of this graph and since $D_{k-1}(u) \cap D_1(z) = \emptyset$ holds for all $z \in \cup_{j=1}^h P_j$, we deduce that $D_1(x) \cap D_1(y) \neq \emptyset$, i.e. $d_G(x, y) = 2$. $\square$

To find the connected components of the graph H_k we construct again a special bipartite graph $B_k = (W, K; U)$. In this graph $W = \{s_1, \ldots, s_{p_{k+1}}\}$ (a vertex s_j represents a set S_j^{k+1}), and K is the vertex set of H_k. A vertex $s_j \in W$ and a vertex $v \in K$ are adjacent in B_k if and only if $v \in N(S_j^{k+1})$. The graph B_k can be constructed in $O(\sum_{v \in N_{k+1}(u) \cup N_k(u)} deg(v))$ time. Note that a vertex $v \in N_k(u)$ belongs to H_k if and only if it has a neighbour in $N_{k+1}(u)$. The connected components of H_k are exactly the intersections of the connected components of B_k with the set K, and thus can be found within the same time bound. All other steps of the procedure can be implemented in the same way as for strongly chordal graphs. Summarizing, the whole procedure for a dually chordal graph G requires

$$O(\sum_{k=1}^{q} (\sum_{v \in N_k(u)} deg(v) + |N_k(u)|)) = O(|V| + |E|)$$

time.

Theorem 3. *Let G be a dually chordal graph. The tree $T = (V, E')$ constructed by the Procedure 2 is a multiplicative 4–spanner as well as an additive 3–spanner of G.*

PROOF IS OMITTED.

References

1. I. ALTHÖFER, G. DAS, D. DOBKIN, D. JOSEPH, and J. SOARES, On sparse spanners of weighted graphs, *Discrete Comput. Geom.*, **9** (1993), 81–100.
2. J.-P. BARTHÉLEMY and A. GUÉNOCHE, Trees and Proximity Representations, *Wiley*, New York, 1991.
3. C. BERGE, Hypergraphs, *North Holland*, Amsterdam, 1989.
4. S. BHATT, F. CHUNG, F. LEIGHTON, and A. ROSENBERG, Optimal simulations of tree machines, in *27th IEEE Foundations of Computer Science*, Toronto, 1986, 274–282.
5. A. BRANDSTÄDT, F. DRAGAN, V. CHEPOI, and V. VOLOSHIN, Dually chordal graphs, *Graph–Theoretic Concepts in Computer Science, Lecture Notes in Computer Science* **790**, J. van Leeuwen, ed., Springer–Verlag, Berlin, New York, 1994, 237–251.
6. A. BRANDSTÄDT, V. CHEPOI, and F. DRAGAN, Clique r-domination and clique r-packing problems on dually chordal graphs, *SIAM J. Discrete Math.* **10** (1997), 109–127.
7. P. BUNEMAN, A characterization of rigid circuit graphs, *Discrete Math.*, **9** (1974), 205–212.
8. L. CAI AND D.G. CORNEIL, Tree spanners, *SIAM J. Disc. Math.*, **8** (1995), 359–387.
9. G.J. CHANG AND G.L. NEMHAUSER, The k–domination and k–stability problems on sun–free chordal graphs, *SIAM J. Alg. Disc. Meth.*, **5** (1984), 332–345.

10. L.P. CHEW, There are planar graphs almost as good as the complete graph, *J. of Computer and System Sciences*, **39** (1989), 205–219.

11. G.A. DIRAC, On rigid circuit graphs, *Abh. Math. Sem. Univ. Hamburg*, **25** (1961), 71–76.

12. F. DRAGAN, C. PRISACARU, and V. CHEPOI, Location problems in graphs and the Helly property (in Russian), *Discrete Mathematics, Moscow*, **4** (1992), 67–73.

13. D.B.A. EPSTEIN, J.W. CANNON, D.F. HOLT, S.V.F. LEVY, M.S. PATERSON AND W.P. THURSTON, Word Processing in Groups, *Jones and Bartlett*, Boston, 1992.

14. M. FARBER, Characterization of strongly chordal graphs, *Discrete Math.*, **43** (1983), 173–189.

15. M. FARBER AND R.E. JAMISON, Convexity in graphs and hypergraphs, *SIAM J. Alg. Discrete Meth.*, **7** (1986), 433–444.

16. E. GHYS AND P. DE LA HARPE, Les Groupes Hyperboliques d'après M. Gromov, *Progress in Mathematics*, Vol. **83**, Birkhauser, 1990.

17. M.C. GOLUMBIC, Algorithmic Graph Theory and Perfect Graphs, *Academic Press*, London, 1980.

18. D. HAREL AND R.E. TARJAN, Fast algorithms for finding nearest common ancestors. *SIAM J. Comput.* **13** (1984), 338–355.

19. HOÀNG-OANH LE, Personal communication.

20. A.L. LIESTMAN AND T. SHERMER, Additive graph spanners, *Networks*, **23** (1993), 343–364.

21. N. LINIAL, E. LONDON AND Y. RABINOVICH, The geometry of graphs and some of its algorithmic applications, *Combinatorica*, **15** (1995), 215–245.

22. M.S. MADANLAL, G. VENKATESAN AND C.PANDU RANGAN, Tree 3–spanners on interval, permutation and regular bipartite graphs, *Information Processing Letters*, **59** (1996), 97–102.

23. M. MOSCARINI, Doubly chordal graphs, Steiner trees and connected domination, *Networks*, **23** (1993), 59–69.

24. R. PAIGE, and R.E. TARJAN, Three partition refinement algorithms, *SIAM J. Comput.*, **16** (1987), 973–989.

25. D. PELEG AND A.A. SCHÄFFER, Graph spanners, *J. Graph Theory*, **13** (1989), 99–116.

26. D. PELEG AND J.D. ULLMAN, An optimal synchronizer for the hypercube, *in Proc. 6th ACM Symposium on Principles of Distributed Computing*, Vancouver, 1987, 77–85.

27. E. PRISNER, Distance approximating spanning trees, in *Proc. of STACS'97, Lecture Notes in Computer Science* **1200**, (R. Reischuk and M. Morvan, eds.), Springer–Verlag, Berlin, New York, 1997, 499–510.

28. P.H.A. SNEATH and R.R. SOKAL, Numerical Taxonomy, *W.H. Freeman*, San Francisco, California, 1973.

29. J.P. SPINRAD, Doubly lexical ordering of dense 0-1– matrices, *Information Processing Letters*, **45** (1993), 229–235

30. D.L. SWOFFORD and G.J. OLSEN, Phylogeny reconstruction, In *Molecular Systematics (D.M. Hillis and C. Moritz, editors)*, Sinauer Associates Inc., Sunderland, MA., 1990, 411–501.

31. J.L. SZWARCFITER, and C.F. BORNSTEIN, Clique graphs of chordal and path graphs, *SIAM J. Discrete Math.*, **7** (1994), 331–336.

Decomposition of Integer Programs and of Generating Sets

G. Cornuejols[1], R. Urbaniak[2], R. Weismantel[3] and L. Wolsey[4]

[1] CMU, USA
[2] ZIB-Berlin, Germany
[3] ZIB-Berlin, Germany
[4] CORE, Belgium

Abstract. In this paper we investigate techniques for decomposing the matrix of coefficients of a family of integer programs. From a more practical point of view, these techniques are useful to design a primal algorithm that solves the integer program via generating sets. In this context our approach is applied to the Frobenius problem and to integer programming instances that seem to be difficult for LP-based integer programming codes. From a theoretical point of view, the techniques for decomposing a matrix that we present in this paper give rise to bounds on the L_1-norm of all the elements in the Hilbert basis of a pointed cone. Moreover, applying our decomposition techniques we can show that any 0/1 linear integer program with a fixed number of constraints and a fixed number of digits to encode each coefficient in the matrix can be solved in polynomial time. A relation of our method to the group theoretic approach exists and is discussed as well.

Keywords: generating set, group theoretic approach, integer programming, test set, scaling, primal method, knapsack problem.

1 Introduction

This paper deals with (families of) integer programs of the form

$$\max \{c^T x : \ d \le Ax \le b, \ x \in \mathbb{Z}^n, \ l \le x \le u\}, \tag{1}$$

where $A \in \mathbb{Z}^{m \times n}$ is a given matrix, $d, b \in \mathbb{Z}^m$, $d \le b$ and $l, u \in \mathbb{Z}^n$, $l \le u$ define feasible ranges and $c \in \mathbb{Z}^n$ is the objective function.

The approach that we propose for solving certain programs of the form (1) uses the following two ingredients:
(i) generating sets for integer programs and
(ii) techniques for decomposing the coefficients of A and b.

Let $X_A(d, b, l, u) = \{x \in \mathbb{Z}^n : \ d \le Ax \le b, \ l \le x \le u\}$ denote the feasible set of the integer program (1).

Let

$$F_A := \{x - y : x \ne y \text{ and there is } (d, b, l, u) \in \mathbb{Z}^{2n+2m} \text{ with } x, y \in X(d, b, l, u)\}.$$

Definition 1. A set of vectors $\{t^1, \ldots, t^k\} \subseteq F_A$ is a *generating set* for F_A if, given any $t \in F_A$, there exist $\lambda_1, \ldots, \lambda_k \in \mathbb{Z}_+$ such that $t = \sum_{j=1}^{k} \lambda_j t^j$ and

(i) $t^+ = \sum_{j=1}^{k} \lambda_j (t^j)^+$,   (ii) $t^- = \sum_{j=1}^{k} \lambda_j (t^j)^-$,

(iii) $(At)^+ = \sum_{j=1}^{k} \lambda_j (At^j)^+$, (iv) $(At)^- = \sum_{j=1}^{k} \lambda_j (At^j)^-$.

Using Gordan's lemma, one can show that such a finite generating set exists: Indeed, Gordan's lemma states that, for any $S \subseteq \mathbb{Z}_+^d$, there exists a unique minimal set of points $p_1, \ldots, p_k \in S$ such that, for every $p \in S$, there exists a p_j such that $p \geq p_j$. By setting $S := \{(z^+, z^-, (Az)^+, (Az)^-) : z \in F_A\}$, Gordan's lemma implies that there exists a unique minimal generating set for F_A. This unique minimal generating set is denoted by G_A.

One can characterize G_A as the set of all irreducible vectors in F_A.

Definition 2. A vector $w \in F_A$ can be reduced by $v \in F_A$ if $v^+ \leq w^+$, $v^- \leq w^-$, $(Av)^+ \leq (Aw)^+$ and $(Av)^- \leq (Aw)^-$. In this situation, we say that we obtain $w - v$ by *reducing* w. If w cannot be reduced by any $v \in F_A$ we say that w is *irreducible*.

Every irreducible vector in F_A must belong to G_A. Conversely, if vector $w \in F_A$ can be reduced by $v \in F_A$, then it is easy to check that $w - v \in F_A$ and $w - v$ reduces w. Now it follows that w is not in the minimal generating set G_A.

The generating set G_A is related to the notion of test set. Given $A \in \mathbb{Z}^{m \times n}$ and $c \in \mathbb{Z}^n$, a *test set* for the family of programs of the form (1) where $(d, b, l, u) \in \mathbb{Z}^{2n+2m}$ is a subset of the integer lattice $\mathbb{Z}^n$ such that for every feasible point x of (1) that is not optimal with respect to c, there exists an element t in the test set with which x can be improved, i.e., $c^T t > 0$ and $x + t$ is again a feasible point of (1).

Lemma 3. *The generating set G_A is a test set for every objective function and every $(d, b, l, u) \in \mathbb{Z}^{2n+2m}$ of (1).*

The generating set G_A contains the Graver basis associated with the matrix A, see Graver [4]. Although there are various algorithms that compute, in principle, such generating sets, see for instance Thomas [7] and the references given there, all these algorithms are still far from being able to compute a test set for a general 0/1 program with 100 variables, say. The main reason for this is that such methods perform, for every pair of vectors from a current set G^i, operations that in the worst case let the set grow by a quadratic factor in the cardinality of G^i. Iterating these steps a few times can exceed a running time limit of several hours on today's workstations.

We show in Section 2 that the generating set G_A can be constructed by considering the rows of A one at a time. This property yields in various cases a significant speed up for computing generating tests. From a theoretical point of view, the row composition algorithm for computing generating sets allows to bound the L_1-norm of the elements of G_A. A consequence of this theorem is that any 0/1 linear integer program with a fixed number of constraints and a fixed number of digits to encode each coefficient in the matrix can be solved in polynomial time. In Section 3, we propose a way to decompose an integer program into a family of problems which all have the same generating set. In various cases, this set can be computed practically and used to solve integer programs that are extremely difficult for LP-based codes. We characterize in Section 4 the generating set for the group relaxation and apply our decomposition method to the Frobenius problem.

2   Decomposing a Generating Set

Let $G_1 := G_{A_1}$ be a minimal generating set for F_{A_1}. Let A_2 be obtained from A_1 by adding a row, namely $A_2 := \begin{pmatrix} A_1 \\ a^T \end{pmatrix}$. We denote by $G_2 := G_{A_2}$ the minimal generating set for F_{A_2}. The sets G_1 and G_2 are related in the following way.

Lemma 4. Let $G_1 = \{t^1, \ldots, t^k\}$. For every $t \in G_2$, there exist $\lambda_1, \ldots, \lambda_k \in \mathbb{Z}_+$ such that $t = \sum_{j=1}^{k} \lambda_j t^j$ and

$$(i) \quad t^+ = \sum_{j=1}^{k} \lambda_j (t^j)^+, \qquad (ii) \quad t^- = \sum_{j=1}^{k} \lambda_j (t^j)^-,$$

$$(iii) \quad (At)^+ = \sum_{j=1}^{k} \lambda_j (At^j)^+, \quad (iv) \quad (At)^- = \sum_{j=1}^{k} \lambda_j (At^j)^-.$$

Lemma 5. Let $G_1 = \{t^1, \ldots, t^k\}$ and $t \in G_2$. There exists a sequence of vectors $z^1, \ldots, z^l$ in G_1 that satisfies Lemma 4 and, for every $j \in \{1, \ldots, l\}$ the following conditions:

(a) $z^1 + \ldots + z^j \in F_{A_1}$ and,
(b) for every $v \in F_{A_1}$ reducing $z^1 + \ldots + z^j$ (with respect to A_1) we have that either $(a^T v)^+ \not\preceq (a^T(z^1 + \ldots + z^j))^+$ or $(a^T v)^- \not\preceq (a^T(z^1 + \ldots + z^j))^-$.

By Lemma 4, every element in G_2 is the nonnegative integer combination of elements in G_1. This property can be used to compute G_2 starting with the set G_1. Initially we set $G := G_1$. Iteratively we take all the sums of two vectors in G, reduce each of these vectors as long as possible by the elements of G (with respect to the matrix A_2), and add all the reduced vectors that are different from the origin to the set G. In pseudo-programming language the algorithm for computing G_2 starting from G_1 may look as follows:

Algorithm 6.

> Set $G_{old} := \emptyset$ and $G := G_1$.
> While $G_{old} \neq G$ repeat the following steps:
> > Set $G_{old} := G$.
> > For all pairs of vectors $v, w \in G$, let $z := v + w$.

As long as there exists $y \in G$ such that $y^+ \leq z^+$, $y^- \leq z^-$ and $(A_2 y)^+ \leq (A_2 z)^+$, $(A_2 y)^- \leq (A_2 z)^-$, update $z := z - y$. If $z \neq 0$, update $G := G \cup \{z\}$.

Theorem 7. *Algorithm 6 terminates in finite time. The set G that is returned by the algorithm contains a generating set of F_{A_2}.*

Proof. The input of Algorithm 6 is the set $G_1 = \{t^1, \ldots, t^k\}$ that generates F_{A_1}. With every $t \in G_2$ we associate a sequence of vectors $Z(t) = (z^1, \ldots, z^l)$ satisfying Lemma 5. Let

$$l_0 := \max\{l : \text{ there exists } t \in G_2 \text{ with } Z(t) = (z^1, \ldots, z^l)\}.$$

l_0 is a number and it defines an upper bound on the number of times we have to perform the While-loop of Algorithm 6, because for every $j \in \{1, \ldots, l\}$ there does not exist any $v \in F_{A_1}$ reducing $z^1 + \ldots + z^j$ (with respect to A_1) and that would satisfy $(a^T v)^+ \leq (a^T(z^1 + \ldots + z^j))^+$ and $(a^T v)^- \leq (a^T(z^1 + \ldots + z^j))^-$. Therefore, Algorithm 6 is finite.

We still have to show that every element $t \in G_2$ is contained in G.

When Algorithm 6 terminates, there exists no $z \in F_{A_1}$ such that

$z = v + w$ with $v, w \in G$ and
$(a^T z)^+ \not\leq (a^T g)^+$ or $(a^T z)^- \not\leq (a^T g)^-$ for some element $g \in G$.

On account of Lemma 5 this situation occurs only if G_2 is contained in G.

We have tested Algorithm 6 against a combinatorial variant of the Buchberger algorithm that works simultaneously with all the rows of A.

There was not a single problem in our series of test examples for which Algorithm 6 required more computation time than the combinatorial variant of the Buchberger algorithm. In fact, with n, there also grows the factor by which Algorithm 6 is faster on average than its competitor. One explanation for this behavior might be that the size of the intermediate sets, that algorithms for computing generating sets create, blow up tremendously during the run of the algorithm when we take, iteratively, sums of vectors from a current set, check for reduction, and add all the vectors that cannot be reduced currently to the set. The more rows one has, the bigger this set might become during the process, because the less powerful reduction becomes. In contrast, at the end only a small fraction of the vectors for which we had to perform operations becomes irreducible.

From a theoretical point of view, the row composition algorithm 6 for computing generating sets can be used to derive in an easy way bounds on the L_1-norm of the elements in the generating set. Theorem 8 makes this precise. It yields a generalization of Theorem 1 in [2], yet the bound given in [2] in this special case is better than the bound that Theorem 8 gives.

Theorem 8. *For $k \in \mathbb{Z}_+$, $A \in \{-k, \ldots, k\}^{m \times n}$ consider the family of integer programs of the form (1). Every element g in the minimal generating set G for F_A satisfies the following bound on the L_1 norm: $|g|_1 := \sum_{i=1}^n |g_i| \leq (2k)^{2^m - 1}$.*

A consequence of Theorem 8 is that the minimal generating set associated with a family of 0/1 programs is polynomial in the number of variables if the number m of constraints of the matrix is fixed and all the coefficients of the matrix belong to the set $\{-k, \ldots, k\}$ with $k \in \mathbb{Z}_+$ fixed.

Corollary 9. *Let $k \in \mathbb{Z}_+$ and $A \in \{-k, \ldots, k\}^{m \times n}$. Consider the family of integer programs of the form (1) such that $l, u \in \{0, 1\}^n$. The cardinality of a minimal generating set G of the set F_A associated with the family of programs of the form (1) with varying $(d, b, l, u) \in Z^{2n+2m}$ is of the size $O(n^{(2k)^{2^m-1}} 3^{(2k)^{2^m-1}})$.*

In the remainder of this section we exhibit how the row composition algorithm 6 can be viewed in case of Hilbert bases. Let A be a $m \times n$ matrix with integral entries only and $C_A := \{x \in \mathbb{R}^n : Ax \leq 0\}$, the cone generated by A.

Definition 10. *Let $\mathcal{O}_j$ be the j-th orthant in $\mathbb{R}^n$ and $\mathcal{H}(C_j)$ the minimal Hilbert basis of the cone $C_j := \{x \in \mathcal{O}_j : Ax \leq 0\}$. The set*

$$\mathcal{G}(C_A) := \bigcup_j \mathcal{H}(C_j).$$

is called the extended Graver set of C_A.

For $a \in \mathbb{Z}^n$ let $C_{A,a} := \{x \in \mathbb{R}^n : Ax \leq 0, a^T x \leq 0\}$. Our goal is to investigate how the extended Graver sets $\mathcal{G}(C_A)$ and $\mathcal{G}(C_{A,a})$ differ when $C_{A,a}$ is a proper subset of C_A.

Let $\{b_1, \ldots, b_k\} = \{a^T h : h \in \mathcal{G}(C_A), a^T h \neq 0\}$ and $G_i := \{h \in \mathcal{G}(C_A) : a^T h = b_i\}$ for $i = 1, \ldots, k$. Define the knapsack cone $C_b := \{x \in \mathbb{R}_+^k : b_1 x_1 + \ldots + b_k x_k \leq 0\}$. C_b is a pointed cone. Therefore C_b has a unique minimal Hilbert basis, $\mathcal{H}(C_b)$ say. With every element $\bar{h} = (\bar{h}_1, \ldots, \bar{h}_k) \in \mathcal{H}(C_b)$ we associate a set of elements $T(\bar{h})$ of $\mathbb{Z}^n$ as follows:

$$T(\bar{h}) := \{h \in \mathbb{Z}^n : \text{ there exist } \alpha^j \in \mathbb{Z}_+^{G_j}, \sum_{h^i \in G_j} \alpha_i^j = \bar{h}_j, j = 1, \ldots, k,$$

$$\text{such that } h = \sum_{j=1}^k \sum_{h^i \in G_j} \alpha_i^j h^i\}.$$

Lemma 11. *If $\bar{h}$ is an element in the Hilbert basis $\mathcal{H}(C_b)$ of the knapsack cone C_b, then $h \in C_{A,a}$ for all $h \in T(\bar{h})$.*

Lemma 12. *For $x, x^1, x^2 \in C_b \cap \mathbb{Z}^k$ such that $x = x^1 + x^2$ and $h \in T(x)$, there exist $h^1 \in T(x^1)$ and $h^2 \in T(x^2)$ with $h = h^1 + h^2$.*

The utility of the set $T(\bar{h})$ will become apparent in the next theorem.

Theorem 13. *The set $\mathcal{G}(C_{A,a})$ coincides with the set of all irreducible elements in*

$$\bigcup_{\bar{h} \in \mathcal{H}(C_b)} T(\bar{h}) \cup \{h \in \mathcal{G}(C_A) : a^T h = 0\}. \tag{2}$$

Proof. Denote the set in (2) by G. Using Lemma 11 we can easily see that $G \subseteq \mathcal{G}(C_{A,a})$. It remains to be show that $\mathcal{G}(C_{A,a}) \subseteq G$.

Let h be in $\mathcal{G}(C_{A,a})$ and assume $h \notin G$. Since $h \in C_A$ there exist elements $h^1, \ldots, h^l$ in $\mathcal{G}(C_A)$ with $h = h^1 + h^2 + \ldots + h^l$. If $a^T h^i = 0$ for all $i = 1, \ldots, l$ we get a contradiction that $h \in \mathcal{G}(C_{A,a})$ because $h^i \in C_{A,a}$ for all $i = 1, \ldots, l$. Therefore, $a^T h^i \neq 0$ for at least one $i \in \{1, \ldots, l\}$. W.l.o.g. we assume that $\{h^1, \ldots, h^j\} = \{h^i : a^T h^i = 0, i = 1, \ldots, l\}$. Let $\{b_1, \ldots, b_k\} = \{a^T h^i : i = j+1, \ldots, l\}$ and $\bar{G}_s := \{h^i : a^T h^i = b_s\}$, $s = 1, \ldots, k$. Set $h' := h^{j+1} + \ldots + h^l$. Then $h = h^1 + \ldots + h^j + h'$ and

$$a^T h = \sum_{h^i \in \bar{G}_1} a^T h^i + \ldots + \sum_{h^i \in \bar{G}_k} a^T h^i$$

$$= |\bar{G}_1| b_1 + \ldots + |\bar{G}_k| b_k \leq 0.$$

It follows that $x = (|\bar{G}_1|, \ldots, |\bar{G}_k|) \in \mathbb{Z}_+^k$ is an element of the knapsack cone C_b. Either x is in $\mathcal{H}(C_b)$ or x can be written as a nonnegative integral combination of the elements in $\mathcal{H}(C_b)$. In the first case we obtain that $h' \in T(x)$ and by Lemma 11 h can be written as a nonnegative integral combination of the elements in $C_{A,a}$. Otherwise $x \notin \mathcal{H}(C_b)$. Then $x = x^1 + x^2$ with $x^1, x^2 \in C_b \cap \mathbb{Z}^k$ and by Lemma 12 $h' = h'^1 + h'^2$ with $h'^i \in T(x^i)$ for $i = 1, 2$. This is a contradiction that $h \in \mathcal{G}(C_{A,a})$.

Notice that $G_- := \{h \in \mathcal{G}(C_A) : a^T h < 0\}$ is a subset of $\{h \in T(\bar{h}) : \bar{h} \in \mathcal{H}(C_b)\}$. This shows essentially

Corollary 14. *The set $G_- \cup G_0$ is the extended Graver set $\mathcal{G}(C_{A,a})$ of the cone $C_{A,a}$ if and only if every element $h \in \bigcup_{\bar{h} \in \mathcal{H}(C_b)} T(\bar{h})$ can be written as a nonnegative integral combination of the elements in $G_- \cup G_0$.*

Related to the question of constructing from a Hilbert basis of $\mathcal{G}(C_A)$ a Hilbert basis of $\mathcal{G}(C_{A,a})$ is the following problem: Let $a := (a_1, \ldots, a_{n-1}, a_n) \in \mathbb{Z}^n$ and G_a the generating set for $F_a = \{x - y : a^T x = a^T y, x \neq y \in \mathbb{Z}_+^n\}$. For $\alpha \in \mathbb{Z}_+$, find the generating set $G_{\bar{a}}$ for $F_{\bar{a}} = \{x - y : \bar{a}^T x = \bar{a}^T y, x \neq y \in \mathbb{Z}_+^n\}$ where $\bar{a} := (a_1, \ldots, a_{n-1}, \alpha a_n)$. Let $\{b_1, \ldots, b_k\} = \{t_n : t \in G_a, t_n \neq 0\}$. We denote by $G_i := \{t \in G_a : t_n = b_i\}$ for all $i = 1, \ldots, k$ and by $\mathcal{H}(C_b)$ the minimal Hilbert basis of the knapsack cone

$$C_b := \{x \in \mathbb{R}_+^{k+1} : b_1 x_1 + \ldots + b_k x_k - \alpha x_{k+1} = 0\}.$$

With every element $h = (h_1, \ldots, h_{k+1}) \in \mathcal{H}(C_b)$ we associate the following set of elements $T(h)$ in $\mathbb{Z}^n$:

$$T(h) := \{t \in \mathbb{Z}^n : \text{ there exist } \alpha^j \in \mathbb{Z}_+^{G_j}, \sum_{h^i \in G_j} \alpha_i^j = h_j, \ j = 1, \ldots, k,$$

$$\text{such that } t = \sum_{j=1}^{k} \sum_{h^i \in G_j} \alpha_i^j h^i \}.$$

Theorem 15. *For every element $t \in T(h)$ with $h \in \mathcal{H}(C_b)$, $h_{k+1} \neq 0$, let*

$$\bar{t}_i \ := \ \begin{cases} t_i, & \text{for } i=1,\ldots,n\text{-}1 \\ \frac{t_i}{\alpha}, & i = n. \end{cases}$$

Then $G_{\bar{a}} = \{t \in G_a : t_n = 0\} \cup \{\bar{t} \in \mathbb{Z}^n : t \in T(h), h \in \mathcal{H}(C_b), \bar{t} \text{ irreducible}\}.$

3 Decomposition of Integer Programs

The decomposition approach that we propose for the integer programming problem

$$\max \{c^T x : \ d \leq Ax \leq b, \ x \in \mathbb{Z}^n, \ l \leq x \leq u\}$$

consists in choosing a row $A_{i\cdot}$ and a positive integer $M \leq max_{j=1,\ldots,n}\{A_{ij}\}$. Write $A_{i\cdot}$ in the form $A_{i\cdot} = a^1 + Ma^2$ and $d_i = d^1 + Md^2$, $b_i = b^1 + Mb^2$ where a^1, a^2 are integer vectors and d^1, d^2, b^1, b^2 are integers. Then, we replace the constraint $d_i \leq A_i^T x \leq b_i$ by the two constraints $(a^2)^T x = \alpha$ and $d_i - M\alpha \leq (a^1)^T x \leq b_i - M\alpha$. By solving the family of these integer programs where α ranges between a lower bound of the exact value min $\{(a^2)^T x : \ d \leq Ax \leq b, \ x \in \mathbb{Z}^n, \ l \leq x \leq u\}$ and an upper bound of the exact value max $\{(a^2)^T x : d \leq Ax \leq b, \ x \in \mathbb{Z}^n, \ l \leq x \leq u\}$, and taking the maximum over all these optimal solutions, we obtain an optimal solution of the original integer program. Every integer program in the family so constructed has the same generating set. The reason this decomposition might be interesting is that the generating set for the decomposed problems can be substantially smaller than the generating set of the original problem:

Let k, M, b and $\gamma < \beta < M$ be positive integers and consider the constraint

$$\beta x_1 + (kM + \gamma)x_2 + (kM + \beta)x_3 = b, \ x \in \mathbb{Z}_+^3.$$

Replacing this constraint by the two constraints

$$kx_2 + kx_3 = \alpha,$$
$$\beta x_1 + \gamma x_2 + \beta x_3 = b - \alpha M, \ x \in \mathbb{Z}_+^3,$$

and computing the generating set for this family of integer programs where α ranges between 0 and $\lfloor \frac{b}{M} \rfloor$, we obtain that the generating set consists of the two vectors $\pm(\beta(e_3 - e_2) - (\beta - \gamma)e_1)$. By contrast, the number of elements in the generating set for the original constraint depends on the data and can be of the size $O(M)$.

The algorithm presented in Section 2 can be used to compute the generating set for the decomposed problem, one constraint at a time. We first construct a generating set for a relaxed problem (obtained by ignoring one or more constraints) and then refine it. In this context we have been performing various computational tests.

We report here about the following 4 subset sum problems in integer variables that we tried to solve by linear programming techniques. Our observations are summarized in the Table 1[5].

$$ex1 \quad 12346x_1 + 14666x_2 + 12366x_3 + 13466x_4 + 13446x_5 + 14566x_6 \leq 685111$$
$$ex2 \quad 12212x_1 + 12214x_2 + 12216x_3 + 24416x_4 + 24418x_5 \qquad\qquad \leq 27123101$$
$$ex3 \quad 12223x_1 + 12224x_2 + 12225x_3 + 12226x_4 + 12227x_5 + 12228x_6 \leq 1123100$$
$$ex4 \quad 1124x_1 + 1366x_2 + 1566x_3 + 2566x_4 + 3566x_5 + 5566x_6 \qquad \leq 148801$$

problem	CPLEX-time	CPLEX-nodes	gen-set	sol-time
ex1	57:43	> 7 mio	0:04	25:47
ex2	> 570 *	> 40 mio	0:01	0:03
ex3	> 600 *	> 40 mio	0:01	0:01
ex4	112:53	>14 mio	0:18	42:52

Table 1

To solve these instances CPLEX searches extensively through the branch and bound tree. The column CPLEX-time shows the CPU-time measured in minutes : seconds on a SUN Sparc 20 that the code spent trying to solve the instances. The column entitled CPLEX-nodes corresponds to the number of branch and bound nodes that the version CPLEX 4.0.3 created. For the instances marked with the symbol *, we stopped CPLEX, because the number of inspected nodes exceeded the threshold of 40 million. The forth column corresponds to the running time to compute the generating set for the decomposed problems. The column entitled "sol-time" shows the time to find an optimal solution of the original subset sum problem with the generating set for the decomposed problem.

We also tested variants of these problems: instead of having one integer variable corresponding to one weight we created 15 0/1-variables with equal weight. With each of these variables we associated a unique objective function value that we defined as 100 times the weight plus an index i corresponding to the index of the variable within the block of all items with the same weight. A typical instance along these lines looks as follows:

---

[5] Notice that in order to run CPLEX on this instances, it is necessary to change the default parameters for the precision

$$\max 222200 \sum_{i=1}^{15} x_i + \sum_{i=1}^{15}(i-1)x_i + 113500 \sum_{i=16}^{30} x_i + \sum_{i=16}^{30}(i-16)x_i +$$
$$555500 \sum_{i=31}^{45} x_i + \sum_{i=31}^{45}(i-31)x_i + 222400 \sum_{i=46}^{60} x_i + \sum_{i=46}^{60}(i-46)x_i +$$
$$333400 \sum_{i=61}^{75} x_i + \sum_{i=61}^{75}(i-61)x_i + 122400 \sum_{i=76}^{90} x_i + \sum_{i=76}^{90}(i-76)x_i$$

$$\text{st..} \quad 2222 \sum_{i=1}^{15} x_i + 1135 \sum_{i=16}^{30} x_i + 5555 \sum_{i=31}^{45} x_i + 2224 \sum_{i=46}^{60} x_i +$$
$$3334 \sum_{i=61}^{75} x_i + 1224 \sum_{i=76}^{90} x_i \leq 16999, \; x_1,\ldots,x_{90} \in \{0,1\}.$$

Problems of this form seem to be extremely difficult for linear programming techniques. In all our test examples of this form we had to stop CPLEX because the code required more than 130 megabytes of memory. For the decomposition approach in conjunction with generating sets problems of this form are easy to solve, because we can aggregate all variables with the same weight to one variable, then decompose the problem and reconstruct from the generating set of this aggregated integer program the generating set of the original problem. This fact follows from Theorem 16.

Theorem 16. *Let $D \in \mathbb{Z}$ and G be a generating set for $F = \{y - z : y, z \in \mathbb{Z}_+^d, \; a^T y = a^T z = \alpha \text{ where } \alpha \in D\}$. Let $N_1, \ldots, N_d$ be a partition of the ground set $\{1, \ldots, n\}$ From G one can construct a generating set W for $H = \{x - w : x, w \in \{0,1\}^n, \; \sum_{j=1}^d \sum_{i \in N_j} a_j x_i = \sum_{j=1}^d \sum_{i \in N_j} a_j w_i = \alpha \text{ where } \alpha \in D\}$ as follows:*

$e_u - e_v \in W$ *for all* $u, v \in N_j$, $j = 1, \ldots, d$;
Let t be any element of G. For all $j = 1, \ldots, d$ and sets $S_j \subseteq N_j$ such that $|S_j| = |t_j|$ we include $\sum_{j=1}^d \sum_{i \in S_j} sign(t_j)e_i$ in the set W.

The generating set for the decomposed, aggregated problem in the example above is computed within 2 minutes. With this test set, an optimal solution of the original instance is found below 2 seconds of CPU time.

There is a clear conclusion that one can draw from our tests with the integer and 0/1-knapsack problems: for knapsack problems such that (1) the number of different coefficients in the inequality is small and (2) the coefficients in the inequality do not differ too much, the decomposition approach of Section 3 makes the computation of generating sets possible, if not effective. In particular, the higher the right hand side of the knapsack inequality, the more effective the combination of decomposition and generating sets becomes, because then, after decomposition, the more right hand sides need to be inspected with just *one* global generating set. Although this inspection of many right hand sides becomes expensive for our approach as well, LP- based codes or dynamic programming algorithms for knapsack will fail to compute an optimal solution of these problems even within several hundreds of CPU minutes on today's workstations.

4 The Group Relaxation and the Frobenius Problem

A decomposition scheme similar to the one introduced in Section 3 can be performed based on the group relaxation for integer programming, as follows.

For a constraint $a^T x = b$, $x \in \mathbb{Z}_+^n$, choose a positive integer M from the set $\{\min_j\{a_j\}, \ldots, \max_j\{a_j\}\}$. For $j = 1, \ldots, n$, write a_j in the form $a_j = a_j^1 + M a_j^2$ where a_j^1 is a nonnegative integer between 0 and $M - 1$. Similarly, we write $b = \beta_1 + M\beta_2$ where β_1 is a nonnegative integer between 0 and $M - 1$. Let $D \subset \mathbb{Z}^2$ denote the family of all vectors of the form $(b_1, b_2) = (\beta_1 + Mx_0, \beta_2 - x_0)$, where $x_0 \in \mathbb{Z}$. Then, we replace the constraint $a^T x = b$, $x \in \mathbb{Z}_+^n$ by the two constraints $(a^1)^T x = b_1$ and $(a^2)^T x = b_2$, $x \in \mathbb{Z}_+^n$. By solving the family of these integer programs where $(b_1, b_2) \in D$ varies, and taking the maximum over all these optimal solutions, we obtain an optimal solution of the original integer program. The first step in constructing a generating set for these two constraints consists in determining the generating set for the family of problems $(a^1)^T x = b_1$ where $b_1 = \beta_1 + Mx_0$, $x_0 \in \mathbb{Z}$ varies. If we treat all the problems in this family simultaneously, we obtain the so called group relaxation $(a^1)^T x = b_1 \bmod M$, see Gomory [3]. In this context, rather than investigating the generators for the set of all difference vectors of feasible points for one program in the family, we deal here with the generating set for $F := \{x - y : \text{ there exists } b_x, b_y \in \mathbb{Z}, b_x \bmod M = b_y \bmod M = \beta_1, \text{ and } (a^1)^T x = b_x, (a^1)^T y = b_y\}$. We denote by G the minimal generating set for F.

The next theorem relates the elements of G to the simple directed cycles of a digraph associated with the group problem (a cycle is *simple* if it contains no repeated nodes). Let $H = (V, F)$ be the digraph with node set $V = \{0, 1, \ldots, M - 1\}$ and arc set defined as follows. For every $v, w \in V$, $v \neq w$, there is an arc (v, w) in F if there exists an index $j \in \{1, \ldots, n\}$ such that $w = (v + a_j^1) \bmod M$ or $v = (w - a_j^1) \bmod M$. Note that the digraph $H = (V, F)$ has parallel arcs when some coordinates of a^1 coincide.

Theorem 17. *To every element in G, there corresponds at least one simple directed cycle in H. Conversely, not every simple directed cycle of H corresponds to an element in G.*

The decomposition scheme for integer programs that we discussed in the previous section can also be applied to the Frobenius problem: Given non-negative integers $a_1, \ldots, a_n$ with $\gcd(a_1, \ldots, a_n) = 1$; find the largest integer b_0 (the Frobenius number) that cannot be expressed as a non-negative integer combination of the weights $a_1, \ldots, a_n$. With elementary number-theoretic arguments one can verify that such a number b_0 exists. In fact, Kannan [5] showed that the Frobenius number can even be computed in polynomial time in the encoding length of the numbers $a_1, \ldots, a_n$ when n is fixed.

Quite a simple algorithm for determining the Frobenius number can be derived from the decomposition scheme in combination with the theorem stated below. Let $a_1 < a_2 < \ldots < a_n \in \mathbb{Z}_+$ and $M \in \mathbb{Z}_+$ such that there exists a vector $y \in \mathbb{Z}_+^n$ with $M = \sum_{i=1}^n a_i y_i$. We write each coefficient a_i as $a_i = \alpha_i M + \beta_i$ with $\beta_i = a_i \bmod M$ and $\alpha_i = \lfloor \frac{a_i}{M} \rfloor$. For each $l \in \{0, \ldots, M - 1\}$, we denote by $f(l)$ the value of the following optimization problem in $n + 1$ integer variables

$x_1, \ldots, x_{n+1}$:

$$f(l) := \min \sum_{i=1}^{n} \alpha_i x_i + x_{n+1} : \sum_{i=1}^{n} \beta_i x_i - M x_{n+1} = l, \ x \in \mathbb{Z}_+^{n+1}. \qquad (3)$$

The value $M f(l) + l$ is the smallest integer congruent to l modulo M that can be expressed as a non-negative integer combination of $a_1, \ldots, a_n$. This number can also be read off from the following integer program:

$$\min x_{n+1} : \sum_{i=1}^{n} a_i x_i = l + M x_{n+1}, \ x \in \mathbb{Z}_+^{n+1}.$$

Noting that $\sum_{i=1}^{n} a_i x_i = M(\sum_{i=1}^{n} \alpha_i x_i) + \sum_{i=1}^{n} \beta_i x_i$, we would decompose this problem as

$$
\begin{aligned}
\min \quad & \lambda, \\
\text{s. t.} \quad & \sum_{i=1}^{n} \alpha_i x_i && = \alpha_0, \\
& \sum_{i=1}^{n} \beta_i x_i && \leq l + \lambda M, \\
& x \in \mathbb{Z}_+^n, \ \lambda \in \mathbb{Z}_+,
\end{aligned}
\qquad (4)
$$

where $\alpha_0 \in \mathbb{Z}$ varies between an appropriate lower bound, B_l say, and an upper bound, B_u say. Denoting by λ_0 the optimal solution of the decomposed problem (4) for a fixed value of α_0, it is clear that $f(l) = \min(\alpha_0 + \lambda_0);$ '$\alpha_0 \in \{B_l, \ldots, B_u\}\}$. From this formula is also clear that, rather than imposing $\sum_{i=1}^{n} \alpha_i x_i = \alpha_0$ with varying $\alpha_0 \in \{B_l, \ldots, B_u\}$ as a constraint in the decomposed problem, one can drop the constraint and make it part of the objective function. This yields formula (3).

The bridge between the formula (3), the decomposition scheme and the Frobenius problem builds

Theorem 18. *For $a_1 < a_2 < \ldots < a_n \in \mathbb{Z}_+$ with $\gcd(a_1, \ldots, a_n) = 1$, and $M \in \mathbb{Z}_+$, $y \in \mathbb{Z}_+^n$ with $M = \sum_{i=1}^{n} a_i y_i$ and $l \in \{0, \ldots, M-1\}$, let $f(l)$ denote the value of the program (3). The Frobenius number associated with $a_1, a_2, \ldots, a_n$ equals the value $\max\{f(l) - M : l \in \{0, \ldots, M-1\}\}$.*

We want to mention that Theorem 18 was proved by Brauer & Shockley [1] when M coincides with one of the numbers $a_1, \ldots, a_n$, see also Selmer [6].

In Table 2 we present 5 instances for which we determined the Frobenius number following this approach. The column "M" contains the values by which we decomposed the computation of the Frobenius number following the approach and notation of Section 3. The column entitled Example shows the data. The columns Frobenius and Time present the Frobenius number for the corresponding example and the time that we spent in determining it, respectively. For none of these examples CPLEX can determine the optimal solution of the corresponding integer subset sum problem when the right hand side equals the Frobenius number.

Example	M	Frobenius	Time
12223, 12224, 36674, 61119, 85569	24447	89643481	6:19
12228, 36679, 36682, 48908, 61139, 73365	12228	89716838	0:47
12137, 36405, 24269, 36407, 84545, 60683	12137	58925134	1:06
13211, 13212, 39638, 66060, 52864, 79268, 92482	26423	104723595	31:51
13429, 26850, 26855, 40280, 40281, 53711, 53714, 67141	13429	45094583	10:25

Table 2

References

1. A. Brauer & J. E. Shockley: On a problem of Frobenius, Journal für reine und angewandte Mathematik **211** (1962) 399-408
2. P. Diaconis & R. Graham & B. Sturmfels: Primitive partition identities, Paul Erdös is 80. Vol. II, Janos Bolyai Society, Budapest (1995) 1-20
3. R. E. Gomory: Some polyhedra related to combinatorial problems, Linear Algebra and its Applications **2** (1969) 451-558
4. J. E. Graver: On the foundations of linear and integer programming I, Mathematical Programming **8** (1975) 207-226
5. R. Kannan: Solution of the Frobenius problem and its generalizations, Manuscript (1991)
6. E. S. Selmer: On the linear diophantine problem of Frobenius, *Journal für reine und angewandte Mathematik* **293/294** (1977), 1-17
7. R. R. Thomas: Gröbner basis methods for integer programming, PhD. Dissertation, Cornell University, (1994)
8. R. Urbaniak & R. Weismantel & G. Ziegler: A variant of Buchberger's algorithm for integer programming, SIAM J. Discrete Math., Vol 1, No 10, (1997) 96 – 108

Acknowledgement:
The second author is supported by the graduate school "Algorithmische Diskrete Mathematik" of the German Science Foundation, grant GRK219/2-96.

The second author is supported by a "Gerhard-Hess-Forschungsförderpreis" of the German Science Foundation (DFG).

Bounded Degree Spanning Trees[*]

(Extended abstract)

Artur Czumaj and Willy-B. Strothmann

Heinz Nixdorf Institute and Department of Computer Science
University of Paderborn, D-33095 Paderborn, Germany
{artur,willy}@uni-paderborn.de

Abstract. Given a connected graph G, let a Δ_T-spanning tree of G be a spanning tree of G of maximum degree bounded by Δ_T. It is well known that for each $\Delta_T \geq 2$ the problem of deciding whether a connected graph has a Δ_T-spanning tree is $\mathcal{NP}$-complete. In this paper we investigate this problem when additionally connectivity and maximum degree of the graph are given. A complete characterization of this problem for 2- and 3-connected graphs, for planar graphs, and for $\Delta_T = 2$ is provided.

Our first result is that given a biconnected graph of maximum degree $2\Delta_T - 2$, we can find its Δ_T-spanning tree in time $O(m + n^{3/2})$. For graphs of higher connectivity we design a polynomial-time algorithm that finds a Δ_T-spanning tree in any k-connected graph of maximum degree $k(\Delta_T - 2) + 2$. On the other hand, we prove that deciding whether a k-connected graph of maximum degree $k(\Delta_T - 2) + 3$ has a Δ_T-spanning tree is $\mathcal{NP}$-complete, provided $k \leq 3$. For arbitrary $k \geq 3$ we show that verifying whether a k-connected graph of maximum degree $k(\Delta_T - 1)$ has a Δ_T-spanning tree is $\mathcal{NP}$-complete. In particular, we prove that the Hamiltonian path (cycle) problem is $\mathcal{NP}$-complete for k-connected k-regular graphs, if $k > 2$. This extends the well known result for $k = 3$ and fully characterizes the case $\Delta_T = 2$.

For planar graphs it is $\mathcal{NP}$-complete to decide whether a k-connected planar graph of maximum degree Δ_G has a Δ_T-spanning tree for $k = 1$ and $\Delta_G > \Delta_T \geq 2$, for $k = 2$ and $\Delta_G > 2(\Delta_T - 1) \geq 2$, and for $k = 3$ and $\Delta_G > \Delta_T = 2$. On the other hand, we show how to find in polynomial (linear or almost linear) time a Δ_T-spanning tree for all other parameters of k, Δ_G, and Δ_T.

1 Introduction

Various problems of constructing spanning trees, or generally spanning subgraphs, that satisfy given constraints have been studied before extensively [1,4,5,10,13–16,19]. Such problems often arise in designing communication networks for the purpose of broadcasting or fault tolerance (e.g., see discussion in [4]). It has been shown that in most of the cases even simple constraints make the problem $\mathcal{NP}$-hard [8,20]. For example, it is well known that for any $\Delta_T \geq 2$ the problem of testing whether a graph has a spanning tree of maximum degree bounded by Δ_T is $\mathcal{NP}$-complete [8].

---

[*] Partially supported by EU ESPRIT Long Term Research Project 20244 (ALCOM-IT), DFG Leibniz Grant Me872/6-1, and DFG Project Me872/7-1.

Let Δ_T be an arbitrary integer greater than 1. A Δ_T-*spanning tree* of G is a spanning tree of G of maximum degree bounded by Δ_T. When can we precisely say that a given graph has a Δ_T-spanning tree, and if it has one, then how efficiently it can be found? Even in the most basic case $\Delta_T = 2$, that is, of verifying whether a graph has a Hamiltonian path, only very few results are known. For example, it is known that when a graph is sufficiently dense or its degree sequence satisfies certain conditions, then it has a Hamiltonian path and one such path can be constructed in polynomial time (see e.g. [3]). In the case of planar graphs, Tutte showed that every 4-connected planar graph has a Hamiltonian cycle and Chiba and Nishizeki [6] provided a linear-time algorithm for the construction. On the opposite side, Garey et al. [9] proved that it is $\mathcal{NP}$-complete to decide whether a 3-connected 3-regular planar graph has a Hamiltonian path.

Even less is known for larger values of Δ_T. Neumann-Lara and Rivera-Campo [19] characterized the values of Δ_T for spanning trees of k-connected graphs as a function of its independence number, and Caro et al. [5] as a function of its minimum degree. Both these results are mainly interesting for dense graphs and are far from being tight for sparse graphs. Barnette [1] showed that every 3-connected planar graph has a 3-spanning tree. Perhaps the most general result was obtained by Fürer and Raghavachari [7] and Win [25]. They showed that the optimal value of Δ_T is approximated within an additive constant term by the inverse of the toughness of the graph. We notice however, that from the algorithmic point of view this result would be not satisfactory, because it is $\mathcal{NP}$-hard to determine the toughness of graphs [2]. Nevertheless, Fürer and Raghavachari [7] were able to estimate the toughness of graphs and designed a polynomial-time algorithm that finds a spanning tree whose maximum degree is within one of optimal.

1.1 New Results

In this paper we investigate the problem of testing whether a graph has a Δ_T-spanning tree when connectivity and maximum degree of the graph are given as parameters. Let G be a k-connected graph of maximum degree $\Delta_G \geq k$ with n vertices and m edges.

We show that increasing the connectivity of graphs provides much weaker conditions on the maximum degree of the graph to ensure the existence and efficient finding of spanning trees of low degree. Our first algorithm is designed for biconnected graphs and runs in $O(m + n^{3/2})$ time. It finds a spanning tree T in which the degree of every vertex v is roughly halved; more precisely, $d_T(v) \leq \lceil d_G(v)/2 \rceil + 1$. In particular, if a graph is of maximum degree at most $2\Delta_T - 2$, then the algorithm finds a Δ_T-spanning tree of G. The second algorithm runs in time $O(n^2 \cdot k \cdot \alpha(n, n) \cdot \log n)$ and finds a Δ_T-spanning tree of any k-connected graph of maximum degree bounded by $k(\Delta_T - 2) + 2$. These results improve significantly upon the trivial lower degree bounds for $k \geq 2$ and/or $\Delta_T \geq 3$.

Then we prove that the Hamiltonian path problem and the Hamiltonian cycle problem are $\mathcal{NP}$-complete for k-connected k-regular graphs, $k \geq 3$. This extends the well-known result for 3-connected graphs [9] and fully characterizes the complexity of the problem for $\Delta_T = 2$ and $k \geq 3$. We can generalize this bound to $\Delta_T > 2$ and prove that (unless $\mathcal{P} = \mathcal{NP}$) our algorithmic results are in a sense best possible for $k \leq 3$ and $\Delta_T > 2$. For that we first construct k-connected graphs of maximum degree $k(\Delta_T - 1)$

General graphs					
$k = 2$		$k = 3$		$k \geq 4$	
$\Delta_G \leq 2\Delta_T - 2$	$\Delta_G > 2\Delta_T - 2$	$\Delta_G \leq 3\Delta_T - 4$	$\Delta_G > 3\Delta_T - 4$	$\Delta_G \leq k(\Delta_T - 2) + 2$	$\Delta_G \geq k(\Delta_T - 1)$
$O(m + n^{3/2})$	$\mathcal{NPC}$	$O(n^2 \alpha(n,n)\log n)$	$\mathcal{NPC}$	$O(n^2 k\alpha(n,n)\log n)$	$\mathcal{NPC}$

Planar graphs					
$k = 1$	$k = 2$		$k = 3$		$k \in \{4,5\}$
$\Delta_G > \Delta_T$	$\Delta_G \leq 2\Delta_T - 2$	$\Delta_G > 2\Delta_T - 2$	$\Delta_T = 2$	$\Delta_T \geq 3$	$\Delta_T \geq 2$
$\mathcal{NPC}$[8]	$O(n \log n)$	$\mathcal{NPC}$	$\mathcal{NPC}$[9]	$O(n)$ [24]	$O(n)$ [6]

Table 1. Summary of results. $\mathcal{NPC}$ means that the problem is $\mathcal{NP}$-complete.

without Δ_T-spanning trees for $k \geq 3$ and $\Delta_T > 2$. Then we prove that verifying whether a k-connected graph of maximum degree $k(\Delta_T - 1)$ has a Δ_T-spanning tree is $\mathcal{NP}$-complete for $k \geq 3$. Finally, we extend these results and show that it is $\mathcal{NP}$-complete to decide whether a 2-connected (planar) graph of maximum degree $2\Delta_T - 1$ has a Δ_T-spanning tree. These results establish a complete characterization of the Δ_T-spanning tree problem for $k \leq 3$ in general graphs.

Our results can be applied to planar graphs. It is known that every 4-connected planar graph has a Hamiltonian path and that such a path can be found in linear time [6]. Barnette [1] showed that every 3-connected planar graph has a 3-spanning tree and one can also find such a tree in linear time [24]. On the other hand, Garey et al. [9] proved that it is $\mathcal{NP}$-complete to decide whether a 3-regular 3-connected planar graph has a Hamiltonian path. One can easily extend this result to show that it is $\mathcal{NP}$-complete to verify whether a connected planar graph of maximum degree $\Delta_T + 1$ has a Δ_T-spanning tree.

In this paper we fill the remaining gap and characterize biconnected planar graphs. Our result for arbitrary graphs implies that every 2-connected planar graph of maximum degree at most $2(\Delta_T - 1)$ has a Δ_T-spanning tree. In the case of planar 2-connected graphs we can design an algorithm that finds a Δ_T-spanning tree in time $O(n \log n)$. Since we can prove that it is $\mathcal{NP}$-complete to decide whether a 2-connected planar graph of maximum degree $2\Delta_T - 1$ has a Δ_T-spanning tree, this result establishes a complete characterization of the Δ_T-spanning tree problem for k-connected planar graphs of maximum degree Δ_G.

Table 1 summarizes the results (it assumes that $\Delta_G > \Delta_T \geq 2$).

Organization of the paper Section 2 provides basic terminology. Then in Sect. 3 an algorithm for finding a Δ_T-spanning tree in a 2-connected graph of maximum degree at most $2(\Delta_T - 1)$ is presented. Section 4 contains a polynomial-time algorithm that finds a Δ_T-spanning tree of k-connected graphs with maximum degree $\Delta_G \leq k(\Delta_T - 2) + 2$ for arbitrary $k \geq 2$. In Sect. 5 we prove that verifying whether a k-connected graph of maximum degree $k(\Delta_T - 1)$ has a Δ_T-spanning tree is $\mathcal{NP}$-complete for every $k \geq 3$, $\Delta_T \geq 2$. In Sect. 6 we provide a characterization of the Δ_T-spanning tree problem in planar graphs.

Some technical details are omitted here but they will appear in the final version.

2 Basic Notation

We use [3] for basic terminology and notation not defined here. The set of vertices adjacent to a vertex v in a graph G is denoted by $\Gamma_G(v)$ and its size by $d_G(v) := |\Gamma(v)|$. For an arbitrary connected graph H let $\mathbb{BCT}(H)$ denote its *block-cutvertex tree* whose vertices correspond to the *articulation points* of H and to the *blocks* (maximal 2-connected subgraphs) of H, and there is an edge between an articulation point v and a block B iff v is in B (see also [3, page 6]). Let v be an articulation point of a connected graph G with $d_{\mathbb{BCT}(G)}(v) = 2$ whose removal disconnects G into two connected components G_1, G_2. The operation *split G at (an articulation point)* v outputs the connected graphs $G_1 \cup \{v\}, G_2 \cup \{v\}$. A *2-tree* is a tree in which all internal vertices are of degree two.

3 Biconnected Graphs of Maximum Degree $\Delta_G \leq 2\Delta_T - 2$

Our algorithm is recursive. On a very high level, in each recursive invocation the algorithm deals with a connected graph G such that $\mathbb{BCT}(G)$ is a 2-tree. If the graph is not biconnected, then every articulation point is of degree two in $\mathbb{BCT}(G)$ and we split the graph into its biconnected components (Lemmas 5 and 6, Corollary 7). Otherwise G is biconnected and we modify G by removing edges (Lemmas 8 and 9) In either case we obtain smaller graphs (with less edges) which are analyzed recursively. A key point is to keep the graphs in a proper shape that can be maintained by the recursive calls and can be used to obtain spanning trees. Therefore we introduce *ws (well-structured) graphs* and design an algorithm that finds a low degree spanning tree in ws-graphs. Finally we apply our construction to any biconnected graph (Theorem 11).

Definition 1. A connected bipartite graph $G = (V_1, V_2, E)$ with labeling function $V_2 \rightarrow \{\boxed{0}, \boxed{1}, \boxed{2}\}$ is *ws (well-structured)* if (i) $d_G(v_1) \in \{1, 2\}$ for all $v_1 \in V_1$ and $d_G(v_2) \geq 1$ for all $v_2 \in V_2$, (ii) vertices with label $\boxed{0}$ are of even degree, those with label $\boxed{1}$ of odd degree, (iii) if there is a vertex with label $\boxed{0}$, then it is the only $\boxed{0}$-vertex and there is at most one other vertex with label $\boxed{1}$, and (iv) if there is no vertex with label $\boxed{0}$, then there are at most three vertices with label $\boxed{1}$.

Definition 2. A *ws-spanning tree* T of a ws-graph G is a spanning tree of G such that every vertex v with label $\boxed{i}$ is of degree at most $\lceil \frac{d_G(v)+i}{2} \rceil$ in T. All unlabeled vertices may be of arbitrary degree in T.

Definition 3. A Δ_G-*ws-graph* is a biconnected ws-graph with every vertex of degree at most Δ_G. A Δ_T-*ws-spanning tree* is a ws-spanning tree of a $(2\Delta_T - 2)$-ws-graph.

Remark 4. Every vertex in a Δ_T-ws-spanning tree is of degree at most Δ_T.

We define a total order $\boxed{0} \prec \boxed{1} \prec \boxed{2}$ on the labels.

Lemma 5. *Let G be a ws-graph. Let $a \in V_2$ be an articulation point with $d_{\mathbb{BCT}(G)}(a) = 2$ and with label in $\{\boxed{1}, \boxed{2}\}$. Let G_1 and G_2 be the connected graphs obtained by splitting G at a. Any of the following cases assigns labels to a in G_1 and G_2 so that G_1, G_2 are ws-graphs and the sum of any ws-spanning trees T_1 and T_2 of G_1 and G_2, respectively, is a ws-spanning tree of G.*
- *If the label of a is $\boxed{2}$ and if $d_{G_1}(a)$ and $d_{G_2}(a)$ are even,*
   then label a in one arbitrary graph (say G_1) with $\boxed{2}$ and in the other (G_2) with $\boxed{0}$.

- *If the label of a is $\boxed{2}$ and if $d_{G_1}(a)$ and $d_{G_2}(a)$ are odd,*
  then label a in both graphs with $\boxed{1}$.
- *If the label of a is $\boxed{2}$ and if $d_{G_1}(a)$ is even and $d_{G_2}(a)$ is odd,*
  then label a in G_1 with $\boxed{2}$ and in G_2 with $\boxed{1}$.
- *If the label of a is $\boxed{2}$ and if $d_{G_1}(a)$ is even and $d_{G_2}(a)$ is odd,*
  then label a in G_1 with $\boxed{0}$ and in G_2 with $\boxed{2}$.
- *If the label of a is $\boxed{1}$, then (w.l.o.g.) $d_{G_1}(a)$ is even and $d_{G_2}(a)$ is odd;*
  label a in G_1 with $\boxed{0}$ and in G_2 with $\boxed{1}$. □

Lemma 6. *Let G be a ws-graph and let $\mathbb{BCT}(G)$ be a 2-tree. If G is not biconnected, then let B_1 and B_2 be the (only) two blocks in G that are incident to only one articulation point. Let s, t, u be a vertex with the smallest, second smallest, and third smallest label in V_2, respectively. Assume that s is not a $\boxed{2}$-vertex, that s is in B_1, and that s is not an articulation point of G. If s is a $\boxed{0}$-vertex and t is a $\boxed{1}$-vertex, then let $x = t$ be in B_2. If s, t, u are all $\boxed{1}$-vertices, then let t or u, say $x = t$, be in B_2. Otherwise, let x be an arbitrary vertex in B_2. Assume further, that x is not an articulation point in G.*

Then we can split G at all articulation points and relabel the articulation points such that the components are biconnected ws-graphs. Additionally, the sum of any ws-spanning trees of the components is a ws-spanning tree of G.

Proof. The proof is by induction on the number of blocks in G.

If there is only one block, then G is a biconnected ws-graph and we are done.

Otherwise, let a be an arbitrary articulation point that separates s from x. Split G at a into G_1 and G_2 such that s is in G_1. Then $\mathbb{BCT}(G_1)$ and $\mathbb{BCT}(G_2)$ are 2-trees. We now show how to assign labels to a in G_1 and G_2 consistently with Lemma 5 such that G_1 and G_2 fulfill all the inductive requirements of the lemma.

Let $y := \{t, u\} - \{x\}$.

If s is a $\boxed{0}$-vertex in G then a has label $\boxed{2}$ in G. Label a in G_1 by at least $\boxed{1}$ and in G_2 by at least $\boxed{0}$.

If a has label $\boxed{1}$ in G then $a = y$ and s is a $\boxed{1}$-vertex in G. Label a in G_1 and G_2 by at least $\boxed{0}$.

Otherwise, a is a $\boxed{2}$-vertex in G and s is a $\boxed{1}$-vertex in G. If G_i contains y, then label a in G_i by at least $\boxed{1}$ and in the other graph G_{3-i} by at least $\boxed{0}$.

According to Lemma 5, G_1 and G_2 are ws-graphs and the sum of their ws-spanning trees is a ws-spanning tree of G. Thus the lemma follows from the inductive hypothesis. □

Corollary 7. *Let G be a connected ws-graph with maximum degree $2\Delta_T - 2$ that fulfills the requirements of Lemma 6. Then we can split G into blocks such that (i) every block is a $(2\Delta_T - 2)$-ws-graph and (ii) the sum of the Δ_T-ws-spanning trees of the blocks is a Δ_T-ws-spanning tree of G.* □

Lemma 8. *Let $G = (V_1, V_2, E)$ be a ws-graph, let $|V_2| > 1$ and let s be a vertex in V_2 with $d_G(s) > 2$. Let x_1 and x_2 be two vertices adjacent to s. Let v_1 and v_2 be the other vertices than s incident to x_1 and x_2, respectively. If the graph obtained by deleting x_1 and x_2 with the incident edges and inserting the new vertex z with the edges (z, v_1) and*

(z, v_2) (all labels remain unchanged) has a ws-spanning tree T^*, then we can construct a ws-spanning tree T of G out of T^*.

Proof. Add the edges $(s, x_1), (s, x_2)$ to T^*. Additionally, if $(v_i, z) \in T^*, i = 1, 2$, then add (x_i, v_i) to T^*. Delete z with all incident edges. The resulting graph has exactly one cycle on which a and s lie. Thus one of the two new edges $(s, x_1), (s, x_2)$, say (s, x_1), lies on the cycle. Removal of (s, x_1) keeps the degree constraint at s and yields a ws-spanning tree T of G. □

We will call the above operation DELETE-AND-COMBINE x_1, x_2 for s and the obtained graph is denoted by $G\langle x_1, x_2, s \rangle$.

We can prove the following lemma:

Lemma 9. *Let G be a $(2\Delta_T - 2)$-ws-graph and s be the vertex with the smallest label in G. Let z be a new vertex. Let $d_G(s) \geq 4$ and let $(s, x_1), (s, x_2), (s, x_3)$ be three edges incident to s. Assume that $G - \{(s, x_1), (s, x_2), (s, x_3)\}$ has exactly one articulation point a of degree 4 in the block-cutvertex tree (Note: all other articulation points are of degree 2).*

If we split $G - \{(s, x_1), (s, x_2)\} \cup \{(z, x_1), (z, x_2)\}$ at a into two $(2\Delta_T - 2)$-ws-graphs G_1, G_2 such that s is in G_1 and z is in G_2, then we can assign labels to s and a in G_1 and to z and a in G_2 so that a Δ_T-ws-spanning tree T of G can be constructed from any Δ_T-ws-spanning trees of G_1 and G_2. □

We present a recursive algorithm that finds a Δ_T-ws-spanning tree in $(2\Delta_T - 2)$-ws-graphs. A high level description is as follows (we assume $|V_2| > 1$):

Algorithm Δ_T-WS-SPANNING-TREE
  Input: $(2\Delta_T - 2)$-ws-graph G
  Output: Δ_T-ws-spanning tree T
  (1) Let s be a vertex with the smallest label in G. If s is a $\boxed{2}$-vertex and $d_G(s)$ is even, then label it $\boxed{0}$. Otherwise label s with $\boxed{1}$.
  (2) If $|V_2| = 2$ then (a); return(T);
  (3) If $d_G(s) = 2$ then (b); return(T);
  (4) Let t be a vertex with the second smallest label and u with the third smallest label. Let $(s, x_i), i = 1, 2, 3$, be three edges incident to s and (x_i, v_i) be the other edge than (s, x_i) incident to x_i. Define $\mathcal{K} := \{v_1, v_2, v_3\}$ and $X := \{x_1, x_2, x_3\}$.
  (5) If $d_G(s) = 3$ then (c); return(T);
  (6) If $|\mathcal{K}| = 1$ then (d); return(T);
  (7) If $|\mathcal{K}| = 2$ then (e); return(T);
  (8) (f); return(T);

Our algorithm considers the following cases (we present a high level description):

(a) This is the base case. Let V_2 be $\{s, v\}$ and $V_1 := \{x_1, \ldots, x_d\}$, where $d := d_G(s)$. If d is even, then the edges $(s, x_1); \ldots; (s, x_{d/2}); (v, x_{d/2}); \ldots; (v, x_d)$ induce a Δ_T-ws-spanning tree T of G, because v has to be a $\boxed{2}$-vertex. If d is odd, then the edges $(s, x_1); \ldots; (s, x_{(d+1)/2}); (v, x_{(d+1)/2}); \ldots; (v, x_d)$ induce a Δ_T-ws-spanning tree T of G, because v is at least a $\boxed{1}$-vertex.

(b) If $d_G(s) = 2$ then let x_1 and x_2 be the two neighbors of s. Let $v_i, i = 1, 2$, be the other neighbor of x_i than s ($v_1 \neq v_2$, because $|V_2| \neq 2$ and G is biconnected). Remove x_1, x_2 and s with the incident edges from G. If one of v_1 or v_2 is a $\boxed{1}$-vertex, say v_1, then relabel v_1 by $\boxed{0}$. Otherwise, relabel v_1 and/or v_2 with $\boxed{1}$ if they are now of odd degree. Observe that we removed the $\boxed{0}$-vertex s and that there was at most one $\boxed{1}$-vertex in G. Thus the obtained graph can be split according to Corollary 7. Let T^* be the sum of the Δ_T-ws-spanning trees of the components. Then $T := T^* \cup \{(s, x_1), (x_1, v_1), (x_2, v_2)\}$ is a Δ_T-ws-spanning tree of G.

(c) W.l.o.g. let $v_1 \neq v_2$.

If $|\{v_1, v_2, v_3\}| = 2$, then $G\langle x_1, x_2, s\rangle$ has two trivial blocks (s, x_3) and (x_3, v_3) and one non-trivial block. Thus $G\langle x_1, x_2, s\rangle$ fullfills the requirements of Corollary 7.

If $|\{v_1, v_2, v_3\}| = 3$, then we can find a vertex x_i, say x_3, such that all $\boxed{1}$-vertices are in the same block in $G - (s, x_3)$. Because t, u are in one leaf-block and v_3 in the other, the requirements of Corollary 7 are satisfied for $G\langle x_1, x_2, s\rangle$.

The construction of the Δ_T-ws-spanning tree T of G out of a Δ_T-ws-spanning tree of $G\langle x_1, x_2, s\rangle$ is done as indicated by Lemmas 6 and 8.

(d) Let $\mathcal{K}$ be $\{v\}$. Delete the vertices x_1 and x_2 together with their incident edges. The resulting graph G^* is biconnected and thus it is a $(2\Delta_T - 2)$-ws-graph. Any Δ_T-ws-spanning tree T^* of G^* can be extended to a Δ_T-ws-spanning tree T of G by adding the edges (s, x_1) and (x_2, v).

(e) Let $\mathcal{K}$ be $\{v_1, v_3\}$ and both v_1 and s be adjacent to x_1 and x_2. DELETE-AND-COMBINE x_2, x_3 for s. The resulting graph is biconnected and hence it is a $(2\Delta_T - 2)$-ws-graph. By Lemma 8, we can construct a Δ_T-ws-spanning tree T of G.

(f) In this case either the graph $G\langle x_1, x_2, s\rangle$ and/or $G\langle x_2, x_3, s\rangle$ is biconnected or $G - \{(s, x_1), (s, x_2), (s, x_3)\}$ has exactly one articulation point a of degree 4 in the block-cutvertex tree. In the first case we use Lemma 8 to get a Δ_T-ws-spanning tree T of G. In the second case we split the separating pair $\{s, a\}$ according to Lemma 9 and get a Δ_T-ws-spanning tree T of G.

Lemma 10. *Every $(2\Delta_T - 2)$-ws-graph G has a Δ_T-spanning tree. One can find a Δ_T-spanning tree of G in time $O(n^{3/2})$.*

Sketch of the proof: One can verify that if the input graph is $(2\Delta_T - 2)$-ws, then the algorithm described above always returns a Δ_T-spanning tree. Thus we only must show that the algorithm can be implemented within the required time. For the proof notice that $n \leq m \leq 2n$. We use three data structures that are dynamically maintained during the run of the algorithm.

The first data structure contains a (real) *representation of the graph*, in which each vertex keeps its incidency list. It is trivial to maintain the representation of the graph under edge deletions and insertions in constant time. Another operation which must be maintained is splitting the graph G at an articulation point a that separates vertex s from vertex x. Let G_1 and G_2 be the resulting graphs. We run two depth first search (DFS) algorithms in parallel in G, one starting from s and another from x. In each of the DFS-algorithms, we assume that the recursion stops every time we are at a. We end when one of the algorithms visits all the edges. This means that all the edges of G_1 or G_2, say G_1, have been visited. Then we create the representation of G_1 and delete the edges of G_1 from the original representation of G (to create the representation of G_2). The running time is $O(m')$, where m' is the number of edges in G_1. Since G_1 has not more edges than G_2, standard amortization argument can be used to show that the overall cost of maintaining the representation of the graph under splitting is $O(n \log n)$.

The second data structure, which is the fully dynamic data structure for maintaining biconnectivity of Rauch [21], is used for *biconnectivity queries*. It enables to perform edge deletion and insertion in amortized time $O(\sqrt{n})$, and answers in constant time the query for an articulation point that separates two specified vertices. The crucial idea is that we do not perform the splitting operation on this representation.

The third data structure is needed for *connectivity queries in the forest*. For this we use the dynamic tree data structure of Sleator and Tarjan [23].

We omit the details, but using these three data structures, one can implement Algorithm Δ_T-WS-SPANNING-TREE to run in $O(n^{3/2})$ time. Actually, the running time is $O(n \log n)$ plus the time for maintaining $O(n)$ biconnectivity updates and queries. $\square$

Theorem 11. *Every biconnected graph $G = (V, E)$ with maximum degree $2\Delta_T - 2$ has a Δ_T-spanning tree. Such a Δ_T-spanning tree can be found in time $O(n^{3/2} + m)$.*

Proof. We first sparsify G. In [11] an $O(n + m)$-time algorithm is given that outputs a spanning subgraph G' of G that contains the same k-connected components as G and has fewer than kn edges. We use this algorithm for $k = 2$.

Now, for each edge $e = (x, y) \in E$, place a new vertex v_e in the middle of e, i.e., remove e and then add two edges (x, v_e) and (v_e, y). Let $G^* = (V^*, E^*)$ be the resulting graph. Since G' has $O(n)$ edges, G^* has $O(n)$ edges too. If we set $V_1 = \{v_e : e \in E\}$, $V_2 = V$, and assign label $\boxed{2}$ to every vertex in V_2, then G^* is a $(2\Delta_T - 2)$-ws-graph. Therefore we can find a Δ_T-spanning tree T^* of G^* by Lemma 10. We construct a Δ_T-spanning tree T of G from T^* by adding an edge $e = \{x, y\}$ iff the edges $\{x, v_e\}$ and $\{v_e, y\}$ are in T^*. One can verify that T is indeed a Δ_T-spanning tree of G. Lemma 10 ensures that the whole construction can be performed in the required running time. $\square$

We finally notice that the running time of our algorithm may be significantly improved if we would use randomization. The fully dynamic data structure for maintaining biconnectivity of King and Rauch [22] could be applied to obtain the running time of $O(m + n \log^4 n)$ with high probability.

4 k-Connected Graphs of Maximum Degree $k(\Delta_T - 2) + 2$

Let G be a k-connected graph of maximum degree $\Delta_G \leq k(\Delta_T - 2) + 2$ and let Δ_T^* be the minimal integer such that G has a Δ_T^*-spanning tree. Fürer and Raghavachari [7] gave an algorithm for finding a spanning tree of G with maximum degree at most $\Delta_T^* + 1$. They designed a polynomial-time algorithm that finds a d-spanning tree T and a set B with vertices of degree $d - 1$ in T that satisfy the following property. Let S be the set of vertices of degree d in T. If F is the forest obtained from T by removing vertices of $S \cup B$, then there are no paths in G through vertices of $V - (S \cup B)$ between different trees in F. In that case T is a spanning tree of maximum degree at most $\Delta_T^* + 1$.

The following lemma extends the result from [7] and shows that the tree found by the algorithm of Fürer and Raghavachari is of degree at most Δ_T.

Lemma 12. *Let T be an arbitrary d-spanning tree of a k-connected graph G of maximum degree $\Delta_G \le k(\Delta_T - 2) + 2$. Let S be the set of vertices of degree d in T and let B be an arbitrary subset of vertices of degree $d - 1$ in T. Let $S \cup B$ be removed from the graph, breaking the tree T into a forest F. Suppose G satisfies the condition, that there are no paths through vertices of $V - (S \cup B)$ between different trees in F. Then $d \le \Delta_T$.*

Proof. Let T^* be the vertex-induced subtree of T consisting of a vertex r of degree d in T and all paths from r to vertices in $S \cup B$. For every edge e of T incident to a vertex in $S \cup B$ that does not belong to any path from T^*, define a *leaf component* that consists of all vertices w of T such that the unique tree path in T from w to r contains the edge e.

There are at least $|S|(d - 2) + |B|(d - 3) + 2$ leaf components (This bound will be reached, if every vertex in T^* of degree at least three is either in S or in B. Otherwise there are more leaf components). Because the graph is k-connected, there must be at least k paths from each leaf component to r, where one path can go over the only vertex of $S \cup B$ incident in T to this leaf component. Hence each leaf component is incident in $G - T^*$ to at least $k - 1$ edges with endpoints in $S \cup B$. Therefore there must be at least $(k - 1)(|S|(d - 2) + |B|(d - 3) + 2)$ edges of $G - T$ incident to vertices in $S \cup B$.

On the other hand, a vertex from S can be incident to at most $\Delta_G - d$ edges in $G - T$ and a vertex from B can be incident to at most $\Delta_G - (d - 1)$ edges in $G - T$. Hence there may be at most $|S|(\Delta_G - d) + |B|(\Delta_G - (d - 1))$ edges in $G - T$ that are incident to vertices in $S \cup B$. Therefore the following inequality must hold:

$$(k - 1)(|S|(d - 2) + |B|(d - 3) + 2) \le |S|(\Delta_G - d) + |B|(\Delta_G - (d - 1))$$

We have assumed that $\Delta_G \le k(\Delta_T - 2) + 2$. One can easily verify that this inequality holds only if $d \le \Delta_T$. □

Combining this result, initial sparsification of an input graph with $O(kn)$ edges [11], and the algorithm of Fürer and Raghavachari [7] we obtain:

Theorem 13. *Every k-connected graph G of maximum degree at most $k(\Delta_T - 2) + 2$ has a Δ_T-spanning tree. One can find such a tree in time $O(n^2\, k\, \alpha(n) \log n)$.* □

5 $\mathcal{NP}$-Completeness in General Graphs

Garey, Johnson, and Tarjan [9] showed that deciding whether a 3-connected 3-regular graph has a Hamiltonian cycle (or a Hamiltonian path) is $\mathcal{NP}$-complete. We extend their result and prove that verifying whether a k-connected $k(\Delta_T - 1)$-regular graph has a Δ_T-spanning tree (for $\Delta_T = 2$ also a Hamiltonian cycle) is $\mathcal{NP}$-complete for every $k \ge 3$, $\Delta_T \ge 2$. [1]

---

[1] We can provide a significantly simpler proof of the $\mathcal{NP}$-completeness if we would assume that $k \ne 5$. However, similarly as in the existential proof for $\Delta_T = 2$ presented in [12], the case $k = 5$ requires more complicated arguments.

In our proof we will analyze the 3-connected 3-regular graphs used in the reduction of Garey, Johnson and Tarjan [9] of 3SAT to the Hamiltonian problem in 3-connected 3-regular graphs. We first briefly characterize all separating edge-triplets that may appear in their construction. Then we will find a special perfect matching M in the graphs constructed in [9] that will be used to build a k-edge connected k-regular graph G_k^M that has a Hamiltonian cycle (respectively a Hamiltonian path) iff the graph constructed by Garey, Johnson and Tarjan has one. By replacing each vertex in G_k^M by a $K_{k,k(\Delta_T-1)-1}$ we will get a k-connected graph that has a Δ_T-spanning tree (respectively a Hamiltonian cycle for $\Delta_T = 2$) iff the graph constructed by Garey, Johnson and Tarjan has one. This will imply the main result.

Using a reduction from 3SAT, Garey, Johnson, and Tarjan [9] proved the $\mathcal{NP}$-completeness of verifying whether a 3-connected 3-regular graph has a Hamiltonian cycle (or a Hamiltonian path). For every 3SAT formula $\mathcal{F}$ they construct a 3-connected 3-regular graph G (abbreviated GJT-graph) such that $\mathcal{F}$ is satisfiable iff G has a Hamiltonian cycle (or a Hamiltonian path).

Each GJT-graph is a composition of four graphs: *the Tutte-graph, the XOR-graph, the 2-OR-graph*, and *the 3-OR-graph* (see Fig. 1–4). For every clause $a \vee b \vee c$, respectively variable x, there is a construction of the form indicated by Fig. 5 and Fig. 6, respectively.

The constructions for clauses (respectively for variables/literals) are connected one after another to form a line. Then the ends of the two lines are connected via a 2-OR-graph (XOR-graph) in the case of the Hamiltonian cycle (Hamiltonian path) problem; this 2-OR- or XOR-graph will be called *the connecting 2-OR-/XOR-graph*. Each literal in each clause is connected via the XOR-graph to a cycle of the literal in the corresponding variable construction, the so-called *literal-cycle*. For an example for the formula $(x \vee y \vee z) \wedge (\bar{x} \vee \bar{y} \vee w) \wedge (y \vee \bar{z} \vee \bar{w})$ see Fig. 7. For further details of the construction of Garey, Johnson, and Tarjan we refer to [9].

We can prove the following lemma that characterizes all edge sets of cardinality three in the GJT composition that separates the graph G and are not all adjacent.

Lemma 14. *Every non-trivial separating edge triplet of a GJT-graph consists of the three a-edges of one induced Tutte-subgraph.* □

We shall use a special perfect matching M in the GJT-graphs.

Lemma 15. *Let $G = (V, E)$ be a GJT-graph. There exists a perfect matching $M \subset E$ in G, such that (i) for every separating edge quadruplet $Q \subset E$ that is minimal: $|Q \cap M| < 4$ and that (ii) for every separating edge triplet (SET) $S \subset E$: $|S \cap M| = 1$* □

Let $G = (V, E)$ be a k-edge connected k-regular graph. For each vertex $v \in V$, let $\Gamma_G(v) = \{\Gamma^1(v), \dots, \Gamma^k(v)\}$ be arbitrarily ordered set of the neighbors of v in G. We define a graph $H = (V_H, E_H)$ as follows. To each vertex $v \in V$ we assign a copy of $K_{k,k(\Delta_T-1)-1}$ that is denoted by $K_{k,k(\Delta_T-1)-1}(v)$. Let $A(v) = \{a_1(v), \dots, a_k(v)\}$ and $B(v) = \{b_1(v), \dots, b_{k(\Delta_T-1)-1}(v)\}$ be the vertices in the k and the $k(\Delta_T - 1) - 1$ elements' vertex set of $K_{k,k(\Delta_T-1)-1}(v)$, respectively. Let $E(v)$ denote the set of the edges in $K_{k,k(\Delta_T-1)-1}(v)$. We define $V_H = \bigcup_{v \in V}(A(v) \cup B(v))$ and $E_H =$

$\{(a_i(v), a_j(u)) : v, u \in V, \Gamma^i(v) = u, \Gamma^j(u) = v\} \cup \bigcup_{v \in V} E(v)$. One can verify that H is of maximum degree $k(\Delta_T - 1)$ and that it is k-connected (see also [18]).

Lemma 16. *G has a Hamiltonian path iff H has a Δ_T-spanning tree.* ☐

Theorem 17. *For every integers k and Δ_T, $k \geq 3$, $\Delta_T \geq 2$, it is $\mathcal{NP}$-complete to verify whether a k-connected graph of maximum degree $k(\Delta_T - 1)$ has a Δ_T-spanning tree.*

Proof. For a graph G, let M be the perfect matching constructed in Lemma 15. For every $k > 2$, define the multigraph G_k^M by:
- if $3 \mid k$: duplicating each edge of G $k/3$ times.
- if $3 \mid (k-1)$: duplicating each edge of M $\lceil k/3 \rceil$ times and the other edges of G $\lfloor k/3 \rfloor$ times.
- if $3 \mid (k-2)$: duplicating each edge of M $\lfloor k/3 \rfloor$ times and the other edges of G $\lceil k/3 \rceil$ times.

One can show that G_k^M is k-edge connected and k-regular. Thus the graph obtained by replacing each vertex in G_k^M by a $K_{k,k(\Delta_T-1)-1}$ is k-connected and of maximum degree $k(\Delta_T - 1)$ [18, Theorem 3]. Additionally it has a Hamiltonian path iff G_k^M has one (Lemma 16). Now, it is easy to see that for the case $\Delta_T = 2$, the obtained graph has a Hamiltonian cycle iff G_k^M has one. ☐

Theorems 13 and 17 yield the following characterization of triconnected graphs.

Theorem 18. *(i) Every triconnected graph of maximum degree at most $3\Delta_T - 4$ has a Δ_T-spanning tree; such a tree can be found in polynomial time. (ii) Verifying whether a triconnected graph of maximum degree $3\Delta_T - 3$ has a Δ_T-spanning tree is $\mathcal{NP}$-complete.* ☐

6 Biconnected Planar Graphs

The result from Sect. 5 does not hold for $k = 2$. Actually, Lemma 10 states that every biconnected graph of maximum degree $2(\Delta_T - 1)$ has a Δ_T-spanning tree. We can show that this bound cannot be improved even for biconnected planar graphs.

Theorem 19. *(i) Every biconnected graph of maximum degree at most $2\Delta_T - 2$ has a Δ_T-spanning tree; such a tree can be found in polynomial time. (ii) Verifying whether a biconnected planar graph of maximum degree $2\Delta_T - 1$ has a Δ_T-spanning tree is $\mathcal{NP}$-complete.* ☐

We finally note that our algorithm from Sect. 3 can be speeded up for planar graphs. For this notice that we have designed our algorithm in Sect. 3 so that all operations can easily preserve the embedding of a plane graph G during the modifications. Apart from operations caused by the fully dynamic biconnectivity algorithm, the algorithm in Sect. 3 requires $O((n + m) \cdot \log(n + m))$ time.

We may use the decremental data structures of [17] for maintaining biconnectivity. The main feature of that data structure is that it can perform the DELETE-AND-COMBINE operation and can determine in a fast way a separating vertex between u and v, if one exists. Thus we get the following results:

Theorem 20. *If a biconnected planar graph G is of maximum degree at most $2\Delta_T - 2$, then one can find a Δ_T-spanning tree of G deterministically in $O(n \log n)$ time using space $O(n^2)$ or in $O(n \log^2 n)$ time using space $O(n)$, and with high probability in $O(n \log n)$ time using space $O(n)$.* □

Acknowledgment   We are grateful to Monika Rauch Henzinger for describing some details of her fully dynamic data structures for maintaining biconnectivity.

References

1. D. Barnette. Trees in polyhedral graphs. *Canadian J. Mathematics*, 18:731–736, 1966.
2. D. Bauer, S. L. Hakimi, and E. F. Schmeichel. Recognizing tough graphs is $\mathcal{NP}$-hard. *Discrete Applied Mathematics*, 28:191–195, 1990.
3. B. Bollobás. *Extremal Graph Theory*. Academic Press, London, 1978.
4. P. M. Camerini, G. Galgiati, and F. Maffioli. Complexity of spanning tree problems, I. *European Journal of Operation Research*, 5:346–352, 1980.
5. Y. Caro, I. Krasikov, and Y. Roditty. On the largest tree of a given maximum degree in a connected graph. *Journal of Graph Theory*, 15:7–13, 1991.
6. N. Chiba and T. Nishizeki. The Hamiltonian cycle problem is linear-time solvable for 4-connected planar graphs. *Journal of Algorithms*, 10:187–211, 1989.
7. M. Fürer and B. Raghavachari. Approximating the minimum-degree Steiner tree to within one of optimal. *Journal of Algorithms*, 17:409–423, 1994. Also in ACM-SIAM SODA 1992.
8. M. R. Garey and D. S. Johnson. *Computers and Intractability: A Guide to the Theory of $\mathcal{NP}$-completeness*. Freeman, New York, 1979.
9. M. R. Garey, D. S. Johnson, and R. E. Tarjan. The planar Hamiltonian circuit problem is $\mathcal{NP}$-complete. *SIAM Journal on Computing*, 5(4):704–714, 1976.
10. M. X. Goemans and D. P. Williamson. A general approximation technique for constrained forest problems. *SIAM Journal on Computing*, 24(2):296–317, 1995.
11. H. Nagamochi and T. Ibaraki. A linear-time algorithm for finding a sparse k-connected spanning subgraph of a k-connected graph. *Algorithmica*, 7:583–596, 1992.
12. B. Jackson and T. D. Parsons. On r-regular r-connected non-Hamiltonian graphs. *Bulletin of Australian Mathematics Society*, 24:205–220, 1981.
13. D. S. Johnson. The $\mathcal{NP}$-completeness column: An ongoing guide. *Journal of Algorithms*, 6:434–451, 1985.
14. S. Khuller and B. Raghavachari. Improved approximation algorithms for uniform connectivity problems. In *Proceedings of the 27 ACM STOC*, pp. 1–10, 1995.
15. S. Khuller, B. Raghavachari, and N. Young. Low degree spanning trees of small weight. *SIAM Journal on Computing*, 25(2):355–368, 1996.
16. S. Khuller and U. Vishkin. Biconnectivity approximations and graph carvings. *Journal of the ACM*, 41(2):214–235, 1994.
17. T. Lukovski and W.-B. Strothmann. Decremental biconnectivity on planar graphs. Manuscript, 1997.
18. G. H. J. Meredith. Regular n-valent n-connected nonHamiltonian non-n-edge-colorable graphs. *Journal of Combinatorial Theory Series B*, 14:55–60, 1973.
19. V. Neumann-Lara and E. Rivera-Campo. Spanning trees with bounded degrees. *Combinatorica*, 11(1):55–61, 1991.
20. C. H. Papadimitriou and M. Yannakakis. The complexity of restricted spanning tree problems. *Journal of the ACM*, 29(2):285–309, 1982.

21. M. Rauch. Improved data structures for fully dynamic biconnectivity. Full version. A preliminary version appeared in *Proceedings of the 26th ACM STOC*, 1994.
22. M. Rauch Henzinger and V. King. Fully dynamic biconnectivity and transitive closure. In *Proceedings of the 36th IEEE FOCS*, pp. 664–673, 1995.
23. D.D. Sleator and R.E. Tarjan. A data structure for dynamic trees. *Journal of Computer and System Sciences* 26:362–391, 1983.
24. W.-B. Strothmann. Constructing 3-trees in 3-connected planar graphs in linear time. Manuscript, 1996.
25. S. Win. On a connection between the existence of k-trees and the toughness of a graph. *Graphs and Combinatorics*, 5:201–205, 1989.

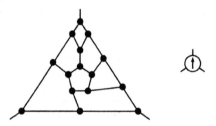

Fig. 1. Tutte-graph and its abbreviation

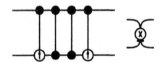

Fig. 2. XOR-graph and its abbreviation

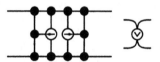

Fig. 3. 2-OR-graph and its abbreviation

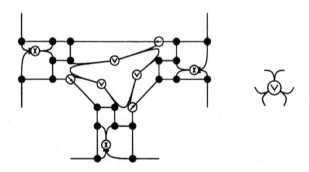

Fig. 4. 3-OR-graph and its abbreviation

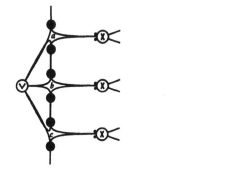

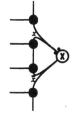

Fig. 5. Construction for a clause $a \vee b \vee c$ **Fig. 6.** Construction for a variable x

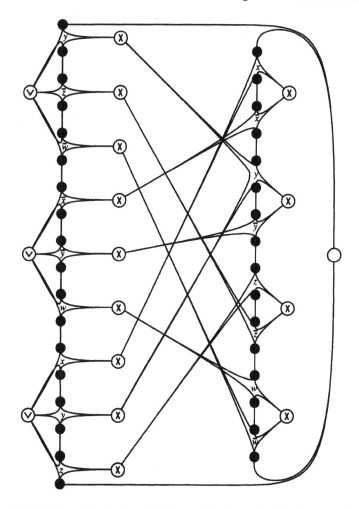

Fig. 7. GJT-graph for the formula $(x \vee y \vee z) \wedge (\bar{x} \vee \bar{y} \vee w) \wedge (y \vee \bar{z} \vee \bar{w})$

Optimal Adaptive Broadcasting with a Bounded Fraction of Faulty Nodes (Extended Abstract)

Krzysztof Diks[1]* Andrzej Pelc[2]**

[1] Instytut Informatyki, Uniwersytet Warszawski, Banacha 2, 02-097 Warszawa, Poland. E-mail: diks@mimuw.edu.pl
[2] Département d'Informatique, Université du Québec à Hull, Hull, Québec J8X 3X7, Canada. E-mail: Andrzej_Pelc@uqah.uquebec.ca

Abstract. We consider broadcasting among n processors, f of which can be faulty. A fault-free processor, called the source, holds a piece of information which has to be transmitted to all other fault-free processors. We assume that the fraction f/n of faulty processors is bounded by a constant $\gamma < 1$. Transmissions are fault free. Faults are assumed of *crash* type: faulty processors do not send or receive messages. We use the *whispering* model: pairs of processors communicating in one round must form a matching. A fault-free processor sending a message to another processor becomes aware of whether this processor is faulty or fault free and can adapt future transmissions accordingly. The main result of the paper is a broadcasting algorithm working in $O(\log n)$ rounds and using $O(n)$ messages of logarithmic size, in the worst case. This is an improvement of the result from [10] where $O((\log n)^2)$ rounds were used. Our method also gives the first algorithm for adaptive distributed fault diagnosis in $O(\log n)$ rounds.

1 Introduction

Broadcasting, also called one-to-all communication, is one of fundamental primitives in network protocols. A processor (node of a network) holds a message which must be transmitted to all other nodes. This task often occurs in distributed computing, e.g., in global processor synchronization and updating distributed databases. Moreover, it is implicit in many parallel computation problems, where data and results are distributed among processors. This happens, e.g., in matrix multiplication, parallel solving of linear systems, parallel computing of Discrete Fourier Transform, or parallel sorting, cf., e.g., [4] and [13].

Two most important measures of performance of broadcasting algorithms are the number of elementary transmissions (*messages*) and the number of rounds (*parallel time*) required. Excellent accounts of literature on broadcasting (and closely related gossiping, i.e., all-to-all communication) focusing on the optimization of these measures can be found in surveys [8] and [12].

---

* This work was done during the author's stay at the Université du Québec à Hull.
** Research supported in part by NSERC grant OGP 0008136.

As communication networks grow in size, they become increasingly vulnerable to component failures. Some links and/or nodes of the network may fail. It becomes important to design communication algorithms in such a way that the desired communication task be accomplished efficiently in spite of these faults, without knowing their location ahead of time. The vast literature on fault-tolerant broadcasting and gossiping has been surveyed in [17]. One of the most common assumptions about faults is that their number is bounded by an integer f and that they are located in a way most detrimental for the communication process. In this case the goal of *fault-tolerant* broadcasting becomes to transmit information from one fault-free node (the *source*) to all other fault-free nodes of the network. Broadcasting algorithms achieving this goal for any location of at most f faults are called f-*tolerant*.

Most of the literature on fault-tolerant broadcasting and gossiping concentrated on transmission faults under assumption that nodes are fault free. Research was focused both on minimizing the number of messages [3] and on minimizing time [9, 11, 18] of the communication process.

The case when nodes can fail was first investigated by Farley [7] whose aim was to minimize time of f-tolerant broadcasting in chordal rings. In [1] the aim was to minimize the number of messages in fault-tolerant gossiping. It was assumed that nodes fail in a Byzantine manner but they have diagnostic capabilities: a fault-free node can correctly diagnose tested nodes, while faulty testers are unpredictable. The authors presented a gossiping algorithm using $3n \log n + O(n)$ message transmissions and working correctly whenever the fault-free part of the network is connected.

The goal of the present paper is simultaneous optimization of the number of messages and the number of rounds in fault-tolerant broadcasting. We assume that there are at most f faulty nodes among n nodes and their fraction f/n is bounded by a constant $\gamma < 1$. Transmissions are fault free. Faults are assumed of *crash* type: faulty nodes do not send or receive messages. We use the *whispering* model: pairs of nodes communicating in one round must form a matching. A fault-free node sending a message to another node becomes aware of whether this node is faulty or fault free and can adapt future transmissions accordingly. The main result of the paper is a broadcasting algorithm working in $O(\log n)$ rounds and using $O(n)$ messages of logarithmic size, in the worst case. (We also count unsuccessful transmissions to faulty nodes.) It should be stressed that although the majority of nodes may be faulty and our algorithm is deterministic, both the number of rounds and the number of messages used have the same order of magnitude as in the best broadcasting algorithm working in the fault-free case.

Our result implies a positive answer to a problem from [10], where broadcasting in a similar model was discussed. (The only difference in assumptions is that faulty processors in [10] could not send but could receive messages and had to get the source message as well. Our algorithm can be easily adapted to this scenario.) The authors showed a fault-tolerant broadcasting algorithm working in $O((\log n)^2)$ rounds and using $O(n(\log n)^2)$ messages. They asked if the number of rounds can be decreased to $O(\log n)$.

The main part of our algorithm is a preprocessing phase in which a constant fraction of fault-free nodes are distributively organized in a connected graph with constant degree and logarithmic diameter. A much simpler preprocessing in which all fault-free nodes were arranged in a list was done in [10] but it used $O((\log n)^2)$ rounds. Speeding up the preprocessing, via a much more intricate method, is a major contribution of the present paper.

As an application of this method we can also obtain an improvement of the result from [5] on simulating computations on faulty parallel machines such as PRAM and OCPC. An important part of the simulation algorithm in [5] was a preprocessing in which fault-free processors were assigned numbers permitting them to share the work of faulty processors. This preprocessing, using $O((\log n)^2)$ rounds in [5], can be done in $O(\log n)$ rounds using our method.

We can also directly apply our method to perform leader election among fault-free processors, provided that only a bounded fraction of processors fail.

Another important application of our preprocessing is an efficient algorithm for adaptive distributed fault diagnosis. We describe the problem and its solution in section 5.

2   The model

We consider n processors (called *nodes*) that are labeled by integers $1,...,n$, know these labels and can communicate with one another. The *source* of broadcasting is assumed fault free. We do not assume that the identity of the source is known to other nodes. At most f nodes are faulty and faults are of *crash* type: faulty nodes do not receive or send any messages. Faults are *static*: the status faulty/fault-free of a node does not change during the execution of broadcasting. We assume that $f/n \leq \gamma$, for some constant $\gamma < 1$.

The elementary procedure in our broadcasting algorithm is a transmission between two nodes u and v, initiated by a fault-free node. All transmissions are fault free. The result of such a transmission is an exchange of information packets between u and v, if both these nodes are fault free. Packets are of size $O(\log n)$. At least one bit signifying "I am fault free" is sent in each direction during a transmission between fault-free nodes. In case when one of the nodes is faulty, the other node becomes aware of it, as it does not receive the expected "alive" bit in prescribed time (a time-out mechanism is used). The algorithm proceeds in *rounds* of prescribed duration synchronized by a global clock.

Our algorithm is *adaptive*: nodes can use information acquired in previous rounds to schedule future transmissions and determine messages to be sent. However, in performing local computations and choosing subsequent transmissions, a node can use only information currently available to it (we do not assume the existence of any central unit monitoring communication).

We use the *1-port* or *whispering* model: each node can be involved in at most one transmission in a given round, i.e., pairs communicating in each round must form a matching. This is one of the most widely used models in the theory of broadcasting and gossiping (cf. [8, 12]). The requirements of this model force us to construct the algorithm in such a way as to avoid *conflicts*: sending to two nodes in the same round, sending to a node which is simultaneously sending to another node, or sending to the same node from two different nodes. While the first type of conflicts is easy to avoid, special care has to be taken to avoid the other two; we will point it out in the description of the algorithm. This requirement is also the only reason to restrict faults to the crash type. If faulty nodes could behave in a Byzantine way, it would be impossible to prevent simultaneous transmissions to the same node, as transmissions are not prescribed in advance but adaptively scheduled by nodes themselves (cf. section 5 where we consider Byzantine node faults).

3 Preliminaries and terminology

For any graph $G = (V, E)$ and any subset $S \subset V$ we denote by $G(S)$ the subgraph of G induced by S. A graph is called *d-regular* if all its nodes have degree d. For any graph $G = (V, E)$ and any set of nodes $S \subset V$ let $N_G(S)$ denote the set consisting of all nodes in S and all their neighbors. (The subscript G will be omitted if there is no ambiguity.) For any graph G and any positive integer c, G^c denotes the graph resulting from G by adding edges between nodes at distance at most c in G (multiple edges are avoided). A graph G is an *expander* if there exists a constant $\beta > 1$ such that $|N(S)| \geq \beta|S|$, whenever $|S| \leq |V|/2$. β is then called the *expansion* of G.

We use the following lemmas proved in [20], concerning the expander constructed in [15].

Lemma 1. *Let k be a number of the form $4m + 2$ for some integer m, where $k - 1$ is prime. There exist positive constants α, μ and d, and an explicitly constructible k-node d-regular graph $G = (V, E)$ such that for every $S \subset V$ of size $|S| \leq \alpha|V|$ the subgraph $G(V \setminus S)$ contains a connected subgraph $H = (U, F)$ with the following properties:*

- *$|U| \geq k - \mu|S|$,*
- *the degree of each node in H is at least $\frac{4}{5}d$,*
- *the diameter of H is $O(\log k)$.*

In Upfal's proof $\alpha = \frac{1}{72}$ and $\mu = 6$. In the sequel we will use these constants. It follows in particular that $|U| > k/2$. In order to show that the diameter of H is logarithmic, Upfal proved that the graph G from lemma 1 has the following expansion property:

Lemma 2. *Let G be the graph from lemma 1. For any node u of G let $M(u)$ be any set of neighbors of u of size at least $\frac{4}{5}d$. Then for every $S \subset V$ of size $|S| \leq k/2$,*

$$|S \cup \bigcup_{u \in S} M(u)| \geq \frac{21}{20}|S|.$$

Lemma 2 implies that the diameter of H is at most $D = \lfloor 2 \log_{\frac{21}{20}} k \rfloor$.

We will use the following theorem of Vizing and a lemma from number theory (cf. [2]):

Theorem 3. *All edges of a graph of degree at most d can be partitioned into at most $d + 1$ disjoint matchings.*

Lemma 4. *For every positive integer n let $f(n)$ denote the largest prime of the form $4m + 1$ not exceeding n. Then $\lim_{n \to \infty} \frac{n - f(n)}{n} = 0$.*

4   The algorithm

We start with a general overview of the algorithm. All nodes are partitioned into k sets of logarithmic size and the algorithm works on these sets, rather than on individual nodes. This is done in order to keep the number of messages linear, instead of $\Theta(n \log n)$. First the *active* sets are identified: those which contain at least two fault-free nodes. One of these nodes, the *chief*, will represent the set in further computations, the other one, the *guard*, is used to avoid conflicts in a later stage.

Next we consider the expander $G = (V, E)$ from lemma 1, built on the sets of the partition. The role of the set S is played by sets which are not active. The goal is to identify a large connected subgraph of G of logarithmic diameter, induced by active sets. This part of the algorithm is executed by chiefs. The existence of such a subgraph follows from lemma 1 but the difficulty lies in distributed labeling of chiefs: they need to "realize" that they belong to the subgraph, learn its size and get consecutive numbers in order to divide subsequent communication work. To do this, leader election is needed.

First each chief identifies chiefs of all sets neighboring in G. Then comes a multistage elimination process identifying *survivors* among chiefs: those chiefs that have more than $k/2$ chiefs at logarithmic distance. This is done to perform fast leader election among the survivors: since survivors are at logarithmic distance from one another, leader election can be done in logarithmically many rounds. Once the leader is elected, it initiates a construction of a BFS tree of logarithmic depth, containing more than $k/2$ chiefs and assigns numbers to its nodes. This concludes the preprocessing phase which is the main part of the algorithm.

To perform broadcasting, the source message is first disseminated inside the tree. Then nodes of the tree share the work of informing all remaining nodes in logarithmically many rounds. It is at this stage that guards prevent conflicts.

We now give a detailed description of the algorithm. All our considerations are carried out for sufficiently large n. In order to use lemmas 1 and 2 we first work under the following additional assumptions that will be removed later:

1.The fraction f/n of faulty nodes is bounded by $\gamma < \frac{1}{72}$.

2.For $l = \lceil \log n \rceil$, the integer $\lceil \frac{n}{l} \rceil$ is of the form $4m + 3$, where $4m + 1$ is a prime number.

Let $l = \lceil \log n \rceil$. Let $k + 1 = \lceil \frac{n}{l} \rceil$ and fix a partition of all nodes into $k + 1$ subsets $X_1,...,X_{k+1}$, such that $X_i = \{(i - 1)l + 1, ..., il\}$, for $i = 1, ..., k$, and $X_{k+1} = \{ki + 1, ..., n\}$. A set X_i is called *active* if it contains at least two fault-free nodes. Otherwise it is called *passive*. The smallest fault-free node in an active set is called its *chief* and the largest is called its *guard*. We omit the easy proof of the following lemma.

Lemma 5. *For any constant γ' such that $1 > \gamma' > \gamma$ and for sufficiently large n, at least $(1 - \gamma')k$ among sets $X_1,...,X_k$ are active.*

Stage 1. In all active sets $X_1,...,X_k$ the chief and the guard are selected in $l - 1$ rounds (in parallel). Let $x_1, ..., x_l$ be an increasing enumeration of X_j. A fault-free node x_i sends messages to nodes $x_{i+1}, x_{i+2},...$, starting in round i, until a fault-free node is found or all nodes x_r, $r > i$ are exhausted. Notice that conflicts are avoided. A fault-free node that did not get a message from a smaller node becomes the chief and a fault-free node that did not find a larger fault-free node becomes the guard. Let C be the set of all chiefs. If a passive set contains only one fault-free node, this node does not get any message and does not find any larger fault-free node hence it learns that its set is passive. Call such a node *singleton*. Stage 1 is executed in $l - 1$ rounds and uses at most n messages, as every node is called at most once.

Let $G = (V, E)$ be the graph from lemmas 1 and 2 with the set of nodes $V = \{X_1, ..., X_k\}$. By assumptions 1 and 2, and in view of lemma 5, the set S of passive sets satisfies the requirements of lemma 1. Let $E_1,...,E_{d+1}$ be a partition of all edges in E into disjoint matchings.

Stage 2. Chiefs build a graph $K = (C, F)$ in which chief c_i of an active set X_i is adjacent to chief c_j of an active set X_j if sets X_i and X_j are adjacent in the graph G. This is done in $d+1$ steps, each consisting of $l-1$ rounds. The rth step, $r = 1, ..., d + 1$, is devoted to communication between chiefs of sets joined in G by edges belonging to E_r. If the edge between X_i and X_j belongs to E_r, where $i < j$, and the set X_i is active, the chief c_i of X_i tries to find the chief of X_j by sending messages to consecutive nodes of X_j (in increasing order), until the chief of X_j is found or it is revealed that X_j is passive. Stage 2 uses $(d+1)(l-1)$ rounds and at most n messages. Upon its completion chiefs adjacent in the graph K know each other.

From now on we identify chiefs with their (active) sets and consider the graph K to be a subgraph of G induced by chiefs. By lemma 1 graph K contains a connected subgraph H of logarithmic diameter and with more than $k/2$ nodes of degree at least $\frac{4}{5}d$. We now want to identify chiefs belonging to this subgraph.

Stage 3. This stage is performed in $D = \lfloor 2 \log_{\frac{21}{20}} k \rfloor$ steps, each consisting of $d + 1$ rounds. Let $F_i = E_i \cap F$. Before step 1 all chiefs are *living*. In consecutive rounds of step i living chiefs communicate with neighbors in graph K. (The rth round is devoted to communication between neighbors whose joining edge is in F_r. Always the chief with lower index sends the message. If the respective neighbor of chief c does not exists, c remains silent in this round. If a living chief did not get the expected message from a neighbor with lower index, it knows that the neighbor is not living.) Every living chief counts those neighbors that were living during step i. If the total number of living neighbors is less than $\frac{4}{5}d$ the chief *dies*, otherwise it remains living in the next step. Stage 3 uses $O(\log k) = O(\log n)$ rounds and $O(kD) = O(n)$ messages.

Lemma 1 implies that after D steps of Stage 3 there exist living chiefs. Call these chiefs *survivors*. We will need the following lemma which can be proved using lemma 2.

Lemma 6. *If v is a survivor then there are more than $k/2$ nodes in graph K at distance $\leq D$ from v.*

The existence of survivors and lemma 6 imply

Corollary 7. *The distance between every pair of survivors in graph K is at most $2D$.*

Stage 4. This stage is the election of a leader among survivors. All chiefs take part in the election and the survivor becoming the leader is the one with smallest index. The election proceeds in $2D$ steps, each consisting of $d+1$ rounds. In the beginning every chief c_i gets a value equal to i if c_i is a survivor, and equal to $k + 1$ otherwise. In every step every chief collects values from its neighbors in the graph K (as usual, in round j of each step chiefs joined by an edge from F_j communicate) and updates its value to the minimum of its own previous value and all values obtained in this step. The chief whose value is equal to its index after $2D$ steps, becomes the leader. In view of corollary 7 exactly one survivor will be elected. Stage 4 uses $O(D) = O(\log n)$ rounds and $O(kD) = O(n)$ messages.

Stage 5. This stage is the construction of a BFS subtree T of graph K. The tree T is of depth D and is rooted at the leader x elected previously. The construction proceeds in D steps, each consisting of $d + 1$ rounds. After the ith step, $i = 0, 1, ..., D-1$, the tree is constructed up to depth i. In consecutive rounds $1,...,d+1$ of the $(i+1)$th step each chief at distance i from the root x communicates with its neighbors in graph K joined by edges from $F_1,..., F_{d+1}$, respectively, adding new nodes as its children. It follows from lemma 6 that the number of nodes in this tree exceeds $k/2$. Stage 5 uses $O(\log n)$ rounds and $O(k) = O(\frac{n}{\log n})$ messages.

Stage 6. In this stage nodes of the tree T are numbered in pre-order. First each node v of the tree counts the size of the subtree rooted at v, proceeding bottom-up in D steps, each consisting of $d+1$ rounds. Next, nodes are numbered in pre-order in a recursive fashion, this time top-down. Simultaneously with the numbering the root broadcasts the size k_0 of the tree T to all nodes. Stage 6 uses $O(\log n)$ rounds and $O(k) = O(\frac{n}{\log n})$ messages.

Stage 7. This is the final stage of preprocessing. $O(\frac{n}{\log n})$ fault-free nodes (elements of the tree T) are already assigned consecutive numbers $1,..., k_0$. Call them **key nodes**. Now they share the work of communicating with all other fault-free nodes. Let $r = \lceil \frac{k+1}{k_0} \rceil$. Thus $r \leq 3$. The jth key node, for $j < k_0$, becomes **responsible** for sets $X_{(j-1)r+1},...,X_{jr}$; the k_0th key node becomes responsible for the remaining sets. The stage proceeds in r steps, each consisting of l rounds. In the ith step each key node v communicates with nodes of the ith set for which it is responsible. In consecutive rounds of this step a key node sends messages to consecutive nodes of this set *from largest to smallest* until the guard of the set is found or until it learns that the set is passive. Now the role of the guard becomes clear. It enables key nodes to operate in parallel eliminating the risk of conflict in which a key node sends a message to another key node which sends a message to some other node simultaneously. Each guard and each singleton found by v (a total of at most 3 nodes) become new children of v. Let T' denote the tree T augmented in this way. Stage 7 uses $O(\log n)$ rounds and at most n messages.

Stage 7 concludes the preprocessing phase. Upon its completion the tree T' (of logarithmic depth and bounded degree) contains guards and singletons from all sets $X_1,...,X_{k+1}$. Now it is easy to complete broadcasting. In l rounds (and using l messages) the source finds the guard of its set X_i and sends the message to it. (If the source is in a passive set it knows already after Stage 1 that it is a singleton and remains idle during these l rounds.) Let s denote the guard in the first case and the source itself in the second case. Next s broadcasts the message in the tree T'. This takes $O(\log n)$ rounds and $O(\frac{n}{\log n})$ messages. Finally, in additional l rounds, each guard sends the message to all nodes of its set. This requires at most n messages.

We have shown that our broadcasting algorithm works in $O(\log n)$ rounds and uses $O(n)$ messages in the worst case. It remains to remove assumptions 1 and 2 stated in the beginning of this section. We first address assumption 1. The fact that the fraction of faulty nodes (and consequently of passive sets) is smaller than $\frac{1}{72}$ was needed to show that the graph G from lemma 1 contains a large connected subgraph of logarithmic diameter. In order to eliminate this assumption, we use the fact that G is an expander (cf. lemma 2).

Lemma 8. *If $G = (V, E)$ is a d-regular expander, for some constant d, then for all constants $0 < \delta_1 < \delta_2 < 1$, there exists a positive integer constant c, such that for every set U of nodes, $|U| \geq \delta_1|V|$ implies $|N_{G^c}(U)| \geq \delta_2|V|$.*

Using lemma 8 it is easy to eliminate assumption 1. Suppose that the number of active sets is $(1-\alpha')k$ for some $\alpha' > \alpha = \frac{1}{72}$. Let U denote the family of active sets. By lemma 8 we find a constant c such that $|N_{G^c}(U)| \geq (1 - \alpha)k$. For every passive set X_i from $N_{G^c}(U)$, the *representative* of X_i is the lowest index active neighbor of X_i in G^c. Now the whole algorithm is modified as follows. Passive sets from $N_{G^c}(U)$ are considered as active and all actions to be performed by the chief of such a set are simulated by the chief of its representative. Since c and d are constants, this simulation increases the number of rounds only by a constant factor and the number of messages remains linear.

In order to remove assumption 2 we use lemma 4. Since the function $n - f(n)$ is $o(n)$ we can use the number $f(k) + 1$ instead of k in the entire algorithm, increasing the number of rounds and messages by at most a constant factor. This concludes the proof of the main result of this paper.

Theorem 9. *If the fraction of faulty processors is at most γ, for some constant $\gamma < 1$, then adaptive broadcasting among n processors can be done in $O(\log n)$ rounds and using $O(n)$ messages of logarithmic size, in the worst case.*

5 Application to fault diagnosis

Fault diagnosis is one of major problems in fault-tolerant computing. Its aim is to locate all faulty processors in a system. The classical approach to fault diagnosis was originated in a seminal paper by Preparata, Metze and Chien [19]. Processors perform predetermined tests on one another and diagnosis is carried out by a central monitoring unit, assumed to be fault free, based on the collection of test results. It is assumed that fault-free processors always give correct test results, while tests conducted by faulty processors are totally unpredictable: a faulty tester can output any test result regardless of the status of the tested processor. Faults are assumed permanent, i.e., the fault status of a processor does not change during testing and diagnosis.

This model and some of its variations have been thoroughly studied in the literature (see the survey [6]). One of the important model modifications is eliminating the assumption that a central monitor exists. In *distributed diagnosis* there is no central monitor and the aim of diagnosis is to identify the fault status of all processors by all fault-free processors in the system. That is, upon completion of diagnosis, every fault-free processor should have a correct binary vector describing which processor is faulty and which is fault free. Every processor considers itself to be fault free. Processors that do not directly test a given processor have to get diagnostic information from others. Messages relayed by faulty intermediaries can be distorted. Distributed diagnosis has been introduced by Kuhl and Reddy [14] and later studied by many authors.

Nakajima [16] was the first to propose a new approach to fault diagnosis, called *adaptive diagnosis*. Instead of fixing all tests in advance, processors can schedule future tests based on the results of previous ones, which is likely to improve efficiency. In [2] (see also numerous references therein) the number of rounds of adaptive diagnosis was investigated, assuming that tests involving disjoint pairs of processors can be conducted in the same round. It was shown that while locating $t < \frac{n}{2}$ faults in the nonadaptive setting requires t rounds in the worst case, adaptive diagnosis can locate less than $\frac{n}{2}$ faults among n in a constant number of rounds. In [2] adaptive diagnosis was performed by a central fault-free monitor which collected test results after each round and scheduled tests for the next round.

We will show how to apply our method to perform distributed fault diagnosis in an adaptive way. Assumptions about test outcomes are the same as in the previously cited papers: fault-free processors are correct testers, while faulty nodes are totally unpredictable. We also assume that fault-free processors faithfully relay messages, while faulty processors can act in a Byzantine way, arbitrarily distorting and/or rerouting relayed messages. Transmissions are assumed fault free.

We are interested in three efficiency criteria: the number of tests, the number of messages and the number of rounds. We assume that communicating processors can exchange all currently available information in one round. The assumption concerning the whispering model has to be revised as follows. Since our diagnosis is distributed and adaptive at the same time, tests are scheduled on-line by processors themselves. Since processors behave in a Byzantine way, it is impossible to prevent several faulty processors from testing the same processor simultaneously or from trying to send messages to the same processor simultaneously. Hence we only assume that a processor cannot test or send messages to two processors in the same round. The latter restriction may be due to network technology. Notice that the above modification was not needed in central adaptive diagnosis (cf. [2]) because test scheduling was done by a central monitor which was assumed fault free. It was not needed in distributed nonadaptive diagnosis either, because then all tests and transmissions were scheduled in advance.

As usual, the aim of distributed diagnosis is that all fault-free processors learn the fault status of all other processors. It turns out that we can efficiently achieve this goal whenever the fraction of faulty processors is bounded by any constant $\gamma < 1$. However, it may happen that fault-free processors have the correct diagnosis while faulty processors have a reverse picture of the system: they diagnose faulty processors as fault free and vice-versa. For $\gamma \geq 1/2$ faulty processors outnumber fault-free processors and such diagnosis is useless to an external user who cannot decide which group of processors should be trusted. Hence, although our method gives efficient distributed diagnosis for any bounded fraction of faulty processors, this diagnosis can be safely used by an external user only if $\gamma < 1/2$. In this case the view of the majority can be used as correct diagnosis.

We now show how to modify the algorithm from section 4 to handle the diagnosis problem. The main difference is that now faults are Byzantine, thus processors cannot trust obtained information. In the sequel, the phrase "x finds a fault-free processor y" will mean that x considers y to be fault free (which does not necessarily reflect reality unless x itself is fault free). In Stage 1 the chief and the guard of each set X_j are selected in two steps, each consisting of $l - 1$ rounds. Let $x_1, ..., x_l$ be an increasing enumeration of X_j and $y_1, ..., y_l$ a decreasing enumeration. In the first phase x_i tests processors $x_{i+1}, x_{i+2}, ...,$ starting in round i, until a fault-free processor is found or all nodes $x_r, r > i$ are exhausted. The second step is a mirror image of the first, using enumeration y_i instead of x_i. A processor that did not find a fault-free processor in either step considers itself to be singleton (in a passive set). A processor that found a fault-free processor only in the first (second) step considers itself a chief (a guard).

Stage 2 is modified by replacing each round in which v sends a message to w by two rounds: in the first round the smaller of these processors tests the larger and in the second the larger tests the smaller. Stages 3-6 remain unchanged. The tree T is formed. In Stage 7 rounds used by key nodes to find all guards and singletons are replaced by pairs of rounds: first a key node v tests processor w and then w informs v if it is a guard or a singleton. If the test result is positive and the answer is yes, w becomes a new child of v in the augmented tree T'. This concludes the (modified) preprocessing.

Now diagnosis is completed as follows. First all guards and singletons test all processors in their sets X_j. Then this information is collected in the root of the tree T'. Next the root broadcasts the complete diagnosis information to all nodes of T'. Finally, each guard broadcasts this information in its set X_j. This completes distributed diagnosis. We have proved the following theorem.

Theorem 10. *If the fraction of faulty processors is at most γ, for some constant $\gamma < 1$, then distributed adaptive diagnosis for n processors can be done in $O(\log n)$ rounds, using $O(n)$ tests and $O(n)$ messages in the worst case.*

All the above orders of magnitude are clearly optimal, in contrast to central adaptive diagnosis which, as we mentioned before, can be done in a constant number of rounds, provided that less than half processors are faulty.

References

1. A. Bagchi and S.L. Hakimi, Information dissemination in distributed systems with faulty units, IEEE Trans. Comp. 43 (1994), 698-710.
2. R. Beigel, W. Hurwood and N. Kahale, Fault diagnosis in a flash, Proc. 36th Symp. on Found. of Comp. Sci. (1995), 571-580.
3. K.A. Berman and M. Hawrylycz, Telephone problems with failures, SIAM J. Alg. Disc. Meth. 7 (1986), 13-17.
4. D.P. Bertsekas and J.N. Tsitsiklis, Parallel and Distributed Computation: Numerical Methods, Prentice-Hall, Englewood Cliffs, NJ, (1989).
5. B.S. Chlebus, L.Gąsieniec and A. Pelc, Fast deterministic simulation of computations on faulty parallel machines, Proc. 3rd Ann. Eur. Symp. on Alg. ESA'95, (1995), 89-101.
6. A.T. Dahbura, System-level diagnosis: A perspective for the third decade, Concurrent Computation: Algorithms, Architectures, Technologies, Plenum Press, New York (1988).
7. A. Farley, Reliable minimum-time broadcast networks, in: Proc. 18th SE Conf. on Combinatorics, Graph Theory and Computing, Congressus Numerantium 59 (1987), 37-48.
8. P. Fraigniaud and E. Lazard, Methods and problems of communication in usual networks, Disc. Appl. Math. 53 (1994), 79-133.
9. L. Gargano and U. Vaccaro, Minimum time networks tolerating a logarithmic number of faults, SIAM J. Disc. Math. 5 (1992), 178-198.
10. L. Gąsieniec and A. Pelc, Adaptive broadcasting with faulty nodes, Parallel Computing 22 (1996), 903-912.
11. L. Gąsieniec and A. Pelc, Broadcasting with linearly bounded transmission faults, Discrete Applied Mathematics, to appear.
12. S.M. Hedetniemi, S.T. Hedetniemi and A.L. Liestman, A survey of gossiping and broadcasting in communication networks, Networks 18 (1988), 319-349.

13. S.L. Johnsson and C.T. Ho, Matrix multiplication on Boolean cubes using generic communication primitives, in: Parallel Processing and Medium-Scale Multiprocessors, A. Wouk (Ed.), SIAM, (1989), 108-156.
14. J.G. Kuhl and S.M. Reddy, Distributed fault-tolerance for large multiprocessor systems, Proc. IEEE Symp. Comp. Architecture (1980), 23-30.
15. A. Lubotzky, R. Philips and P. Sarnak, Explicit expanders and the Ramanujan conjectures, Proc. 18th Ann. ACM Symp. on Theory of Computing (1986), 247-263.
16. K. Nakajima, A new approach to system diagnosis, Proc. 19th Allerton Conf. Commun. Contr. and Computing (1981), 697-706.
17. A. Pelc, Fault-tolerant broadcasting and gossiping in communication networks, Networks 28 (1996), 143-156.
18. D. Peleg and A.A. Schäffer, Time bounds on fault-tolerant broadcasting, Networks 19 (1989), 803-822.
19. F. Preparata, G. Metze and R. Chien, On the connection assignment problem of diagnosable systems, IEEE Transactions on Electron. Computers 16 (1967), 848-854.
20. E. Upfal, Tolerating a linear number of faults in networks of bounded degree, Information and Computation 115 (1994), 312-320.

Weighted Graph Separators
and Their Applications *

Hristo N. Djidjev

Department of Computer Science, Rice University
P.O. Box 1892, Houston, TX 77005, USA
email: hristo@cs.rice.edu

Abstract. We prove separator theorems in which the size of the separa-tor is minimized with respect to non-negative vertex costs. We show that for any planar graph G there exists a vertex separator of total sum of vertex costs at most $c\sqrt{\sum_{v \in V(G)}(cost(v))^2}$ and that this bound is opti-mal to within a constant factor. Moreover such a separator can be found in linear time. This theorem implies a variety of other separation results. We describe applications of our separator theorems to graph embedding problems, to graph pebbling, and to multi–commodity flow problems.

1 Introduction

Background. A separator is a small set of vertices or edges whose removal divides a graph into two roughly equal parts. The existence of small separators for some important classes of graphs such as planar graphs can be used in the design of efficient divide-and-conquer algorithms for problems on such graphs. Formally, a separator theorem for a given class of graphs S states that any n-vertex graph from S can be divided into components of no more than αn vertices, for some constant $\alpha < 1$, by removing a set called an α *separator* of no more than $f(n)$ vertices, where $f(n) = o(n)$. A classical result of Lipton and Tarjan [20] states that a separator theorem with $f(n) = O(\sqrt{n})$ holds for the class of planar graphs and that the separator can be found in linear time. Lipton and Tarjan also described various applications of their separator theorem [21].

Numerous new results on graph separation and many new application have been found since. Separator theorems exist for the class of graphs of bounded genus [6, 10], for the class of chordal graphs [11], for the three-dimensional graphs [24], for the class of graphs that exclude a given minor [1], and for some geo-metrically defined graphs that include many three dimensional finite element graphs [23, 28]. Separators with special structure have been studied, including simple cycle separators [22, 9] and separators consisting of edges [5]. Leighton and Rao [17] have studies approximation algorithms for constructing separators of minimum size.

---

* This work is partially supported by National Science Foundation grant CCR–9409191 and by EPA grant R82-5207-01-0.

Separator theorems have been used for solving large sparse systems of linear equations [19, 12], for developing algorithms for VLSI layout design [4, 18], for shortest path problems [8], in parallel computing [13], and in computational complexity [21].

Our results. In many applications it is useful to associate costs with the vertices or the edges of the graph. For example, the costs might correspond to different sizes of circuit elements to be laid out or different amounts of data that have to be exchanged between communicating processes in a distributed system. For such applications, the size of the separator has to be measured by some function of the costs associated with its vertices (instead by its cardinality).

In this paper we consider graphs with costs and weights on the vertices or edges: the costs are used to evaluate the size of the separator and the weights are used to evaluate the sizes of the components into which the graph is divided. We give tight bounds on the minimum cost of a separator of a planar weighted graph and study related algorithmic problems. Our main theorem states that for any planar graph G with a non-negative real cost $cost(v)$ on each vertex v and non-negative real vertex weights adding to one there exists a vertex set of total cost no more than $\sqrt{8}\sqrt{\sum_{v \in V(G)}(cost(v))^2}$ whose removal leaves no component of weight exceeding $2/3$. For the special case of graphs with equal vertex costs our theorem yields the same bound as the Lipton and Tarjan's planar separator theorem. We show that our bound is tight to within a constant factor and give a linear algorithm for constructing such a separator.

Previously, weights on vertices and edges have been studied mostly for measuring the sizes of the components into which the graph is divided. Separator theorems of such types can usually be proved by generalizing the corresponding non-weighted versions in a straightforward way. However, these generalizations do not readily extend to the case where the size of the separator has also to be measured as a function of the weights. The problem is that by the previous technique a portion of the separator is constructed as a path in the breadth-first tree of the graph. In the unweighted case, all paths of a given length contribute the same number of vertices to the separator. Thus, the objective of most previous algorithms is to find a suitable pair of levels l_1 and l_2 containing small $(O(\sqrt{n}))$ number of vertices such that $l_2 - l_1 = O(\sqrt{n})$. (The second condition implies that the maximum length of simple path in the tree resulting after deleting vertices on levels l_2 and above and shrinking vertices on levels l_1 and below to a single vertex will not exceed $2(l_2 - l_1 - 1) = O(\sqrt{n})$.)

In order to prove a separator theorem for graphs with vertex costs, we define a spanning tree we call a *generalized breadth-first spanning tree* (GBFST) whose levels are real numbers and which generalizes a breadth-first spanning tree in the case of graphs with vertex costs. The property of a GBFST that is essential for separating a planar graph is that, intuitively, if for any real r one cuts along level r all edges either in the original graph, or in its GBFST, the resulting components will have identical vertex sets. Moreover, the vertices of any simple path in the tree joining a vertex on level r_2 with its ancestor vertex on level r_1 will have total cost $r_2 - r_1$.

Our constructive proofs can be transformed into an $O(n \log n)$ algorithm for finding a separator of an n-vertex planar graph in a straightforward manner. We show that it is actually possible to find a separator in $O(n)$ time, by avoiding the additional $O(\log n)$ factor that comes from sorting. Our technique illustrates that in certain cases sorting can be replaced by a binary search of a set of n reals that uses the linear algorithm for finding the median and exploits the structure of the graph.

We prove some other separator theorems that follow from our main result, some of which are needed in the application section. These include a separation into parts of weight not exceeding $1/2$, finding a separator that is a set of edges, and separation of graphs of bounded genus.

Applications. We discuss three applications of our separator theorems: for the problems of embedding planar graphs into various communication structures, for pebbling a planar directed acyclic graph, and for constructing a minimum routing tree.

The graph embedding problem arises in the design of VLSI chips and in the simulation of different interconnection structures. We consider graphs with costs on the edges that model the volumes of information that has to be transfered between different processes. The goal is to minimize the total amount of traffic (viz., the average weighted dilation), if the original graph is implemented on the target structure. We show that any n-vertex planar graph of maximum degree d and non-negative edge costs whose squares sum up to n can be embedded into a binary tree, a hypercube, a butterfly graph, a shuffle-exchange graph, a cube-connected-cycles graph, or a de Brujn graph with $O(\log d)$ average weighted dilation.

Graph pebbling is used to study memory requirements of straight-line programs, for register allocation, and for deriving space-time tradeoffs. We show that any n-vertex planar acyclic directed graph G with non-negative vertex costs can be pebbled so that, at any time, the sum of the costs of all vertices with pebbles be $O(\sqrt{\sum_{v \in V(G)} (cost(v))^2} + I \log n)$. (Here I denotes the maximum sum of costs on the predecessors of any vertex of G.)

Finally, we consider applications of our results to the tree congestion problem. This a version of the multi–commodity flow problem, where one is given a set of demands represented by an edge-weighted graph G and the goal is to construct a network of small cost that can accommodate all demands. In the case studied in the paper, the network is restricted to be a tree and the cost is measured by the maximum edge capacity (called a *congestion*) of the tree. For a general graph G, the problem of finding a tree with a minimum congestion (called a *minimum routing tree*) is NP–complete. We study the case for planar graphs G and provide an estimation on the congestion of minimum routing trees for planar graphs and an $O(n \log n)$ algorithm that constructs such trees.

2 A vertex separator theorem

In this section we prove our main result. As with other separator theorems for graphs embedded on surfaces, the proof relies on the fact that in any planar graph with a spanning tree of small radius there exists a small cycle that divides the graph into two roughly equal parts. In order to use this result for graphs with vertex costs we will define a special type of spanning tree and show how it can be used for slicing the graph into appropriately "thin" pieces. If not stated otherwise, we assume throughout this section that G denotes a connected planar graph with non-negative vertex costs and non-negative vertex weights adding to one.

Constructing a generalized breadth-first spanning tree

Definition 2.1 *Let T be a spanning tree of G with root rt and let for any vertex v of G level(v) denote the sum of the vertex costs on the simple tree path from v to the root of T. T is called a* generalized breadth-first spanning tree (GBFST) *if for any vertex $w \neq rt$ the parent of w is a vertex adjacent to w with a minimum level.*

By definition, T is a GBFST of G iff for any non-tree edge (v, w) of G such that $level(v) \leq level(w)$ and $w \neq rt$

$$level(v) \geq level(parent(w)). \tag{1}$$

The next lemma establishes a close relationship between the GBFST's and the breadth-first spanning trees (BFST's).

Lemma 1. *If all costs of G are unit, then a spanning tree T of G is a GBFST if and only if T is a BFST of G.*

A GBFST can be constructed in linear time by the following algorithm.

Algorithm GBFST
{Constructs a generalized breadth-first spanning tree T of G}

1. Replace any edge (v, w) of G by a pair of directed edges, one from v to w of *length cost(w)* and one from w to v of length $cost(v)$.
2. Choose an arbitrary vertex rt of G to be the root of T. Run the linear time single source shortest path algorithm from [15] on the resulting planar graph, G', with source rt.
3. Define T to be the resulting shortest path tree of G'.

Lemma 2. *Algorithm GBFST constructs a generalized breadth-first spanning tree of G in $O(|G|)$ time, where $|G| = |V(G)| + |E(G)|$.*

Proof: Let (v, w) be a non-tree edge of G with $level(v) \leq level(w)$, where the levels are computed with respect to T. By the definitions of a level and a shortest path tree, the shortest path between rt and w in G' has length $level(w) = level(parent(w)) + cost(w)$. Assume that (1) does not hold for (v, w). Then the path between rt and w consisting of the shortest path between rt and v and the edge (v, w) has length $level(v) + cost(w) < level(parent(w)) + cost(w) = level(w)$. Hence that path is shorter than the tree path between rt and w, contradicting the definition of T in Step 3 of the algorithm. $\square$

Slicing the graph

Given a spanning tree T of G, each non-tree edge induces a (unique) simple cycle. We call such cycle a T-*cycle*. Using Jordan's Curve Theorem, Lipton and Tarjan showed in [20] that if G is triangulated, then at least one T-cycle is a weighted 2/3-separator of G.

For graphs with vertex costs we have the following result.

Lemma 3. *Let G be a triangulated planar graph with non-negative real vertex costs and a GBFST T of maximum level l_{max}. A vertex 2/3 separator of G that is a T-cycle with cost not exceeding $2l_{max} + cost(rt)$ can be found in $O(|G|)$ time, where rt is the root of T.*

Define the *width* of any subgraph of G to be the maximum difference of the levels of any two vertices of G. Our next goal is, by removing a set of vertices of small total cost, to cut from G components of small weights so that the remaining part has appropriately small width (so that Lemma 3 can be efficiently applied on any such component).

For any tree edge $e = (v, w)$ of G assign a non-negative edge cost $cost(e) = |level(v) - level(w)|$ and assign cost 0 to each non-tree edge. Denote by l_{max} the maximum level of G. For any real x define a set

$$E_x = \{(v, w) \in E(G) \mid level(v) \leq x < level(w)\}.$$

Consider the function f defined for $x \geq 0$ by

$$f(x) = \sum_{e \in E_x} cost(e) \tag{2}$$

($f(x) = 0$ if $x > l_{max}$).

Our algorithm for finding a pair of levels whose removal cuts the graph into components with a small width or a small weight is based on the following lemma (see Figure 1).

Lemma 4. *Let $h(l)$ be an integrable non-negative function defined on the interval $[0, L]$ such that $h(0) = 0$. Then*

(i) There exists a number $\hat{l}$ in $[0, L]$ such that

$$h(\hat{l}) + 2(L - \hat{l}) \leq 2\sqrt{\int_0^L h(l)\, dl};$$

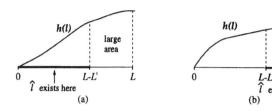

Fig. 1. Illustration of Lemma 4.

(ii) Let $L' \in [0, L]$. If $L'^2 \leq \int_{L-L'}^{L} h(l)\, dl$, then such a number $\hat{l}$ exists in the interval $[0, L-L']$, otherwise such a number $\hat{l}$ exists in the interval $(L-L', L]$.

Proof: Omitted. □

Lemma 5. *If $cost(rt) = 0$ then one can find in $O(|G|)$ time a pair of levels l' and l'' such that*

(i) Both the weight of all vertices of G on levels below l' and the weight of all vertices of G on levels above l'' do not exceed $1/2$;
(ii) $f(l') + f(l'') + 2(l'' - l') \leq \sqrt{8N(G)}$.

Proof: The levels can be found by binary search based on Lemma 4. Details to be given in the final version. □

Next we prove the main theorem of this section.

Theorem 6. *Let G be a planar graph with non-negative vertex weights that sum up to one and non-negative vertex costs whose squares sum up to $N(G)$. There exists a 2/3 vertex separator of G with total vertex cost no more than $\sqrt{8N(G)}$. Such a separator can be found in $O(|G|)$ time.*

Proof: First we will partition the graph into components by removing a set of edges of small cost so that Lemma 3 can be applied to a component of small width. Then we will replace that edge set by a vertex set of low cost.

Embed G in the plane and add new edges (if necessary) in order to triangulate the embedding. We will need to identify a vertex of cost 0 to be the root of a spanning tree of G. If such a vertex rt does not exist in G, we define a new vertex rt of cost 0 in any face (triangle) f of the embedding of G and join it by 3 new edges to the vertices on f. We will use G to denote the resulting graph.

Given a subgraph H of G, let the weight (resp. the cost) of H be the sum of the weights (resp. costs) of the vertices in H. Similarly, we define weight and cost of vertex and edge sets.

Let l' and l'' be a pair of levels satisfying Lemma 5. By definition, the set of edges $E_{sl} = E_{l'} \cup E_{l''}$ has cost $f(l') + f(l'')$ (see (2)). Let K be a component of a maximum weight into which E_{sl} divides G.

Assume that the weight of K is greater than 2/3. Then, by Lemma 5(i), the level of any vertex in K will be in the interval (l', l''). Thus, the difference between the levels of any two vertices of K is less than $l'' - l'$.

Next we will replace E_{sl} by a vertex set. Define

$$M_{sl} = \{v \mid (v, parent(v)) \in E_{sl}\}.$$

By our definition of edge costs, M_{sl} and E_{sl} have equal costs. By the definition of M_{sl}, any *tree* edge from E_{sl} is incident to a vertex from M_{sl}.

Thus, in order to show that M_{sl} can be used instead of E_{sl} for partitioning G, we have also to show that any *non-tree* edge from E_{sl} is incident to a vertex from M_{sl}.

Let (v, w) be a non-tree edge of E_{sl}. Assume w.l.o.g. that $level(v) \leq level(w)$. Then $level(v) \leq l \leq level(w)$ for some $l \in \{l', l''\}$. By (1), $level(v) \geq level(parent(w))$. Then $l \geq level(v) \geq level(parent(w))$ and $level(w) \geq l \geq level(parent(w))$. Hence the tree edge $(w, parent(w)) \in E_{sl}$ and $w \in M_{sl}$.

To summarize, we found a vertex set M_{sl} such that for a pair of levels l' and l'' satisfying Lemma 5

(i) the cost of M_{sl} is $f(l') + f(l'')$;

(ii) if the heaviest component K induced by $V(G) \backslash M_{sl}$ has weight greater than $2/3$, then the width of K is at most $l'' - l'$.

Now apply Lemma 3 to G. Let c be a simple cycle with exactly one non-tree edge and that is a weighted $2/3$-separator of G. Denote by M_c the set of vertices on c that belong to K. Then the cost of M_c is at most $2(l'' - l')$ (we used here the fact that $cost(rt) = 0$).

If the weight of K is less than $2/3$ we define $M_c = \emptyset$.

The set $M = M_{sl} \cup M_c$ is clearly a $2/3$-separator of G. The cost of M is no more than $f(l') + f(l'') + 2(l'' - l')$, which by Lemma 5 does not exceed $\sqrt{8N(G)}$.

By Lemmas 2, 5, and 3, constructing a GBFST, finding levels l' and l'', or constructing a cycle c requires $O(|G|)$ time. All other steps can be implemented in linear time in a straightforward manner. $\qquad\square$

3 Tightness of the theorem

In this section we will show that Theorem 6 is optimal within a constant factor.

Theorem 7. *There exists a constant $c > 0$ such that for any integer $n > 0$ and any n non-negative reals $cost_1, \cdots, cost_n$ there exists a planar n-vertex weighted graph G with costs the given numbers such that any $2/3$ vertex separator of G has total vertex cost at least $c\sqrt{\sum_{i=1}^{n} cost_i^2}$.*

Proof (sketch): The proof is based on the following idea. Let $Q = \sum_{i=1}^{n} cost_i^2$. Define a square R_0 with area Q and partition R_0 into n rectangles with areas $cost_i^2, i = 1, \cdots, n$, and bounded aspect ratio (except some very small rectangles which can have arbitrary aspect ratios). Then define a planar graph G with vertices corresponding to the rectangles and edges corresponding to shared boundaries. If a vertex v corresponds to a rectangle of area A, define $cost(v) = \sqrt{A}$ and $weight(v) = A$. Then the existence of a separator of G of cost $o(\sqrt{Q})$ will imply the existence of a line of length $o(\sqrt{Q})$ which divides R_0 into two roughly equal parts, which is impossible. $\qquad\square$

4   Related separator theorems

The separator theorem from the previous section implies several other separator theorems. The proofs of these theorems are similar to the proofs of the non-weighted version, so we only state the results and omit the proofs. The interested reader is referred to the full version for complete proofs.

Theorem 8. (1/2-separator theorem) *Let G be a planar graph with non-negative vertex weights that sum up to one and non-negative vertex costs whose squares sum up to $N(G)$. There exists a $1/2$ vertex separator of G of cost $O(\sqrt{N(G)})$, which can be found in $O(|G|)$ time.*

Theorem 9. (Edge separator theorem) *Let G be a planar graph of maximum vertex degree d, with non-negative vertex weights that do not exceed $2/3$ and sum up to one, and non-negative edge costs whose squares sum up to $N(G)$. There exists a $2/3$ edge separator of G of total cost not exceeding $4\sqrt{dN(G)}$, which can be found in $O(|G|)$ time.*

Theorem 10. (Bounded-genus separator theorem) *Let G be an n-vertex graph embedded on a surface of genus g with non-negative vertex weights that sum up to one and non-negative vertex costs whose squares sum up to $N(G)$. There exists a $2/3$ vertex separator of G with total vertex cost no more than $\sqrt{(16g + 8)N(G)}$. Such a separator can be found in $O(|G|)$ time, given the embedding of G.*

5   Applications

5.1   Graph embeddings

Many problems in VLSI design and in network simulation can be formulated as graph embedding problems. We show here that, in the case of planar graphs or graphs of bounded genus, one can use our graph separation results to find embeddings with good properties.

Formally, an embedding of a graph G_1 into a graph G_2 is an injective mapping $\mu : V(G_1) \to V(G_2)$. One measure for characterizing an embedding μ is the *average dilation* of μ defined as

$$AD(\mu) = (\sum_{(u,v) \in E(G_1)} d_{G_2}(\mu(u), \mu(v))\,) \,/\, |E(G_1)|,$$

where $d_{G_2}(x, y)$ denotes the distance between vertices x and y in G_2. In the applications, G_1 can model communicating processes of a computer program, a circuit, or a communication network, where vertices denote communicating objects and edges denote existence of communication links. G_2 represents a target structure (e.g., a mesh of processors, or a VLSI chip) and usually has a regular structure. Then $AD(\mu)$ will correspond to the average increase of communication time if G_1 is implemented or simulated by G_2.

In many applications the amounts of information that need to be exchanged between the individual communicating objects are different. For such problems non-negative costs $cost(u, v)$ are assigned to the edges of G_1 and the goal is to minimize the total increase of traffic if G_1 is implemented on G_2. In such a case we use as a measure the *average weighted dilation (AWD)* of μ defined as

$$AWD(\mu) = \sum_{(u,v) \in E(G_1)} d_{G_2}(\mu(u), \mu(v)) \cdot cost(u, v) \, / \, |E(G_1)|.$$

Another parameter that is used to evaluate an embedding μ is the *expansion* of μ defined as $|V(G_2)|/|V(G_1)|$. Good embeddings are required to have $O(1)$ expansion. Using the approach of [5], we will show that weighted separators can be used to construct embeddings of planar graphs with small AWD and expansion one.

Theorem 11. *Every n-vertex connected planar graph G of maximum degree d and with non-negative edge costs whose squares sum up to n can be embedded into a binary tree with average weighted dilation $O(\log d)$ and expansion one.*

Proof (sketch): For any subgraph H of G let $N(H) = \sum_{(u,v) \in E(H)} (cost(u, v))^2$, where $cost(u, v)$ denotes the cost on edge (u, v). Use the following recursive procedure to embed G into a binary tree T.

If G has a single vertex, then define T to be a tree of a single vertex. Otherwise, assign a weight $1/n$ to each vertex of G. Construct in $O(n)$ time a partition A, B of $V(G)$ and an edge separator C of total cost $O(\sqrt{dN(G)}) = O(\sqrt{dn})$ as in Theorem 9. By construction, C will consist of *all* edges incident to the vertices of a certain vertex separator M. Choose any $x \in M$. Embed recursively the subgraph induced by $A \setminus \{x\}$ into a binary tree T_1 and the subgraph induced by $B \setminus \{x\}$ into another binary tree T_2. Make the roots of T_1 and T_2 children of x and denote by T the resulting tree.

Denote $s(G) = \sum_{(u,v) \in E(G)} d_T(\mu(u), \mu(v)) \cdot cost(u, v)$, where μ is the above defined embedding of G into T. Since the endpoints of each edge of C are embedded into vertices at distance $O(\log n)$ (at most twice the height of T), then $s(G) \leq s(G_1) + s(G_2) + O(\sqrt{dN(G)} \log n)$, if $n > 1$.

Solving the recurrence gives $S(G) = O(\sqrt{nN(G)} \log d) = O(n \log d)$. Since G is connected, then it has at least $n - 1$ edges and $AWD(\mu) \leq S(G)/(n-1) = O(n \log d/(n-1)) = O(\log d)$. Thus G can be embedded into a binary tree with AWD $O(\log d)$ and expansion one. $\quad\square$

If the squares of the edge costs of G do not add to n, then Theorem 11 can still be applied on G after scaling of the costs. Note that in case of unit costs Theorem 11 yields a bound on the average dilation that is independent of the size of the graph and, by [5], is optimal within a constant factor.

Corollary 12. *Every n-vertex planar graph G of maximum degree d and with non-negative edge costs whose squares sum up to n can be embedded into a hypercube, a butterfly graph, a shuffle-exchange graph, de Brujn graph, or a cube-connected-cycles graph with average weighted dilation $O(\log d)$ and expansion one.*

Proof: According to [3, 2, 25], any binary tree can be embedded into any of the above graphs with a maximum dilation $O(1)$ and expansion one. $\qquad\square$

Similar results hold for embedding a planar graph into a linear, a 2-dimensional, or a d-dimensional $(d \geq 3)$ array. The proofs of these results are similar to the proof of Theorem 11.

5.2 Pebbling

Pebbling a directed graph is a well studied problem in programming with applications to register allocation and to deriving space-time tradeoffs (see [26, 29] for a survey of results). Pebbling is a one-person game that involves placing pebbles on the vertices of a directed acyclic graph G. During any step, one can remove a pebble from a currently pebbled vertex or place a pebble onto a vertex that currently has no pebble. An additional requirement is that a vertex can be pebbled only if all its predecessors (i.e., all vertices w such that $(w, v) \in E(G)$) are pebbled. The goal is that any vertex be pebbled at least once while minimizing the number of pebbles used at any particular moment. In applications, pebbles correspond to memory units and the objective is to minimize the total memory needed in order to carry out the computation represented by G.

In the weighted version of the pebbling problem all vertices of G have non-negative costs associated with them. The goal is to minimize the maximum sum of the costs on all vertices of G with pebbles placed on them at any given time. We denote this value by $p(G)$.

In this section we generalize the result of Lipton and Tarjan [21], which give a linear algorithm for efficiently pebbling unweighted planar graphs. First, we will need to modify (strengthen) Theorem 6. The next statement shows that one can divide a weighted planar graph by removing a set of vertices of small cost so that both the number of vertices and the weight of each part be small.

Theorem 13. *Let G be a planar n-vertex graph with non-negative weights $weight(v)$ on its vertices that sum up to one and non-negative vertex costs whose squares sum up to $N(G)$. A set C of vertices of total cost $O(\sqrt{N(G)})$ can be found in $O(n)$ time that partitions $V(G)$ into sets A and B such that no edge joins A and B, $weight(A)$, $weight(B) \leq 5/6$, and $|A|, |B| \leq (5/6)n$.*

Proof: Follows from Theorem 6 by divide-and-conquer. $\qquad\square$

We state also the following corollary of Theorem 13 which we will use later and whose proof is similar to the proof of Theorem 9.

Corollary 14. *Let G be a planar n-vertex graph of maximum vertex degree d, with non-negative weights $weight(v)$ on its vertices that do not exceed $2/3$ and that sum up to one, and non-negative edge costs whose squares sum up to $N(G)$. There exists a $2/3$ edge separator of G of total cost not exceeding $O(\sqrt{dN(G)})$ that partitions $V(G)$ into sets A and B such that $weight(A)$, $weight(B) \leq 5/6$, and $|A|, |B| \leq (5/6)n$. The separator can be found in $O(n)$ time.*

Next we describe a pebbling algorithm based on the method from [21]. Let G be an n-vertex planar graph and Q be the sum of the squares of the vertex costs of G. Recall that our goal is to minimize $p(G)$, the maximum sum of the costs on all vertices of G with pebbles placed on them at the same time. Let $in\_cost(G)$ denote the maximum sum of costs on all predecessors of any vertex of G. Assign a weight on each vertex of G equal to the square of the cost on that vertex. Using Theorem 13, find a partition A, B, C of the vertices of G such that the cost of C is $O(\sqrt{Q})$, A and B are edge disjoint, the cost of neither A nor B exceeds αQ, and $|A|, |B| \leq \alpha n$, for $\alpha = 5/6$.

Pebble the vertices of G in topological order as follows. To pebble the next vertex v remove all pebbles except those on vertices of C. Pebble all predecessors w of v recursively. For this end, for any predecessor w of v, apply recursively the same procedure on the graph G_w induced by w and all its unpebbled predecessors. Since the vertices are pebbled in topological order and since pebbles on vertices of C are never removed, the cost of G_w does not exceed αQ. Thus we have the following recurrence for the maximum value $p(Q, n)$ for $p(G)$ over all n-vertex graphs G whose cost squares sum up to Q and for which $in\_cost(G)$ does not exceed some upper bound I

$$p(Q, n) \leq c\sqrt{Q} + I + p(\alpha Q, \lfloor \alpha n \rfloor), \quad \text{if } n > 1,$$

where $c > 0$ is a constant. The solution to the recurrence is $p(Q, n) = O(\sqrt{Q} + I \log n)$.

Thus we proved the following theorem.

Theorem 15. *Let G be an n-vertex planar acyclic directed graph with non-negative vertex costs whose squares sum up to $N(G)$. Let I be the maximum sum of the costs on the predecessors of any vertex of G. The vertices of G can be pebbled so that, at any time, the sum of the costs of all vertices with pebbles be $O(\sqrt{N(G)} + I \log n)$.*

5.3   Designing multi–commodity flow trees

The multicommodity flow problem is related to optimally routing several different commodities in a network in order to satisfy a given set of demands. The total amount of flow through any of the edges should not exceed the capacity of that edge. Approximation algorithms exist that find whether there exists a *feasible flow* (flow that satisfies the demands and obeys all capacity constraints) [16]. We consider a related version of the problem where, given the set of demands, one has to construct a network for which a feasible flow exists and whose maximum edge capacity (called a *congestion of the network*) is small. The case where the network is required to be a tree is called the *tree congestion problem* [14].

Formally, let $G = (V, E)$ be a digraph with non-negative real weights (*demands*) $dem(v, w)$ on its edges. The goal is to find a tree T with vertex set V, degree 3 of internal vertices and capacities on the edges such that: (i) a feasible flow satisfying the demands determined by the edges of G exists, and (ii) the congestion is minimized over the congestions of *all* trees that meet the requirements. For example, the demands may correspond to the expected volume of phone calls between corresponding pairs of locations and the goal will be to minimize the maximum bandwidth required [27]. Such a tree T is called a *minimum routing tree*. If T is fixed (but not its capacities), one can determine the smallest possible capacity $load(e)$ on any edge e of T as follows. Delete e from T and let T_1 and T_2 be the resulting trees. Let E_1 and E_2 be the sets of edges of G with both endpoints in $V(T_1)$ and $V(T_2)$, respectively. Then

$$load(e) = \sum_{(v,w)\in E(G)\setminus(E_1\cup E_2)} dem(v, w).$$

For a general weighted demands digraph G, the tree congestion problem is NP–complete [27]. A polynomial approximation algorithm with a logarithmic approximation factor is described in [14]. Seymour and Thomas [27] show that exact solution can be found in polynomial $(O(|G|^4))$ time if G is planar and all demands are integers.

Using our separator theorems, we will provide the first known non-trivial upper bound on the worst-case congestions of minimum routing trees for the class of planar digraphs and will also give an $O(n \log n)$ algorithm that, given any n-vertex planar digraph, constructs a routing tree with congestion not exceeding the claimed bound.

Theorem 16. *For any n-vertex planar digraph G of maximum degree d and with non-negative edge demands whose squares sum up to n there exists a balanced binary routing tree of G with congestion $O(\sqrt{dn})$.*

Proof: Let G be an n-vertex planar graph of maximum degree d with non-negative edge demands whose squares sum up to n. (We ignore the orientation of the edges of the original graph.) Denote for any subgraph H of G $N(H) = \sum_{(u,v)\in E(H)}(dem(u, v))^2$, where $dem(u, v)$ denotes the demand on edge (u, v). Apply to G the following algorithm.

The routing algorithm

If G consists of a single vertex, then define T to be the tree of that vertex. Otherwise, assign a weight to each vertex v of G equal to the sum of the squares of the demands on all edges incident to v divided by $2N(G)$ (so that the sum of all weights is one). Construct in $O(n)$ time a partition A, B of $V(G)$ and an edge set C of total demand $O(\sqrt{dN(G)}) = O(\sqrt{dn})$ (as in Corollary 14) such that any edge joining A and B belongs to C, the weight of neither A nor B exceeds $\alpha N(G)$, and $|A|, |B| \leq \alpha n$, for $\alpha = 5/6$. Assume w.l.o.g. that $|A| \geq |B|$. Choose a vertex x in A with a minimum weight. Embed recursively the subgraph induced by $A \setminus \{x\}$ into a binary tree T_1 and the subgraph induced by B into another binary tree T_2. Make the roots of T_1 and T_2 children of x and denote by T the resulting tree. Let edges e_1 and e_2 connect x to the roots of T_1 and T_2, respectively.

Analysis

For any edge e in T_1, the load on e with respect to G is not more than the load on e in G_1 plus the sum σ of the demands on all edges of G that join a vertex in T_1 to a vertex in T_2. Similar argument is valid for the load on any edge in T_2. From the choice of x, the load on e_1 or e_2 does not exceed $N(G)/|T_1| + \sigma$. Recall that T_1 has a vertex set $A \setminus \{x\}$. Since $|T_1| \geq n/2 - 1$ and $N(G) = n$, then $N(G)/|T_1| = O(1)$. Furthermore, $\sigma = \sum_{(u,v) \in C} (dem(u,v))^2 = O(\sqrt{dn})$.

Thus the maximum congestion $cong(n)$ of the routing tree constructed by this algorithm for any n vertex planar graph satisfies the recurrence

$$cong(n) \leq \max\{cong(n_1), cong(n_2) \mid n_1 + n_2 = n, \ n_1, n_2 \leq \alpha n\} + O(\sqrt{dn})$$

$$= cong(\alpha n) + O(\sqrt{dn}),$$

whose solution is $cong(n) = O(\sqrt{dn})$. □

We formulated the results from this section (Theorems 11, 15, and 16) for the class of planar graphs. The generalizations for the class of graphs of bounded genus are straightforward.

References

1. Noga Alon, Paul Seymour, and Robin Thomas. A separator theorem for graphs with an excluded minor and its applications. *Proceedings of the 22nd Symp. on Theory of Computing*, pages 293–299, 1990.

2. S. Bhatt, F. Chung, T. Leighton, and A. Rosenberg. Optimal simulations of tree machines. In *Proc. 27th IEEE Symposium on Foundations of Computer Science (FOCS)*, pages 274–282, 1986.

3. S. N. Bhatt, F. R. K. Chung, J. W. Hong, F. T. Leighton, and A. L. Rosenberg. Optimal simulations by butterfly networks. In *Proc. 20th ACM Symposium on Theory of Computing (STOC)*, pages 192–204, 1988.

4. S.N. Bhatt and F.T. Leighton. A framework for solving VLSI graph layout problems. *Journal of Computer and System Sciences*, 28:300–343, 1984.

5. Krzystof Diks, Hristo N. Djidjev, Ondrej Sykora, and Imrich Vrto. Edge separators of planar and outerplanar graphs with applications. *J. Algorithms*, 14:258–279, 1993.

6. Hristo N. Djidjev. A separator theorem. *Compt. rend. Acad. bulg. Sci.*, 34:643–645, 1981.

7. Hristo N. Djidjev and Shankar Venkatesan. Reduced constants for simple cycle graph separation. *Acta Informatica*, 1995. in print.

8. G.N. Frederickson. Fast algorithms for shortest paths in planar graphs, with applications. *SIAM Journal on Computing*, 16:1004–1022, 1987.

9. Hillel Gazit and Gary L. Miller. Planar separators and the Euclidean norm. In *SIGAL 90, Lecture Notes in Computer Science, vol. 450*, pages 338–347. Springer-Verlag, Berlin, Heidelberg, New York, Tokio, 1990.

10. John R. Gilbert, Joan P. Hutchinson, and Robert E. Tarjan. A separator theorem for graphs of bounded genus. *J. Algorithms*, 5:391–407, 1984.

11. John R. Gilbert, Donald J. Rose, and Anders Edenbrandt. A separator theorem for chordal graphs. *SIAM Journal on Algebraic and Discrete Methods*, pages 306–313, 1984.

12. John R. Gilbert and Robert E. Tarjan. The analysis of a nested dissection algorithm. *Numerische Mathematik*, 50:377–404, 1987.
13. Michael T. Goodrich. Planar separators and parallel polygon triangulation. *Proceedings of 24th Symp. on Theory of Computing*, pages 507–516, 1992.
14. Samir Khuller, Balaji Raghavachari, and Neal Young. Designing multi-commodity flow trees. *Information Processing Letters*, 50:49–55, 1994.
15. P. Klein, S. Rao, M. Rauch, and S. Subramanian. Faster shortest-path algorithms for planar graphs. In *26th ACM Symp. Theory of Computing*, pages 27–37, 1994.
16. Leighton, Makedon, Plotkin, Stein, Tardos, and Tragoudas. Fast approximation algorithms for multicommodity flow problems. *Journal of Computer and System Sciences*, 50:228–243, 1995.
17. F. Thomas Leighton and Satish Rao. An approximate max-flow min-cut theorem for uniform multicommodity flow problems with applications to approximation algorithms. In *Proceedings of the 29th IEEE Symposium on the Foundations of Computer Science*, pages 422–431, 1988.
18. C.E. Leiserson. Area efficient VLSI computation. In *Foundations of Computing*. MIT Press, Cambridge, MA, 1983.
19. Richard J. Lipton, D. J. Rose, and Robert E. Tarjan. Generalized nested dissection. *SIAM J. Numer. Anal.*, 16:346–358, 1979.
20. Richard J. Lipton and Robert E. Tarjan. A separator theorem for planar graphs. *SIAM J. Appl. Math*, 36:177–189, 1979.
21. Richard J. Lipton and Robert E. Tarjan. Applications of a planar separator theorem. *SIAM Journal on Computing*, 9:615–627, 1980.
22. Gary L. Miller. Finding small simple cycle separators for 2-connected planar graphs. *Journal of Computer and System Sciences*, pages 265–279, 1986.
23. Gary L. Miller, Shang-Hua Teng, and Stephen A. Vavasis. A unified geometric approach to graph separators. *Proceedings of the 32nd FOCS*, pages 538–547, 1991.
24. Gary L. Miller and William Thurston. Separators in two and three dimensions. *Proceedings of the 22nd Symp. on Theory of Computing*, pages 300–309, 1990.
25. B. Monien and H. Sudborough. Comparing interconnection networks. In *Symposium on Mathematical Foundations of Computer Science*, volume 320, pages 138–153, 1988.
26. N. Pippenger. Advances in pebbling. In *Annual International Colloquium on Automata, Languages and Programming*, pages 407–417, 1982.
27. P. D. Seymour and R. Thomas. Call routing and the ratcatcher. *Combinatorica*, 14:217–241, 1994.
28. Shang-Hua Teng. Points, spheres, and separators: A unified geometric approach to graph partitioning. Ph.D.-Thesis CMU-CS-91-184, Carnegie Mellon University, Pittsburgh, 1991.
29. H. Venkateswaran and M. Tompa. A new pebble game that characterizes parallel complexity classes. *SIAM Journal on Computing*, 18:533–549, 1989.

A New Exact Algorithm for General Orthogonal D-Dimensional Knapsack Problems

Sándor P. Fekete, Jörg Schepers*

Center for Parallel Computing, Universität zu Köln
D–50923 Köln, GERMANY
[fekete,schepers]@zpr.uni-koeln.de

Abstract. The *d-dimensional orthogonal knapsack problem* (OKP) has a wide range of practical applications, including packing, cutting, and scheduling. We present a new approach to this problem, using a graph-theoretical characterization of feasible packings. This characterization allows us to deal with classes of packings that share a certain combinatorical structure, instead of single ones. Combining the use of this structure with other heuristics, we develop a two-level tree search algorithm for finding exact solutions for the *d*-dimensional OKP. Computational results are reported, including optimal solutions for all two–dimensional test problems from recent literature.

1   Introduction

The problem of cutting a rectangle into smaller rectangular pieces of given sizes is known as the *two–dimensional cutting stock problem*. It arises in many industries, where steel, glass, wood, or textile materials are cut, but it also occurs in less obvious contexts, such as machine scheduling or optimizing the layout of advertisements in newspapers. The three-dimensional problem is important for practical applications as container loading or scheduling with partitionable resources. It can be thought of as packing boxes into a container. We refer to the generalized problem in $d \geq 2$ dimensions as the *d-dimensional orthogonal knapsack problem (OKP-d)*. Being a generalizition of the bin packing problem, the OKP-d is $\mathcal{NP}$-complete in the strict sense. (Note that we consider the constrained problem, where bounds on the number of rectangles of each size are imposed.) The vast majority of work done in this field refers to a restricted problem, where only so–called *guillotine patterns* are permitted. This constraint arises from certain industrial cutting applications: guillotine patterns are those packings that can be generated by applying a sequence of edge-to-edge cuts. The recursive structure of these patterns makes this variant much easier to solve than the *general* or *non–guillotine* problem.

Relatively few authors have dealt with the exact solution of the non–guillotine problem. All of them focus on the problem in two dimensions. Biró and Boros

---

* Supported by the German Federal Ministry of Education, Science, Research and Technology (BMBF, Förderkennzeichen 01 IR 411 C7).

(1984) [4] gave a characterization of non–guillotine patterns using network flows but derived no algorithm. Dowsland (1987) [6] proposed an exact algorithm for the case that all boxes have equal size. Arenales and Morabito (1995) [1] extended an approach for the guillotine problem to cover a certain type of non–guillotine patterns. So far, only two exact algorithms have been proposed and tested for the general case. Beasley (1985) [2] and Hadjiconstantinou and Christofides (1995) [12] have given different 0–1 integer programming formulations of this problem. Even for small problem instances, they have to consider very large 0–1 programs, since the number of variables depends on the size of the container that is to be packed. The largest instance that was solved in either article has 9 out of 22 boxes packed into a 30 × 30 container. After an initial reduction phase, Beasley gets a 0-1 program with more than 8000 variables and more than 800 constraints; the program by Hadjiconstantinou and Christofides still contains more than 1400 0–1 variables und over 5000 constraints. From Lagrangean relaxations, they derived upper bounds for a branch–and–bound algorithm, which are improved using subgradient optimization. The process of traversing the search tree corresponds to the iterative generation of an optimal packing.

In this paper, we describe a different approach to characterizing feasible packings and constructing optimal solutions. We use a graph-theoretic characterization of the relative position of the boxes in a feasible packing. This allows us a much more efficient way to construct an optimal solution for a problem instance: combined with good heuristics for dismissing infeasible subsets of boxes, we develop a two-level tree search. This exact algorithm has been implemented; it outperforms previous methods by a wide margin.

The rest of this paper is organized as follows: after describing the OKP-d and the related *d-dimensional orthogonal packing problem (OPP-d)*, we introduce the novel concepts of packing classes (Section 2.2) and conservative scales (Section 3). Section 4 contains a solution strategy for the OPP-d. This method is used as a building block for the exact OKP-d algorithm presented in Section 5. We conclude the paper by reporting computational results for OKP-2 test problems from literature as well as new test instances.

2 Problem formulation and mathematical approach

Suppose we are given a finite set V of d-dimensional rectangular boxes with "sizes" $w(v) \in \mathbb{R}_0^{+d}$ and "values" $c(v) \in \mathbb{R}_0^+$ for $v \in V$. When arranging these boxes in a container C, we have to preserve the orientations of the boxes; this constraint usually arises from considerations for stability ("this side up") in packings, from asymmetric texture of material in cutting-stock problems, or from different types of coordinates in scheduling problems. The objective of the *d-dimensional orthogonal knapsack problem (OKP-d)* is to maximize the total value of a subset $V' \subseteq V$ fitting into the container C and to find a complying packing. Closely related is the *d-dimensional orthogonal packing problem (OPP-d)*, which is to decide whether a given set of boxes B fits into a unit size container, and to find a complying packing whenever possible.

2.1 Modeling the Problem

The hardness of finding optimal solutions for an OKP instance is compounded by the difficulty of giving a useful problem formulation: once we have placed a box in a container, the remaining feasible space is no longer convex. As a consequence, standard methods for integer program modeling are inedequate. Previous attempts to overcome this difficulty have used a discretization of the set of possible box positions; the number of variables (basically one 0-1 variable for each combination of a grid position and a box type) in the resulting integer program is exponential in the size of the input coordinates. Consequently, the problem instances for which Beasley [2], and Hadjiconstantinou and Christofides [12] were able to use this approach are relatively small. It seems unlikely that even moderately sized three-dimensional problems can be solved with this method.

In the following, we will describe a different way of modeling feasible packings. The basic idea is to use the combinatorial information induced by relative box positions. For this purpose, we consider projections along the different coordinate axes; overlap in these projections defines an interval graph for each coordinate. Using properties of interval graphs and additional conditions, we can tackle two fundamental problems of exact enumeration algorithms:

1. How can we prove in short time that a particular subset of boxes is infeasible for packing?
2. How do we avoid treating equivalent cases more than once?

2.2 Packing classes

Instead of dealing with single packings we handle classes of packings that share a certain combinatorial structure. This structure arises from the way different boxes in a packing can "see" each other orthogonal to one of the coordinate axes. (So-called *box visibility graphs* have been considered as means for representing graphs, see [7].) We will see that transitive orientations of certain classes of graphs correspond to possible packings if and only if specific additional conditions are satisfied. This allows us to make use of a number of powerful theorems [11, 14] to prune our branch-and-bound tree.

For a d-dimensional packing, consider the projections of the boxes onto the d coordinate axes x_i. Each of these projections induces a graph G_i: two boxes are adjacent in G_i, if and only if their x_i projections overlap. (See Figure 1 for a two-dimensional example.) By definition, the G_i are interval graphs, thus they have algorithmically useful properties. (See [11, 14].)

A set of boxes $S \subseteq V$ is called x_i-*feasible*, if the boxes in S can be lined up along the x_i-axis without exceeding the x_i-width of the container. Then we get the following straightforward statements:

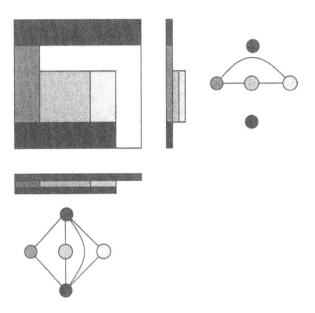

Fig. 1. A two-dimensional packing pattern E and the interval graphs G_1, G_2 induced by the projections

Theorem 1. *Given a packing, then the graphs $G_i = (V, E_i), i = 1, \ldots, d$ as defined have the following properties:*

$P1$: *the graphs $G_i := (V, E_i)$ are interval graphs.*

$P2$: *each stable set S of G_i is x_i-feasible.*

$$P3: \bigcap_{i=1}^{d} E_i = \emptyset.$$

A set $E = (E_1, \ldots, E_d)$ of edges is called a *packing class* for (V, w), if and only if it satisfies the conditions $P1$, $P2$, $P3$. For any given $G_i = (V, E_i)$, we denote by G_i^C the complement graph for G_i. If G_i arises from a packing, any edge in G_i^C corresponds to two boxes with non-overlapping x_i-projection, hence we can think of G_i^C as the comparability graph of the boxes in direction x_i. For any packing class E, we will consider transitive orientations F_i of the comparability graph $G_i{}^C$, and call $F = (F_1, \ldots, F_d)$ a *transitive orientation* of the packing class E. In the following, we will show that for any transitive orientation F of a packing class, there is a packing. For this purpose, define a mapping $p_i^F : V \to \mathbb{R}_0^{+d}$ by

$$p_i^F(v) := \max\{p_i(u) + w_i(u) | \vec{uv} \in F_i\} \tag{1}$$

for $v \in V$, $i \in \{1, \ldots, d\}$.

Without loss of generality, it is sufficient to consider only packings that are generalized "bottom-left" justified, that is any lower bounding coordinate of a box is at zero or at the upper bounding coordinate of another box (without requiring the two boxes to touch). More formally, we say that a packing has *solid projections*, if

$$\forall i \in \{1, \ldots, d\} \; \forall v \in V : \quad p_i(v) = 0 \quad \lor \quad \exists u \in V : p_i(v) = p_i(u) + w_i(u).$$

Then we have the following theorem:

Theorem 2. *A mapping $p : V \to \mathbb{R}_0^{+d}$ is a packing with solid projections, iff there is a packing class E with a transitive orientation F, such that $p = p^F$.*

Proof. First consider a packing p with solid projections. Let F be the canonical orientation of p defined for any $i = \{1, \ldots, d\}$ by:

$$F_i = \{\vec{uv} \mid p_i(u) + w_i(u) \leq p_i(v)\} \, .$$

Note that F is a transitive orientation of the packing class obtained from the G_i as described above. We show $p = p^F$ by induction over the number of elements in the partial order F_i. Let $v \in V$. If v is the only element in F_i, hence a minimal element of the partial order F_i, then $p_i(v) = 0$ by definition and the claim holds. Therefore consider $p_i(v) > 0$ and assume with the induction hypothesis that for all u with $\vec{uv} \in F_i$, we have $p_i(u) = p_i^F(u)$. Let t be the box where the maximum of p_i^F is attained. Since $\vec{tv} \in F_i$, we have

$$p_i^F(v) = p_i^F(t) + w_i(t) = p_i(t) + w_i(t) \leq p_i(v).$$

Since p has solid projections, there must be a t' with $p_i(v) = p_i(t') + w_i(t')$. Then $\vec{t'v} \in F_i$, hence

$$p_i(v) = p_i(t') + w_i(t') = p_i^F(t') + w_i(t) \leq p_i^F(v).$$

To prove the converse, consider a packing class E with a transitive orientation F. Place all boxes v at the positions given by $p^F(v)$. To show that p^F is a packing, we have to show that

1. All boxes are within the boundaries of the container.
2. No two boxes overlap.

Let $v \in V, i \in \{1, \ldots, d\}$. By construction of p^F, the circle-free digraph (V, F_i) contains a directed path $(v^{(0)}, \ldots, v^{(r)})$ with $\deg_{in}(v^{(0)}) = 0$ and $v^{(r)} = v$, such that

$$p_i^F(v) = \sum_{k=0}^{r-1} w_i(v^{(k)}).$$

Since F_i is transitive, $S := \{v^{(0)}, \ldots, v^{(r)}\}$ induces a clique in G_i^C, implying that S is a stable set in G_i. Hence by condition $P2$

$$p_i^F(v) + w_i(v) = \sum_{k=0}^{r} w_i(v^{(k)}) \leq 1, \text{ implying 1.}$$

To see 2., consider $u, v \in V$, $u \neq v$. By $P3$, there is an $i \in \{1, \ldots, d\}$ with the (undirected) edge $\overline{uv} \notin E_i$, hence $\overline{uv} \in E_i{}^C$. Since F_i is an orientation of E_i^C, either $\vec{uv} \in F_i$ or $\vec{vu} \in F_i$. By construction of p^F, we get

$$p_i^F(v) \geq p_i^F(u) + w_i(u) \quad \vee \quad p_i^F(u) \geq p_i^F(v) + w_i(v).$$

In both cases, u and v can be separated by an x_i-orthogonal hyperplane. Hence, the boxes u and v are disjoint and p^F describes a packing. By construction of p^F, the packing has solid projections.

This completes the proof. $\square$

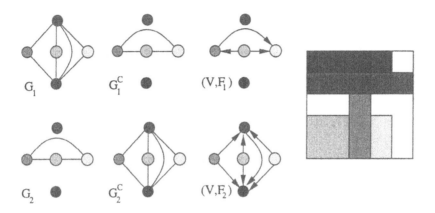

Fig. 2. Constructing a packing pattern for a two-dimensional packing class E

The complements of the component graphs E_1 and E_2 from the example in Figure 2 each have six transitive orientations. Hence there are 36 transitive orientations of E. Figure 3 shows the corresponding packings, constructed by virtue of (1).

If a set of boxes S has a total sum of x_i-widths exceeding k times the x_i-width of the container, then any feasible packing must have an x_i-orthogonal cut that intersects at least $k + 1$ boxes. Using the terminology of packing classes, we get an algorithmically useful formulation of this condition:

Theorem 3. *Let E be a packing class for (V, w), $i \in \{1, \ldots, d\}$, $G_i := (V, E_i)$ and $S \subseteq V$. Then $G_i[S]$ contains a clique of size $\left\lceil \frac{\sum_{s \in S} w_i(s)}{w_i(C)} \right\rceil$.*

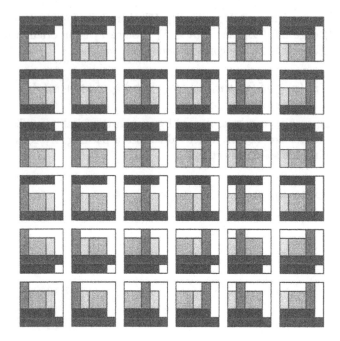

Fig. 3. Packing patterns of the packing class E

3 Conservative scales

In this section, we describe an additional way to reduce the number of cases that we have to consider when looking for an optimal subset.

When defining the packing classes, the size of the boxes only matters when defining the i-feasible sets in P2. Therefore, an OPP-d instance $P = (V, w)$ is characterized by the families

$$\mathcal{Z}(V, w_i) := \{S \subseteq V | w_i(S) \le w_i(C)\}, \quad i \in \{1, \ldots, d\}.$$

In particular, P2 remains valid when enlarging these families. Let $P = (V, w)$ and $Q = (V, w')$ be instances of OPP-d. If

$$\mathcal{Z}(V, w_i) \subseteq \mathcal{Z}(V, w_i'), i \in \{1, \ldots, d\} \tag{2}$$

holds, then any packing class for P is also a packing class for Q, thus any packing for P is a packing for Q. We call such a w' a *conservative scale* for P.

The following theorem provides an easy construction method for conservative scales.

Theorem 4. *Consider an OPP-d instance given by (V, w) and $k \in \mathbb{R}_0^{+d}$. Define w' by*

$$w_i'(v) := \frac{w_i(C) \left(\left\lceil \frac{(k_i+1)w_i(v)}{w_i(C)} \right\rceil - 1 \right)}{k_i}.$$

for all $v \in V$, $i \in \{1, \ldots, d\}$ Then (V, w') satisfies (2).

In [8], we use a modfied version of this rounding technique for the fast generation of a class of lower bounds for the one-dimensional bin packing problem. See also [15].

Conservative scales can be used in several ways:

- A simple rejection heuristic for the OPP-d tries to find a conservative scale w' for (V, w), so that the total (d-dimensional) volume of V regarding w' exceeds the volume of the unit container. In this case it is shown that no packing for (V, w') and therefore for the original problem (V, w) can exist.
- A relaxation of the OKP-d ist the corresponding one–dimensional knapsack problem where the box volumes serve as costs. This is used in Beasley's reduction test *Area Program*. The box voluminas regarding conservative scales can either be used as additional costs for this program, turning it into a multidimensional knapsack problem, or as costs for additional one–dimensional relaxations.
- Theorem 3 can be strengthened:

Theorem 5. *Given the assumptions of Theorem 3 and a conservative scaling of (V, w). Then $G_i[S]$ contains a clique of size $\left\lceil \frac{\sum_{s \in S} w_i'(s)}{w_i(C)} \right\rceil$.*

4 Solution of OPP-d

In our search for an optimal subset of boxes, we may have to decide whether a specific subset has a feasible packing. Our solution strategy for this Orthogonal Packing Problem consists of three steps. Their ordering is due to increasing computational effort.

1. Try to disprove the existence of a packing with the heuristic rejection method described above. Stop in case of success.
2. Try to find a packing using an efficient packing heuristic. Stop in case of success.
3. Try to build up a packing class by the following tree search algorithm. In case of failure we can state that no packing for the given OPP–instance exists.

We give only the ideas of the tree search method. A fully detailed description can be found in [16].

The search tree is traversed by DFS. The algorithm branches by fixing $(b, c) \in E_i$ or $(b, c) \notin E_i$. After each assignment, it is checked if new edges forced by one of the defining properties of a packing class or by Theorem 5 must be added to some E_i or if this condition yields a contradiction. The use of P3 is clear. Properties P1 and P2 are hereditary, we have to avoid three types of forbidden induced subgraphs:

1. C_4's,
2. a generalization of odd antiholes; see [16] for more details,
3. infeasible stable sets.

Each time we detect such a fixed subgraph, we can abandon the search on this node. Each time we detect such a fixed subgraph, except for a set of "equivalent edges", we can fix one of them.

These tests as well as the test if the fixed edges already form a packing class are based on comparability graph recognition and on the calculation of maximal weighted cliques in comparability graphs, so they can be performed efficiently [11].

5 A tree search algorithm for OKP-d

The subsets of V that are candidates for an optimal solution are enumerated in a search tree, where each node corresponds to the OKP-d restricted by upper and lower bounds on the number of boxes used from each *box type*.

Box types are classes of the partition $V = \sum_{j=1}^{m} V_j$ that arise as the union of boxes with equal size and equal value. For $j \in \{1, \ldots, m\}$ let $n_j = |V_j|$ and order $V_j = \{b_{j,1}, \ldots, b_{j,n_j}\}$. $\overline{n}_j$ and $\underline{n}_j$ denote the upper and lower bounds on box type V_j.

The initial tree node starts with the original OKP-d. (For $j \in \{1, \ldots, m\}$: $\overline{n}_j = n_j$, $\underline{n}_j = 0$.)
Consider a node N given by $\overline{n} = (\overline{n}_j)_{j=1,\ldots,m}$ and $\underline{n} = (\underline{n}_j)_{j=1,\ldots,m}$.
If $\overline{n} = \underline{n}$ holds, N is a leaf. The corresponding restricted OKP-d is equivalent to the OPP-d $(\underline{V}, w)$ with

$$\underline{V} := \sum_{\substack{j = 1 \\ \underline{n}_j > 0}}^{m} \{b_{j,1}, \ldots, b_{j,\underline{n}_j}\}.$$

We apply the solution strategy from Section 4.
Otherwise we choose j with $\overline{n}_j > \underline{n}_j$, and branch to the tree nodes given by $(\underline{n}_1, \ldots, \underline{n}_{j-1}, \nu, \underline{n}_{j+1}, \ldots, \underline{n}_m)$ and $(\overline{n}_1, \ldots, \overline{n}_{j-1}, \nu, \overline{n}_{j+1}, \ldots, \overline{n}_m)$ for all $\nu \in \{\underline{n}_j, \ldots, \overline{n}_j\}$.
The tree is traversed by best–first search. In order to keep it small, we implemented the following pruning and reduction techniques:

- The reduction tests *Overlapping Pieces, Free Area, Free Value* and *Area Program* from [2] are applied to reduce the gap between $\bar{n}$ and $\underline{n}$. *Area Program* is used with eight different conservative scales, constructed according to Theorem 4. In addition, *Area Program* supplies an upper bound for the OKP-d.
- Define $\bar{V}$ on the analogue of $\underline{V}$. If $\bar{V}$ can be packed into C using a simple heuristic, we have an optimal packing for this subtree.
- The lower bound $\underline{n}$ demands that $\underline{V}$ is covered by each optimal solution of the restricted OKP-d. So if the OPP-d $(\underline{V}, w)$ can be shown to have no packing, the search on this subtree can be abandoned. Again we apply the method from Section 4.

If V contains "large" boxes, the following theorem provides an additional reduction method.

Theorem 6. *Let $\iota \in \{1, \ldots, d\}$ be fixed. Consider a disjoint partition $V = V_1 \cup V_2 \cup V_3$ with $w_i(v_1) + w_i(v_2) > w_i(C)$ for $v_1 \in V_1$, $v_2 \in V_1 \cup V_2$ and $\iota \neq i$. Let C' be a container with size $w_i(C)$ in all directions except of ι, where it has size $\sum_{v_1 \in V_1} w_\iota(v_1)$.*
If V fits into C and $V_1 \cup V_3$ fits into C', then there exists a packing for (V, w) where the boxes in $V_1 \cup V_3$ and the boxes in V_2 are separated by the hyperplane $x_i = \sum_{v_1 \in V_1} w_i(v_1)$.

If the assumption holds, we can restrict our attention to the solution of the OKP-d given by (V_2, w) with the container C'' of size $w(C)$ in all directions except for x_ι, where it has size $w_\iota(C) - \sum_{v_1 \in V_1} w_\iota(v_1)$, then merge the optimal packing with the packing of $(V_1 \cup V_3, C')$.

6 Computational Results

The algorithm was implemented in C++ and run on a SPARCserver 1000. The code was compiled using the gcc compiler. Besides the running times we list the number of nodes in the OKP search tree, the number of calls of the OPP search tree, and the total number of OPP nodes.

We solved the test problems beasley 1–12 from the OR–Library [2, 3]. The problem instances chrhad 3,8,11,12 come from [12]. These are all published test instances for the general OKP. All optimal solutions were found in less than 0.3 s. For solutions to chrhad8 and chrhad11 we could give the first proof of optimality. For comparison, we show the running times reported in [2, 12].

The instances wang 20 and chrwhi 62 come from [17] and [5]. They were originally meant as test instances for the guillotine-constrained problem.

The new instances okp 1-5 are random instances generated in the same way as beasley 1-12 (see [2]) after applying initial reduction.

Table 1. Computational results. The columns "B85" and "HC95" show the running times as reported in [2,12].

problem	container size	box types	# boxes	OKP nodes	OPP calls	OPP nodes	time/s	B85	HC95	opt. sol.
beasley1	(10, 10)	5	10	8	0	0	0.03	0.9		164
beasley2	(10, 10)	7	17	5	0	0	0.04	4.0		230
beasley3	(10, 10)	10	21	21	3	35	0.09	10.5		247
beasley4	(15, 10)	5	7	1	0	0	0.01	0.1	0.04	268
beasley5	(15, 10)	7	14	1	0	0	0.01	0.4		358
beasley6	(15, 10)	10	15	34	4	5	0.13	55.2	45.2	289
beasley7	(20, 20)	5	8	0	0	0	0.01	0.5	0.04	430
beasley8	(20, 20)	7	13	18	3	149	0.09	218.6		834
beasley9	(20, 20)	10	18	3	0	0	0.03	18.3	5.2	924
beasley10	(30, 30)	5	13	1	0	0	0.02	0.9		1452
beasley11	(30, 30)	7	15	45	10	10	0.13	79.1		1688
beasley12	(30, 30)	10	22	46	6	63	0.26	229.0	>800	1865
chrhad3	(30, 30)	7	7	1	0	0	0.01		532	1178
chrhad8	(40, 40)	10	10	3	0	0	0.02		>800	2517
chrhad11	(30, 30)	15	15	45	1	47	0.20		>800	1270
chrhad12	(40, 40)	15	15	2	0	0	0.02		65.2	2949
wang20	(70, 40)	20	42	630	69	9235	12.51			2726
chrwhi62	(40, 70)	20	62	398	87	17431	7.47			1860
okp1	(100,100)	15	50	986	79	4937	11.60			27718
okp2	(100,100)	30	30	9355	2677	3413	116.24			22502
okp3	(100,100)	30	30	5799	140	148	73.03			24019
okp4	(100,100)	33	61	1610	0	0	50.09			32893
okp5	(100,100)	29	97	1557	159	12038	40.14			27923

Table 2. The new problem instances okp1–okp5.

Problem okp1: container = (100,100), 15 box types (50 boxes)
  size = [(4,90),(22,21),(22,80),(1,88),(6,40),(100,9),(46,14),(10,96),
        (70,27),(57,18),(10,84),(100,1),(2,41),(36,63),(51,24)]
value = [838,521,4735,181,706,2538,1349,1685,5336,1775,1131,129,179,
        6668,3551]
    n = [5,2,3,5,5,5,3,1,3,1,1,5,5,2,4]

Problem okp2: container = (100,100), 30 box types (30 boxes)
  size = [(8,81),(5,76),(42,19),(6,80),(41,48),(6,86),(58,20),(99,3),(9,52),
        (100,14),(7,53),(24,54),(23,77),(42,32),(17,30),(11,90),(26,65),
        (11,84),(100,11),(29,81),(10,64),(25,48),(17,93),(77,31),(3,71),
        (89,9),(1,6),(12,99),(33,72),(21,26)]
 value= [953,389,1668,676,3580,1416,3166,537,1176,3434,676,1408,2362,
        4031,1152,2255,3570,1913,1552,4559,713,1279,3989,4850,299,
        1577,12,2116,2932,1214]
    $n_j = 1, \quad j \in \{1, \ldots, 30\}$

Problem okp3: container = (100,100), 30 box types (30 boxes)
  size = [(3,98),(34,36),(100,6),(49,26),(14,56),(100,3),(10,90),(23,95),
        (10,97),(50,47),(41,45),(13,12),(19,68),(50,46),(23,70),(28,82),
        (12,65),(9,86),(21,96),(19,64),(21,75),(45,26),(19,77),(5,84),
        (16,21),(23,69),(5,89),(22,63),(41,6),(76,30)]
 value= [756,2712,1633,2332,2187,470,1569,4947,2757,4274,4347,396,3866,
        5447,2904,6032,1799,929,5186,2120,1629,2059,2583,953,1000,
        2900,1102,2234,458,5458]
    $n_j = 1, \quad j \in \{1, \ldots, 30\}$

Problem okp4: container = (100,100), 33 box types (61 boxes)
  size = [(48,48),(6,85),(100,14),(17,85),(69,20),(12,72),(5,48),(1,97),
        (66,36),(15,53),(29,80),(19,77),(97,7),(7,57),(63,37),(71,14),(3,76),
        (34,54),(5,91),(14,87),(62,28),(6,7),(20,71),(92,7),(10,77),(99,4),
        (14,44),(100,2),(56,40),(86,14),(22,93),(13,99),(7,76)]
value = [5145,874,2924,3182,2862,1224,531,249,6601,1005,6228,3362,907,
        473,6137,1556,313,4123,581,1999,5004,2040,3143,795,1460,841,
        1107,280,5898,2096,4411,3456,1406]
    n = [1,2,1,1,1,1,3,3,2,1,3,1,1,2,2,1,3,1,2,1,2,1,3,3,1,1,2,3,2,3,2,1,1,3,3]

Problem okp5: container = (100,100), 29 box types (97 boxes)
  size = [(8,81),(5,76),(42,19),(6,80),(41,48),(6,86),(58,20),(99,3),(9,52),
        (100,14),(7,53),(24,54),(23,77),(42,32),(17,30),(11,90),(26,65),
        (11,84),(100,11),(29,81),(10,64),(25,48),(17,93),(77,31),(3,71),
        (89,9),(1,6),(12,99),(21,26)]
value = [953,389,1668,676,3580,1416,3166,537,1176,3434,676,1408,2362,
        4031,1152,2255,3570,1913,1552,4559,713,1279,3989,4850,299,
        1577,12,2116,1214]
    n = [3,4,4,4,1,5,5,5,5,4,5,1,1,5,5,4,2,3,1,1,2,1,4,1,5,4,5,2,5]

References

1. M. Arenales and R. Morabito. An AND/OR–graph approach to the solution of two–dimensional non–guillotine cutting problems. *European Journal of Operations Research*, **84**, 1995, pp. 599–617.

2. J. E. Beasley. An exact two–dimensional non–guillotine cutting stock tree search procedure. *Operations Research*, **33**, 1985, pp. 49–64.

3. J. E. Beasley. OR-Library: distributing test problems by electronic mail. *Journal of the Operations Research Society*, **41**, 1990, pp. 1069–1072.

4. M. Biró and E. Boros. Network flows and non–guillotine cutting patterns. *European Journal of Operations Research*, **16**, 1984, pp. 215–221.

5. N. Christofides and C. Whitlock. An algorithm for two–dimensional cutting problems. *Operations Research*, **25**, 1977, pp. 31–44.

6. K. A. Dowsland. An exact algorithm for the pallet loading problem. *European Journal of Operations Research*, **31**, 1987, pp. 78-84.

7. S. P. Fekete and H. Meijer. Rectangle and box visibility graphs in 3D. To appear in *International Journal of Computational Geometry and its Applications*. Available at `ftp://ftp.zpr.uni-koeln.de/pub/paper/zpr96-224.ps.gz`.

8. S. P. Fekete and J. Schepers. A new classs of lower bounds for bin packing problems. ZPR Report 97-265. Available at `ftp://ftp.zpr.uni-koeln.de/pub/paper/zpr97-265.ps.gz`.

9. S. P. Fekete, J. Schepers, and M. Wottawa. PACKLIB: a library of packing problems. Under construction.

10. G. Galambos and G. J. Woeginger. On-line bin packing – a restricted survey. *Zeitschrift für Operations Research*, **42**, 1995, pp. 25–45.

11. M. C. Golumbic. *Algorithmic graph theory and perfect graphs*. Academic Press, New York, 1980.

12. E. Hadjiconstantinou and N. Christofides. An exact algorithm for general, orthogonal, two–dimensional knapsack problems. *European Journal of Operations Research*, **83**, 1995, pp. 39–56.

13. M. R. Jerrum. A data structure for systems of orthogonal, non-overlapping rectangles. Internal Report CSR-239-87, Department of Computer Science, Edinburgh University, August 1987.

14. N. Korte and R. H. Möhring. An incremental linear–time algorithm for recognizing interval graphs. *Siam Journal of Computing*, **18**, 1989, pp. 68-81.

15. S. Martello and P. Toth. *Knapsack Problems – Algorithms and Computer Implementations*, Wiley, Chichester, 1990.

16. J. Schepers. *Exakte Algorithmen für orthogonale Packungsprobleme*. Doctoral thesis, Universität zu Köln, in preparation for 1997.

17. P. Y. Wang. Two algorithms for constrained two–dimensional cutting stock problems. *Operations Research*, **31**, 1983, pp. 573-586.

Dynamic Data Structures for Realtime Management of Large Geometric Scenes *
(Extended Abstract)

M. Fischer, F. Meyer auf der Heide, and W.-B. Strothmann

Heinz Nixdorf Institute and Department of Computer Science
University of Paderborn, D-33095 Paderborn, Germany
{mafi,fmadh,willy}@uni-paderborn.de

Abstract We present a data structure problem which describes the requirements of a simple variant of fully dynamic walk-through animation: We assume the scene to consist of unit size balls in $\mathbb{R}^2$ or higher dimensions. The scene may be arbitrarily large and has to be stored in secondary memory (discs) with relatively slow access. We allow a visitor to walk in the scene, and a modeler to update the scene by insertions and deletions of balls. We focus on the realtime requirement of animation systems: For some t (specified by the computation power of (the rendering hardware of) the graphic workstation) the data structure has to guarantee that the balls within distance t of the current visitor's position are presented to the rendering hardware, 20 times per second. Insertions and deletions should also be available to the visitor with small delay, independent of the size of the scene. We present a data structure that fulfills the above task in realtime. Its runtime is output-sensitive, i.e. linear in a quantity close to the output size of the query. We further present (preliminary) experimental results indicating that our structure is efficient in practice.

1   Introduction

The aim of this paper is to present fully dynamic data structures that allow navigation in, and manipulation of arbitrarily large geometric scenes in realtime. Our goal in this paper is to focus on the realtime requirement of such systems. This means that we have to answer queries in time independent of the overall size of the scene. We want to design data structures that have provably the realtime property, and can be used in practice. For this purpose we also include (very preliminary) experimental results, performed on our experimental platform which is based on a commercial system for supporting virtual reality applications.

The class of scenes we consider in this paper is still restricted: We assume the scene to consist of an arbitrary number of non-overlapping simple objects of the same size. Further, we assume that only objects within a fixed distance to

---
* Partially supported by EU ESPRIT Long Term Research Project 20244 (ALCOM-IT), DFG Leibniz Grant Me872/6-1 and DFG Grant Me872/7-1.

the visitors position are visible for the visitor. This distance is dictated by the speed of the graphic workstation; it has to be able to render all visible objects in time $\approx \frac{1}{20}$ second in order to guarantee a smoothly moving scene. This distance is large enough so that single objects farer away are very small from the visitor's viewpoint.

Our system does not (yet) allow simplification of many objects very close together, although they could be visible as one bigger object. State of the art animation systems can partially handle such extensions, but such systems do not guarantee realtime behavior in our strong sense, especially not in case of dynamic changes of the scene. They all need computation-intensive preprocessing to set up the data structure like octrees [26,27] to present the scene. In this paper we propose dynamic variants of graph spanners for this task. Possible extensions of our system to allow objects of different sizes and simplification are discussed towards the end of the paper.

1.1 Known Results

We first give a (very short) insight into the current state of algorithms development for walk-through animation. Then we survey results from computational geometry related to our approach.

Data structures used for walk-through animation of large scenes It is not possible in this extended abstract to describe the state of art of walk-through animation. We only focus on the underlying basic data structures for representing and manipulating the scene, and for searching in the scene. The graphics pipeline of a graphic workstation expected a scene consisting of polygons. From this it has to set up a picture. For a realistic optical impression, a frame rate of 20 pictures per second is necessary. In following table the rendering capability of up-to-date graphic workstation is shown.

SGI	Triangles
O2	382K/sec
Indigo^2 (Solid IMPACT)	1 M/sec
OCTANE	1.82M/sec

As scenes are much larger then the rendering capacity, Clark [9] has introduced a hierarchical representation of scenes. The scenes consist of objects, which are again components of (simpler) objects, and so on, until a level of polygonal description is reached. Standard data structures to represent the objects is the BSP-tree [15] or the octree [26,27]. Objects are of different size and number of polygons. In order to reduce rendering time, for an object exists multiple descriptions, each with a different number of polygons. For faraway polygons we render simple descriptions, and for near objects descriptions with more polygons (level of detail).

If we have polygonal approximations of groups of objects for nodes at higher levels of the hierarchy, we render the subtree with fewer polygons and reduce the rendering time (replacement). Additionally, the hierarchical organization offers to cull parts of the model that are not visible for the viewer if we know the extent of a scene part in the hierarchy.

Animations systems which use octrees are presented in [18,22], for BSP-trees in [29]. Another possibility where the scene is partitioned into a spatial hierarchy is shown in [5,6]. For large architectural models e.g. in [30,17,21,2,16] algorithms are described which computes an overestimation of the portion of the scene visible from the visitors viewpoint. The underlying data structure is a visibility graph which connects parts of the scene (cells) from which a viewer can see other cells. An approach to separate the management of a complex model from the application is described in [28].

In the context of the walk-through of large scenes the simplification of polygonal objects for reducing the rendering time is important. The aim is to calculate a good approximation (e.g. minimize divergences, same appearance) with a reduced number of polygons of a model, see e.g. [14]. In the last years a very popular approach for representing shape and analyzing features at multiple levels of detail are wavelets, see e.g. [13,20,10].

Related work in computational geometry As stated in the introduction we have to find for a viewpoint x all objects in distance at most t, refered to as a (circular) range query [23,1]. There are known data structures with linear space and output sensitive query time for the two dimensional case [7,8]. Extensions to higher dimension are still open [1]. We are also interested in dynamic changes of the scene that should only result in a local update in the range query data structure.

For this we will adapt a construction previously used for the construction of spanners (for a survey of recent results see [4]). An undirected edge-weighted subgraph G' of a graph G is a $\triangledown$-spanner if for every pair of vertices u, v the distance between u and v in G' is at most $\triangledown$ times their distance in G. A $\triangledown$-spanner guarantees informally in our context that we we can find all balls in distance t from x by exploring the $t \cdot \triangledown$ neighborhood of x and their neighbors in the spanner. As we are looking for output sensitive query time, the number of neighbors of a vertex should be constant.

Keil and Gutwin [19,3] present a directed spanner for $k \geq 9$. In [25], Ruppert and Seidel present a similar spanner construction. We will introduce this construction in Section 2, because it is a basis for our range searching problem underlying our animation system. Our goal, however, is different: rather than bounding the path length between vertices v, w (as in case of spanners), we want to minimize the maximum Euclidean distance from v to any vertex on the path from v to w. For this (weaker) condition we get stronger bounds than shown for spanners in [25].

1.2 Our models, results and techniques

We will describe our approach for scenes in 2D, extensions to higher dimensions are discussed in Section 3.

Our abstract fully dynamic animation problem We assume that the scene in 2D consists of an arbitrary number m of simple objects (balls) of unit size. These balls are arbitrarily distributed without overlaps. The scene S is stored in secondary memory (discs). The queries come from a *visitor* moving through the scene. She is sitting in front of a graphic workstation, the *rendering machine*, and sees the part $V_t(x)$ of the scene within distance t of her current position x, i.e. $V_t := \{b \in S \mid d(b,s) \le t\}$. Because we have balls of unit size we can use for $d(\cdot, \cdot)$ the Euclidean distance function, i.e. if $U_t(x) := \{y \mid \|x - y\|_2\}$ and S are the centers of the balls $V_t(x) = U_t(x) \cap S$. (In fact she only sees a $45° - 90°$ viewport of $V_t(x)$, but this is not important for our considerations.) The distance t is chosen such that the graphic pipe of our graphic workstation is able to render the sub-scene in $V_t(x)(\approx t^2$ balls of unit size) in $\frac{1}{20}$ seconds. This makes sure that a frame rate of 20 pictures per second, i.e., an impression of a smoothly moving scene is guaranteed. In order to reach this frame rate, we have to make sure that it is particularly simple for the workstation to access the balls in $V_t(x)$. They should be contained in the portion of the scene in main memory of the rendering machine, and easily be extractable given x.

Changes of the scene are done by a *modeler*, sitting in front of (another) graphic workstation. He moves through the scene like the visitor. At any time he can insert a ball into the scene at his current position, or he can *move* to a ball and remove it from the scene. The update should be available for the visitor within very little time.

The architecture of our animation system As we assume the scene to be very large, we have to cope with the fact that it is stored on secondary memory, i.e. on discs. Thus we cannot assume that $V_t(x)$ is directly extracted from the scene 20 times per second, because accessing a disc is to slow. Instead we assume a two level access: We load large portions of the scene, like areas $V_T(y)$ for $T \gg t$, from the disc. This is necessary only in time intervals lasting a few seconds. From $V_T(y)$, now stored in main memory, we extract the sets $V_t(x)$20 times per second. In order not to load too much work on the visitor's graphic workstation, we use another workstation to communicate with the disc.

Conceptually, the system consists of three connected workstations, the *rendering machine*, the *modeling machine* and the *manager*. The modeler and the manager have access to discs storing the scene. The rendering machine is a graphic workstation used by the visitor. It can render a scene consisting of up to A balls, guaranteeing a frame rate of 20 pictures per second. The modeling machine is used by the modeler. It has the same characteristics as the rendering machine. The manager is a (not necessarily graphic) workstation connected both to the modeling and the rendering machine. (It may be convenient to have, or to think about having, two managers, one that serves the rendering machine, one that serves the modeling machine). We describe the manager's service for the rendering machine:

Assume the time being partitioned in constant length *time intervals*. Let x_i denote the position of the visitor at the end of time interval i. Then the manager has to make sure that the rendering machine gets a description of $V_T(x_i)$ at the end of time interval $i + 1$. (The size of T is discussed below.) For this purpose, the rendering machine sends x_i to the manager at the beginning of time interval $i + 1$. The remaining time of the time interval $i + 1$ is used by the manager to compute necessary information that allows a simple computation of $V_T(x_i)$ from $V_T(x_{i-1})$. Also within interval $i + 1$, this information has to be sent to the rendering machine, where $V_T(x_{i-1})$ is then updated to $V_T(x_i)$.

T has to be chosen large enough so that $V_t(y) \subset V_T(x_{i-1})$ for arbitrary positions y reachable by the visitor in time interval $i+1$ from x_{i-1}. Thus we need $T \geq 2d + t$, where d denotes the distance the visitor can move in a time interval.

The modeling machine gets the same support from the manager as the rendering machine. Updates of the scene, insertions or deletions of balls at the modeler's current position, are then executed by modifying the data structure on the discs. The updates should be available by the manager within few time intervals. "Few" means an absolute constant independent on m, d, t, T.

Requirements for our data structure Our data structure has to store an arbitrary number of balls of unit size in 2D space. The positions of the balls are arbitrary, but the balls do not overlap. We want to support the following operations: Let x be a position in 2D (not necessarily a center of a ball), $t \geq 1, T \gg t$.

- SEARCH(x, T) reports all balls in $V_T(x)$ (this is sometimes refered to as *circular range query*) as a data structure $D(x)$ such that, for each position y with $d(x, y) \leq T - t$, all balls in $V_t(y)$ can be reported very fast. (This is a crucial property, because the rendering machine has to do this 20 times per second!) Very fast means, that the time needed is close to the output size $|V_t(y)| \leq t^2$.
- UPDATE(x, T, y) reports all balls in $V_T(x) \backslash V_T(y)$ and in $V_T(y) \backslash V_T(x)$. They have to be organized as a data structure $U(x, T, y)$ such that, given $D(x)$ and $U(x, T, y)$, $D(y)$ can be computed very fast. (This is important because the rendering machine has to do the update within each time interval. This loads additional work on this machine but reduces the communication between the manager and the rendering machine.) Here "very fast" means that the time needed is close to the output size.
- INSERT(x) inserts a ball at position x.
- DELETE(x) deletes the ball with center x.

The UPDATE operation is interesting because typically the positions x_{i-1} and x_i of the visitor at the ends of time intervals $i - 1$ and i are not too far apart. Thus the update might be much smaller than $V_T(x_i)$, which decreases the computation time to produce and the communication time to submit the update.

Our main concern is that the above operations are executed in realtime, i.e., in time independent of the overall number of balls currently in the scene. SEARCH and UPDATE should run in time proportional to their output size. INSERT and DELETE should be similarly fast as SEARCH, because we want to have finished such an operation within a small, constant number of time intervals.

2 Survey of our techniques and results

We first describe a static data structure which supports SEARCH, if the position x of the visitor is the center of an existing ball. For this purpose adopt a data structure proposed in [25].

Fix an integer k. Let $\gamma = \frac{2\pi}{k}$. The γ-*sectors* of a position x in $\mathbb{R}^2$ are defined as follows: Draw k rays from x such that they form angles $\frac{2\pi(i-1)}{k}, i = 1, \ldots k$, with the vertical line through x. These rays subdivide $\mathbb{R}^2$ into k sectors around x (we assume that the left ray bounding a sector belongs to it). The γ-*angle graph* on a (finite) set $M \subseteq \mathbb{R}^2$ has vertex set M. Each $x \in M$ has directed edges to the closest balls from M lying in the k γ-sectors around x. Thus the graph has out-degree k. For our purpose it is convenient not to use the Euclidean distance $d(\cdot, \cdot)$ for defining the closest point. Instead we define the distance $\tilde{d}_\gamma(x, y)$ from x to y as follows:

Consider the γ-*sector* S of x containing y. Then $d_\gamma(x, y)$ is the (Euclidean) distance from x to the line L through y being orthogonal to the line halving the angle in S at x. Note: If $k \equiv 2 \pmod 4$, then L is also a boundary of a γ-*sector* of y. By using k scan line algorithms we can compute the γ-angle graph in time $\mathcal{O}(km \log m)$, where m denotes the number of balls. The graph of [19] in its undirected version, is a $(\frac{1}{\cos(\gamma)} \cdot \frac{1}{1-\tan(\gamma)})$-spanner for the Euclidean metric, if $k \geq 9$. Our γ-angle graph is a $(\frac{1}{1-2\sin(\gamma/2)})$-spanner, as shown in [25], if $k \geq 7$. We show:

Theorem 1. *Let $k \geq 6, \gamma = \frac{2\pi}{k}, \Delta(\gamma) := \max\{\sqrt{1 + 48 \sin^4 \frac{\gamma}{2}}, \sqrt{5 - 4\cos\gamma}\}$. For x being the center of a ball of the 2D-scene, each ball of $V_T(x)$ is reachable via a directed path from x in the γ-angle graph. One such path contains no balls farer away than $T' = \Delta(\gamma) \cdot T$ from x. The bound for $\Delta(\gamma)$ is tight for even k.*

As noted above, also the results for spanners from [19] and [25] would yield analogous to Theorem 1, but with worse Δ-values. E.g., for $k = 9$ (10,11), the result from [19] yields Δ-values 8.11 (4.52, 3.32), the one from [25] 3.16 (2.61, 2.29), whereas we obtain 1.39 (1.32, 1.27). As our Δ is defined for $k \geq 6$, we can use a graph with small out-degree for our implementation.

Proof of Theorem 1: First assume that $k \equiv 2 \pmod 4$. Let x, y be balls in the scene with $d(x, y) = t$. We assume that an adversary tries to place balls we have to choose as our neighbor in order to lead us far away from x before (if at all) we find y. Our strategy to reach y is as follows: When having reached a ball z, we choose as next the edge going in that sector around z containing y. The strategy of choosing the balls, and the choices of the adversary is shown in Figure (b). Figure (a) shows the simpler, special case where $\gamma = \frac{\pi}{3} (\hat{=} 60°)$ and $k = 6$. (Note that, for $k \equiv 2 \pmod 4$, the rays from z_i through y, $i = 1, 2, \ldots$, are boundaries of γ-sectors of z_i.) It is easy to check that y is always reached, and that always z_2 or z_3 is the ball on the path farest away from x ($\Delta(\gamma) = \max\{d(x, z_2), d(x, z_3)\}$). With $u = 2t \sin(\frac{\gamma}{2})$ and the cosines theorem we get for $d(x, z_2)$:

$$d(x, z_2) = \sqrt{u^2 + t^2 - 2ut \cos(90° + \frac{\gamma}{2})} = t \cdot \sqrt{5 - 4\cos(\gamma)}$$

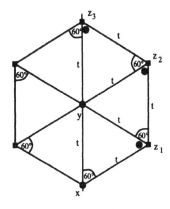
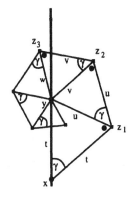

(a) Search environment for $k = 6$. (b) Search environment for any $k > 6$.

And with $w = 8t \sin^3(\frac{\gamma}{2})$ and the cosines theorem we get for $d(x, z_3)$:

$$d(x, z_3) = \sqrt{w^2 + t^2 - 2wt \cos(90° + \frac{3}{2}\gamma)} = t\sqrt{1 + 48 \sin^4(\gamma/2)}$$

Note that the above construction also shows the optimality of our bound for $\Delta(\gamma)$, if k is even. If k is odd, then $\Delta(\gamma)$ is an upper bound. Executing SEARCH(x) is now easy: We find all balls from $V_t(x)$ by starting breadth first search (BFS) from x, without ever leaving $U_{T'}(x)$. For each ball y found during the search we check whether $d(x, y) \leq t$.

Thus, SEARCH(x, T) can be done in time $\mathcal{O}(l) = \mathcal{O}(T^2)$, where l denotes the number of balls in $V_{T'}(x)$. Note that $\Delta(\gamma) \to 1$ with $\gamma \to 0$. We get the smallest out-degree $k = 6$ for $\gamma = 60°$ and $\Delta(60°) = 2$.

The main problem of this result, even for the static case, lies in the fact that it only computes $V_T(x)$, if x is the center of a ball. For other positions, the problem is to find "nearby" balls in order to start searching in the graph. But if the current position is far away from any ball, i.e. the visitor crosses a "desert", it seems complicated to figure out when balls from the other side of the desert" become visible. We therefore first show how to handle scenes "without deserts".

Fix an angle γ. We call a scene (γ, c)-*crowded*, if for each point $z \in \mathbb{R}^2$ (not necessarily a center of a ball), each γ-sector around z contains either no ball or a ball within distance at most c from z.

For (γ, c)-crowded scenes we are able to efficiently compute SEARCH(y, T) and UPDATE(x, T, y), if y is sufficiently close to x and $V_T(x)$ is given. We also can efficiently support the DELETE operation. (A general SEARCH(x, T) or INSERT(x) is still complicated, because we have to locate x in the scene and thus find at least some ball close to x, which seems not to be possible in realtime; it seems to need time $\log(m)$.)

Theorem 2. *Consider a (γ, c)-crowded scene of m balls in $\mathbb{R}^2$, organized as a γ-angle graph, $k = \frac{2\pi}{\gamma} \in \mathbb{N}, k \geq 6$. Let $c \leq t \ll T, T' = T + \Delta(\gamma) \cdot c$. For positions x, y with $r = d(x, y) \leq T - c$, let $l_1 := |V_{T'}(x) \backslash V_T(y)|, l_2 := |V_{T'}(y) \backslash V_T(x)|$.*

- *UPDATE(x, T, y) can be done in time $\mathcal{O}(l_1 + l_2) = \mathcal{O}(T \cdot (r + c))$, if $V_T(x)$ is given.*
- *DELETE(x) can be done in time $\mathcal{O}(c^2 \log(c))$; if $k \equiv 2 \pmod 4$, time $\mathcal{O}(c^2)$ suffices.*

This proof is sketched in Section 3.

There is a simple way to turn any scene into a (γ, c)-crowded scene: Just add "dummy balls", e.g., on all positions $(x_1, x_2), x_1, x_2$ being multiples of $c/(1 + 1/2 \tan(\gamma/2))$. (These balls may overlap with "real" balls). The advantage of this structure is that we even can perform SEARCH(x, T) efficiently, because we can locate dummy balls close to x in constant time. The main disadvantage is the huge amount of storage overhead. We present a way to reduce this overhead by only storing "important" dummy balls and use dynamic perfect hashing [12], [11] to find out whether a dummy ball is important or not.

Theorem 3. *Consider an arbitrary scene of m balls in $\mathbb{R}^2$. Using the notations from Theorem 2, and $l := |V_{T'}(x)|$, the following holds:*

- *Our data structure needs space $\mathcal{O}(m)$.*
- *SEARCH(x, T) can be done in time $\mathcal{O}(l + (\frac{T}{c})^2) = \mathcal{O}(T^2)$.*
- *UPDATE(x, T, y) can be done in time $\mathcal{O}(l_1 + l_2 + \frac{T(r+c)}{c}) = \mathcal{O}(T \cdot (r + c))$, if $V_T(x)$ is given.*
- *DELETE(x) can be done in time $\mathcal{O}(c^2 \log(c))$, w.h.p.; if $k \equiv 2 \pmod 4$, time $\mathcal{O}(c^2)$ suffices, w.h.p. [1].*
- *INSERT(x) can be done in time $\mathcal{O}(c^2)$, w.h.p.*

Further, the result of SEARCH(x, T) is a data structure $D(x)$ that allows to execute SEARCH(z, t) in $D(x)$ deterministically in time $\mathcal{O}(l' + (\frac{t}{c})^2) = \mathcal{O}(t^2)$, with $l' = |V_{t + \Delta(\gamma) \cdot c}(z)|$. The result of UPDATE$(x, T, y)$ is a data structure $U(x, T, y)$ that allows to compute $D(y)$ using $D(x)$ deterministically in time $\mathcal{O}(l_1 + l_2 + \frac{T(r+c)}{c}) = \mathcal{O}(T(r + c))$.

This proof is sketched in Section 4.

3 The fully dynamic structure for (γ, c)-crowded scenes

In this section we sketch a proof of Theorem 2. Let M be a (γ, c)-crowded scene of size m. Thus, in each γ-sector around each point z in $\mathbb{R}^2$, there is either no ball, or a ball within distance at most c to z. Assume M to be organized as a γ-angle graph G, $k = \frac{2\pi}{\gamma} \in \mathbb{N}, k \geq 6$. For $z \in \mathbb{R}^2, t \geq 0$, let $U_t(z) := \{y \in \mathbb{R}^2 \mid d(z, y) \leq t\}$. Thus $V_t(x) = U_t(x) \cap S$. We first note that for (γ, c)-crowded scenes we can strengthen Theorem 1.

---

[1] w.h.p. means "with high probability", i.e. with probability $1 - m^{-f}$. f can be chosen as an arbitrarily large constant.

Corollary 4. *Let a (γ, c)-crowded scene be given. For balls x, y of the scene with $d(x, y) = t$, there is a path from x to y in G that never leaves $U_{t'}(x)$, for $t' = t + \Delta(\gamma) \cdot c$.*

The proof follows directly from that of Theorem 1.

We first show how to execute UPDATE(x, T, y). We assume that $V_T(x)$ is represented as $D(x)$, the subgraph of G induced by the balls found during the BFS from x. By Corollary 4, BFS finds all balls from $V_T(x)$ and some balls from $V_{T'}(x) \backslash V_T(x), T' = T + \Delta(\gamma) \cdot c$. Let $B(x)$ be a linear list of those nodes of $D(x)$ that have a neighbor outside $D(x)$ in G. Given $D(x), B(x)$ and a position y being $r \leq T - c$ units away from x, we do the following for each $z \in B(x)$:

- Compute a BFS tree $R(z)$ rooted in z containing all balls in $V_{r+c\Delta(\gamma)}(z)$.
- Compute $D(y), B(y)$ from $D(x), B(x)$ and $R(z), z \in B(x)$.

We first note:

Lemma 5. *All balls from $D(y)$ are contained in $D(x) \cup \bigcup_{z \in B(x)} R(z)$.*

The proof is omitted (follows from the (γ, c)-crowdedness of the scene).

Thus the second step of the algorithm can be done. The time needed is bounded by

$$\mathcal{O}(\sum_{z \in B(x)} |R(z)|) = \mathcal{O}(T \cdot (r + c)^2).$$

If $r \ll T$, this is faster than $\mathcal{O}(T^2)$, the worst case time bound we get for constructing $D(y)$ from scratch. Clearly, the $R(z)'$s may greatly overlap. In the full paper, we describe the UPDATE operation exactly and show the time bound claimed in Theorem 2.

In order to execute DELETE(x), for x being the center of a ball, we have to delete the ball x and its outgoing edges from the graph (which is easy). In addition we have to find the set U of balls that have an edge pointing to x, and redirect these edges.

As the scene is (γ, c)-crowded, $U \subseteq V_c(x)$. This implies that $|U| \leq c^2$. In fact, even $|U| = \mathcal{O}(c)$ holds, because at most $\mathcal{O}(c)$ balls in $V_c(x)$ can have x as their closest neighbor in one of the k directions. (The worst case would be to cluster the border of $U_c(x)$ with balls.) Now we have to find the new closest balls for the balls in U. It is clear that a new neighbor n for a ball $u \in U$ is at most at distance $2c \sin \alpha$ from x or there is no other ball in the corresponding sector since the scene is (γ, c)-crowded.

Using the preprocessing procedure we get a DELETE-time bounded by the time to find U plus the time to find the new potential neighbors N plus k scan line algorithms (see Section 2). This sums up to time $\mathcal{O}(c^2 \log(c))$. In case of $k \equiv 2 \pmod 4$, this can be improved to $\mathcal{O}(c^2)$. This can be shown as follows:

Lemma 6. *If $k \equiv 2 \pmod 4$, then for every $u \in U$ there is either no other ball other than x in the corresponding sector, or there is a closest neighbor n of u in S that is also in U.*

We can further improve the DELETE-time, if we store for each ball not only the k outgoing edges, but also the list of incoming edges. We have to use dynamic memory management. In this case we can get U in time $|U| = \mathcal{O}(c)$ which leads to a DELETE-time of $\mathcal{O}(c \log c)$.

4 The fully dynamic structure for arbitrary scenes

In this section we sketch a proof of Theorem 3. Now we fix $c \leq t \ll T$, and add dummy balls at positions $\{(ac', bc'), a, b \in \{-q, \ldots q\}\}$, for $c' = c/(1 + 1/2 \tan(\gamma/2))$. q is chosen much larger than the real scene M will ever be. We build our γ-angle graph on the balls from M and the dummy balls. This yields a (γ, c)- crowded scene.

Thus we get UPDATE- and DELETE-times as described in Theorem 3 by the algorithm described in the last section. SEARCH(x, T) is done by computing the dummy ball closest to x (in constant time) and start the search from there. Also, INSERT is easy: We execute SEARCH(x, c). Only these $\approx c^2$ balls are candidates for (undirected) neighbors of x_i, each one is handled in constant time. The overheads in the runtimes result from the fact that we also visit dummy balls.

Thus, it only remains to show how to reduce the space requirement from $\mathcal{O}(m + q^2)$ to $\mathcal{O}(m)$. For this purpose we observe the following: If m is small compared to q^2, then there are large "deserts" where only dummy balls exist. In the "inner of a desert" these balls have only dummy balls as neighbors in the γ-angle graph. As we know this, we can try to avoid storing them explicitly. This idea is made precise below. A dummy ball is called *essential*, if at least one of its undirected neighbors is a real ball.

Lemma 7. *There are only $\mathcal{O}(m)$ essential dummy balls.*

Proof: The worst scene consists of m balls with pairwise large distance. In this case each of these balls causes $\mathcal{O}(k)$ (recall : k is the out-degree of vertices) dummy balls to be essential.

Our data structure now only holds essential dummy balls. In addition to being included in the γ-angle graph explicitly, the essential dummy balls are maintained in a dynamic perfect hash list as described in [11]. This is a randomized data hash structure that guarantees lookups, insertions, deletions in constant time, and needs linear space. The time bounds for INSERT and DELETE are very reliable. They are guaranteed with probability $1 - \frac{1}{m^f}$, where f can be made arbitrarily large. The space bound and the time for UPDATE and SEARCH are worst case.

Using this representation of the γ-angle graph on M and the dummy balls, all operations can be done as fast as before, except for the fact that we may have to check in the hash table, whether a newly found dummy ball is essential. For inserts and deletes, it may happen that dummy nodes change their status, non-essential to essential or vice versa. This demands insertions or deletions in the hash table, which needs constant time, w.h.p.

5 Extensions to higher-dimensional spaces

γ-angle graphs can be generalized to scenes in $\mathbb{R}^d$. In this case, we have to partition the space around a position x into sectors that are cones with maximum angle γ. In [24] Rogers has shown that $k(\gamma) = \mathcal{O}(d^{3/2} \log \frac{d}{\sin(\gamma/2)} \sin^{-d} \frac{\gamma}{2})$ many such cones suffice. These cones cover $\mathbb{R}^d$, but overlap. It can be show that the fact that they overlap does not cause a problem for our data structure. We can get similar results to what we have for 2D, the degree of our graph becomes $k(\gamma)$. For practical use, we need the structure for the 3-dimensional case. Further, for efficient implementations we should have very simple, easily computable, disjoint cones describing the sectors.

The most elegant cones, all of identical structure, can be described by using the Platonian Soloids *Icosahedron* and *Dodecahedron*: The sectors around x are defined by drawing such a body with center x, and defining the sectors as the positive hulls of x with the 2-dimensional faces of the body. In case of a Dodecahedron we get degree 12. In case of the icosahedron we get degree 20. Unfortunately γ is in both cases bigger than 60°, i.e., too large for our purposes. Another regular body is the "football" – an Icosahedron where each vertex is cut off and is replaced by a 5-gon. Here we can guarantee $\gamma < 60°$ and out-degree 32.

The notion of c-crowdedness and dummy nodes can be extended to d-dimensional space in a straight forward way yielding a d-dimensional version of our data structure. Also the analogues of our theorems hold (replacing time bounds by $\mathcal{O}(T^d)$ for SEARCH and by $\mathcal{O}(T^{d-1}(r + c))$ for UPDATE).

6 First preliminary experimental results

Until now we have implemented the 60°-angle graph in 2D on a set S of "real balls" and dummy balls, and our SEARCH-algorithm. The dummy balls are placed on the vertices of the grid with given edge length c'. We have inserted dummy balls explicitly (no hashing) and a set S of randomly placed balls of size one. As such a placement does not have large deserts, we have made experiments also on graphs where we have enforced deserts.

Our experiment is as follows: Fix the environment with $t = 40$. We have measured the running time needed for answering the SEARCH(x, t) queries, for a position x and different values of $|S|, c$ and the extension of the scene. In the following table we see the measured running times. The implementation is not optimized until now.

balls	scene	$c = 37.3$		$c = 18.6$	
		$|V_t(x)|$	time [s]	$|V_t(x)|$	time [s]
25000	400x400	872	0.047	851	0.0495
50000	400x800	847	0.046	823	0.048
100000	800x800	826	0.045	775	0.048
200000	800x1600	853	0.0475	799	0.0485

If we increase $|S|$, we also increase the extension of the scene. Because of the random placement of balls the values of $|V_t(x)|$ are roughly equal. Our scene still fits in main memory. Thus, as expected, the search time does not depend on $|S|$. As expected, the query time does not much differ, if we double both the scene and the number of balls. Still we can guarantee time $\leq \frac{1}{20}$ sec. for output sizes up to 850 balls on a SUN Ultra 2 (200 MHz UltraSPARC) Workstation. Because of the additional dummy balls of the scene with $c = 18.6$ the measured running time is slightly larger than as the running time for $c = 37.3$. In the implementation of our animation system we have to balance the costs for search and delete.

The following table shows measured running times for building the γ-angle graph from scratch:

balls	time [s]	balls	time [s]	balls	time [s]	balls	time [s]	balls	time [s]
50	0.0052	300	0.0379	625	0.08	20000	3.48	640000	141.89
100	0.0114	350	0.0448	1250	0.18	40000	7.37	1280000	305.94
150	0.0179	400	0.0519	2500	0.37	80000	15.44		
200	0.0245	450	0.0588	5000	0.78	160000	32.23		
250	0.0311	500	0.0664	10000	1.62	320000	67.59		

We use an algorithm similar to one from [25] which uses plane sweeps to get a running time of $\mathcal{O}(n \log n)$. The implementation can be improved by using only sorted sequences for the balls reached by the sweepline instead of a more complex orthogonal range query. This works because of the following observation: The balls reached by the sweepline (without nearest neighbor) form a linear order of the projection of the points to the vertical line of the bisector of a sector.

The efficient implementation of the preprocessing algorithm is important for the DELETE(x) operation in order to find the new nearest neighbors of the nearest neighbors of the deleted ball x. For our strong realtime requirements we demand running times of at most $1/20s$. In the table we see that this can be fulfilled for up to 350 balls. In later implementations of our animation system, we have to adjust this with the parameter c.

7 Directions of future research

We have proposed a rigorous approach to a management of arbitrarily large scenes in realtime. For this we have described a very simple version of the management problem (identical objects, no simplification) and designed fully dynamic data structures that are provably efficient. We gave preliminary experimental evidence that they are very efficient also in practice, when implemented on our architecture consisting of three (or more) workstations. We consider the state of our research described in this paper as a first step:

- We plan to realize a 2D- and 3D-version of our system and data structure in order to demonstrate that it works in practice. An obstacle here might be the use of perfect hashing. It is an interesting problem to develop deterministic versions of our general structure. Ideas from spanner constructions with bounded in- and out-degree might help.

- An interesting problem is how to handle objects of different sizes. The problem now is that the distance measure has to be modified: The distance from x to a ball B around y with radius r should be measured by the "visibility of B from x". This can, e.g., be measured by the angle α with $sin(\alpha/2) = \frac{r}{\|x-y\|_2}$. This makes the definition of an analogue to the γ-angle graph difficult. A simple way out would be to partition the objects in classes of roughly equal size and to handle the classes separately.
- Further, it is of great importance to realize a version of simplification. E.g., one could try to define an area "densely clustered with objects" as one simplified object which has to be added to the scene. It can be handled like a bigger object.

References

1. Pankaj K. Agarwal. Range Searching. Technical Report CS-1996-05, Duke University, Department of Computer Science, 1996.
2. John M. Airey, John H. Rohlf, and Jr Frederick P. Brooks. Towrads Image Realism with Interactive Update Rates in Complex Virtual Environments. In *Proceedings of the ACM SIGGRAPH Symposium on Interactive 3D Graphics 1990*, pages 41 – 50, 1990.
3. Ingo Althöfer, Gautam Das, David Dobkin, Deborah Joseph, and José Soares. On Sparse Spanners of Weighted Graphs. *Discrete & Computational Geometry*, 9:81 – 100, 1993.
4. Sunil Arya, Gautam Das, David M. Mount, Jeffrey S. Salowe, and Michiel Smid. Euclidian Spanners: Short, Thin, and Lanky. In *27th Annual ACM Symposium on Theory of Computing*, pages 489 – 498, 1995.
5. Bradford Chamberlain, Tony DeRose, Dani Lischinski, David Salesin, and John Snyder. Fast Rendering of Complex Environments Using a Spatial Hierarchy. In *Proceedings of the Graphics Interface '96*, 1996.
6. Bradfort Chamberlain, Tony DeRose, Dani Lischinski, David Salesin, and John Snyder. Fast Rendering of Complex Environments Using a Spatial Hierarchy. Technical Report UW-CSE-95-05-02, University of Washington, July 1995.
7. B. Chazelle and H. Edelsbrunner. Optimal Solutions for a Class of Point Retrieval Problems. *Journal of Symbolic Computation*, 1:47 – 56, 1985.
8. B. Chazelle and H. Edelsbrunner. Linear Space Data Structures for Two Types of Range Search. *Discrete & Computational Geometry*, 2:113 – 126, 1987.
9. James H. Clark. Hierarchical Geometric Models for Visible Surface Algorithms. *Communications of the ACM*, 19(10):547 – 554, October 1976.
10. Tony DeRose, Michael Lounsbery, and Joe Warren. Multiresolution Analysis of Arbitrary Topological Type. Technical Report 93-10-05, University of Washington, Seattle, WA, 98195, October 1993.
11. M. Dietzfelbinger and F. Meyer auf der Heide. Dynamic Hashing in Real Time. In *Informatik: Festschrift zum 60. Geburtstag von Günter Hotz*. Teubner, Stuttgart, 1992.

12. M. Dietzfelbinger, A. Karlin, F. Meyer auf der Heide, H. Rohnert, and R.E. Tarjan. Dynamic Perfect Hashing: Upper and Lower Bounds. *SIAM Journal on Computing*, 23(4):748 – 761, 1994.

13. Matthias Eck, Tony DeRose, Tom Duchamp, Hugues Hoppe, Michael Lounsbery, and Werner Stuetzle. Multiresolution Analysis of Arbitrary Meshes. In *Proceedings of the SIGGRAPH '95*, pages 173–182, August 1995.

14. Carl Erikson. Polygonal Simplification: An Overview. Technical Report 96-016, University of North Carolina at Chapel Hill, Department of Computer Science, 1996.

15. Henry Fuchs, Zvi M. Kedem, and Bruce F. Naylor. On Visible Surface Generation by a Priori Tree Structures. In *Computer Graphics*, 1980.

16. Thomas A. Funkhouser and Carlo H. Sequin. Adaptive Display Algorithm for Interactive Frame Rates During Visualisation of Complex Virtual Environments. In *Proceedings of the SIGGRAPH '93*, pages 247 – 254, 1993.

17. Thomas A. Funkhouser, Carlo H. Sequin, and Seth J. Theller. Management of Large Amounts of Data in Interactive Building Walkthroughs. In *Proceedings of the SIGGRAPH '91*, pages 11 – 20, 1992.

18. Ned Greene, Michael Kass, and Gavin Miller. Hierarchical Z-Buffer Visibility. In *Proceedings of the SIGGRAPH '93*, pages 231–238, 1993.

19. J. Mark Keil and Carl A. Gutwin. Classes of Graphs Which Approximate the Complete Euclidian Graph. *Discrete & Computational Geometry*, 7:13 – 28, 1992.

20. Michael Lounsbery. *Multiresolution analysis for surfaces of arbitrary topological type*. PhD thesis, Dept. of Computer Science and Engineering, University of Washington, 1994.

21. David Luebke and Chris Georges. Portals and Mirrors: Simple, Fast Evaluation of Potentially Visible Sets. In *Proceedings of the ACM SIGGRAPH Symposium on Interactive 3D Graphics 1995*. ACM Press, April 1995.

22. Paulo W. C. Maciel and Peter Shrirley. Visual Navigation of Large Environments Using Textured Clusters. In *Proceedings of the ACM SIGGRAPH Symposium on Interactive 3D Graphics 1995*, 1995.

23. Jiří Matoušek. Geometric Range Searching. *ACM Computing Surveys*, 26:421 – 461, 1994.

24. C.A. Rogers. Covering a Sphere with Spheres. *Mathematika*, 10:157 – 164, 1963.

25. J. Ruppert and R. Seidel. Approximating the d-dimensional complete Euclidean graph. In *3rd Canadian Conference on Computational Geometry*, pages 207 – 210, 1991.

26. Hanan Samet and Robert E. Webber. Hierarchical data structures and algorithms for computer graphics, part I. *IEEE Computer Graphics and Applications*, 8(3):48–68, May 1988.

27. Hanan Samet and Robert E. Webber. Hierarchical data structures and algorithms for computer graphics, part II. *IEEE Computer Graphics and Applications*, 8(4):59–75, July 1988.

28. Jonathan Mark Sewell. *Managing Complex Models for Computer Graphics*. PhD thesis, University of Cambridge, Queens' College, March 1996.

29. Jonathan Shade, Dani Lischinski, David H. Salesin, Tony DeRose, and John Snyder. Hierarchical Image Caching for Accelerated Walkthroughs of Complex Environments. In *Proceedings of the SIGGRAPH '96*, August 1996.

30. Seth J. Teller and Carlo H. Sequin. Visibility Preprocessing For Interactive Walkthroughs. In *Proceedings of the SIGGRAPH '90*, volume 25, pages 61 – 69, 1991.

Solving Rectilinear Steiner Tree Problems Exactly in Theory and Practice

Ulrich Fößmeier   Michael Kaufmann

Universität Tübingen, Wilhelm-Schickard-Institut, Sand 13,
72076 Tübingen, Germany,
email: { foessmei / mk } @informatik.uni-tuebingen.de

Abstract. The rectilinear Steiner tree problem asks for a shortest tree connecting given points in the plane with rectilinear distance. The best theoretically analyzed algorithm for this problem with a fairly practical behaviour bases on dynamic programming and has a running time of $O(n^2 \cdot 2.62^n)$ (Ganley/Cohoon). The best implementation can solve random problems of size 35 (Salowe/Warme) within a day. In this paper we improve the theoretical worst-case time bound to $O(n \cdot 2.38^n)$, for random problem instances we prove a running time of less than $O(2^n)$. In practice, our ideas lead to even more drastic improvements. Extensive experiments show that the range for the size of random problems solvable within a day on a workstation is almost doubled. For exponential time algorithms, this is an enormous step.

1   Introduction

The Steiner tree problem is one of the most famous combinatorial problems next to the traveling salesman problem. Let $S = \{s_0, .., s_{n-1}\}$ be a set of points in the plane which are called *terminals*. A *Steiner tree* is a tree in the plane which contains S. The Steiner tree problem is to find a Steiner tree of minimum length (SMT). There are several versions of the problem, extensively described in the literature [6,7,13]. They mainly differ by the underlying distance metric. In the *rectilinear* Steiner tree problem, the distance between two points is defined within the L_1 metric, it is the sum of the differences of their $x-$ and $y-$coordinates. This classical combinatorial problem received new importance in the development of techniques for VLSI routing [11,12]. In this area, it is highly desirable to achieve optimum solutions. One percent increase of the wire length could mean a considerable loss of performance of the corresponding chip. Further, knowing exact solutions for subproblems is important for many heuristic approaches. The problem is known to be *NP*-complete [6], so it has been used as a test field for many approximation algorithms appeared in the last two decades [1,2,16]. The known algorithms that achieve a guaranteed optimum perform extremely poorly from the analytical point of view as well as from the implementation side. Our progress is twofold: We show how to find *optimum* rectilinear Steiner trees in a 'reasonable' computing time in theory as well as in practice. We give:

1. an improved worst-case analysis
2. new bounds for random point sets
3. new insights for better implementations
4. a program that solves problems of more than 55 points in a day.

A rectilinear Steiner tree T consists of horizontal and vertical segments connecting the terminals. Those endpoints of the segments which are not terminals are called *Steiner points*. T is called a *full Steiner tree* if all terminals are leaves in T. At terminals with degree more than one we can split T into edge-disjoint full Steiner trees, which have only the split terminals in common. Those trees are called *full components*. In Fig. 1 vertices are drawn as filled circles, and every three line segments joining at the same place define a Steiner point there; in the left lower corner of the drawing a full component consisting of six terminals is displayed, in the right lower corner there is a full component of two terminals.

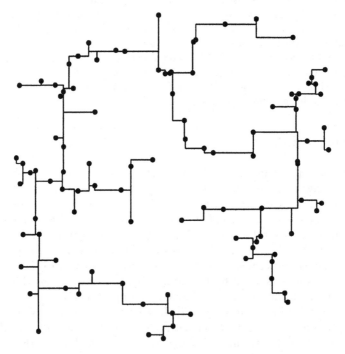

Fig. 1. A Steiner minimum tree for 100 terminals

All previous approaches to compute rectilinear SMT's (see [5] and [9] for a review) use a two-phases scheme developed by Pawel Winter: First a sufficient number of full components is computed and in the second phase a Steiner minimum tree is constructed using dynamic programming or Branch & Bound. For both approaches the number of full components constructed in phase 1 is the main parameter for the total running time; thus all algorithms try to manage with a small number of full components.

The experiences in [5] and [9] show that the dynamic programming approach leads to the best theoretic results, while a Branch & Bound algorithm [14] has the best practical behaviour. The theoretically best algorithm ([15]) with a complexity of $n^{O(\sqrt{n})}$ is highly impractical because of the large constant in the exponent. The most advanced work on dynamic programming [5] gives simple algorithms that perform quite poorly in practice (27 random points in a day). Here Steiner minimum trees are constructed for all possible subsets of S with increasing sizes by putting together a full component FC and a previously computed Steiner minimum tree ST for a smaller subset if FC and ST have exactly one terminal in common. If two trees for the same subset are constructed then only the shorter tree is stored. The total time complexity is $\sum_{m=3}^{n} \binom{n}{m} \cdot m \cdot f(m)$, where $f(m)$ denotes the number of full components within a terminal set of size m. In [5], $f(m)$ is bounded by 1.62^m and a time bound of $O(n^2 \cdot 2.62^n)$ follows.

The known Branch & Bound algorithms [14] compute lower bounds for Steiner trees containing a certain set $Comp$ of full components. If this lower bound is larger than an already known Steiner tree for S then the full components in $Comp$ may not arise together in the optimum. Otherwise a new full component is added to $Comp$. It is clear that the practical behaviour of this procedure depends highly on the number of full components; though it works well in practice it can only guarantee a double exponential running time.

We considerably reduce the number of necessary full components for both approaches and the number of all subsets of S to be considered for the dyn. prog.

To demonstrate the enormous importance of the number of full components for both the dynamic programming and the Branch & Bound approach we give two concrete examples in Table 1 (with 20 resp. 31 terminals). All experiments in this paper are performed on Sparc Ultra 140 workstations. To all CPU times we must add the time needed for phase 1 which is comparatively small for large problems (about 5 minutes for 55 terminal problems).

terminals	full components Salowe/Warme	CPU Branch&Bound	full components this paper	CPU Branch&Bound	CPU dyn. prog.
20	106	23 sec	50	1 sec	11 sec
31	168	42452 sec	63	30 sec	2530 sec

Table 1. The influence of full components in B & B and dyn. prog.

In the next section we review some basics on rectilinear Steiner tree problems, we derive new properties for full components which could be part of the SMT. Further we give statistics for the enormous profit we gain. In Section 3 we propose an alternative way to put the components together such that many subsets do not need to be considered and give theoretical bounds as well as practical results for this approach. Finally, we demonstrate the implementation of our approaches giving the results of about one thousand experiments. Following a tradition of [5,14] to show a Steiner minimum tree of relatively big size on page 2 of the paper, we display in Fig. 1 an SMT with 100 terminals. The theoretical results used in this paper will be summarized in the poster session of [4].

2 Basics and Full Components

A classical Lemma of Hwang [8] restricts the topology of the full components:

Lemma 1. *Let $n > 4$. Suppose that the SMT of a given set of terminals is a full Steiner tree. Then there exists an SMT that either consists of a single line with $n - 1$ alternating incident segments, or of a corner with $n - 3$ alternating segments being incident to one leg and a single segment being incident to the other leg.*

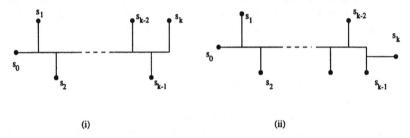

(i)   (ii)

Fig. 2. Two shapes of a full Steiner component

In Fig. 2 we show the two shapes described in Lemma 1. The single middle line is called *Steiner chain*, the only terminal on the Steiner chain is the *root* of the full component. Note that the second shape is similar to the first one, except of the last leg which is oriented in the same way as the Steiner chain. From now on we only consider full components of type (i) with a horizontal Steiner chain having the root at the left end. The number of components of type (ii) can be assumed not to exceed the number of components of type (i) ([3]).

At the beginning of phase 1 of the algorithm every subset of the terminals is computed that has one of the shapes of Fig. 2; this set of full components is sufficient to construct a Steiner minimum tree. In many cases we can easily *prove* that some full component cannot be part of a Steiner minimum tree.

2.1 Reducing the Number of Full Components

2.1.1 Direct Tests

To get better bounds for the number of full components we need to explore more properties of relevant full components.

We say that a full component is defining *empty rectangles* if for every pair of consecutive terminals of the component the smallest rectangle containing them does not contain any other terminal of the given set. From [2] we know

Lemma 2. *For any set of terminals there is a Steiner minimum tree using only full components defining empty rectangles.*

Corollary 1. *When we fix the set of terminals above (below) the Steiner chain, then the set of terminals below (above) the chain is determined automatically, namely the terminals with the smallest distance to the chain in the corresponding vertical stripes.*

Let n' be the number of terminals to the right of a fixed root t; we check on which side of the chain there are more terminals (say below). Hence at most $n'/2$ terminals lie above of the chain. Choose $k/2$ out of them. They completely determine the full component of size k. This can be done in $\binom{n'/2}{k/2}$ ways. Hence there are at most

$$4 \cdot \sum_{n'=1}^{n} \sum_{k=1}^{n'} \binom{n'/2}{k/2} \leq 4 \cdot \sum_{n'=1}^{n} 2 \cdot \sum_{k=1}^{n'/2} \binom{n'/2}{k} \leq 4 \cdot \sum_{n'=1}^{n} 2 \cdot 2^{n'/2} \leq 32 \cdot 2^{n/2}$$

full components for n terminals.

Lemma 3. *For a set S of n terminals, there are only $O(2^{n/2}) \approx O(1.42^n)$ full components defining empty rectangles.*

To improve the characterization we introduce a new concept for full components. A full component FC with Steiner points $c_1, \ldots, c_k$ is called a *tree star* if the minimum spanning tree of the point set $S \cup \{c_1, \ldots, c_k\}$ contains every edge of FC (i.e. the connections of terminals to Steiner points and the connections between two Steiner points) and if FC defines empty rectangles. In [3], we show

Lemma 4. *For any set of terminals S there is a Steiner minimum tree containing only tree stars as full components.*

Corollary 2. *In a tree star the triangles defined by $45°$-diagonals on both sides of each segment do not contain any other terminal of S (see Fig. 3).*

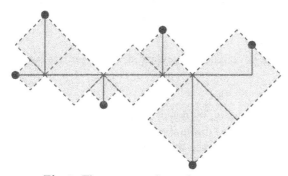

Fig. 3. The empty regions of a tree star.

2.1.2 Component Augmentation

Removing full components that do not fulfill the constraints described so far (Hwang's shapes, empty rectangles, tree stars) leads to a set of full components with a cardinality similar to the candidates of Salowe/Warme. Our next step considerably decreases the number of full candidates further by a factor of approximately two in practice and is responsible for our good results: We can delete a full component FC if we can show that FC must be a part of a larger full component in the optimum.

Lemma 5. *Let* FC *be a full component,* u *a terminal not belonging to* FC, dist *the shortest distance between* u *and* FC *and* d *a point on an edge of* FC *with distance* dist *to* u. *Let* G *be the complete graph having as vertices all terminals and all Steiner points arising in any full component of the candidate set and having edge lengths according to the rectilinear metric. If every path in this graph connecting* u *and* d *and not using the edge* (u, d) *contains an edge being longer than* dist *then* FC *cannot be a full component of a Steiner minimum tree.*

Proof. If FC would belong to a Steiner minimum tree T, u would be connected to FC via a path p; let e be the longest edge along p; adding the segment (u, d) to T creates a cycle containing e; since e is longer than *dist* the deletion of e leads to a Steiner tree being shorter than T, a contradiction. ◇

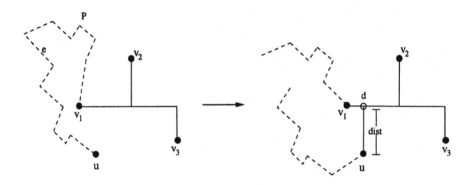

Fig. 4. A tree star is extended to a larger tree star.

E.g. the tree star $FC = (v_1, v_2, v_3)$ in Fig. 4 will be extended to the tree star $FC' = (u, v_1, v_2, v_3)$ if the terminal u cannot be connected to FC via a path consisting of edges being shorter than the distance between u and FC.

Note that the size of the (in Lemma 5 complete) graph G can be considerably reduced, since the paths will only contain edges between terminals and Steiner points or between Steiner points adjacent in the same full component. Furthermore, the Steiner points of full component FC^* do not need to be included in G, if the Steiner chains of FC^* and FC cross or they have more than one terminal in common, since FC^* cannot arise together with FC in an SMT in that case. A careful implementation using this fact yields extremely good results.

Further techniques include the simultanous augmentation test for a component together with several terminals, and the application of dynamization techniques since the graph G does not change too much for different subsequent tests. These ideas are currently implemented and are not included in the statistics.

2.2 Practical Results

In practical examples the number of tree stars is small: In [14] a very large number of examples is considered and it can be seen that the number of their full components is less than $4n$ (linear!). Our experiments confirm these values (see the rightmost column of Table 2) Our implementation can delete about half of them, therefore we have to deal with about $1.7n$ full components for randomly generated examples (see the second line of Table 4). Table 2 shows the effect of our tests for some typical 'average case' examples (each one has exactly the average number of full components shown in Table 4): The number of generated full components, i.e. full components defining empty rectangles, is reduced by applying first some simple fast tests, then the tree star test and finally the augmentation test.

n	# FC	simple tests	tree star	augmen- tation	Salowe Warme
30	1507	547	109	46	114
35	2726	910	123	56	138
40	2959	936	164	67	168
45	4770	1490	167	76	172
50	5507	1607	188	87	186
55	6607	1755	192	90	184
60	8341	2545	231	111	233
65	11086	3202	255	117	232

Table 2: The number of full components.

2.3 Theoretical Bounds

2.3.1 Worst-Case Instances

In this section we analyse the concept of tree stars theoretically. From Corollary 1 we know that only the terminals at one side of the Steiner chain are evident. Say that these terminals ($S \uparrow$) are above the chain; they are referred to as *upper terminals* in the following, terminals below the chain are called *lower terminals*. We will show that not every subset of $S \uparrow$ can define a tree star. If in the situation of Fig. 5 l and r are consecutive upper terminals of some tree star τ and l' is the lower terminal between l and r belonging to τ, then at least one of any four subsets $S_1, S_2, S_3, S_4 \subseteq (S \uparrow)$ with $a \notin S_1, b \in S_1, a \in S_2, b \in S_2$ $b \in S_3, c \in S_3$, $b \in S_4, c \notin S_4$ cannot define a tree star. The proof for this fact makes use of Corollary 2 which implies that the triangles indicated in Fig 5 must be empty; it can be found in [3].

A careful analysis of the consequences of this result (see [3]) leads to

Lemma 6. *The number of tree stars is not larger than the number of binary strings of length $n/2$ with: There is no sequence of at least five '0' in the string, each '100001'-sequence has its first '0' at a position p with $(p \bmod 3) = 0$ and each '10001'-sequence has its first '0' at a position p with $(p \bmod 3) \neq 2$.*

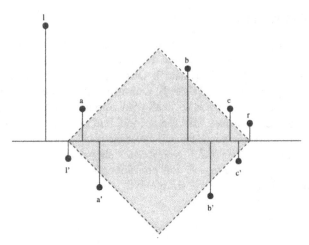

Fig. 5. Not every star is a tree star.

Each binary string represents a subset of the upper terminals (thus having a length of $n/2$). A bit '1' in the string can be interpreted as a terminal that is part of the corresponding tree star, whereas a '0' stands for a terminal that does not belong to the tree star. We are now ready to formulate the main result of this section:

Theorem 1. *In a rectilinear Steiner tree problem with n terminals there are $O(1.384^n)$ tree stars.*

Proof. The number of strings fulfilling the conditions of Lemma 6 (call them *good* strings) follows the rules of the following recurrence:
$$F(0) = 1; \qquad F(1) = 2; \qquad F(2) = 4; \qquad F(3) = 8;$$
$$F(k) = F(k-1) + F(k-2) + F(k-3) + F(k-4) + F(k-5), \text{ if } (k \bmod 3) = 0.$$
$$F(k) = F(k-1) + F(k-2) + F(k-3) + F(k-4), \text{ if } (k \bmod 3) = 2.$$
$$F(k) = F(k-1) + F(k-2) + F(k-3), \text{ if } (k \bmod 3) = 1.$$
This can be seen in the following way: A k-bit string with a '1' at the end is good if and only if the $(k-1)$-bit string without the final '1' is good. A k-bit string with '10' at the end is good if and only if the $(k-2)$-bit string without the final '10' is good, and so on. A k-bit string with '1000' at the end is good if and only if the $(k-4)$-bit string without the final '1000' is good and the first '0' of the '1000' is at a position p with $(p \bmod 3) \in \{0, 1\}$; thus $(k \bmod 3) \in \{2, 0\}$. A k-bit string with '10000' at the end is good if and only if the $(k-5)$-bit string without the final '10000' is good and the first '0' of the quadruple is at a position p with $(p \bmod 3) = 0$; thus $(k \bmod 3) = 0$.
Using automatic systems for resolving recurrences we get that $F(n) \in O(1.384^n)$ for a fixed root; following the argumentation for Lemma 3 we can prove that the overall number of full components exceeds this number only by a constant factor, so we have $O(1.384^n)$ tree stars overall. It is worth to note that the factor in the 'O' is small (about 0.045). $\diamond$

We have shown that applying the tree star test can reduce the number of full components to be considered; nevertheless the concept of tree stars does not lead to a polynomial number of necessary full components: In [3] we give an RST problem with $\Omega(1.32^n)$ tree stars.

2.3.2 Random Instances

Let the n terminals be randomly distributed within the unit square. Salowe/Warme proved a bound of $O(n^2 \cdot 2^k \cdot k^k)$ for full components of size k in this setting. By Chernoff bounds, we can see that the maximum empty diamond has side length $O(\sqrt{\log n/n})$ with high probability. Then by Corollary 2, all terminals of a tree star have distance $O(\sqrt{\log n/n})$ from the Steiner chain. Since there are only $O(\sqrt{n \cdot \log n})$ terminals within this distance of the Steiner chain, there are only tree stars of at most this size. Hence,

Theorem 2. *The number of tree stars in a random problem instance is*

$$O(\textstyle\sum_{k \geq 1} n \cdot \binom{\sqrt{n \cdot \log n}}{k/2})) \leq O(n \cdot 2^{\sqrt{n \log n}}).$$

3 Phase 2 in Dynamic Programming: The Merge – Step

In our previous analysis of the running time, we assumed that for *all* subsets R an SMT(R) will be constructed. But we only need to compute the subsets R that could be extended to an SMT for the total point set S by subsequently adding other full components. There are many subsets R that do not fulfill this condition and that can be excluded: If e.g. two terminals in R lie at opposite corners and the terminals between them are not in R, the tree SMT(R) will obviously not appear as a part of SMT(S) and we do not need to consider R. We now make this observation more precise.

Let $Q(a, b)$ be the axisparallel square defined by the points a, b such that a and b lie at opposite sides of $Q(a, b)$ and the center of $Q(a, b)$ lies in the middle of the segment $[ab]$ (see Fig. 6(i)). We define EQ to be the graph (V, E) with V to be the set of terminals and $E = \{(a, b); Q(a, b)$ does not contain any terminals$\}$ (see Fig. 6(ii) for an example). The black vertices in Fig. 6(ii) show a subset of terminals that the algorithm should not consider (one of the terminals lies inside the fat cycle, the other one outside). In [3] we prove

Lemma 7. *Let R be any subset of the terminals, $\overline{R} = V - R$ and $EQ(\overline{R})$ is the induced subgraph of $\overline{R}$. If $EQ(\overline{R})$ contains a simple cycle such that there are vertices of R inside and outside of the cycle then we do not need to consider the subset R of terminals.*

We can show that an algorithm that makes use of this property has to consider less than α^n subsets of terminals in random problem instances for $\alpha < 2$. Together with Theorem 2, we can only prove that for random terminal sets the running time is smaller than β^n for a constant $\beta < 2$. There are some worst-case examples where the number of subsets cannot be reduced by the technique of

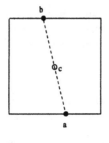 

Fig. 6.   (i) $Q(a,b)$ (ii) An example for a 'not reasonable' subset

graph EQ, but for these cases only a small fraction of them will be considered by other reasons.

We summarize the theoretical analyses and combine the last results from this section, Theorem 2 and 1:

Theorem 3. *We can solve the rectilinear Steiner tree problem for n points in the plane in time $O(n \cdot 2.384^n)$ in the worst case, and for n randomly distributed points we only need $O(\beta^n)$ time, with $\beta < 2$.*

In practical examples the effect that not every subset of S is considered is much clearer. We combine the EQ-concept with an idea that cannot decrease the worst-case bound, but leads to obvious improvements in practice: Determine any terminal $s_0 \in S$ as a start point. Construct only subsets ST containing s_0. Clearly this strategy reduces the number of constructed subsets because no subset ST with $s_0 \notin ST$ will ever be considered. Since there is a way to construct a Steiner minimum tree T using this strategy (namely starting with the full component of T containing s_0 and adding one by one just the full components belonging to T) the correctness of the algorithm follows.

Table 3 shows the number of considered subsets R for some typical examples. This number is proportional to the total running time.

problem size	total number = x^n
25	$1916 = 1.35^n$
25	$3349 = 1.38^n$
30	$40829 = 1.42^n$
30	$26927 = 1.38^n$
35	$19645 = 1.33^n$
35	$179115 = 1.41^n$
40	$628230 = 1.40^n$
40	$8701481 = 1.49^n$

Table 3: How many STs are computed?

We summarize our results by the following algorithm:

(1) **for** $m = 2$ **to** $|S|$ **do**
(2) $L(m) := \{T \in F(S); |T| = m, s_0 \in T\};$
(3) $F(m) := \{T \in F(S); |T| = m\};$
(4) **od;**
(5) **for** $m = 2$ **to** $|S|$ **do**
(6)   **for** $i = 2$ **to** $|S|$ **do**
(7)   **for all** $A \in F(i)$ **and** $B \in L(m)$ **with:**
(8) $|A \cap B| = 1$ **and** the component A does
(9)   not cross an edge $(a, b) \in EQ(\overline{A \cup B})$
(10)   **do** $R := A \cup B;$
(11) $L(m + i - 1) := L(m + i - 1) \cup R;$
(12) $l(R) := min\{l(R), l(A) + l(B)\};$
(13) **od; od; od;**
(14) **return** $l(entry(L(|S|)));$

The F-sets contain the set of full components computed in phase 1. In the $L(m)$-sets we only store sets of size m that contain terminal s_0 and could have been constructed by combining full components under the conditions in lines (8) and (9). Note that the first condition in line (8) means that sets A and B can be combined, and the second in line (9) concerns the conditions given by Lemma 7. So by far, not all subsets of size m will be considered.

4 Practical Results

4.1 Overview

Our approach has two main advantages compared with former algorithms:

1. We use a smaller number of full components.
2. We do not consider all possible subsets of the set of terminals.

We carried out experiments with two different algorithms: Our dynamic programming approach closely follows the description given in the previous section. Our best results have been achieved by combining our phase 1 with the Branch & Bound approach. Therefore, we will mainly concentrate on the results of this *combined algorithm*. We run 100 random problem instances for each of the problem sizes 30, 35, 40, 45, 50, 55 and 60.

Table 4 summarizes

- the average number of full components used by Salowe/Warme (for comparisons).
- the average number of full components used by the combined algorithm.
- the average CPU time for 80% of the examples (in order to avoid an excessive influence of some few extreme examples we omitted the 10% fastest and the 10% slowest examples).
- the time when 90% of the examples were finished.

- the influence of the number of full components on the running time. To that purpose we subdivided the set of examples E into two equal-sized sets E_{low} and E_{high} such that the examples in E_{low} need less full components than the examples in E_{high}. We show the minimum and the average time needed by the examples in E_{high} as well as the maximum and the average time needed by the examples in E_{low}.

The graphical plot in Fig. 7 shows the average running time for the 'middle' 80% of the examples (thus every point stands for 80 examples) for the combined algorithm and for the algorithm of Salowe and Warme. In Fig. 8 we give (logarithmic scaled) the time $t(x)$ for every percentage x when $x\%$ of the examples were finished.

Table 5 shows the average length of the SMT (the terminals are randomly distributed on a 10000×10000 grid) for our examples as well as the average relative difference, given in percent ($\frac{MST-SMT}{SMT} \times 100$).

The statistics for the examples of size 60 concern only the 50 problems in E_{low}; problems in E_{high} cannot be finished within one day in most of the cases.

	30	35	40	45	50	55	60
average number of full components (Salowe/W.)	107	124	149	164	190	205	229
average number of full components (combined alg.)	46.7	55.9	67.3	76.1	86.9	90	111
average CPU-time of the 80% middle examples	0.67	3.77	31.9	229	780	6916	–
90% were finished after (seconds)	4.0	31.4	366	2502	13191	42123	–
minimum CPU-time in E_{high} (seconds)	0.2	0.4	1.1	3	2	9	–
average CPU-time in E_{high} (seconds)	4.0	28.3	356	3262	9129	61083	–
maximum CPU-time in E_{low} (seconds)	0.4	1.3	48.5	164	2341	1724	92604
average CPU-time in E_{low} (seconds)	0.2	0.4	4.3	126	114	259	9168

Table 4. Statistics

	30	35	40	45	50	55	60
length of SMT	41394	43893	47771	50238	52938	55290	–
length of SMT in E_{high}	42564	44466	48244	51011	53415	57232	–
length of SMT in E_{low}	40585	43538	47286	49502	52499	53537	56986
rel. difference MST vs. SMT	12.65	12.28	12.77	12.39	12.93	12.41	–
rel. difference in E_{high}	13.48	13.09	13.56	12.88	12.99	12.69	–
rel. difference in E_{low}	11.83	11.47	11.98	11.89	12.88	12.14	11.93

Table 5. Average lengths of SMT's and MST's.

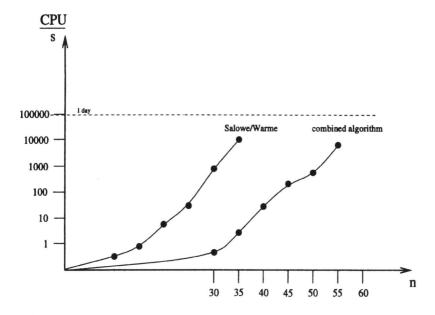

Fig. 7. Average CPU-times for Salowe/Warme and the combined algorithm

4.2 Discussion and Remarks

Our dynamic programming implementation can solve problem instances of size 42 in one day and has thus a larger power than the Branch & Bound algorithm of Salowe and Warme. Besides the better worst case running time it has an important advantage compared to every Branch & Bound algorithm: After a short while it is possible to predict the total running time while this is impossible using Branch & Bound. But only looking at the total running time Branch & Bound yields to much better results. Furthermore the much higher amount of space for phase 2 turned out to be the main bottleneck for the dynamic programming approach.

The *combined algorithm* using our phase 1 and the Branch & Bound approach of Salowe and Warme for phase 2 now can solve problems of size 55 in a day (and a large percentage of larger problems). Note that for algorithms with exponential running time this implies an enormous improvement. The reduction of the number of full components by a factor of more than 2 (from $4n$ to $1.7n$) does not quite mean a doubling of the solvability range for the problem size, but it comes near. The number of full components is not directly proportional to the running time of the Branch & Bound approach in practice, but it is a good indicator for it.

It is worth to mention that randomly generated problems are often easier to solve than problems where the terminals are equally distributed on the plane.

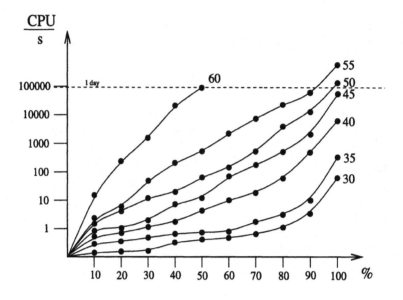

Fig. 8. Which percentage of the examples can be solved in what time?

An indication for this fact is our observation (line 2 and 3 of Table 5) that the Steiner trees in hard cases (E_{high}) are longer than in simpler cases (E_{low}). Large total length roughly means the presence of less short edges which again means that the terminals are less clustered. They are better distributed.

The relative difference seems to be largely independent from the number of terminals and lies between 12% and 13%. Line 5 and 6 of Table 5 indicate a relation between the number of full components and the relative difference; if the relative difference is small (i.e. there is no large difference between the SMT and the MST) then the number of full components is small (i.e. the problem is 'easy'). In an extreme situation, if MST = SMT, there is obviously no relevant component of size three or more, and thus the number of full components is equal to the number of edges of the MST, namely $n - 1$.

In some cases even large randomly generated examples are easy enough to be solved. Fig. 1 shows a Steiner minimum tree for a random set of 100 points that was computed by our program in only 15 minutes. For this example, 146 full components survived the tests of our phase 1.

These results can directly be applied even to large problem sizes, as they usually occur in VLSI-applications. Starting from a not-so-good solution produced by a standard heuristic, we can subdivide the tree into small subtrees of size say less than 50. Solving these subtrees optimally can be done with our approach within reasonable time and the solutions are composed finally to a final Steiner tree of seemingly very good quality. Note that the subproblems can be handled independently such that a parallel approach seems natural.

A critical point is the correctness of our programs. We do not know any efficient check of the optimality of the Steiner trees computed by our programs. We only compared them to some heuristics (like 1-Steiner heuristic [10]) and always found that our solutions have been better than the solution by the heuristics. This is not a proof but only an indication for the correctness of our implementation.

Acknowledgements Thanks go to Carsten Hess and Bernd Schatz for their contributions to the full component problem, and to Job Sibeyn and Klaus Reinhardt for the contributions to the merge-step. Bernd Schatz did all the implementations.

References

1. S. Arora, *Polynomial Time Approximation Schemes for Euclidian TSP and other Geometric Problems*. In Proc. of 37th Annual IEEE Symp. on Foundations of Computer Science, 2–11, 1996.
2. P. Berman and V. Ramaiyer. *Improved approximations for the Steiner tree problem*. Proc. of 3d ACM-SIAM Symp. on Discrete Algorithms, 325–334, 1992.
3. U. Fößmeier and M. Kaufmann. *On Exact Solutions for the Rectilinear Steiner Tree Problem*. Technical Report WSI-96-09, Universität Tübingen, 1996.
4. U. Fößmeier and M. Kaufmann. *On Exact Solutions for the Rectilinear Steiner Tree Problem*. ACM Conference on Computational Geometry: Poster Session, 1997 (3 pages).
5. Ganley J. L., J. P. Cohoon, *Optimal rectilinear Steiner minimal trees in $O(n^2 2.62^n)$ time*, Proc. 6th Canad.Conf.Comput.Geom., 308-313, 1994.
6. M. R. Garey, D. S. Johnson. *The Rectilinear Steiner Problem is NP-Complete*. SIAM J. Appl. Math., 32, 826-834, 1977.
7. M. Hanan. *On Steiner's Problem with Rectilinear Distance*. SIAM J. Appl. Math.,14, 255-265, 1966.
8. F. K. Hwang. *On Steiner Minimal Trees with Rectilinear Distance*. SIAM J. Appl. Math.,30, 104-114, 1976.
9. F. K. Hwang, D. S. Richards and P. Winter. *The Steiner Tree Problem*. Annals of Disc. Math. 53, North-Holland, 1992.
10. A. B. Kahng and G. Robins. *A new class of iterative Steiner tree heuristics with good performance*. IEEE Trans. Comp.-Aided Design 11, 893-902, 1992.
11. B. Korte, H. J. Prömel, A. Steger. *Steiner Trees in VLSI-Layouts*. In Korte et al.: Paths, Flows and VLSI-Layout, Springer, 1990.
12. Th. Lengauer. *Combinatorial Algorithms for Integrated Circuit Layout*. John Wiley, 1990.
13. D. Richards. *Fast Heuristic Algorithms for Rectilinear Steiner Trees*. Algorithmica, 4, 191-207, 1989.
14. J. S. Salowe and D. M. Warme. *Thirty-Five Point Rectilinear Steiner Minimal Trees in a Day*. Networks Vol. 25, 69-87, 1995.
15. W. D. Smith. *How to find Steiner minimal trees in Euclidian d-space*. Algorithmica 7 (1992), 137-177.
16. A. Z. Zelikovsky. *An 11/8-approximation Algorithm for the Steiner Problem on Networks with Rectilinear Distance*. In Sets, Graphs and Numbers. Coll. Math. Soc. J. Bolyai 60: 733-745, 1992.

Dynamically Switching
Vertices in Planar Graphs*
(Extended Abstract)

Daniele Frigioni[1]   Giuseppe F. Italiano[2]

[1] Dipartimento di Informatica e Sistemistica, Università di Roma "La Sapienza", via Salaria 113, I–00198 Roma, Italy. frigioni@dis.uniroma1.it.
[2] Dipartimento di Matematica Applicata ed Informatica, Università "Ca' Foscari" di Venezia, Venice, Italy. italiano@unive.it, http://www.dsi.unive.it/~italiano/.

Abstract. We consider graphs whose vertices may be in one of two different states: either *on* or *off*. We wish to maintain dynamically such graphs under an intermixed sequence of updates and queries. An update may reverse the status of a vertex, by switching it either *on* or *off*, and may insert a new edge or delete an existing edge. A query tests properties of the subgraph induced by the vertices that are *on*. We give efficient algorithms that maintain information about connectivity on planar graphs in $O(\log^3 n)$ amortized time per query, insert, delete, switch-on and switch-off operation over sequences of at least $\Omega(n)$ operations, where n is the number of vertices of the graph.

1 Introduction

In the last years research in dynamic graph algorithms has been a blossoming field (see e.g. [1–11,13–18,20,21,23]). The main dynamic model that has been considered in the literature is the following. We are given a graph $G = (V, E)$, and we wish to maintain some property $\mathcal{P}$ in G during edge deletions and edge insertions. We refer to this as the *dynamic edge model*. If the graph represents a communication network, for instance, the edge update operations reflect the network changes as *links* go up and down throughout the lifetime of the network. The dynamic edge model, however, does not capture the whole set of possible updates in a dynamic graph. Indeed some important changes, such as the failure and repair of *processors* (rather than links) in the previous example, do not fall directly in this model. In order to account for these updates, one could consider a *dynamic vertex model* instead, where the following modifications are allowed:

* Work supported in part by EU ESPRIT Long Term Research Project ALCOM-IT under contract no. 20244, and by the Italian MURST Project "Efficienza di Algoritmi e Progetto di Strutture Informative". The research of the second author was supported in part also by a Research Grant from University "Ca' Foscari" of Venice and by the German–Italian Program "Vigoni 1997".

a vertex of G may be removed together with all its incident edges, or a removed vertex may be added back to G together with its incident edges.

The dynamic edge model and the dynamic vertex model can be combined into a more general model, in which a certain property $\mathcal{P}$ is maintained on a graph G subject edge insertions, edge deletions, plus removal and reinsertions of vertices together with their incident edges. We refer to this as the *complete dynamic model*. Since algorithms for the dynamic edge model are usually able to support insertions and deletions of disconnected vertices, one could implement removing a vertex v as follows: first delete all the edges incident to v, and then remove the disconnected vertex v. Similarly, inserting v with all its incident edges can be implemented as a sequence of multiple updates in the dynamic edge model. We are interested in *efficient* algorithms in the complete dynamic model, namely algorithms that are faster than these simple–minded applications of a dynamic edge algorithm. Despite a rich research in dynamic graph algorithms, to the best of our knowledge no efficient complete dynamic algorithm was previously known.

In this paper, we start the study of efficient algorithms in the complete dynamic model. Assume that we are given an undirected graph G, whose vertices can be in two different states: either *on* or *off*. In communication networks, vertices that are *on* correspond to active processors, and vertices that are *off* to faulty processors. An active processor may be switched *off* because of a fault, while a faulty processor may be repaired and turned *on* again in the future. We wish to perform on G updates of the following kinds:

switch–on(v):   Turn vertex v *on*;

switch–off(v):   Turn vertex v *off*;

insert(x, y):   Insert an edge between x and y;

delete(x, y):   Delete the edge between x and y;

Let G_{on} be the subgraph of G induced by the vertices that are switched *on*. Namely, G_{on} contains all the vertices that are *on* and all the edges whose endpoints are both *on*. If G models a communication network, at any time G_{on} represents the portion of the network that is operational. Note that, differently from insert and delete operations, each switch–off and switch–on operation can cause many changes in G_{on}, since it may remove or add as many as $\Theta(n)$ different edges. The query operations we would like to perform relate to some property $\mathcal{P}$ in G_{on}. For instance, we may be interested in checking whether any two given vertices x and y are joined by a path consisting solely of vertices that are switched *on* (i.e., whether a connection through active processors can be established between x and y), or whether x and y are joined by at least two edge– or vertex–disjoint such paths (i.e., whether processors x and y can survive the failure of one active link or one active processor). A path that contains only vertices that are switched *on* is referred to in the following as a *on–path*.

In this paper, we consider the problem of maintaining information about connectivity on undirected graphs subject to complete dynamic operations, and we prove that this can be done very efficiently on planar graphs. Namely, we show how to maintain connectivity on a planar graph under an intermixed sequence of switch–on, switch–off, insert, delete, and query operations. In order to put our results in perspective, we mention the bounds for query, switch–on and switch–off operations attainable by means of existing algorithms. If we record only whether each vertex is *on* or *off*, we can carry out each update in $O(1)$ time. However, a query operation could require performing a visit of the whole planar graph G in the worst case, and thus would have a worst–case running time of $O(n)$. On the other hand, we could apply an edge dynamic algorithm such as the one by Eppstein *et al.* [7], which would support queries faster (i.e., in at most $O(\log n)$ time). However, since switching a vertex *on* or *off* could cause as many $\Theta(n)$ edges to be deleted or inserted, each switch–on or switch–off could require $O(n \log n)$ time in the worst case. In this paper, we show how to adapt the separator based sparsification technique of Eppstein *et al.* [5,6] to this problem, and give algorithms and data structures for solving the problem in $O(\log^3 n)$ amortized time per insert, delete, switch–on, switch–off, and query operations over sequences of $\Omega(n)$ operations. This improves sharply on previous approaches.

Our algorithms exploit separator properties of planar graphs. Namely, we compute a recursive decomposition of a planar graph G into edge–disjoint connected subgraphs. This decomposition is obtained by using suitably large separators, which contain all the high–degree vertices of G, and supports connectivity queries in a quite natural fashion. However, it can be changed dramatically during updates. In order to perform updates faster, we maintain the properties of the decomposition in a lazy fashion. Many details, proofs and boundary cases are omitted from this extended abstract for lack of space.

2 Balanced Piece Decompositions

Let G be an n–vertex undirected planar graph. A *separator* of G is a set of vertices $S \subseteq V(G)$ whose removal disconnects G into two or more subgraphs roughly of the same size. The *size* of a separator S is given by the number of vertices in S. Assume that the vertices of G have non–negative costs summing to no more than one. The separator theorem of Lipton and Tarjan [19] states that the vertices of G can be partitioned into three sets A, B, and S such that no edge joins a vertex in A with a vertex in B, neither A nor B has total cost exceeding $\frac{2}{3}$, and S contains no more than $2\sqrt{2n}$ vertices. The set S is a separator, and we call a separator of size at most $2\sqrt{2n}$ a *small separator*. As proved by Lipton and Tarjan [19], a small separator can be found in $O(n)$ time. This process can be repeated recursively, yielding a *separator decomposition* of a graph G. Each level of the separator decomposition partitions the edges of G: for $k \geq 0$, G is split into at most 2^k edge–disjoint subgraphs, with the partition at level k being a refinement of the partition at level $(k - 1)$. The deeper we go in the separator decomposition, the smaller is the size of the subgraphs and the larger becomes the size of the separator. Eventually, we will end up decomposing the graph G into a collection of constant–size edge–disjoint subgraphs by using a suitably large separator. Since a planar separator can be found in linear time [19], the total time required to compute this decomposition is easily seen to be $O(n \log n)$. Goodrich showed how to speed up this computation to $O(n)$ time [12].

Let S be any separator of G. The edges of G are partitioned by S into equivalence classes as follows: two edges are in the same class if and only if they are in a same path that contains no vertex of S (except as endpoints). These equivalence classes are called the *separation classes of G with respect to S*. A class consisting only of one edge is called a *trivial separation class*. Let $E_1, E_2, \ldots, E_p$ be the separation classes of G with respect to S. Each class E_j defines a *bridge* $B_j = (V_j, E_j)$ of S, where B_j is the subgraph of G induced by the edge set E_j (that is, $v \in V_j$ if and only if there is an edge of E_j incident on v). An edge (u, v), with u and v both in S, is called a *trivial bridge*. The vertices of a bridge which are elements of S are called its *attachments*. Note that by definition any two bridges B_i and B_j, for $i \neq j$, can share at most their attachments.

In this paper, we will consider planar graphs with n vertices and at most $\mu_0 n$ multiple edges, for some constant $\mu_0 > 0$. More precisely, we will start with a planar graph without multiple edges and apply transformations to this graph so that there can be multiple edges; however, we will place the restriction that there can be at most $\mu_0 n$ multiple edges at any time.

Lemma 1. *Let G be an n–vertex planar graph with at most $\mu_0 n$ multiple edges, for some constant μ_0, and let $\Delta > 0$ be a given integer. Let $V_\Delta \subseteq V(G)$ be the set of vertices in G whose degree is greater than Δ. Then $|V_\Delta| < (2/\Delta)(\mu_0 + 3)n$.*

Proof. The number of edges incident to vertices in V_Δ is at least $(\Delta |V_\Delta|)/2$. By Euler's formula, the number of edges of a planar graph with at most $\mu_0 n$ multiple edges is $m \leq 3n - 6 + \mu_0 n$. Hence, $(\Delta |V_\Delta|)/2 \leq m \leq (\mu_0 + 3)n - 6$, and the lemma follows.

Fix $\Delta_0 = 124(\mu_0 + 3)$. Given a planar graph G, we refer to vertices of degree greater than Δ_0 as *high-degree* vertices, and refer to the remaining vertices as *low-degree* vertices. As a consequence of Lemma 1, a n–vertex planar graph with no more than $\mu_0 n$ multiple edges has less than $\frac{n}{62}$ *high-degree* vertices.

Definition 2. Let G be a planar graph with n vertices. A *piece decomposition* of G is defined by a separator $S \subseteq V(G)$ and a collection of subgraphs (or *pieces*) P_i, having the following properties:

(1) Each piece P_i such that $|S \cap P_i| > 2$ is a bridge of S;
(2) Each piece P_i such that $|S \cap P_i| \leq 2$ is the union of bridges of S sharing the same attachment(s);
(3) Each P_i has size bounded by a constant (not depending on n).

In what follows, we refer to a piece decomposition of G as $\prec S, P_i \succ$. Among all piece decompositions, we will work with *c-balanced piece decompositions*, which are defined as follows:

Definition 3. Let G be a planar graph, and let c be a given constant, $1/62 < c < 1$. A *c-balanced piece decomposition* of G is a piece decomposition $\prec S, P_i \succ$ of G such that:

(1) S contains all of the high–degree vertices of G;

(2) $|S| \leq c|V(G)|$; and

(3) The sum, over all pieces P_i, of $|S \cap P_i|$, is less than $15|S|$.

By Lemma 1, $c > 1/62$ is required to make room for high–degree vertices in S. Using some results from [5], we can prove the following lemma:

Lemma 4. *Let G be a planar graph, and let c be a given constant, $1/62 < c < 1$. A c–balanced piece decomposition $\prec S, P_i \succ$ of G can be computed in linear time.*

3 Certificates

Balanced piece decompositions are the first ingredient of our data structure. Another ingredient needed is the notion of certificate.

Definition 5. Let graph property $\mathcal{P}$ be fixed, and assume that we are given a graph G with a set $X \subseteq V(G)$. A *global certificate for X in G* is a graph $\mathcal{C}$, with $X \subseteq V(\mathcal{C})$, such that for any H with $V(G) \cap V(H) \subseteq X$, $V(\mathcal{C}) \cap V(H) \subseteq X$, $G \cup H$ has property $\mathcal{P}$ if and only if $\mathcal{C} \cup H$ has the property.

According to this definition, if G is a planar graph that is split into G_1 and G_2 by a certain separator X, we can maintain information about property $\mathcal{P}$ for G also after replacing G_1 with a global certificate $\mathcal{C}_1$ for X in G_1. Definition 5 is actually stronger. Indeed, let $\mathcal{P}$ be connectivity, and let graph G_1 be fixed. Let $\mathcal{C}_1$ be a global certificate for X in G_1: then, for *any* graph H such that $V(H) \cap V(G_1) \subseteq X$, $\mathcal{C}_1 \cup H$ is connected if and only if $G_1 \cup H$ is connected.

There are some properties that can be described either as global graph properties or in terms of pairs of vertices. More generally, let $\mathcal{P}$ a property of graphs: we say that $\mathcal{P}$ is *local* if it can be defined with respect to a particular pair $\langle x, y \rangle$ of vertices in the graph. A query related to property $\mathcal{P}$ is referred to as *global* if $\mathcal{P}$ is meant for the entire graph, and is referred to as *local* if $\mathcal{P}$ is meant for a particular pair $\langle x, y \rangle$. Examples of such properties are connectivity, and edge and vertex connectivity, which are defined both for the entire graph (*global property*) as well as for any pair of vertices in the graph (*local property*). While maintaining a graph under sequences of updates on the edges/vertices, at any time we might ask, for instance, queries on whether the entire graph is connected (*global query*), or rather we might want to ask queries on whether any two given vertices are connected (*local query*). For local properties, we need a slightly different notion of certificate, which we call a *local certificate*.

Definition 6. Let $\mathcal{P}$ be a local property of graphs, and let G be a graph with a set $X \subseteq V(G)$. A graph $\mathcal{C}$ is a *local certificate* of $\mathcal{P}$ for X in G if and only if for any H with $V(H) \cap V(G) \subseteq X$, and any x and y in $V(H)$, $\mathcal{P}$ is true for $\langle x, y \rangle$ in $G \cup H$ if and only if it is true for $\langle x, y \rangle$ in $\mathcal{C} \cup H$.

Note that a local certificate C has to preserve the behavior of the property not only with respect to the vertices in X, but also with respect to all vertices in H. For instance, let $\mathcal{P}$ be connectivity, and let C be a local certificate for X in G: then, for any H such that $V(H) \cap V(G) \subseteq X$, and any $x, y \in V(H)$, x and y are connected in $C \cup H$ if and only if they are connected in $G \cup H$. In this paper we use *compressed* certificates whose definition is given in the following and is borrowed from [5]:

Definition 7. Let graph property $\mathcal{P}$ be fixed, and let G be a given graph with $X \subseteq V(G)$. A *compressed certificate for X in G* is a certificate C for X in G, such that C has at most $O(|X|)$ vertices and edges.

As an example, we give here a local compressed certificate C for connectivity in G. This certificate will be used throughout the paper. The idea is to replace each connected component of G with a compressed spanning tree for that component. More formally, let G and $X \subseteq V(G)$ given, and let $C_1, C_2, \ldots, C_p$, with $p \geq 1$, be the connected components of G. For $1 \leq i \leq p$, define $X_i = X \cap V(C_i)$. Note that for $i \neq j$, $X_i \cap X_j = \emptyset$, and thus $\sum_{i=1}^{p} |X_i| = |X|$.

Form a graph C as follows. First compute a spanning tree $ST(C_i)$ for each connected component C_i. Then, compute a tree $T_X(C_i)$ in two phases, as follows. The first phase is called the *pruning* phase, and applies a sequence of the following operations on $ST(C_i)$: remove from $ST(C_i)$ a leaf $z \notin X$ together with its incident edge. After the pruning phase, we obtain a tree T whose leaves are all in X, and this property will remain true throughout the second phase. The second phase is called the *shortcutting* phase, and applies a sequence of the following operations on T: find in T a vertex $v \notin X$ adjacent to exactly two neighbors, say u and w, remove v from T and replace edges (u, v) and (v, w) with a single edge (u, w). Let $T_X(C_i)$ be the tree obtained at the end of the shortcutting phase, i.e., when no more such transformations can be applied. As $T_X(C_i)$ is a tree in which all vertices not in X_i have degree three or more, it is easily seen that the number of vertices of $T_X(C_i)$ is at most $2|X_i| - 2$, which is $O(|X_i|)$. Each vertex $x \in C_i$ may be included in $T_X(C_i)$ in four different ways, as follows:

(a) x belongs to $V(T_X(C_i))$;

(b) x lies in a path of $ST(C_i)$ that has been recursively contracted to edge (v, w) of $T_X(C_i)$ during the shortcutting phase. In this case, we say that x is *hidden inside edge (v, w)* and we refer to (v, w) as a *path–edge for x*.

(c) x lies in a path of $ST(C_i)$ that has been recursively contracted in a vertex v of $T_X(C_i)$ during the pruning phase. Namely, the path between x and v belongs to a subtree of $ST(C_i)$ which is rooted at v and which contains no vertex of X. In this case, we say that x is *hidden inside vertex v*, and we refer to v as a *subtree–vertex for x*.

(d) x lies in a path of $ST(C_i)$ that has been recursively contracted to vertex z of $ST(C_i)$ during the pruning phase, and z lies in a path of $ST(C_i)$ that has been recursively contracted to edge (v, w) of $T_X(C_i)$ during the shortcutting phase. In this case, we say that x is *hidden inside edge* (v, w) and we refer to (v, w) as a *subtree–edge* for x.

We define the above vertices as type (a), (b), (c), and (d), respectively. We say that a type (a) vertex is *preserved* in $T_X(C_i)$, while a type (b), (c), or (d) vertex is *absorbed* inside $T_X(C_i)$. The local compressed certificate C of connectivity for X in G is defined as the union of all $T_X(C_i)$, for $i = 1, 2, \ldots, p$.

Lemma 8. *Let G be a graph with n vertices and m edges, and let $X \subseteq V(G)$. The graph C defined above is a local compressed certificate of connectivity for X in G, and can be computed in $O(m + n)$ time.*

4   Recursive Piece Decompositions

Our data structure hinges upon a recursive piece decomposition of a planar graph, suitably combined with a careful choice of local compressed certificates. Informally, given a graph G, we first find a balanced piece decomposition of G, and replace the (constant–size) pieces in this decomposition with some properly chosen local compressed certificates. This will produce a graph G_1, which has two important properties:

1. The local connectivity of G_1 is related to the local connectivity of G because we are replacing subgraphs of G with their *local certificates*.

2. The size of G_1 is at most a constant fraction of the size of G, because of the size constraints of the balanced decomposition, and the fact that the certificates are *compressed*.

Then, we proceed recursively on G_1. As we reduce by a constant factor the size of the resulting graph at each level, this recursive decomposition will have at most logarithmic depth.

We now define in more details this recursive decomposition. Let c be a given constant, $1/62 < c < 1$. We first find a c–balanced piece decomposition $\prec S, P_i \succ$ of G, according to Definition 3. Namely, separator S has size at most $c|V(G)|$ and contains all of the high–degree vertices of G, the pieces P_i have constant size, and the sum, over all pieces P_i, of $|S \cap P_i|$, is less than $15|S|$. We next replace each piece P_i with a local compressed certificate of connectivity for $S_i = S \cap V(P_i)$ in G: we choose this certificate so that vertices of S_i appear in the leaves of the certificate. This is always possible, as P_i is either a bridge of S, or it is the union of bridges with at most two attachments (sharing the same attachments). Furthermore, a certificate like this can be computed in linear time. We refer to this certificate as $C(P_i)$.

We denote by $C_1(G)$ the graph obtained after replacing each piece P_i with $C(P_i)$: it can be shown that $C_1(G)$ is a certificate of G for local connectivity, and that $|V(C(P_i))| \leq 2|S_i| - 2$. Thus, the total number of vertices in $C_1(G)$ can be bounded as follows:

$$|V(C_1(G))| \leq |S| + \sum_{i=1}^{p} |V(C(P_i))| \leq |S| + 2 \sum_{i=1}^{p} |S_i| - 2.$$

Since $|S| \leq c|V(G)|$ and $\sum_{i=1}^{p} |S_i| \leq 15|S|$ by Definition 3, $|V(\mathcal{C}_1(G))| < 31c|V(G)|$. A suitable choice of constant c in the range $1/62 < c < 1/31$ yields that $\mathcal{C}_1(G)$ has at most $c_1|V(G)|$ vertices, for some constant c_1, $1/2 < c_1 < 1$. Thus, in linear time we have reduced the size of G by at least a constant fraction while preserving information about local connectivity. We now build recursively a similar data structure for the graph $\mathcal{C}_1(G)$, and so on at higher levels. Namely, at level 0, we have $\mathcal{C}_0(G) = G$. At level $\ell > 0$, we have a graph $\mathcal{C}_\ell(G)$, whose size is a constant fraction of the size of $\mathcal{C}_{\ell-1}(G)$, and whose vertex set is a subset of the vertices of $\mathcal{C}_{l-1}(G)$ and hence of G. We stop when the graph at the top level has a constant number of vertices.

In the following we give some details of our data structure. As it is rather involved, we refer the interested reader to the full paper for a complete description of the low–level details. For each vertex v in G we denote by $piece(v)$ the lowest level in the recursive piece decomposition in which v does not belong to the separator at that level. Namely, from level 0 to level $(piece(v) - 1)$ of the recursive piece decomposition v always belongs to the separators at those levels, while v does not belong to the separator at level $piece(v)$. As all of the high-degree vertices are contained in the separator, v must be of constant degree at this level. Furthermore, from level $(piece(v) + 1)$ to the topmost level the degree of v is never increased, and thus v is of constant degree from level $piece(v)$ up, independently on whether it belongs or not to the separators at those levels.

For each vertex v in G we denote by $absorb(v)$ the lowest level in the recursive piece decomposition for which the certificate at that level does not contain vertex v. Namely, for level i, $0 \leq i \leq absorb(v) - 1$, v always belongs to $\mathcal{C}_i(G)$, while v does not belong to $\mathcal{C}_j(G)$ for $j \geq absorb(v)$. Note that $absorb(v) \geq piece(v)$, as at any level vertices in the separator are automatically preserved in the certificate. Furthermore, we maintain links between adjacent levels as follows: for each vertex z at each level j of the decomposition we store a pointer to vertex z itself at level $(j-1)$, and a pointer to the item at level $(j+1)$ inside which vertex z is mapped (this might be either a vertex or an edge which absorbs z at level $(j+1)$, or z itself, if z appears also at level $(j+1)$).

The main difficulty with our recursive decomposition is that it can change quite dramatically during a sequence of updates, and it indeed turns out that maintaining explicitly a recursive piece decomposition can be very expensive. To circumvent this problem, we proceed in a lazy fashion. Namely, we initialize our data structure with a recursive piece decomposition of G. Throughout a sequence of operations, we tolerate graphs and separators at any level in this decomposition to grow, as long as all of the high–degree vertices are in the separator, each piece is connected, and the size of the separator and the graph is at most a constant number of times its initial value. Any time that either the separator or the whole graph at some level k in the decomposition increases its size by a certain constant factor, we rebuild our data structures from scratch starting from level k up. Besides recomputing separators and certificates, this rebuilding takes care of updating the information on $piece(v)$ and $absorb(v)$ for each affected vertex v, and the pointers between adjacent levels.

5 Implementing the Operations

For lack of space, we only concentrate here on sequences of query, switch–off and switch–on operations. However, our data structures can be extended so as to support insert and delete operations as well within the same time bounds. We recall that a switch–off(v) operation corresponds to deleting all the edges in G incident to v, and consequently all the paths passing through those edges. A switch–on(v) operation consists of restoring all the edges incident to v (which were previously deleted by a switch–off(v)). We first describe extract(v), a primitive which is used by all other operations.

5.1 Operation extract(v)

The goal of *extract*(v) is to make vertex v part of the certificates at all levels of the decomposition. Let $\ell = absorb(v)$ be the lowest level in which v is not part of the certificate at that level. As separator vertices are part of the certificate, $v \notin S_\ell$ and thus by Definition 3 v is of constant degree in $C_i(G)$, $i \geq \ell$.

In the following we describe how to uncompress v in level ℓ. Note that by definition vertex v does not belong to $C_\ell(G)$, and thus it cannot be a vertex of type (a). Consequently, only the following cases may arise, according to the type of vertex v inside $C_\ell(G)$. Let us denote as $SF(G)$ a spanning forest of a graph G, and let P be the piece at level ℓ containing vertex v.

1. Vertex v is of type (b) in $C(P)$, i.e., it is hidden inside an edge (x, y), which is a path–edge of $C(P)$ for v. We replace (x, y) with two edges (x, v) and (v, y) in $C(P)$ (an thus $C_\ell(G)$), corresponding to the two subpaths on either side of v.

2. Vertex v is of type (c) in $C(P)$, i.e., it is hidden inside a vertex x, which is a subtree–vertex of $C(P)$ for v. In this case we add edge (x, v) to $C(P)$ (and thus $C_\ell(G)$).

3. Vertex v is of type (d) in $C(P)$, i.e., it is hidden inside edge (x, y), which is a subtree–edge (x, y) of $C(P)$ for v. In this case, we add a new vertex z and replace edge (x, y) with three edges (x, z), (z, y) and (z, v).

4. Vertex v belongs to a subtree of $SF(C_\ell(G))$ which is connected with no other vertex of $C(P)$. In this case we make v an isolated vertex of $C(P)$ (and thus $C_\ell(G)$).

In summary, we have to replace at any level some edge e by a star of two or three edges (case 1 or 3 above), or we have to add a new single edge to the graph at that level (case 2 above), or we have to add a new isolated vertex to the graph at that level (case 4 above). As we do not recompute explicitly the piece decomposition, a sequence of these operations can cause the size of pieces at any level to grow arbitrarily, well beyond their initial constant size. Differently from [5], we adapt the dynamic trees of Sleator and Tarjan [22] to maintain the compressed certificate of a piece, and thus all of the above operations require

$O(\log n)$ time to be performed at each level. Next, we show how to propagate those transformations at higher levels in the recursive decomposition. Let e be an edge to be replaced at level k. If e already belongs to the certificate $C_k(G)$ of G at level k, then the changes we perform at level k are analogous to the changes described above for level ℓ. If edge e does not belong to the certificate $C_k(G)$, then e must be in some piece P at the considered level, and again a number of cases may arise.

1. If e belongs to a path edge (x, y) of $C(P)$ for v, we replace edge (x, y) in $C(P)$ with a star of two or three edges, according to the replacements performed at the lower level.

2. If e is part of a subtree of $SF(P)$ contracted to a single vertex z in $C(P)$, then we add edge (v, z) to $C(P)$; that edge corresponds to the path from v to z in $SF(P)$.

3. If e belongs to a subtree of $SF(P)$ with no connections either with path edges, or with vertices of $C(P)$, then we add vertex v as an isolated vertex to $C(P)$.

4. If e does not belong to $SF(P)$ and it has been replaced by two or three edges, then $SF(P)$ must contains all but one of those edges. Let w be the only endpoint of e touching $SF(P)$. If w is on a path of $SF(P)$ contracted to a path edge (x, y) in $C(P)$, we replace edge (x, y) by a star of three edges (x, w), (w, y) and (w, v). If w is a type (c) vertex that has been contracted to a single vertex z in $C(P)$, we add edge (v, z) to $C(P)$. Finally if w belongs to a tree of $SF(P)$ not containing any point of $C(P)$, then we add w to $C(P)$ as an isolated vertex.

During the execution of the $extract(v)$ operation, at each level $i \geq \ell = absorb(v)$, we can insert at most two new vertices in $C_i(G)$, where each of them has degree at most three. Furthermore, we can replace an edge with at most three edges or we can add a new edge. Finally, we can increase by one the degree of at most one vertex z at level i. If z was previously low–degree and now becomes high–degree, there are exactly $(\Delta_0 + 1)$ edges incident on z: we remove z from the piece and include it in the separator at level i. This is done by deleting the $(\Delta_0 + 1)$ edges incident to z, finding replacements for them, and then by inserting these edges again so that the newly separator vertex z will be a leaf in its certificates.

At a generic level k, each deletion is performed as follows. If the deleted edge is not part of a piece we simply remove the edge from $C_k(G)$, and pass the deletion on to higher levels. Otherwise, let P be the piece containing the deleted edge before the uncompression step. If necessary, we find a replacement for that edge in P, and include the replacement in $C_k(G)$ as an insertion, propagating such insertion recursively at higher levels. Note that the insertion of a replacement edge at level k cannot create new high–degree vertices at level k, as the replacement edge was already part of the graph at level k. If the deleted edge was part of a path corresponding to an edge already in $C_k(G)$, we delete the edge recursively.

If no piece becomes disconnected, each edge deletion determines in the worst case a constant number of operations at each recursive level: another deletion and an insertion (that of the replacement edge). An edge insertion requires at most $O(\log n)$ time to be performed at each level. So, if δ is the constant number of deletions performed at level $piece(v)$, then at level $(piece(v)+1)$ we perform at most δ deletions and δ insertions, and so on at higher levels. Also an edge deletion takes $O(\log n)$ time at each level. The total number of operations performed after a deletion is thus bounded by the quantity $\delta+2\delta+3\delta+\cdots+O(\log n)\delta = O(\log^2 n)$.

The most delicate part of this process is when a piece becomes disconnected: in this case, we have to compute the new pieces (bridges) after the piece is broken. This can cause many new edges to be recursively inserted at the higher levels. With some care, we are able to maintain the following invariant for any level k: between any two successive rebuildings of level k at most $6|S_k|$ edge insertions are propagated from level k to higher levels because of the breaking of pieces. Here, S_k denotes the separator vertices at level k immediately before the second rebuilding. Another important property that we exploit relates to controlling the growth of the separators S_k throughout any sequence of switch–on and switch–off operations. Using these properties and a sophisticated amortized analysis, we can prove the following lemma.

Lemma 9. *Operation extract can be implemented in $O(\log^3 n)$ amortized time over sequences of at least $\Omega(n)$ operations.*

5.2 Switching a Vertex Off

We now give a high–level description of a switch–off(v) operation. We first make v part of the certificates of G at all the levels of the decomposition in $O(\log^3 n)$ amortized time by performing extract(v). This uncompresses the certificates of G at the various levels of the data structure to make room for v. Next, we perform the following operations. First of all we simply mark v as *off* at each level i, $0 \leq i \leq piece(v) - 1$. Note that vertex v belongs to the separator S_i at all these levels, and that v is of constant degree in all graphs $C_i(G)$, $i \geq piece(v)$. We then make vertex v part of the separator at each level higher than $(piece(v) - 1)$. Next, we delete all edges adjacent to vertex v at level $piece(v)$, leaving v isolated, and propagate the constant number of performed deletions at higher levels. As we may need this information during a subsequent switch–on operation, we still store these deleted edges in v, and consider v of the same degree as before the switch–off operation.

Deletions are performed as described for extract(v). Again, if no piece becomes disconnected, the total number of operations performed after a switch–off(v) is bounded by $O(\log^2 n)$. If some piece becomes disconnected, then we recompute the new pieces (i.e., bridges) and propagate the insertions of some edges higher up. Again, this contributes to at most a total of $6|S_k|$ edge insertions between any two successive rebuildings of level k. Using these properties, we can prove that the time required to perform a switch–off(v) operation is $O(\log^3 n)$ amortized over sequences of at least $\Omega(n)$ operations.

5.3 Switching a Vertex On

A switch–on(v) operation is essentially the reversal of a switch–off(v). Namely, we first make v appear in all levels of the data structure by performing extract(v). This will cost $O(\log^3 n)$ amortized time. Next, we proceed from the lowest level to the topmost level as follows: as long as v is a high–degree vertex at that level, we simply mark v as *on* (note that v must already be in the separator at that level by Definition 3). Let ℓ be the lowest level in which v appears as a low–degree vertex. We make v a separator vertex at level ℓ and insert the constant number of edges incident to v. Some of these edges will be inserted in certificates at level ℓ, and we have to make sure that v is a leaf in all the certificates where it appears. We next proceed recursively to higher levels: if an edge incident to v is inserted in the certificate of some piece at level ℓ, its insertion is propagated to the next level of the decomposition. As no piece is broken, at most a constant number of edges are inserted per level, and each edge can be inserted in at most $O(\log n)$ time. As we spend $O(\log n)$ time per level, the total cost of switch–on(v) is dominated by the initial extract(v) operation, yielding an $O(\log^3 n)$ amortized bound.

5.4 Performing Queries

It remains only to describe how to perform queries about the existence of a path between two arbitrary vertices x and y, using our data structure. Consider a n–vertex planar graph G, and its *separator decomposition* built as described previously. Suppose we are interested in checking whether there exists a on–path between vertices x and y in G, after that an arbitrary intermixed sequence of switch–off and switch–on operations on the vertices of G has been performed.

First of all we verify if x and y are both switched *on*. If at least one of them is *off* clearly there cannot be a path between x and y in G_{on}, and the query returns *no*. If both the vertices are switched *on* we first make x and y appear in the certificates at all levels of the decomposition, as in the update operations, by extracting them in $O(\log^3 n)$ amortized time. Next, we check whether x and y are connected through a path that is switched *on* in the topmost level. Note that the topmost level is kept always of constant size throughout the sequence of operations, as it is of constant size after each rebuilding, and it is recomputed each time its size becomes a constant factor larger than its initial size.

The total cost of a query operation is thus dominated by the cost of performing the initial extract operations, yielding an $O(\log^3 n)$ amortized bound.

Theorem 10. *We can maintain information about connectivity for an n–vertex planar graph G, under an arbitrary sequence of switch–on, switch–off and query operations, in $O(\log^3 n)$ amortized time per operation over sequences of at least $\Omega(n)$ operations.*

6 Conclusions and Open Problems

In this paper, we have presented algorithms for maintaining information about connectivity on planar graphs in $O(\log^3 n)$ amortized time per query, insert, delete, switch-on and switch-off operation over sequences of at least $\Omega(n)$ operations. This is the first efficient algorithm that we know of in the complete dynamic model. The large constant factors involved in our asymptotic time bounds make our algorithm mainly of theoretical interest and rather unsuitable for practical applications. Can we simplify our method without sacrificing polylogarithmic time bounds?

Acknowledgments

We are grateful to the anonymous reviewers for many helpful comments.

References

1. D. Alberts and G. Cattaneo and G. F. Italiano. An empirical study of dynamic graph algorithms. *Proc. 7th ACM-SIAM Symp. Discrete Algorithms*, 1996, 192–201.
2. D. Alberts and M. R. Henzinger. Average case analysis of dynamic graph algorithms. *Proc. 6th ACM-SIAM Symp. Discrete Algorithms*, 1995, 312–321.
3. G. Amato, G. Cattaneo, G. F. Italiano. Experimental analysis of dynamic minimum spanning tree algorithms. *Proc. 8th ACM-SIAM Symp. on Discrete Algorithms*, 1997, 314–323.
4. D. Eppstein, Z. Galil, G.F. Italiano, and A. Nissenzweig. Sparsification—A technique for speeding up dynamic graph algorithms. *Proc. 33rd IEEE Symp. on Foundations of Computer Science*, 1992, 60–69.
5. D. Eppstein, Z. Galil, G. F. Italiano, T. H. Spencer, "Separator based sparsification I: planarity testing and minimum spanning trees", *Journal of Computer and System Science*, Special issue of STOC 93, vol. 52, no. 1 (1996), 3–27.
6. D. Eppstein, Z. Galil, G. F. Italiano, T. H. Spencer, "Separator based sparsification II: edge and vertex connectivity", *SIAM J. Comput.*, to appear.
7. D. Eppstein, G.F. Italiano, R. Tamassia, R.E. Tarjan, J. Westbrook, and M. Yung. Maintenance of a minimum spanning forest in a dynamic plane graph. *J. Algorithms* 13 (1992), 33–54.
8. G.N. Frederickson. Data structures for on-line updating of minimum spanning trees, with applications. *SIAM J. Comput.* 14 (1985), 781–798.
9. G.N. Frederickson. Ambivalent data structures for dynamic 2-edge-connectivity and k smallest spanning trees. *Proc. 32nd IEEE Symp. Foundations of Computer Science*, 1991, 632–641.
10. D. Frigioni, A. Marchetti-Spaccamela, U. Nanni. Fully dynamic output bounded single source shortest path problem. *Proc. 7th ACM-SIAM Symp. Discrete Algorithms*, 1996, 212–221.
11. Z. Galil, G.F. Italiano, and N. Sarnak. Fully dynamic planarity testing. *Proc. 24th ACM Symp. Theory of Computing*, 1992, 495–506.
12. M.T. Goodrich. Planar separators and parallel polygon triangulation. *Proc. 24th ACM Symp. Theory of Computing*, 1992, 507–516.

13. M. Rauch Henzinger and V. King. Randomized dynamic graph algorithms with polylogarithmic time per operation. *Proc. 27th ACM Symp. on Theory of Computing*, 1995, 519–527.

14. M. Rauch Henzinger and V. King. Fully dynamic biconnectivity and transitive closure, *Proc. 36th IEEE Symp. Foundations of Computer Science*, 1995.

15. M. Rauch Henzinger and V. King. Maintaining minimum spanning trees in dynamic graphs. *Proc. 24th Int. Coll. on Automata, Languages and Programming*, 1997.

16. M. Rauch Henzinger and J. A. La Poutré. Certificates and fast algorithms for biconnectivity in fully dynamic graphs. *Proc. 3rd European Symp. on Algorithms.* Springer-Verlag LNCS 979 1995, 171–184.

17. M. Rauch Henzinger and M. Thorup. Improved sampling with applications to dynamic algorithms. *Proc. 23rd Int. Coll. on Automata, Languages and Programming.* Springer-Verlag LNCS 1099, 1996, 290–299.

18. J. Herschberger, M. Rauch and S. Suri. Fully dynamic 2-connectivity on planar graphs. *Proc. 3rd Scandinavian Workshop on Algorithm Theory.* Springer-Verlag LNCS 621, 1992, 233–244.

19. R. J. Lipton, R. E. Tarjan. A separator theorem for planar graphs, *SIAM J. Appl. Math.* 36 (1979), 177–189.

20. M. Rauch. Fully dynamic biconnectivity in graphs. *Proc. 33rd IEEE Symp. Foundations of Computer Science*, 1992, 50–59.

21. M. Rauch. Improved data structures for fully dynamic biconnectivity. *Proc. 26th ACM Symp. on Theory of Computing*, 1994, 686–695.

22. D. D. Sleator and R. E. Tarjan. A data structure for dynamic trees. *J. Comput. System Sci.*, 26 (1983), 362–391.

23. M. Thorup. Decremental dynamic connectivity. *Proc. 8th ACM-SIAM Symp. on Discrete Algorithms*, 1997, 305–313.

A New Family of Randomized Algorithms for List Accessing

Theodoulos Garefalakis

Department of Computer Science
University of Toronto
Toronto, Canada
M5S 1A4
e-mail: theo@cs.toronto.edu

Abstract. Sequential lists are a frequently used data structure for implementing dictionaries. Recently, self-organizing sequential lists have been proposed for "engines" in efficient data compression algorithms. In this paper, we investigate the problem of *list accessing* from the perspective of *competitive analysis*. We establish a connection between randomized list accessing algorithms and Markov chains, and present *Markov-Move-To-Front*, a family of randomized algorithms. To every finite, irreducible Markov chain corresponds a member of the family. The family includes as members well known algorithms such as *Move-To-Front*, *Random-Move-To-Front*, *Counter*, and *Random-Reset*.

First we analyze *Markov-Move-To-Front* in the standard model, and present upper and lower bounds that depend only on two parameters of the underlying Markov chain. Then we apply the bounds to particular members of the family. The bounds that we get are at least as good as the known bounds. Furthermore, for some algorithms we obtain bounds that, to our knowledge, are new.

We also analyze *Markov-Move-To-Front* in the paid exchange model. In this model, the cost of an elemant transposition is always paid, and costs d. We prove upper and lower bounds that are relatively tight. Again, we apply the bounds to known algorithms such as *Random-Move-To-Front* and *Counter*. In both cases, the upper and lower bounds match as the parameter d tends to infinity.

1 Introduction

In this paper we consider the *static list accessing* (also known as the *list update*) problem. The problem is to maintain an unsorted list of items in such a way that the cost of successive accesses is kept small. More specifically, an initial list of items is given, and a request sequence is generated. A request specifies an item in the list, and is serviced by accessing the item. In the on-line setting, each request has to be serviced before the next request is made known. The cost incurred by an access to an item is equal to the position of the element in the current list (as maintained by the on-line algorithm). In order to reduce the cost of future requests, the algorithm is allowed to reorganize the list between requests. One can view this reorganization as a sequence of exchanges between consecutive items in the list. The cost of such a rearrangement is measured in terms of the minimum number of exchanges needed for the rearrangement. We will consider two different cost models for the exchanges that are performed in order to rearrange the list. In the standard model, the algorithm may move the item just accessed to a position closer to the front of the list *free of charge*. Those exchanges are called *free*. All other exchanges are called *paid*, and cost 1 each. In the paid exchnage model every exchange has a cost. One can justify the absence of free transposition, if one thinks of the list as an unsorted array. In order to move an item in the list, we have to perform a series of transpositions. The constant d is typically greater than 1, and reflects the fact that link traversals and exchanges are different operations, and therefore may have a different cost.

1.1 Competitive Analysis

For the analysis of on-line algorithms, we use *competitive analysis* [9]. In this framework, the on-line algorithm is compared with an optimal off-line algorithm. An optimal off-line algorithm knows the entire request sequence in advance, and can make optimal choices incurring minimum cost. The performance measure of the on-line algorithm is the *competitive ratio*.

Definition 1. A deterministic on-line algorithm ALG is said to be c-competitive, if there exists a constant α such that for every request sequence σ,

$$ALG(\sigma) \leq c \cdot OPT(\sigma) + \alpha.$$

where $OPT(\sigma)$ is the cost incurred by an optimal off-line algorithm with full knowledge of the request sequence σ. The competitive ratio of ALG, denoted by $R(ALG)$ is the infimum of c, such that ALG is c-competitive.

A useful way to view the problem of analyzing an on-line algorithm is as a game between an *on-line player* and a cruel *adversary* [3]. The on-line player runs the on-line algorithm that services the request sequence created by the adversary. The adversary, on the other hand, based on knowledge of the algorithm used by the on-line player, constructs the input that maximizes the competitive ratio. Usually, we identify the adversary and the off-line algorithm, and think of it as the *off-line player*. For randomized algorithms, the knowledge on which the adversary bases the construction of the cruel sequence is important. In this work we consider only *oblivious* adversaries, which are allowed to examine the on-line algorithm, but have to generate the entire request sequence in advance, without any knowledge of the random bits of the on-line player. The adversary services the request sequence off-line, incurring the optimal off-line cost.

Definition 2. A randomized on-line algorithm ALG is said to be c-competitive against an oblivious adversary, if there exists a constant α such that for every request sequence σ generated as described above,

$$E[ALG(\sigma)] \leq c \cdot OPT(\sigma) + \alpha$$

where OPT(σ) is the cost incurred by an optimal off-line algorithm on σ, and the expectation is taken over the random choices made by ALG. The infimum of c, such that ALG is c-competitive, is called the competitive ratio of ALG against an oblivious adversary, and is denoted by $R_{OBL}(ALG)$.

1.2 Competitive Algorihtms

Sleator and Tarjan [9] have shown that the well-known deterministic *Move-To-Front* algorithm (MTF) is 2-competitive. Karp and Raghavan then observed that no deterministic list accessing algorithm can be better than 2-competitive, and therefore MTF is competitive-optimal. Later, Albers [1] presented *Timestamp*, and proved that it is also 2-competitive. In 1996, El-Yaniv [5] presented an infinite family of optimal daterministic algorithms, and showed that MTF and *Timestamp* are members of this family.

While the best performance of deterministic on-line algorithms is known, this is not the case for randomized algorithms against oblivious adversaries. Irani [6, 7] gave the first randomized list-accessting algorithm, SPLIT, and proved it to be $\frac{15}{8}$-competitive, thus beating the deterministic lower bound. Then Reingold, Westbrook, and Sleator [8] presented two families of randomized algorithms, *Counter* and *Random-Reset*. The best member of the two families achieves a competitive ratio of $\sqrt{3}$, and for several years it was the best randomized list accessing algorithm known. Later, Albers [1] presented *Timestamp(p)*, a randomized algorithm whose competitive ratio is the golden ratio $\frac{1+\sqrt{5}}{2}$. Finally, in 1995, Albers *et al.* [2] gave the *Combination* algorithm that is 1.6-competitive, the best upper bound known so far. The best lower bound known for the problem is due to Teia [10]. He proved that no randomized algorithm can be better than 1.5-competitive.

The paid exchange model, which was also defined in [9], received much less attention than the standard model. Most of the results concerning this model were given by Reingold, Westbrook, and Sleator in [8]. The best known deterministic algorithm, due to Reingold, Westbrook, and Sleator is 5-competitive, independent of d, while the best known deterministic lower bound is 3, [8]. In the same paper, the authors considered the *Random-Reset* algorithm in the paid exchange model, and were able to beat the deterministic lower bound. As expected, for each value of d, there is a different member of the family that has a better performance. Quite surprisingly though, the competitive ratio starts from 2.64 (for $d = 1$), and drops as d grows. No lower bound against oblivious adversaries is known for this model.

1.3 Results

The work in [8] suggests that a greedy algorithm, in the sense that the decision "where to place" the requested item is made without considering any other item, can attain a good competitive ratio. In the present work we further exploit such greedy algorithms. We establish a connection between randomized list accessing algorithms and Markov chains. To every finite, irreducible Markov chain corresponds a list accessing algorithm. We name this family of randomized algorithms *Markov-Move-To-Front*(MMTF). We consider *Markov-Move-To-Front* in both the standard and the paid exchange models, and prove upper and lower bounds for its competitive ratio (against an oblivious adversary) that depend only on two parameters of the underlying Markov chain: π_0, the stationary probability of state 0, and S, the expected time to hit state 0, when in the stationary distribution.

The *Markov-Move-To-Front* family includes many well-known algorithms such as *Move-To-Front*, BIT, *Random-Move-To-Front*, the *Counter* and the *Random-Reset* families. The bounds we achieve are at least as good as the known bounds, thus in this aspect *Markov-Move-To-Front* presents a unified analysis for a wide class of interesting algorihtms. Furthermore, the application of the general bounds to specific members of the family – such as *Random-Move-To-Front* and *Random-Reset* – yield some new, to our knowledge, results.

2 The MMTF Family

We introduce a family of algorithms that simplify the decision *where to place* the item just requested to *when to move it to front*. The engine of each algorithm is a Markov Chain M with a finite set of states $S_M = \{0, 1, ..., s\}$. A copy of the Markov chain is associated with each item in the list. The initialization of the Markov chains is made according the stationary distribution, and therefore, they will remain at the stationary distribution thereafter.

MMTF algorithm: *Upon a request for item x, serve the request, and then make a step in the Markov chain associated with x. If the state at which the algorithm just arrived is 0 move x to front; otherwise do nothing.*

Note that MMTF is *not* a mixed strategy, i.e., it is *not* a distribution over a set of deterministic algorithms, but it is rather a behavioral randomized algorithm [4].

3 The Standard Model

In this section we analyse the MMTF family in the standard model, and obtain upper and lower bounds against oblivious adversaries.

3.1 Upper Bound

For all the upper bounds in this paper we use the potential function method. Potential function arguments are quite common in the analysis of deterministic algorithms. In the randomized setting the same argument holds with some small adjustments. Let Φ be a potential function, and consider an event sequence $e_1, e_2, \ldots, e_n$. Let Φ_i be the value of Φ after the ith event. Φ_0 is the value of

Φ before the first event. Also let ALG and OPT be the on-line and optimal off-line algorithms respectively. Denote by ALG_i (OPT_i) the cost incurred by ALG (OPT) during the ith event, and define the *amortized cost* of ALG during the ith event to be $a_i = ALG_i + \Phi_i - \Phi_{i-1}$. Then we can prove the following lemma, where all the expectations are taken over the random choices made by the on-line algorithm[1].

Lemma 3. *Suppose there exists a constant c such that, with respect to all possible event sequences, for each event e_i, $E[a_i] \leq c \cdot OPT_i$, and $E[\Phi]$ is lower bounded by some constant. Then ALG is c-competitive against an oblivious adversary.*

Let M be an irreducible Markov Chain with set of states $S_M = \{0, 1, ..., s\}$, and transition probabilities $P = (p_{ij})$ that has a stationary distribution $\pi = (\pi_0, \pi_1, ..., \pi_s)$. For $i, j \in S_M$, h_{ij} is the hitting time from state i to state j in M. In our analysis, the hitting time *to* state 0 is of special interest, therefore for brevity we denote by h_i the hitting time h_{i0} [2]. By S we denote the expected hitting time *to* state 0, i.e, $S = \sum_{i=0}^{s} \pi_i h_i$.

Theorem 4. *Let M be an irreducible Markov Chain that has a steady-state distribution $\pi = (\pi_0, \pi_1, ..., \pi_s)$, and transition probabilities $P = (p_{ij})$. The MMTF algorithm that operates on M has competitive ratio that is upper bounded by $\max\{1 + \pi_0 S, S\}$.*

Proof. We start by defining the notion of an inversion. An inversion is an ordered pair $\langle y, x \rangle$ of items such that x appears before y in OPT's list, but it appears after y in the list maintained by MMTF (i.e., the pair indicates the relative order of the items in the on-line algorithm's list). We define the type $T(x)$ of an item x, whose Markov chain is at state i, to be the hitting time h_i. The type of an inversion $\langle y, x \rangle$ is then defined to be the type of x. As our potential function, we define $\Phi = \sum T(x)$, where the sum is taken over all inversions $\langle y, x \rangle$.

Consider now a request for item x, and assume that x is in position k in OPT's list. We will distinguish between two types of events. One is the service of the request by OPT and MMTF, including the possible free exchanges. The other is a paid exchange made by OPT.

Event 1: The cost of OPT for this event is k. The amortized cost of MMTF is $a = MMTF + \Delta\Phi \leq k + R + A + B + C$, where R is the number of inversions $\langle y, x \rangle$ at the time of the access, A is the change in the potential due to new inversions created, B is the change in the potential due to old inversions destroyed, and C is the change in the potential due to old inversions that change type. For the rest of the proof we will fix the value of R to r. Suppose that the state of x

---

[1] We consider finite request sequences of any length n. The algorithm uses a constant number, say c, of random bits to serve each request, therefore the total number of random bits needed to serve the whole request sequence is cn. The expectations are taken over those random strings of length cn.

[2] Recall that h_{ii} is the expected time to return to state i. Therefore $h_0 = h_{00} > 0$.

before the access is $I = i$, and after the access it is $J = j$. There are two cases to consider.

Case 1: If $j = 0$ then x is moved to front. In that case, exactly r inversions $\langle y, x \rangle$ are destroyed. The type of x was h_i. So $B = -rh_i$. Since there are no old inversions that change type, $C = 0$ and therefore

$$E[R + B + C \mid I = i, J = 0] = r - rh_i.$$

Case 2: If $j \neq 0$ then x is not moved to front. In that case, no inversions are destroyed, so $B = 0$. But because of the change of x's type, some inversions $\langle y, x \rangle$ might change type. The change in the potential due to this change is $C = r(h_j - h_i)$, so that

$$E[R + B + C \mid I = i, J = j \neq 0] = r + r(h_j - h_i).$$

And therefore, we can compute $E[R + B + C]$ as

$$\sum_{i=0}^{s} \sum_{j=0}^{s} E[R + B + C \mid I = i, J = j] Pr(I = i, J = j)$$

$$= \sum_{i=0}^{s} E[R + B + C \mid I = i, J = 0] Pr(I = i, J = 0) +$$

$$\sum_{i=0}^{s} \sum_{j=1}^{s} E[R + B + C \mid I = i, J = j] Pr(I = i, J = j)$$

$$= \sum_{i=0}^{s} \pi_i p_{i0} r (1 - h_i) + \sum_{i=0}^{s} \pi_i \sum_{j=1}^{s} p_{ij} r (1 + h_j - h_i)$$

$$= r \sum_{i=0}^{s} \pi_i (p_{i0} - p_{i0} h_i + \sum_{j=1}^{s} p_{ij} + \sum_{j=1}^{s} p_{ij} h_j - \sum_{j=1}^{s} p_{ij} h_i)$$

$$= r \sum_{i=0}^{s} \pi_i (1 - h_i + \sum_{j=1}^{s} p_{ij} h_j)$$

$$= r(1 - S + \sum_{j=1}^{s} h_j \sum_{i=0}^{s} \pi_i p_{ij})$$

$$= r(-S + \sum_{j=0}^{s} \pi_j h_j)$$

$$= 0$$

We turn now to the estimation of $E[A]$. Assume that after servicing the request, OPT moves x forward to position k'. Let y_i, $i = 1, ..., k - 1$ be the items that preceded x in OPT's list before the access. Finally, let Y_i be a random variable that measures the change in the potential due to each pair $\{y_i, x\}$. Again we consider the same two cases as above.

Case 1: If $j = 0$, i.e, x is moved to front, then $k' - 1$ new inversions are created

$(\langle x, y_1 \rangle, ..., \langle x, y_{k'-1} \rangle)$. Since each element in the list has its own copy of the Markov chain, each of the above inversions has expected type $\sum_{i=0}^{s} \pi_i h_i = S$, so

$$E[Y_j] = \begin{cases} S \text{ for } 1 \le j \le k' - 1 \\ 0 \text{ for } k' \le j \le k - 1 \end{cases}$$

In that case we can compute

$$E[A \mid case1] = \sum_{j=1}^{k-1} E[Y_j] = \sum_{j=1}^{k'-1} S = (k' - 1)S$$

Case 2: If $j \ne 0$, i.e., x is not moved, then at most $k - k'$ new inversions are created due to the move by OPT $(\langle y_{k'}, x \rangle, ..., \langle y_{k-1}, x \rangle)$. The expected type of x is $\sum_{i=1}^{s} \pi_i h_i = S - 1$, so

$$E[Y_j] = \begin{cases} 0 & \text{for } 1 \le j \le k' - 1 \\ S - 1 & \text{for } k' \le j \le k - 1 \end{cases}$$

$$E[A \mid case2] \le \sum_{j=1}^{k-1} E[Y_j] = \sum_{j=k'}^{k-1} (S - 1) = (k - k')(S - 1)$$

Case 1 occurs with probability π_0, and case 2 with probability $1 - \pi_0$. Hence, $E[A]$ is at most

$$\pi_0(k' - 1)S + (1 - \pi_0)(k - k')(S - 1)$$
$$= (1 + 2\pi_0 S - S - \pi_0)k' + (1 - \pi_0)(S - 1)k - \pi_0 S$$

$E[A]$ is a linear function of k' and achieves a maximum either for $k' = 1$, or for $k' = k - 1$ (depending on the coefficient of k'). Therefore, $E[A] \le \max\{(1 - \pi_0)(S-1), \pi_0 S\} \cdot k$. We are able now to compute an upper bound for the amortized cost for the request for x.

$$\begin{aligned} E[a] &\le E[k + R + A + B + C] \\ &= k + E[R + B + C] + E[A] \\ &\le (1 + \max\{(1 - \pi_0)(S - 1), \pi_0 S\}) \cdot k \\ &= \max\{1 + (1 - \pi_0)(S - 1), 1 + \pi_0 S\} \cdot OPT \end{aligned}$$

Event 2: A paid exchange made by OPT can create at most one inversion. The type of the inversion is $\sum_{i=0}^{s} \pi_i h_i = S$ on the average. On the other hand OPT pays 1 for the exchange. Hence

$$E[a] = S \cdot 1 = S \cdot OPT$$

Combining event 1 and event 2 we obtain

$$R_{OBL}(MMTF) \le \max\{1 + (1 - \pi_0)(S - 1), 1 + \pi_0 S, S\}.$$

In order to eliminate the term $1 + (1 - \pi_0)(S - 1)$ notice that

$$S \ge 1 \quad \Longrightarrow \quad S \ge 1 + (1 - \pi_0)(S - 1)$$

Therefore, $R_{OBL}(MMTF) \le \max\{1 + \pi_0 S, S\}$.

3.2 Lower Bound

We will determine now a lower bound for MMTF that operates an any Markov chain M. As it will become clear, the lower bound is non-trivial only for Markov chains with $S \geq \frac{3}{2}$.

Theorem 5. *Let MMTF operate on a Markov chain M. Then $\forall \epsilon > 0$, $R_{OBL}(MMTF) > S - \epsilon$.*

Proof. We will prove the theorem by describing a request sequence that enforces the above lower bound. More specifically, we will show that for any given $\epsilon > 0$, there exists a sufficiently large list, such that $E[MMTF(\sigma)] > (S - \epsilon) \cdot MTF(\sigma)$.

Assume that the initial configuration of MTF's list is $\langle x_1, x_2, ..., x_\ell \rangle$ with x_1 at the front, and at the same time MMTF's list is a permutation ρ, i.e., for each i, $1 \leq i \leq \ell$ the item x_i is at position $\rho(i)$ in MMTF's list. Let k be some integer whose value will be determined later, and consider the following request sequence:

$$\sigma = (x_1)^k, (x_2)^k, ..., (x_\ell)^k.$$

The cost incurred by MTF is

$$MTF(\sigma) = (1 + 2 + ... + \ell) + \ell(k - 1) = \frac{\ell(\ell+1)}{2} + \ell(k - 1).$$

MMTF, on the other hand, will move each requested item to front every S requests on the average, so its expected cost $E[MMTF(\sigma)]$ is at most

$$(\rho(1)S + \rho(2)S + ... + \rho(\ell)S) + \ell(k - S)$$
$$= S \cdot \frac{\ell(\ell+1)}{2} + \ell(k - S)$$

because, when MMTF services a request for x_i, the fact that it is (eventually) moved to front does not affect future requests for elements that are behind x_i in the list, but it does make a difference for elements that are positioned in front of x_i. The cost to access those elements increases by one because of the move-to-front action. Thus the cost of $\rho(i)$ attributed to the S first requests for x_i is clearly a lower bound of the expected cost. Note, that the right hand side of the above inequality is independent of the initial configuration of MMTF's list. Then

$$\frac{E[MMTF(\sigma)]}{MTF(\sigma)} = \frac{S\ell(\ell+1) + 2\ell(k - S)}{\ell(\ell+1) + 2\ell(k - 1)}$$
$$= \frac{S + 2\frac{k-S}{\ell+1}}{1 + 2\frac{k-1}{\ell+1}}$$

If we choose $\ell \gg k$, formally if we let $\frac{k}{\ell} \longrightarrow 0$, then the ratio goes to S

$$\frac{E[MMTF(\sigma)]}{MTF(\sigma)} \longrightarrow S \tag{1}$$

By definition, 1 means that

$$\forall \epsilon > 0 \quad S - \epsilon < \frac{E[MMTF(\sigma)]}{MTF(\sigma)} < S + \epsilon$$

The left inequality gives the lower bound. To complete the proof, it remains to say that this procedure can be repeated enough times to outweigh any additive constant.

It is interesting to mention here that the above result is not only expected, but also occurs with high probability. In order for MMTF to incur a (significantly) different cost, it has to move the requested elements to front with significantly different rate, than once every S requests. That is, the time T_0 to hit state 0 has to deviate from its expected value $E[T_0] = S$. From Chebyshev's inequality for the random variable T_0 we get the high probability result, namely

$$Pr(|T_0 - S| \geq \frac{S}{\sqrt{\delta}}) \leq \delta.$$

3.3 Some Members of the Family

In this section, we present some well-known algorithms that are members of the MMTF family, and we apply the genaral bounds of the previous sections.

Random-Move-To-Front(p). RMTF(p) algorithm works as follows. Upon a request for an item x, RMTF(p) serves the request, and then with probability p moves x to front. For $p = 0$, items are never moved, so RMTF(0) clearly does not achieve a bounded competitve ratio. We can prove the following theorem[3].

Theorem 6. Let $0 < p \leq 1$. Then $\frac{1}{p} \leq R_{OBL}(RMTF(p)) \leq \max\{2, \frac{1}{p}\}$.

Proof. RMTF(p) can be formulated as an MMTF algorithm that operates on the simple two-state Markov Chain shown in Figure 1.
The transition matrix of the chain is the following

$$P = \begin{pmatrix} p & 1-p \\ p & 1-p \end{pmatrix}$$

It is not difficult to see that for this Markov chain $(\pi_0, \pi_1) = (p, 1 - p)$, and $h_0 = h_1 = \frac{1}{p}$. Therefore, $S = \pi_0 h_0 + \pi_1 h_1 = \frac{1}{p}$. We can now apply Theorem 4 and Theorem 5 to obtain the result.

Counter(k, {0}) family. The *Counter(k,{0})* family, presented by N. Reingold, J. Westbrook, and D. Sleator [8], is a special case of the MMTF family. The upper bound of Theorem 4 when applied to this subclass of algorithms gives the same bounds as computed by Reingold *et al.*

---

[3] J. Westbrook has obtained a better bound for RMTF, namely $\max\{\frac{1}{p}, \frac{2}{2-p}\}$.

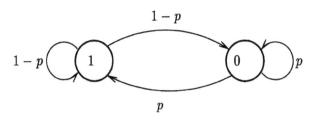

Fig. 1. The Markov chain for RMTF

A good two-state algorithm. Consider now the general form of a two-state algorithm, shown in Figure 2, with transition matrix

$$P = \begin{pmatrix} 1-q & q \\ p & 1-p \end{pmatrix}$$

The vector of the stationary probabilities of the above chain is $(\pi_0, \pi_1) = (\frac{p}{p+q}, \frac{q}{p+q})$, and the hitting times are $h_0 = \frac{p+q}{p}$, and $h_1 = \frac{1}{p}$, therefore $S = 1 + \frac{q}{p(p+q)}$. Using Theorem 4, we can express now an upper bound on the competitive ratio as a function of p and q. Interestingly enough, it turns out that the optimal[4] values are $p = q = 1$, i.e., under our analysis the best two-state algorithm is BIT (intoduced by Reingold, Westbrook, and Sleator [8]), which achieves a competitive ratio of 1.75. Recall that BIT is randomized because the starting state is randomly chosen. BIT is a good example of a mixed randomized strategy. It is a distribution over two deterministic algorithms: one that starts at state 0, and moves an element to front every second time it is requested, and one that starts at state 1, and moves an element to front every second time it is requested. RMTF(p), on the other hand, is a behavioral randomized algorithm. The behavior of the two algorithms is quite different [5] , and, as shown in the previous sections, the mixed strategy (i.e., BIT) achieves a better competitive ratio than the behavioral algorithm (i.e., RMTF(p) for $p \leq \frac{1}{2}$). For a discussion on the differences between mixed and behavioral randomized strategies, the reader is referred to [4].

Random-Reset. In [8], Reingold *et al.*, argued that the best possible member of their *Random-Reset* family has three states, and they determined the best choice for the probabilities to reset to state 1 and to state 2. Subsequently, they proved an upper bound of $\sqrt{3}$ for the competitive ratio of this particular member. For many years this was the best randomized algorithm for list accessing known (see Figure 3). It is interesting to see now, that this algorithm is a member of the MMTF family, and its analysis follows from the general analysis. In particular, for this algorithm $S = \sqrt{3}$, and also $1 + \pi_0 S = \sqrt{3}$, and therefore,

---

[4] optimal under this analysis.
[5] J. Westbrook was the first to point out the difference in the behavior of the two algorithms, and proved that for $p = 2$ the competitive ratio of RMTF is lower bounded by 2, and thus worse than the competitive ratio of BIT.

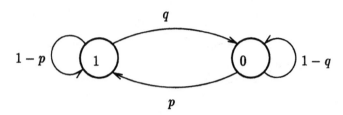

Fig. 2. The MMTF with two states

from theorems 4 and 5 we conclude that $\sqrt{3}$ is both an upper and a lower bound, and therefore the competitive ratio of this algorithm.

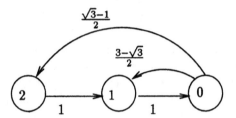

Fig. 3. The optimal *Random-Reset* algorithm

4 Paid Exchange Model

It is possible to prove similar upper and lower bounds for the performance of MMTF in the paid exchange model. Recall that in this cost model, every exchange has a cost d, where d is a constant typically greater than 1. Again, as in the standard model, the cost of accessing the ith item in the list costs i.

4.1 Upper Bound

Let M be an irreducible Markov chain with transtion porbabilities $P = (p_{ij})$, and a stationary distribution $\pi = (\pi_0, \pi_1, ..., \pi_s)$. $S = \sum_{i=0}^{s} \pi_i h_i$ as in the previous section. Then we can prove the following theorem.

Theorem 7. *The MMTF algorithm that operates on M has competitive ratio that is upper bounded by* $\max\{1 + \pi_0(2d + S), 1 + \frac{S}{d}\}$.

Proof. The analysis is very similar to that for the standard model. The notions of the inversion and the type of an inversion are defined as in the proof of Theorem 4. We define now our potential function to be $\Phi = \sum(d + T(x))$, where the sum is taken over all inversions $\langle y, x \rangle$. Intuitively, $T(x)$ pays for the increased access cost because of the inversion, and d pays for the cost of removing the inversion, when the item is moved to front.

Consider a request for item x, which is at position k in OPT's list. In this model, since all exchanges are paid, the types of events we consider are slightly different. The first type is the service of a request by MMTF or OPT, and the exchanges made by MMTF. The second type is an exchange made by OPT.

Event 1: OPT will pay k to service this request. The amortized cost for MMTF is $a = MMTF + \Delta\Phi \leq k + R + A + B + C + D$, where R is the number of inversions of the form $\langle y, x \rangle$ at the time of the access, A is the change in the potential due to new inversions created, B is the change in the potential due to old inversions destroyed, C is the change in the potential due to old inversions that change type, and D is the cost of (paid) exchanges made by MMTF. For the rest of the proof we will fix the value of R to r.

Suppose that the state of x is $I = i$ before the access and $J = j$ after the access. We consider the following two cases:

Case 1: $j = 0$, so x is moved to front. In this case exactly r inversions of the form $\langle y, x \rangle$ are destroyed. The type of x was h_i, so $B = -r(d + h_i)$. There are not any old inversions that change type, so $C = 0$. Also, $D \leq (k + r - 1)d$. Finally, at most $k - 1$ new inversions of the form $\langle x, y \rangle$ are created. Each such inversion has expected type $\sum_{i=0}^{s} \pi_i h_i = S$, so $A \leq (k - 1)(d + S)$. Therefore,

$$E[R + A + B + C + D \mid I = i, J = 0] \leq r(1 - h_i) + (k - 1)(2d + S)$$

Case 2: $j \neq 0$, so x is not moved to front. In this case no new inversions are created, and no old inversions are destroyed, so $A = 0$ and $B = 0$. However, x might change type, causing a change of type to the old inversions of the form $\langle y, x \rangle$. The change in the potential due to this change is $C = r(h_j - h_i)$. Also, $D = 0$, since MMTF does not perform any exchanges.

$$E[R + A + B + C + D \mid I = i, J = j \neq 0] = r + r(h_j - h_i)$$

We can now see that the expected value of $R + A + B + C + D$ equals

$$\sum_{i=0}^{s} E[R + A + B + C + D \mid I = i, J = 0]\pi_i p_{i0} +$$

$$\sum_{i=0}^{s} \sum_{j=1}^{s} E[R + A + B + C + D \mid I = i, J = j]\pi_i p_{ij}$$

$$= \sum_{i=0}^{s} \pi_i p_{i0}(r(1 - h_i) + (k - 1)(2d + S)) +$$

$$\sum_{i=0}^{s} \pi_i \sum_{j=1}^{s} p_{ij} r(1 + h_j - h_i)$$

$$= r(\sum_{i=0}^{s} \pi_i p_{i0}(1 - h_i) + \sum_{i=0}^{s} \pi_i \sum_{j=1}^{s} p_{ij}(1 + h_j - h_i)) +$$

$$\sum_{i=0}^{s} \pi_i p_{i0} (k-1)(2d+S)$$

Note that the expression r is multiplied with is the same as in the previous section, and is identically 0, therefore

$$E[R + A + B + C + D] = \sum_{i=0}^{s} \pi_i p_{i0} (k-1)(2d+S)$$

$$= (k-1)(2d+S) \sum_{i=0}^{s} \pi_i p_{i0}$$

$$= \pi_0 (k-1)(2d+S)$$

We can now bind the amortized cost of MMTF for this event:

$$E[a] \leq k + E[R + A + B + C + D] \leq k[1 + \pi_0(2d+S)]$$

Event 2: A transposition by OPT can create at most one inversion. The expected change in the potential due to that inversion is at most

$$\sum_{i=0}^{s} \pi_i (d + h_i) = d + \sum_{i=0}^{s} \pi_i h_i = d + S$$

On the other hand, OPT pays d. Therefore,

$$E[a] = E[\Delta\Phi] \leq d + S = (1 + \frac{S}{d}) \cdot d = (1 + \frac{S}{d}) \cdot OPT$$

From the above we conclude that the competitive ratio of MMTF against an oblivious adversary is upper bounded by $\max\{1 + \pi_0(2d+S), 1 + \frac{S}{d}\}$.

4.2 Lower Bound

For the lower bound we describe two request sequences, the first of which enforces the competitive ratio to be at least $1 + \frac{S-1}{d+1}$, and the second enforces it to be at least $1 + \frac{d}{S}$.

Theorem 8. *Let MMTF operate on a Markov chain M. Then $R_{OBL}(MMTF) \geq \max\{1 + \frac{d}{S}, 1 + \frac{S-1}{d+1}\}$.*

Proof. For the first sequence we will compare MMTF with MTF. Let $\langle x_1, x_2, ..., x_\ell \rangle$ be the MTF's list, and let item x_1 be in position $\rho(i)$ in the list maintained by MMTF. Consider the request sequence $\sigma = (x_1)^k, (x_2)^k, ..., (x_\ell)^k$. The value of k will be determined later. The cost of MTF to service σ is

$$(1 + 2 + ... + \ell) + \ell(k-1) + d(0 + 1 + ... + (\ell - 1))$$

$$\leq (d+1)\frac{\ell(\ell+1)}{2} + \ell(k-1)$$

On the other hand, MMTF will move each requested item to front once every S request, on the average. Also notice that a move-to-front action does not affect the elements that are behind the moved element, but will increase the access cost of elements that are in front of it in the list and have not been requested yet. Therefore, to assume that item x_i is in position $\rho(i)$ when it is first requested gives a lower bound to the access cost incurred by MMTF. Therefore, $MMTF(\sigma)$ is at least

$$(\rho(1) + \rho(2) + ... + \rho(\ell))S + \ell(k - S) + (\rho(1) + \rho(2) + ... + \rho(\ell) - \ell)d$$
$$= (d + S)\frac{\ell(\ell + 1)}{2} + \ell(k - S) - \ell d$$

If we choose $\ell \gg k$, formally $\frac{k}{\ell} \longrightarrow 0$, then we get

$$\frac{E[MMTF(\sigma)]}{MTF(\sigma)} \longrightarrow \frac{d + S}{d + 1}$$

which by definition means that

$$\forall \epsilon > 0 \quad \frac{d + S}{d + 1} - \epsilon < \frac{E[MMTF(\sigma)]}{MTF(\sigma)} < \frac{d + S}{d + 1} + \epsilon$$

The left inequality gives the lower bound of $\frac{d+S}{d+1}$.

Consider now the request sequence $\sigma = (x_1, x_2, ...x_l)^k$. The cost incurred by OPT is at most the cost incurred by the algorithm that services the request sequence without moving any item, i.e., $OPT(\sigma) \le k\frac{\ell(\ell+1)}{2}$. MMTF will move an item once every S requests, on the average, so its cost on σ is

$$(\rho(1) + ... + \rho(\ell))k + \frac{k}{S}(\rho(1) + ... + \rho(\ell) - \ell)d = k\frac{\ell(\ell + 1)}{2}(1 + \frac{d}{S}) - \ell d$$

Taking $\frac{k}{\ell} \longrightarrow 0$, the ratio approaches $1 + \frac{d}{S}$.

It remains to say that both request sequences can be repeated sufficiently many times to outweigh any additive constant.

4.3 Some Members of the Family

In this section, we apply the general bounds determined for the paid exchange model to some well known members of the family. Those bounds, as expected, depend on the scaling parameter d, and their limit behavior, as $d \longrightarrow \infty$ is of interest.

Counter(k, $\{0\}$). In [8], Reingold et al., analyzed the Counter(k, $\{0\}$) algorithm in the paid exchange model, and were able to determine an upper bound of its competitive ratio in terms of the maximum counter value k and d. Subsequently, they observed that as d tends to infinity, the best ratio (i.e., the ratio that corresponds to the best choice of k for the particular d) tends to $(5+\sqrt{17})/4$.

In this section we will show that the above limit is tight. We begin by describing *Counter*$(k, \{0\})$ as an MMTF algorithm. As shown in Figure 4, *Counter*$(k, \{0\})$ is a simple directed cycle of $k + 1$ nodes. In the stationary distribution, $\pi_0 = \pi_1 = \ldots = \pi_k = \frac{1}{k+1}$. Also, $h_0 = k + 1$, and for $1 \le i \le k$, $h_i = i$. Therefore, $S = \frac{k+2}{2}$ and by Theorem 7 we get an upper bound for the competitive ratio $R_{OBL}(COUNTER)$:

$$R_{OBL}(Counter) \le \max\{1 + \frac{k+2}{2d}, 1 + \frac{1}{k+1}(2d + \frac{k+2}{2})\}$$

Now fix d, and set $1 + \frac{k+2}{2d} = 1 + \frac{1}{k+1}(2d + \frac{k+2}{2})$. This equation has one non-negative solution

$$k = -\frac{3}{2} + \frac{d}{2} + \frac{1}{2}\sqrt{17d^2 + 2d + 1}$$

Since $R_{OBL}(Counter)$ has only one local minimum, which is also global, we conclude that the optimal choice for k is either the floor or the ceiling of the above expression. If we take the limit now, as d tends to infinity, $R_{OBL}(Counter)$ is upper bounded by

$$\lim_{d \to \infty} (1 + \frac{1}{2d}\lceil -\frac{3}{2} + \frac{d}{2} + \frac{1}{2}\sqrt{17d^2 + 2d + 1}\rceil) = \frac{5 + \sqrt{17}}{4} \qquad (2)$$

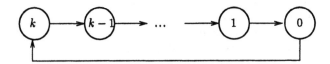

Fig. 4. The *Counter* algorithm

Consider now the lower bound for the competitive ratio for the above, optimal choice of k

$$\lim_{d \to \infty} (1 + \frac{1}{d+1}(\lfloor -\frac{3}{2} + \frac{d}{2} + \frac{1}{2}\sqrt{17d^2 + 2d + 1}\rfloor - 1)) = \frac{5 + \sqrt{17}}{4} \qquad (3)$$

From 2 and 3 we conclude that

$$\lim_{d \to \infty} R_{OBL}(Counter) = \frac{5 + \sqrt{17}}{4}$$

Random-Move-To-Front(p). Consider now the RMTF(p) algorithm. In Section 3.3 we determined that for RMTF(p), $S = \frac{1}{p}$. It is not difficult to compute now the optimal choice of the probability p, given the value of d. From Theorem 7 we get that

$$R_{OBL}(RMTF) \le \max\{1 + \frac{1}{pd}, 2 + 2pd\}$$

For a particular d, the best choice for p is when the two expressions are equal, that is

$$1 + \frac{1}{pd} = 2 + 2pd \quad \Longleftrightarrow \quad p = -\frac{1}{d} \quad or \quad \frac{1}{2d} \tag{4}$$

As $p \geq 0$, the only choice is $p = 1/(2d)$, and then the competitive ratio is 3, independent from d. The lower bound, on the other hand, depends on d.

$$R_{OBL}(RMTF) \geq \max\{1 + \frac{1}{pd}, 1 + \frac{\frac{1}{p} - 1}{d+1}\}$$
$$= \max\{1.5, \frac{3d}{d+1}\} \tag{5}$$

From 4 and 5 we get

$$\frac{3d}{d+1} \leq R_{OBL}(RMTF) \leq 3$$

which is relatively tight, and clearly

$$\lim_{d \to \infty} R_{OBL}(RMTF) = 3$$

5 Concluding Remarks

This paper leaves open several questions. One is to determine the correct competitive ratio for the *Markov-Move-To-Front* family. This would imply the correct competitive ratio for a number of interesting algorithms such as *Random-Move-To-Front* and BIT. Although the presented family includes a wide variety of algorithms, we were not able to find one that outperforms *Random-Reset*. It is interesting to further investigate the family, and see if a better member exists, or alternatively prove that *Random-Reset* is the best member of the family. In this work, we required that each element in the list has a copy of the *same* Markov chain. It is interesting to consider an algorithm, where each element is allowed to have a different chain. This would allow "important" items to move to front more frequently. Although in an adversarial setting this might not improve the performance, it is possible that in practice, or in a model that captures *locality of reference*, such an algorithm would have a better performance.

6 Acknowledgments

The author would like to thank Allan Borodin, Adi Rosén and Mike Molloy for many intresting discussions.

References

1. Albers, S.: Improved Randomized On-line Algorithms for the List Update Problem. Proceedings of the 6th Annual ACM-SIAM Symposium on Discrete Algorithms (1995) 412-419
2. Albers, S., von Stengel, B., Werchner, W.: A Combined BIT and TIMESTAMP Algorithm for the List Update Problem. TR-95-039 (1995), International Computer Science Institute, Berkeley
3. Borodin, A., El-Yaniv, R.: Online Computation and Competitive Analysis. Draft (1996)
4. Borodin, A., El-Yaniv, R.: On Randomization in Online Computation. To appear in 12th IEEE Conference on Computational Complexity (1997).
5. El-Yaniv, R.: There are Infinitely Many Competitive-Optimal Online List Accessing Algorithms. Submitted to SODA 97.
6. Irani, S.: Two Results on the List Update Problem. Information Processing Letters **38** (6) (1991) 202-208.
7. Irani, S.: Corrected Version of the SPLIT Algorithm for the List Update Problem. (1996) ICS Department, U.C. Irvine. Technical Report 96-53. Note: Corrected version of the SPLIT algorithm appearing in [6].
8. Reingold, N., Westbrook, J., Sleator, D.: Randomized Competitive Algorithms for The List Update Problem. Algorithmica **11** (1994) 15-32.
9. Sleator, D., Tarjan, R.: Amortized Efficiency of List Update and Paging Rules. Communications of the ACM **28** (2) (1985) 202-208.
10. Teia, B.: A Lower Bound for Randomized List Update Algorithms. Information Processing Letters **47** (1993) 5-9.

On-Line Construction of Two-Dimensional Suffix Trees *

Raffaele Giancarlo** and Daniela Guaiana***

Dipartimento di Matematica ed Applicazioni, Università di Palermo
Via Archirafi, 34 - 90123 Palermo - ITALY

Abstract. We present a new technique, which we refer to as *implicit updates*, based on which we obtain: (a) an algorithm for the *on-line* construction of the Lsuffix tree of an $n \times n$ matrix A - this data structure, described in [13], is the two-dimensional analog of the suffix tree of a string; (b) simple algorithms implementing primitive operations for **LZ1-type** *on-line lossless* image compression methods. Those methods, recently introduced by Storer [35], are generalizations of **LZ1-type** compression methods for strings (see also [24, 31]). For the problem in (a), we get nearly an order of magnitude improvement over algorithms that can be derived from known techniques [13]. For the problem in (b), we do not get an asymptotic speed-up with respect to what can be done with known techniques, e.g. [13, 28], but a major simplification in the implementation of the primitive operations. To the best of our knowledge, our technique is the first one that effectively addresses problems related to the *on-line* construction of two-dimensional suffix trees.

1 Introduction

The suffix tree T_x of a string x [26, 37] is a very useful data structure with applications ranging from String Matching [3] to Computational Biology [5, 21]. For the RAM and PRAM models of computation [1, 11], we know how to build it optimally [4, 6, 9, 8, 17, 21, 26, 30, 33, 37].

Ukkonen [36] has recently added to the above collection a simple and elegant linear time algorithm for the *on-line* construction of the suffix tree. The algorithm is *on-line* because, after i steps, it knows only the first i symbols of x and it has built the suffix tree for $x[1, i]$. The algorithm by Ukkonen [36] is based on a clever observation about the encoding of information on the edges of the suffix tree. Due to its simplicity, the algorithm is very useful in application areas where the string x is provided *on-line* and, after each new symbol is read, we need to know the suffix tree for the string "we have seen so far". One area is

---

* Work supported in part by Grants from the Italian Ministry of Scientific Research and from the Italian National Research Council
** Part of this work was done while visiting Bell Labs of Lucent Technologies, U.S.A.. Email: raffaele@altair.math.unipa.it
*** Email: daniela@altair.math.unipa.it

given by the fast implementation of *on-line lossless* **LZ1-type** text compression methods [10, 22, 28]. A text compression method is classified as **LZ1** (see [34]), if it uses the basic ideas presented by Ziv and Lempel in their seminal work on text compression [23, 38].

The state of the art about the construction of two-dimensional data structures that are analogous to the suffix tree of a string is not so rich. Those data structures find natural applications in Low Level Image Processing [29], Image Data Compression [35] and Visual Data Bases [19]. Let A be an $n \times m$ matrix with entries defined over a finite alphabet Σ and assume that $n \geq m$. Informally, a two-dimensional analog of the suffix tree for A is a tree data structure storing all submatrices of A. Let's call this data structure *an index* for A. It must support a wide variety of queries (see [12] for a list), the most important of which is the following: given a pattern matrix $P[1:s_1, 1:s_2]$, find all submatrices of A that are equal to P. Those submatrices are called *occurrences* of P in A. An index can support this kind of query in time that depends *only* on the size of P and the number of occurrences of P in A.

Early results investigating the construction of index data structures for matrices are reported in [16]. Recently, it has been shown that any index requires $\Omega(nm^2)$ time to be built [12] and efficient algorithms have been proposed both for the RAM and PRAM model of computation [7, 12, 14]. All those algorithms are a polylogarithmic factor away from optimal time and work. Because of the lower bound stating that it is not possible to build index data structures for an $n \times m$ matrix in time "close" to nm, those data structures may not have wide applicability. Fortunately, for the important special case in which A is an $n \times n$ matrix, much better results are available. Indeed, the Lsuffix tree of a square matrix has been recently proposed [13, 15]. It is an index that compactly stores all *square submatrices* of A. An example of this data structure is provided in Fig. 1 and a formal definition is given in Section 2. It can be built in $O(n^2 \log n)$ time and takes $O(n^2)$ space. Moreover, it can support many queries in optimal time (see [13] for a list).

Both for the case in which the index represents all submatrices of the given matrix and for the special case of the Lsuffix tree, all the algorithms we have mentioned so far are *off-line*, i.e., they must know the matrix ahead of time. However, in applications, it is very important to be able to build the index *on-line*, i.e., while the matrix is being provided in input. In what follows, we restrict attention to square matrices and to the Lsuffix tree. Our model of computation is the RAM [1]. We present a new technique, which we refer to as *implicit updates*, that allows us to obtain the results described in detail in Sections 1.1-1.2.

As for the analogy between the *on-line* algorithm by Ukkonen [36] for the construction of the suffix tree of a string and our technique, we point out that we make use of Ukkonen's clever observation mentioned earlier, but that is only a very small part of our technique.

1.1 On-line Suffix Trees for Square Matrices

Let $W_p = A[1:p, 1:p]$. An $n \times n$ matrix A is read *on-line* as a sequence of matrices $W_1, W_2, \cdots, W_n$. At time p, we know W_p and nothing else about A. At time $p+1$, we get W_{p+1} by getting in input subrow $A[p+1, 1:p+1]$ and subcolumn $A[1:p+1, p+1]$. We remark that the particular input ordering we have chosen for A makes the description of the algorithms easier. However, W_p could be analogously defined when A is given in input in row or column major order and our algorithms would work also in those cases. Let LT_p be the Lsuffix tree for matrix W_p. Again, it is a data structure that compactly stores all square submatrices of W_p (see Fig. 1). Since A is given *on-line* as a sequence of matrices W_p, $1 \le p \le n$, the Lsuffix tree for A can be built as a sequence of trees $LT_1, LT_2, \cdots, LT_n$. We have new algorithms such that:

(a1) The time taken to build the entire sequence of trees is $O(n^2 log^2 n)$. That is, each tree can be obtained from the other in $O(n \log^2 n)$ amortized time.

(a2) Assume that, when we have seen only a part of A, say W_p, we are given a pattern $PAT[1:m, 1:m]$ and we want to check whether PAT occurs in W_p. We can answer that query in $O(m^2 \log |\Sigma|)$ time. Such a query can be extended to report all *occ* occurrences of PAT in W_p in $O(m^2 \log |\Sigma| + occ \log p)$ time.

Remark. Our *on-line* algorithm is a $\log^2$ factor away from optimal time and nearly an order of magnitude improvement with respect to algorithms based on known techniques [13]. Indeed, we can modify the *off-line* algorithm in [13] so that it works *on-line* obtaining an $O(n^3 \log^2 n)$ time bound. Moreover, (a2) implies that we can efficiently check for occurrences of patterns in A, while the matrix is being provided *on-line*.

1.2 LZ1-Type Compression Methods for Images

Storer [35] has recently proposed a new family of methods for *on-line lossless* compression of images. They are generalizations to images of **LZ1-type** compression methods for text (for discussions of those latter methods, the reader is referred to [23, 34, 38]). Related work on extending **LZ1-type** methods from text to images is presented in [25, 31]. We briefly mention results quantifying how well those new methods are expected to compress images. Sheinwald, Lempel and Ziv [25, 31] show that, asymptotically, methods in this new family provide an optimal compression ratio with respect to what can be done by "finite state encoders". Experiments by Storer [35] show that the compression ratio obtained by those methods is 70% of the one obtained by the best available compression methods for images, i.e., JBIG. Those results are considered encouraging [35].

Here we confine our attention to the primitive operations and data structures needed to implement the compression methods in the new family and provide efficient algorithms for those operations. We describe the simplest compression method in this new family for the reason of pointing out the primitive operations needed to implement it. We remark that the same primitive operations can be used to support all other methods in the family.

Assume that the image is an $n \times n$ matrix A, with entries drawn from a finite alphabet Σ. Assume that we want to compress A *on-line*, i.e., while the matrix is given as input. In analogy with **LZ1-type** methods for strings [10, 28, 34, 38], we use a window that slides over the matrix sweeping all of it. Consider the "L-shaped" window of height h covering part of the matrix (see Fig. 2). In analogy with the sweeping order described in [31], the window slides along the main diagonal of A and, initially, it is an $h \times h$ square covering the upper left corner of A. Let $h(p) = p + h - 1$ and let $FW_{h(p)}$ be the "L-shaped" window that starts in row and column p and ends in row and column $p + h - 1$ of A (see Fig. 2), for $1 \leq p \leq n - h + 1$. We refer to the entries on row and column $p + h$ of A, just outside the border of $FW_{h(p)}$, as the *neighbors* of $FW_{h(p)}$.

A is given *on-line*. Assume that when the window has reached row $h(p)$ of A, we have seen A up to row $h(p + h)$, i.e., $W_{h(p+h)}$ is available. Moreover, assume that we have compressed $W_{h(p)}$. The next "compression step" works as follows. For each neighbor position of $FW_{h(p)}$, find the largest square matrix originating in that position and that appears within the window (one such match is reported in Fig. 2). Let $Match(FW_{h(p)})$ be the query that reports all those maximal matrices. The area S of the matrix A covered by the results of the query (see Fig. 3) is encoded in compressed form: we replace the matrices in S by pointers to their occurrences in the window. There are several methods and strategies currently under investigation to generate the encoding from the results of $Match(FW_{h(p)})$ and, probably, the best method can be identified only experimentally (see [35]). Once we have compressed the area outside the window, we slide it by some amount and repeat the process.

The compression method we have outlined has the following problem as one of its main computational bottlenecks:

- Maintain a data structure that represents all square submatrices of A that are within the window $FW_{h(p)}$. Since the window slides over A, the data structure must be dynamically changed. Moreover, it must support the query $Match$.

From now on, as far as Data Compression is concerned, we concentrate on the above subproblem. The data structure that seems to naturally fit the description is the Lsuffix tree of a square matrix [13]. It can be defined for the part of A covered by the window. We have new algorithms such that:

(b1) While the window slides over A, they can maintain the Lsuffix tree of the window in a total of $O(n^2 \log^2 n)$ time. For each $FW_{h(p)}$, $1 \leq p \leq n - h + 1$, the total space used is $O(|FW_{h(p)}|) = O(h^2 + ph)$. $Match(FW_{h(p)})$ can be answered in $O(S \log^2 n)$ time, where S is the total area covered by the matrices in $Match(FW_{h(p)})$ (see Fig. 3).

Remark. We point out that we can use other types of windows getting results analogous to the ones reported in **(b1)** for the L-shaped window of Fig. 2. In particular, we can use rectangular windows of fixed size that sweep matrix A in row or column major order or according to the Peano-Hilbert space filling curve [18, 27].

Remark. While the window slides over A, maintaining its Lsuffix tree is simple: we continuously delete and insert nodes in the data structure. Implementation of $Match(FW_{h(p)})$ is also very simple. With known techniques, we can obtain the same asymptotic results as in **(b1)**, but more complicated algorithms. Indeed, using the ideas in [28] for text compression together with the algorithm for the *off-line* construction of the Lsuffix tree [13], we can implement the query $Match(FW_{h(p)})$ using two Lsuffix trees that compactly store submatrices from two overlapping parts of the window. However, due to the sliding of the window, at some point we have to discard one of those data structures and start building a new one. Moreover, the implementation of $Match(FW_{h(p)})$ with two Lsuffix trees is complicated.

The remainder of this paper is organized as follows. In Section 2 we briefly recall from [13] the definition of Lsuffix tree of a matrix. In Section 3 we state some facts about that data structure. The basic algorithm, i.e., the one that transforms LT_p in LT_{p+1}, is outlined in Section 4. Due to space limitations, we will omit its detailed description as well as presentation of its application to data compression. A journal version giving a full account of the results claimed here is available from the authors.

2   The LSuffix Tree of a Matrix

Intuitively, the Lsuffix tree of a matrix is a data structure that compactly stores all square submatrices of A. Its presentation is organized as follows. We first define Lstrings, which are a suitable linear representation of square matrices. Then, we define the Lsuffix tree for one Lstring. The Lsuffix tree for a matrix A will be defined as the Lsuffix tree for a collection of Lstrings. The material in this section is presented in a more detailed form in [13].

For $1 \leq j \leq n$, submatrix $A[j:n, j:n]$ is the j-th *suffix* of A and submatrix $A[1:j, 1:j]$ is the j-th *prefix* of A. Note that any square submatrix of A whose upper left corner lies on the main diagonal of A can be described as a prefix of a suffix of A.

We need a suitable linear representation of matrices, i.e., a representation of matrices as "strings". Rather than being formal, we define it in intuitive terms. (This representation, denoted Lstring, has been introduced in [13] and, with a different formalism, by Amir and Farach [2].) Given $A[1:n, 1:n]$, we point out that we can divide A into n L-shaped characters, the i-th being composed of row $A[i, 1:i-1]$ and column $A[1:i, i]$. We refer to those characters as *Lcharacters*. Let us write down those L-shaped characters in one dimension and in the order given by their top-down appearance in A. We get a representation of A in terms of a string of L-shaped characters, which we call *Lstring*. We also need for Lstrings the notion of a chunk, which is the analog of the notion of substring for strings. A *chunk* is obtained by writing down the L-shaped characters of A in one dimension, in the order given by their top-down appearance in A, starting at row k and ending at row j. Notice that Lstrings are intended to represent matrices while chunks are intended to represent L-shaped pieces of matrices centered around the main diagonal.

The notion of Lstring outlined above can be stated formally by suitably defining that object as a string over an alphabet of Lcharacters $L\Sigma = \cup_{i=1}^{\infty} \Sigma^{2i-1}$ (for details see [13]). We consider each of the strings of $L\Sigma$ as an atomic item (composed of subatomic parts, which are the characters of Σ). Two Lcharacters are *equal* if and only if they are equal as strings over Σ.

Once that we have a way of representing matrices in one dimension, i.e., as strings, it is easy to define tries that represent them (see [20] for the definition of trie). Indeed, a trie representing a set of matrices can be defined as a trie over the alphabet $L\Sigma$, representing the set of Lstrings corresponding to the matrices. The edges of a trie for matrices are labeled with Lcharacters. Tries for matrices (and therefore Lstrings) can have nodes of outdegree one. We can compact them by compressing chains of nodes of outdegree one. Labels on the edges of the compressed structure are now chunks.

The *Lsuffix tree* for the *Lstring corresponding to* matrix $A[1:n, 1:n]$ is a compacted trie over the alphabet $L\Sigma$ representing all suffixes of $A[1:n]$ (see Fig. 1 for an example). Notice that there is no one-to-one correspondence between the leaves of the Lsuffix tree and the suffixes of A (this is in contrast with [13]). Let us number each diagonal of A by d, if its elements are $A[i, j]$, with $i - j = d$, $0 \leq |d| < n$. Let A_d be the square submatrix of A whose main diagonal is d. The *Lsuffix tree for matrix* A is the Lsuffix tree of the Lstring corresponding to A_d, for $0 \leq |d| \leq n - 1$ (see Fig. 1). Since each square submatrix of A is prefix of some suffix of some A_d, $0 \leq |d| \leq n - 1$, one can easily show that the Lsuffix tree for A represents all square submatrices of A [13].

Let $W_p = A[1:p, 1:p]$, for $1 \leq p \leq n$. For the diagonals of W_p, we follow the same numbering used for the diagonals of A. Let D_p be the set of those diagonal indices. For $d \in D_p$, let $W_{p,d}$ be the square submatrix of W_p of longest side whose main diagonal is d. The Lsuffix tree LT_p for W_p is the Lsuffix tree for the Lstrings corresponding to $W_{p,d}$, for $d \in D_p$.

3 Some Useful Facts

We start by stating some assumptions and terminology. It is convenient to introduce a dummy matrix $W_{p,d}\$$, for each $d \in D_p$. Each $W_{p,d}\$$ has $W_{p,d}$ as second suffix and it is different from all other $W_{p,d}\$$'s. We also introduce dummy leaves in LT_p, each of which is associated to a distinct $W_{p,d}\$$. Each dummy leaf is connected to the root of LT_p by a dummy edge.

Given a compacted trie LT over the alphabet $L\Sigma$, a node u is the *locus* of a matrix B if and only if the concatenation of the labels on the path from the root of LT to u is equal to B. Note that a matrix may not have a locus in LT. The *extended locus* of B is the locus of the shortest matrix (if any) with locus in LT and that has B as prefix. It can be easily shown that the locus and extended locus of a matrix are unique nodes in LT.

Given a node u in LT_p, let $\ell(u)$ be the side of the matrix having locus in u. We refer to $\ell(u)$ as the *distance* of u from the root of LT_p. Consider the matrix obtained by concatenating the labels on the path from the root of LT_p to a leaf f. That matrix can be suffix of more than one $W_{p,d}\$$, $d \in D_p$. Therefore, all suffixes of the $W_{p,d}\$$'s that "end" at f are an equivalence class, which we denote by $CLASS[f]$. In what follows, $M(f)$ denotes an arbitrarily chosen matrix in $CLASS[f]$. We also need the following definitions and observations.

- **Extension Matrices.** We now define extension matrices (an example is given in Fig. 4). Assume that $W_{p,d_1}, \cdots, W_{p,d_g}$ have their suffixes of side q equal. Denote this suffix matrix by M. Assume also that no other matrix, among the $W_{p,d}\$$'s, has a suffix equal to M. Now, for each W_{p+1,d_e}, $1 \le e \le g$, take its suffix of side $q+1$. Among those suffixes, keep only one copy of equal matrices. The result is the *set of extensions* of M. Each matrix in the set is referred to as an *extension matrix* of M. Notice that M has at least one extension. Let $\tilde{M}_1, \cdots, \tilde{M}_r$, be the extensions of M. We point out that they all differ on their last Lcharacter because they all have M as prefix.

- **Blossom Forest of LT_p.** We now define a useful tool for the fast transformation of LT_p into LT_{p+1} (see Section 4 for an outline of its use). Let f be a leaf of LT_p and assume that $M(f)$ is such that its second suffix has extended locus in a node u. We have a suffix link from f to u in LT_p. Notice that the definition of suffix link does not follow the one in [26, 36] because here we are not granted that, given a node u of LT_p locus of a matrix B, then there exists a node u' in LT_p that is locus of the second suffix of B (see [13] for a discussion illustrating this fact). We divide suffix links in external and internal. A suffix link from f to u is *external* if and only if u is a leaf and it is the locus of the second suffix of $M(f)$. Otherwise, it is *internal*. Notice that the suffix link leaving a dummy leaf f (corresponding to some $W_{p,d}\$$) points to the leaf f' such that $W_{p,d} \in CLASS[f']$. Therefore, all suffix links departing from dummy leaves are external. Consider all external suffix links in LT_p as edges and the leaves at their endpoints as nodes. It can be easily shown that the graph so obtained is a forest of trees. Each node of a given tree has the direction of the edge from parent to offspring reversed. We refer to this forest as the *blossom forest* of LT_p. Recalling that all $W_{p,d}\$$'s are distinct, one can easily show that there is a one-to-one correspondence between those matrices and the leaves in the forest.

- **LT_p can be stored in $O(|W_p|) = O(p^2)$ space.** Indeed, one can easily show that LT_p has $O(p^2)$ nodes. Moreover, the label, i.e., a chunk, on each of its edges can be represented by means of a quadruple of integers, as we now explain (the ideas are as in [13, 26]). Consider an edge (u, v) of LT_p and assume that it is labeled with the chunk corresponding to the rows and columns from the q-th to the g-th in the l-th suffix of $W_{p,d'}\$$. We represent this chunk by the quadruple $(c, c', \ell(u), d')$, where c and c' are the projections of row q and row g of the l-th suffix of $W_{p,d'}\$$ on the rows of W_p. Given the quadruple, we can easily access in constant time the subrows and subcolumns of W_p that correspond to the given chunk. For details, see [13].

• **Encoding of the labels on the edges of LT_p entering leaves.** For this kind of label, we use a special form of the encoding described earlier. As outlined in Section 4, this encoding will allow us to save a substantial amount of work in transforming LT_p into LT_{p+1}. Let (u, v) be the same edge as above, labeled with the same chunk, but assume that v is a leaf. In that case, we are sure that the matrix M "spelled out" on the path from the root of LT_p to v is equal to a suffix of $W_{p,d'}\$$. Since that suffix "touches" one of the boundaries of W_p, i.e., either row or column p, we can think of M as touching the same boundary too. But then, since the chunk labeling (u, v) is equal to an L-shaped part of M ending in its lower right corner, we can think of that chunk as touching the same boundary of W_p as M. We use the encoding $(c, \infty, \ell(u), d')$ rather than $(c, c', \ell(u), d')$. ∞ stands for the fact that the given chunk touches one of the boundaries of W_p. Notice that we can easily compute the row c' of W_p where that chunk ends. Indeed, when $d' \geq 0$, it is p; else it is $p + d'$. This encoding is analogous to the one devised by Ukkonen [36] for his *on-line* algorithm for the construction of the suffix tree of a string. We point out that, through the encoding of the label entering leaf f, we can access in constant time an occurrence of $M(f)$ in W_p, i.e., the topmost and leftmost entry of a submatrix of W_p equal to $M(f)$.

We maintain LT_p as a dynamic tree [32]. Moreover, in order to obtain the time bounds claimed for our algorithms in the Introduction, we need to maintain some additional data structures. Due to space limitations, we omit their description.

4   From LT_p to LT_{p+1}-Main Components

When W_p is extended into W_{p+1}, each matrix $W_{p,d}\$$ is extended by one Lcharacter to become a matrix $W_{p+1,d}\$$, for $d \in D_p$. Moreover, two new matrices are created: $W_{p+1,p}\$$ and $W_{p+1,-p}\$$. We want to transform LT_p into LT_{p+1}. Now, the insertion of the two new matrices in LT_p is easy and can be done in constant time (we omit the details). From now on, we concentrate on how the extension of $W_{p,d}\$$ into $W_{p+1,d}\$$, for $d \in D_p$, affects the structure of LT_p.

Notice that, since $W_{p+1,d}\$$ has $W_{p,d}\$$ as longest proper prefix, each suffix of $W_{p+1,d}\$$ can be obtained by appending an Lcharacter at the end of the "corresponding" suffix in $W_{p,d}\$$. But, by definition, LT_p "stores" all suffixes of $W_{p,d}\$$ and LT_{p+1} "stores" all suffixes of $W_{p+1,d}\$$, for all $d \in D_p$. Therefore, the transformation of LT_p into LT_{p+1} consists of extending the suffixes in LT_p by one Lcharacter. We give an example of the possible changes in LT_p that those extensions may cause.

Let M be a matrix that is suffix of, say, $W_{p,d'}\$$, $d' \in D_p$, and with extended locus z in LT_p. For sake of discussion, let us assume that M "ends" on the edge (v, z), i.e., z is not the locus of M in LT_p. Consider one of its extensions $\tilde{M}$. If $\tilde{M}$ has an extended locus in LT_p (which must be z), we do nothing. Else, we create the locus of M in LT_p (by breaking the edge (v, z)) and we insert a new leaf (locus of $\tilde{M}$) as offspring of the new node.

In general, for each M that is suffix of $W_{p,d}\$$, $d \in D_p$, the changes in LT_p are dictated by the extension matrices of M that *do not* have extended locus in LT_p. In terms of the topology of LT_p, we may have: (i) leaves that become internal nodes by generating new leaves; (ii) internal nodes that generate new leaves; (iii) edges that are broken by the insertion of new internal nodes and leaves. We use our new technique to identify and carry out those changes. It works in two phases: *Frontier Expansion* and *Internal Structure Expansion*. The first phase takes care of the changes in (i) and some other changes that do not modify the topology of the tree. We outline this part in Section 4.1. The second phase takes care of changes in (ii)-(iii) and it is outlined in Section 4.2.

4.1 Frontier Expansion-Outline

Consider a leaf u of LT_p. Recall that $M(u)$ is the matrix having locus in u and that $M(u)$ is suffix of some $W_{p,d}\$$, $d \in D_p$. Obviously, all extension matrices of $M(u)$ have no extended locus in LT_p. In order to establish which changes in LT_p are caused by those extension matrices, we have to consider two cases: (a) $M(u)$ has more than one extension matrix; (b) the complement. For case (a), the changes are:

Fact 1. *Let $u \in LT_p$ be a leaf. Assume that $M(u)$ has $r \geq 2$ extensions $\tilde{M}_1, \cdots, \tilde{M}_r$. We create r new leaves (in LT_{p+1}, each leaf is the locus of one extension of $M(u)$). Those leaves become offsprings of u in LT_{p+1} and u becomes an internal node.*

Consider the case in which $M(u)$ has only one extension matrix. We show that we can avoid to perform explicitly the changes caused by the only extension of $M(u)$. That is, we can ignore u. Let $f = parent(u)$ in LT_p. In order to represent $\tilde{M}$ in LT_{p+1}, we need to append one Lcharacter at the end of the path from the root of LT_p to u. Indeed, since $M(u)$ is the longest proper prefix of $\tilde{M}$, we can represent $\tilde{M}$ in LT_p by creating a new leaf u', u' becomes offspring of u and the edge (u, u') is labeled with the last Lcharacter of $\tilde{M}$. But, since among the suffixes of $W_{p,d}\$$'s, $M(u)$ is the only matrix that has $\tilde{M}$ as extension and that is also its only extension, we have that in LT_{p+1} the path $((f, u); (u, u'))$ is unary and must be compressed. We can obtain the same result by appending the required Lcharacter at the end of the path from the root of LT_p to u. As already anticipated, we can perform such a change *implicitly*, i.e., we do nothing, by using the special encoding that we have for the labels of edges entering leaves. The idea is best explained by means of an example.

Let the matrix M be the one in Fig. 4 and assume that $M_1 = M_2 = M_3$. So, M has only one extension $\tilde{M}$. Assume that the chunk on the edge (f, u) is encoded in terms of the occurrence of M as a suffix of $W_{p,7}$, so it is of the form $(c, \infty, \ell(u), 7)$, where ∞ stands for p. Now, let us "decode" that quadruple with respect to W_{p+1}. In that case, ∞ stands for $p + 1$. Therefore, the quadruple, when referred to W_{p+1}, corresponds to the chunk of M labeling the edge (f, u) in LT_p plus the new Lcharacter that we need to append to get $\tilde{M}$ in LT_{p+1}. In general, we have:

Fact 2. *Let $u \in LT_p$ be a leaf. Assume that $M(u)$ has only one extension $\tilde{M}$. The concatenation of the labels on the path from the root of LT_p to u gives exactly $\tilde{M}$, when those labels are "decoded" with respect to the matrices $W_{p+1,d}\$$'s. Therefore, u is the locus of $\tilde{M}$ in LT_{p+1} and we need not modify u when LT_p is transformed in LT_{p+1}.*

As implied by the above two facts, the identification of which leaf u of LT_p generates new leaves in LT_{p+1} reduces to the computation of the extension matrices of $M(u)$. *Frontier Expansion* performs such a computation by identifying the leaves that actually generate new leaves, i.e., the ones satisfying Fact 1. The leaves that can be updated implicitly, i.e., the ones satisfying Fact 2, are ignored. *Frontier Expansion* can be thought of as a sweep of the leaves of LT_p in order of decreasing distance from the root, starting from the dummy leaves of LT_p. The sweep is implemented through a bottom-up visit of all trees in the blossom forest of LT_p. During the visit of a given tree T, once we reach one of its nodes, we may be able to skip over part of the path from that node to the root of T (the nodes being skipped are part of the leaves satisfying Fact 2). We have criteria that, when satisfied, establish precisely how much of the "upward path" we can skip.

We now outline the processing of one tree T in the blossom forest of LT_p. Let a node $u' \in T$ be *marked* if and only if we have already computed all extensions of $M(u')$ and (consequently) all new leaves it generates in LT_{p+1}. For the nodes of T we have the following:

Processing Rule: initially, only the leaves of T_p are *ready* and no internal node is *marked*. An internal node is *ready* if and only if all of its offsprings are *marked*. At a given time instant, we arbitrarily pick a node that is ready and process it. We are finished when we have reached the root of T.

Processing of a node u that is ready consists of computing the extension matrices of $M(u)$. We reduce that problem to the computation of a partition into equivalence classes of the extension matrices of the offsprings of u in T (those nodes are marked). In turn, such a partition can be efficiently computed through a suitable lexicographic sorting of strings. Based on the number of equivalence classes in the resulting partition, we can also compute how much of the "upward" path from u to the root of T we can skip.

4.2 Internal Structure Expansion-Outline

Here we consider the cases in which internal nodes of LT_p generate new leaves or edges of LT_p are broken by the insertion of new internal nodes and leaves. Again, those changes are ruled by extension matrices. We need two preliminary observations. The first one states when we need to change the "internal structure" of LT_p and what kind of changes are needed. The second one is a condition that is used to eliminate extension matrices as candidates for changes in LT_p.

Fact 3. *Let $\tilde{M}$ be a matrix that has no extended locus in LT_p. Assume that $\tilde{M}$ is an extension of some suffix M of some $W_{p,d}\$$ and v is the extended locus of M in LT_p. v can be a leaf only when it is not the locus of M. We create a new leaf $\hat{f}$, which is the locus of $\tilde{M}$ in LT_{p+1}. If v is the locus of M in LT_p, $\hat{f}$ becomes offspring of v in LT_{p+1}. Else, we break the edge $(parent(v), v)$ in LT_p to create a new node v' that will be the locus of M in LT_{p+1}. In this case, the new leaf $\hat{f}$ becomes offspring of v' in LT_{p+1}.*

Fact 4. *Let $\tilde{M}$ be a matrix that has an extended locus v in LT_p. Assume that $\tilde{M}$ is an extension of some suffix M of some $W_{p,d}\$$, $d \in D_p$. v can be a leaf only when it is not the locus of M. Then, all the suffixes of $\tilde{M}$ have extended locus in LT_p.*

Fact 4 states that once we have identified an extension matrix that has already extended locus in LT_p, we can ignore that matrix and all of its suffixes (which are also extension matrices) since all of those matrices are already in LT_p (and therefore in LT_{p+1}).

The matrices considered by *Internal Structure Expansion* are suffixes of $W_{p,d}\$$'s that do not have a locus at a leaf of LT_p. Starting from an initial set identified at the end of *Frontier Expansion*, *Internal Structure Expansion* consists of dynamically maintaining a set of extension matrices for which we are sure Fact 3 holds. A matrix is discarded, together with its suffixes, as soon as Fact 4 holds.

More specifically, we keep in a priority queue a set of extension matrices that are either part of the initial configuration of the queue (to be specified shortly) or have already generated new leaves in LT_p. The matrix with highest priority is the one with "longest" side and ties are broken arbitrarily. The matrices initially in the queue are all the extensions of matrices whose locus in LT_p corresponds to the root of a tree in the blossom forest of LT_p.

We now outline the processing of the matrix $\tilde{M}$ of highest priority in the queue at an arbitrary time instant. It consists of checking whether its second suffix $\tilde{M}'$ generates a new leaf. That reduces to check whether $\tilde{M}'$ has an extended locus in LT_p. If it has an extended locus, then by Fact 4 (applied to $\tilde{M}'$) neither $\tilde{M}'$ nor its suffixes will generate new leaves in LT_p. So, we can ignore those matrices and, as a result, we do not insert $\tilde{M}'$ in the queue. If $\tilde{M}'$ does not have an extended locus in LT_p, then by Fact 3 (applied to $\tilde{M}'$) that matrix generates a new leaf in LT_{p+1}. We insert the new leaf in LT_{p+1} (as described by Fact 3) and $\tilde{M}'$ in the queue (because its suffixes may generate new leaves).

Acknowledgements

The authors are deeply indebted to Jim Storer for suggesting some problems from Data Compression that led to this research and to Bruno Carpentieri, Shimon Even and Roberto Grossi for helpful discussions.

References

1. A.V. Aho, J.E. Hopcroft, and J.D. Ullman. *The Design and Analysis of Computer Algorithms.* Addison-Wesley, Reading, MA., 1974.
2. A. Amir and M. Farach. Two-dimensional dictionary matching. *IPL*, 44:233–239, 1992.
3. A. Apostolico. The myriad virtues of subword trees. In A. Apostolico and Z. Galil, editors, *Combinatorial Algorithms on Words*, pages 85–96, Berlin, 1984. NATO ASI Series F, Vol 12, Springer-Verlag.
4. A. Apostolico, C. Iliopoulos, G. Landau, B. Schieber, and U. Vishkin. Parallel construction of a suffix tree with applications. *Algorithmica*, 3:347–365, 1988.
5. W.I. Chang and E.L. Lawler. Approximate string matching in sublinear expected time. In *Proc. 31st Symposium on Foundations of Computer Science*, pages 116–124. IEEE, 1990.
6. M. T. Chen and J. Seiferas. Efficient and elegant subword-tree construction. In A. Apostolico and Z. Galil, editors, *Combinatorial Algorithms on Words*, pages 97–107, Berlin, 1984. NATO ASI Series F, Vol 12, Springer-Verlag.
7. Y. Choi and T.W. Lam. Two-dimensional pattern matching on a dynamic library of texts. In *COCOON '95, D-Z. Du and M. Li eds*, pages 530–538. LNCS-Springer-Verlag, 1995.
8. M. Farach. Optimal suffix tree construction with large alphabets. Technical report, DIMACS Tr. Number 96-48, 1996.
9. M. Farach and S. Muthukrishnan. Optimal logarithmic time randomized suffix tree construction. *icalp96*, 1996.
10. E.R. Fiala and D.H. Green. Data compression with finite windows. *Communication of the ACM*, 32:490–505, 1988.
11. S. Fortune and J. Wyllie. Parallelism in random access machines. In *Proc. 10th Symposium on Theory of Computing*, pages 114–118. ACM, 1978.
12. R. Giancarlo. An index data structure for matrices, with applications to fast two-dimensional pattern matching. In *Proc. of Workshop on Algorithms and Data Structures*, pages 337–348. LNCS-Springer-Verlag, 1993.
13. R. Giancarlo. A generalization of the suffix tree to square matrices, with applications. *Siam J. on Computing*, 24:520–562, 1995.
14. R. Giancarlo and R. Grossi. Parallel construction and query of suffix trees for two-dimensional matrices. In *Proc. of the 5-th ACM Symposium on Parallel Algorithms and Architectures*, pages 86–97, 1993.
15. R. Giancarlo and R. Grossi. On the construction of classes of suffix trees for square matrices: Algorithms and applications. *Information and Computation*, Vol.130, pages 151–182, 1996.
16. G.H. Gonnet. Efficient searching of text and pictures- Extended Abstract. Technical report, University Of Waterloo- OED-88-02, 1988.
17. R. Hariharan. Optimal parallel suffix tree construction. In *Proc. the 26th Symposium on Theory of Computing*, pages 290–299. ACM, 1994.

18. D. Hilbert. Uiber stetige abbildung einer linie auf ein flachenstuck. *Mathematiche Annalen*, 38:459–460, 1891.
19. R. Jain. Workshop report on visual information systems. Technical report, National Science Foundation, 1992.
20. D.E. Knuth. *The Art of Computer Programming, VOL. 3: Sorting and Searching*. Addison-Wesley, Reading, MA., 1973.
21. R.S. Kosaraju. Real-time pattern matching and quasi-real-time construction of suffix trees. In *Proc. the 26th Symposium on Theory of Computing*, pages 310–316. ACM, 1994.
22. N. Jesper Larsson. Extended application of suffix tree to data compression. In *Proc. Data Compression Conference*, pages 190–199. IEEE, 1996.
23. A. Lempel and J. Ziv. On the complexity of finite sequences. *IEEE Trans. on Information Theory*, IT-22:75–81, 1976.
24. A. Lempel and J. Ziv. Compression of two-dimensional images. In A. Apostolico and Z. Galil, editors, *Combinatorial Algorithms on Words*, pages 141–154, Berlin, 1984. NATO ASI Series F, Vol 12, Springer-Verlag.
25. A. Lempel and J. Ziv. Compression of two-dimensional data. *IEEE Trans. on Information Theory*, IT-32:2–8, 1986.
26. E.M. McCreight. A space economical suffix tree construction algorithm. *J. of ACM*, 23:262–272, 1976.
27. G. Peano. Sur une courbe qui remplit toute une aire pliane. *Mathematiche Annalen*, 36:157–160, 1890.
28. M. Rodeh, V.R. Pratt, and S. Even. Linear algorithm for data compression via string matching. *J of ACM*, 28:16–24, 1981.
29. A. Rosenfeld and A.C. Kak. *Digital Picture Processing*. Academic Press, 1982.
30. S.C. Sahinalp and U. Vishkin. Simmetry breaking for suffix tree construction. In *Proc. the 26th Symposium on Theory of Computing*, pages 300–309. ACM, 1994.
31. D. Sheinwald, A. Lempel, and J. Ziv. Two-Dimensional encodings by finite state encoders. *IEEE Trans. on Communications*, 38:341–347, 1992.
32. D.D. Sleator and R.E. Tarjan. A data structure for dynamic trees. *J. of Computer and System Sciences*, 26:362–391, 1983.
33. A. O. Slisenko. Detection of periodicities and string matching in real time. *J. of Soviet Mathematics, Plenum Publishing Co.*, 22:1316–1386, 1983.
34. J. Storer. *Data Compression: Methods and Theory*. Computer Science Press, Rockville, MD, 1988.
35. J. A. Storer. Lossless image compression using generalized LZ1-type methods. In *Proc. Data Compression Conference*, pages 290–299. IEEE, 1996.
36. E. Ukkonen. On-line construction of suffix trees. *Algorithmica*, 14:249–260, 1995.
37. P. Wiener. Linear pattern matching algorithms. In *Proc. 14th Symposium on Switching and Automata Theory*, pages 1–11. IEEE, 1973.
38. J. Ziv and A. Lempel. A universal algorithm for sequential data compression. *IEEE Trans. on Information Theory*, IT-23:337–343, 1977.

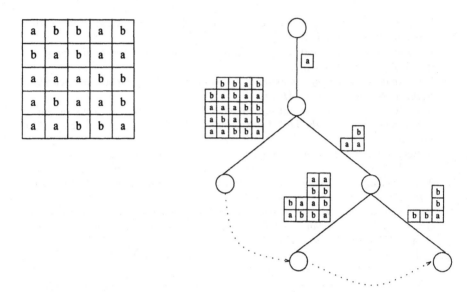

Fig. 1. The Lsuffix tree for the Lstring corresponding to the matrix in the figure. Dotted lines are the suffix links connecting leaves. Since $A = A_0$, this is also part of the the Lsuffix tree for the entire matrix.

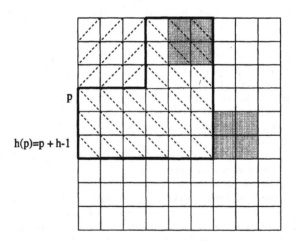

Fig. 2. The L-shaped region within bold lines is the window FW_6 with $h = 3$. The 3×3 matrix in the upper left corner of A is known, but it is not accessible any longer. Notice that this latter matrix plus the window gives W_6. The Lsuffix tree for FW_6 (not shown) represents all square submatrices of A that are also in the window. The light shaded part outside the window is a maximal match for the neighbor position (5,7) of FW_6. It matches the light shaded part in the window.

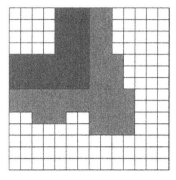

Fig. 3. The dark shaded part is the window FW_7, with $h = 3$. The light shaded part is the area S covered by the (possibly overlapping) matrices reported by $Match(FW_7)$.

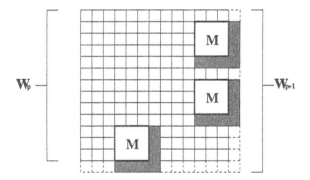

Fig. 4. Assume that matrix M (of side 3) appears only as suffix of $W_{p,7}, W_{p,-4}, W_{p,-9}$. Consider the three suffixes (of side 4) M_1, M_2, M_3 of $W_{p+1,7}, W_{p+1,-4}, W_{p+1,-9}$, respectively. Each of them is obtained by appending the dark shaded Lcharacter to each of the occurrences of M. Assume that $\tilde{M}_1 = M_1 = M_2 \neq M_3 = \tilde{M}_2$. The set of extensions of M is given by $\{\tilde{M}_1, \tilde{M}_2\}$.

Scheduling Multiclass Queueing Networks on Parallel Servers: Approximate and Heavy-Traffic Optimality of Klimov's Priority Rule

Kevin D. Glazebrook[1] and José Niño-Mora[2]

[1] Department of Statistics, Newcastle University, Newcastle upon Tyne NE1 7RU, UK, kevin.glazebrook@newcastle.ac.uk

[2] CORE, Université catholique de Louvain, 34 voie du Roman Pays, 1348 Louvain-la-Neuve, Belgium, jnimora@alum.mit.edu

Abstract. We address the problem of scheduling a multiclass queueing network on M parallel servers to minimize time-average linear holding costs. We analyze a heuristic priority-index rule, based on Klimov's solution to the single-server model: Compute the indices given by Klimov's adaptive greedy algorithm and, when a server becomes free, select a customer with largest index. We present closed-form performance guarantees for this heuristic, with respect to (1) the optimal cost in the original parallel-servers network, and (2) the optimal cost in a "corresponding" single-server network, attended by a server working M times faster. Simpler expressions are derived for the special case that there is no customer feedback, where the heuristic becomes the $c\mu$-rule. Our analysis is based on a primal-dual approach: We compare the cost of the heuristic to the value of (the dual of) a strong linear programming (LP) relaxation, which represents the optimal cost for the "corresponding" single-server network. This relaxation follows from a set of approximate conservation laws (ACLs) satisfied by the network. Our proof of these laws relies on the first set of work decomposition laws known for this model, which we obtain from a classical flow conservation law.

1 Introduction

Can we match the performance of a fast processor (with speed M) with a set of M slow parallel processors (with speed 1)? Clearly not, due to the inefficiencies inherent in parallel processing: the slow-parallel-processors system may process work at a rate strictly less than M, when some processor is idle, whereas the fast-single-processor system always works at rate M. How close, then, can we get to matching the performance of the fast processor with the "corresponding" set of slow processors, and how should we schedule the parallel processors to achieve their best performance? These issues are significant in the design and operation of complex service systems, such as flexible manufacturing systems and computer-communication networks. In this paper we address those issues in the context of a versatile service system model: a multiclass queueing network (MQNET) with parallel servers. Our results indicate that understanding of single-server systems yields considerable insight into the behaviour of their parallel-servers counterparts: In the model we study, the known optimal scheduling policy to the single-server model yields a performance bound together with a provably good heuristic for the parallel-servers model, and both the bound and the heuristic are asymptotically optimal in heavy-traffic.

In a landmark paper, Klimov (1974) solved the problem of scheduling dynamically a MQNET on a single server to minimize the time-average holding cost: compute an index for each class, by running Klimov's algorithm, and then service at each time a customer with largest available index. No optimal policy, however, is known for the corresponding parallel servers model. Still, Klimov's solution naturally suggests a simple heuristic for the extended model: when a server becomes available, assign to it a customer with the largest available Klimov's index. Although it had been conjectured that Klimov's index heuristic should exhibit a good performance (see, e.g., Weiss (1990, 1995)), no analysis of it was previously available. We present in this paper the first exact and asymptotic analyses of Klimov's heuristic for scheduling a MQNET on parallel servers. Our results include exact performance guarantees, which imply its asymptotic optimality in the heavy-traffic limit.

Weiss (1990, 1992) has analyzed the performance of Smith's rule for a parallel machines stochastic scheduling model with general processing time distributions, but with no arrivals or feedback. He also gave in Weiss (1995) an account of Gittins index policies for a more general model involving the preemptive scheduling of a batch of stochastic jobs on parallel machines. In Weiss (1992, 1995) he argued the importance of proceeding to more complex models, including those incorporating job arrivals and networks of queues.

This paper develops a methodology which is effective in evaluating the performance of priority-index rules in the context of such models, focusing on a single-station MQNET model with parallel servers. In a companion paper (see Glazebrook and Niño-Mora (1997b)) we study the performance of priority-index rules for scheduling multi-station MQNETs. Reflection on Weiss' work (1990, 1992, 1995) on Smith's and Gittins' rules for scheduling a batch of jobs on parallel servers will influence how we approach the analysis. Weiss found that the index rules which were the subject of his analysis, and which may be thought of as policies whose aim is to drive down fastest the rates at which waiting jobs incur costs, were in general suboptimal due to an "end effect" related to the loss of processing efficiency when the number of machines exceeds the number of jobs available. Weiss was able to quantify this effect, and he obtained $O(1)$ bounds on the difference between the expected cost of the index rule and the optimal policy. These results implied the asymptotic optimality of the heuristic in the limit as the batch size grows to infinity. In our MQNET model with job arrivals and feedback, these considerations lead us naturally toward investigations of asymptotic optimality in a heavy traffic limit in which the relative importance of machine deployment when not all machines are busy might be expected to decline to zero.

Our methodology is based on formulating strong LP relaxations of the MQNET region of achievable performance. The performance guarantees we derive follow from physical properties of the network (Glazebrook and Garbe's (1996) approximate conservation laws), which yield tight polyhedral relaxations of its performance region. The approximate conservation laws are established by identifying and applying the first work decomposition laws known for the parallel-server model, which are derived from a set of linear equations on performance measures that formulate flow conservation laws. Our analysis recovers the optimality of Klimov's rule in the single-server model, as well as in the special case of the parallel-server model in which all service times are i.i.d. and exponential (see Federgruen and Groenevelt (1988)).

Our approach extends to a hard dynamic and stochastic optimization problem a solution technique that has proven fruitful in the domain of hard combinatorial optimization problems: the primal-dual method for approximation algorithms (see, e.g., Nemhauser and Wolsey (1988)). In this method one seeks to construct simultaneously heuristic solutions to an optimization problem and to the dual of an LP relaxation of it. A performance guarantee follows by comparing the values of both solutions.

The primal-dual approach we pursue is based on the framework developed by Bertsimas and Niño-Mora (1996a) and extended by Glazebrook and Garbe (1996), in which the notion of *conservation laws* plays a central role. Bertsimas and Niño-Mora (1996a) presented a framework for constructing exact LP formulations of dynamic and stochastic scheduling problems that satisfy *generalized conservation laws*. This framework yields the optimality of *priority-index* scheduling policies in a variety of models, including certain single-server MQNETs and multi-armed bandit problems. Glazebrook and Garbe (1996) extended its scope to encompass more complex systems that satisfy conservation laws only in an *approximate* sense. They show that for a system that satisfies *approximate conservation laws* priority-index policies are still *approximately optimal*, by providing explicit performance guarantees. They further apply this framework to several models, including certain classes of discounted multi-armed bandits.

The rest of the paper is structured as follows: Section 2 introduces the MQNET model of interest, formulates the corresponding scheduling problem and describes the proposed priority-index heuristic. Section 3 gives an appropriate reformulation of Glazebrook and Garbe's approximate conservation laws framework on which our main results are based. In Section 4 the analysis of the index heuristic is carried out. We show that work decomposition results yield approximate conservation laws, which are the basis of our approximate and heavy-traffic optimality results. Finally, Section 5 ends the paper with some concluding remarks.

For proofs of the results presented here we refer the reader to the full version of this paper: Glazebrook and Niño-Mora (1997a).

2 The MQNET Model

We consider a single-station MQNET model with N customer classes and M identical parallel servers. Customers of class $i \in \mathcal{N} = \{1, \ldots, N\}$ (or i-customers) arrive at the network from outside as a Poisson process with rate α_i. They may be processed by any server during a time drawn from an exponential distribution with rate μ_i. Upon completion of his service, an i-customer may be routed for further service as a j-customer, with probability p_{ij}, or it may leave the network, with probability $p_{i0} = 1 - \sum_{j \in \mathcal{N}} p_{ij}$. The matrix $\mathbf{I} - \mathbf{P}$ is assumed to be invertible. We further assume that all customer arrival processes, service times and routing events are mutually independent.

The *total arrival rate of j-customers*, denoted λ_j, is defined by the solution of

$$\lambda_j = \alpha_j + \sum_{i \in \mathcal{N}} \lambda_i p_{ij}, \qquad \text{for } j \in \mathcal{N}.$$

The *traffic intensity of j-customers*, denoted ρ_j, is defined by

$$\rho_j = \frac{\lambda_j}{\mu_j}, \qquad \text{for } j \in \mathcal{N}.$$

The *network traffic intensity*, denoted ρ, is given by

$$\rho = \sum_{j \in \mathcal{N}} \rho_j.$$

Given a subset of job classes $S \subseteq \mathcal{N}$, we define similarly the *traffic intensity for S-customers*, denoted $\rho(S)$, by

$$\rho(S) = \sum_{j \in S} \rho_j.$$

We further require the notion of *mean S-workload of a j-customer*, denoted V_j^S, which is the mean remaining service time a current j-customer receives until he leaves classes in subset S for the first time after completing his current service. The V_j^S's are computed by solving

$$V_i^S = \frac{1}{\mu_i} + \sum_{j \in S} p_{ij} V_j^S, \qquad \text{for } i \in \mathcal{N},\, S \subseteq \mathcal{N}.$$

We further define the *external traffic intensity for S-customers*, denoted $\rho^0(S)$, by

$$\rho^0(S) = \sum_{j \in S} \alpha_j V_j^S.$$

The network evolution is governed by a *scheduling policy*, which specifies how servers are assigned to customers. We define *admissible* scheduling policies to be (1) *nonanticipative* (scheduling decisions are not based on future information), (2) *stable* (the time-average number of customers in the network is finite), (3) *preemptive* (the service of a customer may be interrupted at any time), (4) *server-symmetric* (decisions do not use server label information), and *nonidling* (no server stays idle when there are customers present).

The network *state* at time t is given by the following random variables:

$L_j(t)$: number of j-customers in system at time t.
$B_j^m(t)$: 1 if server m is busy with a j-customer at time t; 0 otherwise.
$B_j(t)$: 1 if a j-customer is in service at time t; 0 otherwise.
$B^m(t)$: 1 if server m is busy at time t; 0 otherwise.

It is known that any nonanticipative and nonidling policy is stable for this model if the condition

$$\rho < M$$

holds, and therefore we shall assume it in what follows.

We consider the optimal scheduling problem

$$Z_{parallel}^{OPT} = \min\{c_1 E[L_1] + \cdots + c_N E[L_N]\}, \tag{1}$$

where the minimum in (1) is taken over all admissible policies. We shall further denote Z_{single}^{OPT} the minimum cost rate (1) in a "corresponding" single-server MQNET model.

This problem has only been solved in the single-server case, for which Klimov (1974) showed that a priority-index rule is optimal: for each class i one computes a corresponding *priority-index* γ_i, such that it is optimal to serve at each time a customer with the largest available index. The vector of optimal indices $\gamma = (\gamma_j)_{j \in \mathcal{N}}$ is computed by running Klimov's N-step *adaptive greedy* algorithm on input $(\mathbf{c}, \mathbf{V})$, where $\mathbf{c} = (c_j)_{j \in \mathcal{N}}$ and $\mathbf{V} = (V_j^S)_{j \in S, S \subseteq \mathcal{N}}$.

Although Klimov's rule is no longer optimal for $M \geq 2$, it provides a simple heuristic.

Klimov's priority-index heuristic:

1. Compute Klimov's priority-indices γ_i by running Klimov's algorithm (shown in Figure 1) on input $(\mathbf{c}, \mathbf{V})$.
2. Schedule customers by assigning higher preemptive priority to classes with higher index γ_i.

We shall denote $Z_{parallel}^{INDEX}$ the time-average cost rate (1) achieved by Klimov's rule for the network model with parallel servers described above.

3 Approximate Optimality via Approximate Conservation Laws (ACLs)

The approximate/asymptotic optimality results of Section 4 below will be established by casting the corresponding optimal scheduling problem into the general framework of *approximate conservation laws*, developed by Glazebrook and

Input: $(\mathbf{c}, \mathbf{V})$, where $\mathbf{c} = (c_j)_{j \in \mathcal{N}}$ and $\mathbf{V} = (V_j^S)_{j \in S, S \subseteq \mathcal{N}}$.

Output: $(\boldsymbol{\pi}, \overline{\mathbf{y}}, \boldsymbol{\gamma})$, where $\boldsymbol{\pi} = (\pi_1, \ldots, \pi_N)$ is a permutation of $\mathcal{N}$, $\overline{\mathbf{y}} = (\overline{y}(S))_{S \subseteq \mathcal{N}}$ and $\boldsymbol{\gamma} = (\gamma_1, \ldots, \gamma_N)$.

Step 0. Set $S_1 = \mathcal{N}$; set $\overline{y}(S_1) = \min \left\{ \dfrac{c_i}{V_i^{S_1}} : i \in S_1 \right\}$;

    pick $\pi_1 \in \mathrm{argmin} \left\{ \dfrac{c_i}{V_i^{S_1}} : i \in S_1 \right\}$;

    set $\gamma_{\pi_1} = \overline{y}(S_1)$.

Step k. For $k = 2, \ldots, N$:

    set $S_k = S_{k-1} \setminus \{\pi_{k-1}\}$; set $\overline{y}(S_k) = \min \left\{ \dfrac{c_i - \sum_{j=1}^{k-1} V_i^{S_j} \overline{y}(S_j)}{V_i^{S_k}} : i \in S_k \right\}$;

    pick $\pi_k \in \mathrm{argmin} \left\{ \dfrac{c_i - \sum_{j=1}^{k-1} V_i^{S_j} \overline{y}(S_j)}{V_i^{S_k}} : i \in S_k \right\}$;

    set $\gamma_{\pi_k} = \gamma_{\pi_{k-1}} + \overline{y}(S_k)$.

Step N+1. For $S \subseteq \mathcal{N}$: set

$$\overline{y}(S) = 0, \qquad \text{if } S \notin \{S_1, \ldots, S_N\}.$$

Fig. 1. Klimov's adaptive greedy algorithm.

Garbe (1996), which extends the theory of *generalized conservation laws* of Bertsimas and Niño-Mora (1996a). We review in this section the approximate conservation laws framework, in an appropriately reformulated form.

Consider a *service system* with a set of servers that serve a set $\mathcal{N} = \{1, \ldots, N\}$ of customer classes. The system evolution is governed by a *scheduling policy*. Under each admissible policy the system performance is given by a corresponding *performance vector* $\mathbf{x} = (x_j)_{j \in \mathcal{N}}$, where x_j is a performance measure for class j-customers.

Given a subset S of customer classes, we say that a policy gives *priority* to S-customers (whose class is in S) over S^c-customers if a server never services an S^c-customer when it has an S-customer available.

Let $\mathbf{V} = (V_i^S)_{i \in S, S \subseteq \mathcal{N}}$ be a matrix with all $V_i^S > 0$, and let $b(S)$ and $\Phi(S)$ be nonnegative set functions defined on subsets S of customer classes, with $b(S) > 0$ for all $S \subseteq \mathcal{N}$.

Definition 1 Approximate conservation laws (ACLs). We say that performance vector $\mathbf{x}$ satisfies *approximate conservation laws* with parameters $\mathbf{V}$, b and Φ if the following conditions hold:

(i) Under any admissible scheduling policy,

$$\sum_{j \in S} V_j^S x_j \geq b(S), \qquad \text{for } S \subseteq \mathcal{N}. \tag{2}$$

(ii) Under any admissible nonidling policy that gives *priority* to S-customers over S^c-customers,

$$\sum_{j \in S} V_j^S x_j \leq b(S) + \Phi(S), \qquad \text{for } S \subseteq \mathcal{N}. \tag{3}$$

When a performance vector satisfies ACLs we can come close to solving optimal scheduling problems on linear performance objectives. Consider the *optimal scheduling problem* that consists of (1) finding an admissible policy that minimizes the objective

$$c_1 x_1 + \cdots + c_N x_N, \tag{4}$$

for a given cost vector $\mathbf{c} = (c_j)_{j \in \mathcal{N}} \geq 0$, over all admissible policies, and (2) computing the corresponding optimum cost, Z^{OPT}. We consider the *performance region $\mathcal{X}$* achievable by performance vector $\mathbf{x}$ under *all* admissible scheduling policies. We can now formulate this optimal scheduling problem as the mathematical program

$$Z^{OPT} = \min\{c_1 x_1 + \ldots + c_N x_N : \mathbf{x} \in \mathcal{X}\}. \tag{5}$$

Suppose that performance vector $\mathbf{x}$ satisfies ACLs (2) and (3). Then performance region $\mathcal{X}$ is contained in the polyhedron

$$\mathcal{P} = \left\{ \mathbf{x} \in \Re_+^N : \sum_{j \in S} V_j^S x_j \geq b(S), S \subseteq \mathcal{N} \right\},$$

which yields the following *LP relaxation* of problem (5):

$$Z^{LP} = \min\{c_1 x_1 + \ldots + c_N x_N : \mathbf{x} \in \mathcal{P}\}. \tag{6}$$

Our approach for solving problem (5) approximately corresponds precisely to the *primal-dual method* for approximation algorithms, which has a rich history in the field of combinatorial optimization (see, e.g., Nemhauser and Wolsey (1988)): we will construct simultaneously a heuristic solution to the problem and a feasible solution to the dual of an LP relaxation (given by (6)); a performance guarantee will follow by comparing the values of both solutions.

We shall construct feasible solutions to problems (5) and the dual of (6) as follows:

1. Run Klimov's adaptive greedy algorithm on input $(\mathbf{c}, \mathbf{V})$ to obtain an output $(\pi, \overline{\mathbf{y}}, \gamma)$.
2. The proposed heuristic solution for problem (5) is the performance vector of the priority-index rule that gives higher priority to classes with higher index γ_i. We denote Z^{INDEX} the value (4) achieved by this policy.
3. The proposed feasible solution for the dual of LP relaxation (6) is given by $\overline{\mathbf{y}}$ (see Bertsimas and Niño-Mora (1996a) for a proof that $\overline{\mathbf{y}}$ is indeed dual feasible.)

We shall assume in what follows that the permutation π returned by Klimov's algorithm is $\pi = (1, \ldots, N)$, so that

$$\gamma_1 \leq \cdots \leq \gamma_N.$$

In Bertsimas and Niño-Mora (1996a) it is shown that the value of dual feasible solution $\overline{y}$ is given by

$$Z^D = \gamma_1 b(\{1, \ldots, N\}) + (\gamma_2 - \gamma_1)b(\{2, \ldots, N\}) + \cdots + (\gamma_N - \gamma_{N-1})b(\{N\}). \quad (7)$$

Notice that, by weak LP duality,

$$Z^D \leq Z^{LP} \leq Z^{OPT} \leq Z^{INDEX}. \quad (8)$$

The next result, which reformulates and extends the original result in Glazebrook and Garbe (1996), establishes performance guarantees for the value Z^{INDEX} and the lower bound Z^D.

Theorem 2 Performance guarantees. *Suppose that performance vector* $\mathbf{x}$ *satisfies ACLs (2) and (3) with parameters* $\mathbf{V}$, b *and* Φ. *Then*
(a)

$$Z^{INDEX} \leq Z^{OPT} + \epsilon \quad (9)$$

and

$$Z^D \geq Z^{OPT} - \epsilon, \quad (10)$$

where

$$\epsilon = \gamma_1 \Phi(\{1, \ldots, N\}) + (\gamma_2 - \gamma_1)\Phi(\{2, \ldots, N\}) + \cdots + (\gamma_N - \gamma_{N-1})\Phi(\{N\}). \quad (11)$$

(b) *Furthermore,*

$$Z^{INDEX} \leq rZ^{OPT} \quad (12)$$

and

$$Z^D \geq \frac{1}{r}Z^{OPT}, \quad (13)$$

where

$$r = 1 + \max_{S \subseteq \mathcal{N}} \frac{\Phi(S)}{b(S)}. \quad (14)$$

4   Analysis of Klimov's Index Heuristic

Our analysis approach will be to obtain a *work decomposition* law in Section 4.1 via flow conservation principles. This, along with other constraints on achievable performance, will be used to establish ACLs as in Section 4.2. From these we deduce a range of approximate/heavy-traffic optimality results in Section 4.3.

4.1 Work Decomposition Laws for a MQNET with Parallel Servers

In a variety of single-server multiclass queueing systems researchers have identified *work decomposition laws*, which describe a linear relation between the mean number in the system from each class at an arbitrary time and at an arbitrary time during an interval when the server is idle. Such work decomposition results have played a major role in the performance analysis of vacation and polling models (see, e.g., Fuhrmann and Cooper (1985) and Boxma (1989)). Extensions of those work decomposition laws to MQNETs with one or multiple single-server stations have been developed by Bertsimas and Niño-Mora (1996b, 1996c), who have further applied them to yield new performance bounds. To the best of our knowledge, no work decomposition law has previously been presented for a MQNET with parallel servers.

The main result in this section is a family of new work decomposition laws (Theorem 5) for the parallel server MQNET model studied in this paper. These work decomposition laws will play a central role in our analysis, as they will yield the performance bounds required to cast the heuristic analysis problem into the ACLs framework.

To establish the new work decomposition laws we shall pursue the following approach:

1. Formulate a polyhedral relaxation of the network performance region by *lifting* it into an extended space that involves *auxiliary performance variables*. The key constraints that define this relaxation formulate *flow conservation laws*, and we shall refer to it as the *flow conservation formulation*.
2. Express the constraints in the above relaxation to obtain a *workload reformulation*, in which the key constraints formulate *work decomposition laws*.

We consider the auxiliary performance measures

$$- \; x_{ij} = E\left[L_j \mid B_i^1 = 1\right]; \mathbf{X} = (x_{ij})_{i,j \in \mathcal{N}}.$$
$$- \; x_j^0 = E\left[L_j \mid B^1 = 0\right]; \mathbf{x}^0 = (x_j^0)_{j \in \mathcal{N}}.$$

The flow conservation relaxation. In the next two results we present corresponding families of linear constraints on performance measures $\mathbf{x}$, $\mathbf{X}$ and $\mathbf{x}^0$ that define the *flow conservation relaxation*.

Proposition 3 Conditioning constraints. *Under any admissible scheduling policy performance measures $\mathbf{x}$, $\mathbf{X}$ and $\mathbf{x}^0$ satisfy the following relations:*

$$x_j = \sum_{i \in \mathcal{N}} \frac{\rho_i}{M} x_{ij} + (1 - \frac{\rho}{M}) x_j^0, \qquad for \; j \in \mathcal{N}. \tag{15}$$

It was first shown by Niño-Mora (1995) (see also Bertsimas and Niño-Mora (1996b)) that in a variety of station MQNETs application of the classical *flow conservation law* $L^- = L^+$ of queueing theory yields a set of linear equations on performance measures analogous to that which appears in (16) below as part of the statement of Proposition 4. Extend this technique to our MQNET model with parallel servers yields the following result. Let $\boldsymbol{\alpha} = (\alpha_j)_{j \in \mathcal{N}}$ and $\Lambda = \mathrm{Diag}(\boldsymbol{\lambda})$, where $\boldsymbol{\lambda} = (\lambda_j)_{j \in \mathcal{N}}$.

Proposition 4 Flow conservation laws. *Under any admissible scheduling policy performance measures* **x** *and* **X** *satisfy the following linear equations:*

$$-\alpha \mathbf{x}' - \mathbf{x}\alpha' + (\mathbf{I} - \mathbf{P})'\Lambda\mathbf{X} + \mathbf{X}'\Lambda(\mathbf{I} - \mathbf{P}) \; = \; (\mathbf{I} - \mathbf{P})'\Lambda + \Lambda(\mathbf{I} - \mathbf{P}). \quad (16)$$

The *flow conservation relaxation* of the network performance region is the polyhedron in $(\mathbf{x}, \mathbf{X}, \mathbf{x}^0)$-space defined by constraints (15), (16) and the nonnegativity constraints

$$\mathbf{x}, \mathbf{X}, \mathbf{x}^0 \geq \mathbf{0}. \quad (17)$$

The workload reformulation. Proposition 4 allows us to establish a new work decomposition result for the parallel server MQNET model of interest. The proof involves straightforward algebraic manipulation and follows similar lines to those in Niño-Mora (1995) for single-server station models.

Theorem 5 Work decomposition laws. *Under any admissible policy performance measures* **x**, **X** *and* $\mathbf{x}^0$ *satisfy the following linear constraints: For* $S \subseteq \mathcal{N}$,

$$\sum_{j \in S} V_j^S x_j \; = \; \frac{\sum_{j \in S} \rho_j V_j^S}{M - \rho^0(S)} + \sum_{i \in S^c} \sum_{j \in S} \frac{\lambda_i V_i^S V_j^S}{M - \rho^0(S)} x_{ij} + \frac{M - \rho}{M - \rho^0(S)} \sum_{j \in S} V_j^S x_j^0. \quad (18)$$

The proof of Theorem 5 is based on reformulating a classical flow conservation law in terms of workloads, via Palm calculus.

4.2 Approximate Conservation Laws

In order to use Theorem 5 to develop ACLs for the network we require ways of bounding the last two terms on the right hand side of (18) under nonidling policies that give priority to S-customers. The appropriate results are now given as Propositions 6 and 7.

Proposition 6 Nonidling constraints. *Under any admissible nonidling policy performance vector* $\mathbf{x}^0$ *satisfies the inequality constraint*

$$\sum_{j \in \mathcal{N}} x_j^0 \leq (M - 1) \min \left(1, \frac{\rho}{M - \rho} \right). \quad (19)$$

Proposition 7 Priority constraints. *Under any admissible nonidling policy that gives preemptive priority to S-customers:*

$$\sum_{j \in S} x_{ij} \; \leq \; M - 1, \qquad for \; i \in S^c.$$

The ACLs given in Theorem 8 below follow simply from Theorem 5, Propositions 6-7 and the nonnegativity constraints. The reader is referred to definition 1.

As before, performance vector $\mathbf{x} = (x_j)_{j \in \mathcal{N}}$ is given by

$$x_j = E[L_j], \quad \text{for } j \in \mathcal{N}.$$

Let set functions b and Φ be given by

$$b(S) = \frac{\sum_{j \in S} \rho_j V_j^S}{M - \rho^0(S)} \tag{20}$$

and

$$\Phi(S) = \left\{ V_{\max}^S \sum_{i \in S^c} \lambda_i + (M - \rho) \min\left(1, \frac{\rho}{M - \rho}\right) \right\} \frac{M - 1}{M - \rho^0(S)} V_{\max}^S, \tag{21}$$

where

$$V_{\max}^S = \max_{i \in \mathcal{N}} V_i^S,$$

for $S \subseteq \mathcal{N}$.

Theorem 8 ACLs. *Performance vector $\mathbf{x}$ satisfies ACLs (2) and (3) with parameters $\mathbf{V}$, b and Φ.*

We obtain results which are both stronger and simpler in the special case of our model in which $\mathbf{P} = \mathbf{0}$, i.e., there is no customer feedback. The resulting model is a multiclass M-server queue with independent Poisson arrival streams and exponential service times which are i.i.d. within each class. Klimov's rule reduces in this case to the celebrated $c\mu$-rule.

Let us define matrix $\hat{\mathbf{V}} = (\hat{A}_i^S)_{i \in S, S \subseteq \mathcal{N}}$ by

$$\hat{A}_i^S = \frac{1}{\mu_i}, \quad \text{for } i \in S, S \subseteq \mathcal{N},$$

and let set functions $\hat{b}$ and $\hat{\Phi}$ be given by

$$\hat{b}(S) = \frac{\sum_{j \in S} \rho_j / \mu_j}{M - \rho(S)}$$

and

$$\hat{\Phi}(S) = (M - 1) \min\left(1, \frac{\rho}{M - \rho}\right) \max_{j \in S} \frac{1}{\mu_j}, \tag{22}$$

for $S \subseteq \mathcal{N}$.

Theorem 9 Approximate conservation laws; no feedback model. *Performance vector $\mathbf{x}$ satisfies approximate conservation laws (2) and (3) with parameters $\hat{\mathbf{V}}$, $\hat{b}$ and $\hat{\Phi}$.*

Remark: We further show in Glazebrook and Niño-Mora (1997a) that in the special case that $\mu_j = \mu$, for $j \in \mathcal{N}$ (with $M > 1$) an ACLs analysis recovers the optimality of the $c\mu$-rule, first established by Federgruen and Groenevelt (1988).

4.3 Approximate and Heavy-Traffic Optimality of Klimov's Index Rule

Having established ACLs, we proceed to obtain performance guarantees for Klimov's heuristic based on Theorem 2. We shall retain the customer numbering in which 1 and N are the customer classes with smallest and largest indices, respectively, and we further assume that $\mathbf{c} > 0$.

As before, $Z^{INDEX}_{parallel}$ and $Z^{OPT}_{parallel}$ denote the cost rate achieved by Klimov's heuristic and the optimal cost rate for the model with M parallel servers, and Z^{OPT}_{single} denotes the optimal cost rate for the "corresponding" single-server model, where the server works M times faster than each server in the parallel model. The next result establishes relative performance guarantees for Klimov's heuristic, and also relates the optimal performances achieved in the single and parallel server models.

Theorem 10 Performance guarantees for Klimovs' rule. *We have*

$$Z^{INDEX}_{parallel} \leq r Z^{OPT}_{parallel} \tag{23}$$

and

$$Z^{OPT}_{parallel} \geq Z^{OPT}_{single} \geq \frac{1}{r} Z^{OPT}_{parallel}, \tag{24}$$

where

$$r = 1 + (M-1)(M-\rho)V^{\mathcal{N}}_{\max} \frac{(\gamma_N - \gamma_1)V^{\mathcal{N}}_{\max} \sum_{i \in \mathcal{N} \setminus \{N\}} \lambda_i + \left(M - \rho^0(\mathcal{N} \setminus \{1\})\right)\gamma_N}{\left(M - \rho^0(\mathcal{N} \setminus \{1\})\right)\gamma_1 \sum_{j \in \mathcal{N}} \rho_j V^{\mathcal{N}}_j}.$$

We next investigate the behaviour of Klimov's rule in heavy traffic, as $\rho \to M$. To this end we consider a sequence of models in which only the external arrival rates vary. We thus suppose that the vector of external arrival rates $\boldsymbol{\alpha}$ varies according to a convergent sequence $\{\boldsymbol{\alpha}^n\}_{n=1}^{\infty}$ with limit $\boldsymbol{\alpha}^*$ such that, in an obvious notation,

$$\rho(\boldsymbol{\alpha}^n) \to \rho(\boldsymbol{\alpha}^*) = M \quad \text{as } n \to \infty$$

and

$$\rho_j(\boldsymbol{\alpha}^*) > 0 \quad \text{for } j \in \mathcal{N}. \tag{25}$$

Corollary 11 Heavy-traffic optimality. *The following relations hold:*

$$\frac{Z^{INDEX}_{parallel}(\boldsymbol{\alpha}^n)}{Z^{OPT}_{parallel}(\boldsymbol{\alpha}^n)} = 1 + O(M - \rho(\boldsymbol{\alpha}^n)) \to 1 \quad \text{as } n \to \infty$$

and

$$\frac{Z^{OPT}_{parallel}(\boldsymbol{\alpha}^n)}{Z^{OPT}_{single}(\boldsymbol{\alpha}^n)} \to 1 \quad \text{as } n \to \infty.$$

We now focus on the simple model with no feedback analyzed in Theorem 9. The index policy reduces in this case to the $c\mu$-rule, for which

$$\gamma_i = c_i \mu_i, \qquad \text{for } i \in \mathcal{N}.$$

The performance guarantee in Theorem 12 below is an elegant consequence of the theory. Let $Z^{c\mu}$ denote the expected cost achieved by the $c\mu$-rule heuristic.

Theorem 12 Performance guarantee for $c\mu$-rule.

$$Z^{c\mu}_{parallel} - Z^{OPT}_{parallel} \leq (M-1) \sum_{j \in \mathcal{N}} c_j.$$

The form of $\hat{\Phi}$ in Theorem 9 implies that for a heavy-traffic limit result in the model with no feedback we only require a sequence of traffic intensities $\{\rho^n\}_{n=1}^{\infty}$ with limit M. No further conditions need be imposed. The proof of Theorem 13 below involves straightforward calculations and is omitted.

Theorem 13 Heavy-traffic optimality of $c\mu$-rule.

$$1 \leq \frac{Z^{c\mu}_{parallel}(\rho^n)}{Z^{OPT}_{parallel}(\rho^n)} \leq 1 + \frac{(M-1)(M-\rho^n)\sum_{j \in \mathcal{N}} c_j}{c_1 \mu_1 \sum_{j \in \mathcal{N}} \frac{\rho^n_j}{\mu_j}}$$

$$\to 1 \qquad as\ n \to \infty.$$

Notice the presence of the factor $M-1$ in all of the above performance guarantees. This ensures that the known optimality of the corresponding index rule in the $M = 1$ case is recovered from our analyses.

5 Concluding Remarks

Three ideas that emerge from our analysis are the following: (1) understanding of a single-server system has yielded useful insights into the behaviour of its parallel-servers counterpart; (2) understanding the fundamental laws of a complex parallel-server model has yielded the key to its analysis; in the model we addressed, these fundamental laws were the flow conservation, work decomposition and approximate conservation laws we identified and applied; (3) studying strong LP relaxations of a complex stochastic optimization problem has yielded an exact and asymptotic analysis of a heuristic which had resisted traditional approaches. We believe that these ideas, which guided our approach, should prove fruitful for addressing other complex stochastic optimization problems.

6 Acknowledgements

The work of the first author was supported by the Engineering and Physical Sciences Research Council by means of grant no. GR/K03043. The work of the second author was partially supported by a CORE Research Fellowship and by EC Marie Curie Research Fellowship no. ERBFMBICT961480. Part of this research was carried out during the second author's visit to the Department of Statistics at Newcastle University.

References

Bertsimas, D. and Niño-Mora, J.: Conservation laws, extended polymatroids and multi-armed bandit problems; a polyhedral approach to indexable systems. *Math. Oper. Res.* **21** (1996a) 257-306

Bertsimas, D. and Niño-Mora, J.: Optimization of multiclass queueing networks with changeover times via the achievable region approach: Part II, the multi-station case. Working paper, Operations Research Center, MIT, (1996b)

Federgruen, A. and Groenevelt, H.: Characterization and optimization of achievable performance in general queueing systems. *Oper. Res.* **36** (1988) 733-741

Fuhrmann, S.W. and Cooper, R.B.: Stochastic decompositions in the $M/G/1$ queue with generalized vacations. *Oper. Res.* **33** (1985) 1117-1129

Glazebrook, K.D. and Garbe, R.: Almost optimal policies for stochastic systems which almost satisfy conservation laws. Working paper, Department of Statistics, Newcastle University, (1996)

Glazebrook, K.D. and Niño-Mora, J.: Scheduling multiclass queueing networks on parallel servers: Approximate and heavy traffic optimality of Klimov's rule. CORE Discussion Paper 9710, Université catholique de Louvain, (1997a)

Glazebrook, K.D. and Niño-Mora, J.: An LP approach to stability, optimization and performance analysis for Markovian multiclass queueing networks. CORE Discussion Paper, Université catholique de Louvain, (1997b)

Klimov, G.P.: Time sharing service systems I. *Theory Probab. Appl.* **19** (1974) 532-551

Nemhauser, G.L. and Wolsey, L.A.: *Integer and Combinatorial Optimization.* Wiley, New York, (1988)

Niño-Mora, J. *Optimal Resource Allocation in a Dynamic and Stochastic Environment: A Mathematical Programming Approach.* PhD Dissertation, Sloan School of Management, MIT, (1995)

Weiss, G.: Approximation results in parallel machines stochastic scheduling. *Ann. Oper. Res.* Special Volume on Production Planning and Scheduling, M. Queyranne, ed., **26** (1990) 195-242

Weiss, G.: Turnpike optimality of Smith's rule in parallel machines stochastic scheduling. *Math. Oper. Res.* **17** (1992) 255-270

Weiss, G.: On almost optimal priority rules for preemptive scheduling of stochastic jobs on parallel machines. *Adv. in Appl. Probab.* **27** (1995) 821-839

Optimal Reconstruction of Graphs Under the Additive Model

Vladimir Grebinski and Gregory Kucherov

INRIA-Lorraine and CRIN/CNRS,
615, rue du Jardin Botanique, BP 101,
54602 Villers-lès-Nancy, France,
e-mail: {grebinsk,kucherov}@loria.fr

Abstract. We study the problem of combinatorial search for graphs under the additive model. The main result concerns the reconstruction of *bounded degree* graphs, i.e. graphs with the degree of all vertices bounded by a constant d. We show that such graphs can be reconstructed in $O(dn)$ non-adaptive queries, that matches the information-theoretic lower bound. The proof is based on the technique of separating matrices. In particular, a new upper bound is obtained for d-separating matrices, that settles an open question stated by Lindström in [17]. Finally, we consider several particular classes of graphs. We show how an optimal non-adaptive solution of $O(n^2 / \log n)$ queries for general graphs can be obtained.

1 Introduction and Definitions

Combinatorial Search studies problems of the following general type: determine an unknown object by means of indirect questions about this object. Perhaps the most common example of combinatorial search is the variety of problems of determining one or several counterfeit coins in a set using scales of some kind. Many of these problems still lack an optimal general solution.

Each instance of a Combinatorial Search problem has two main components: a finite *domain of objects* $\mathcal{M}$ and a *class of queries* $\mathcal{Q}$, which is a family of functions from the domain of objects to a domain $\mathcal{A}$ of *answers*. Given $\mathcal{M}$ and $\mathcal{Q}$, the combinatorial search problem is to find a sequence of queries $(q_1, q_2, \ldots, q_k)$, $q_i \in \mathcal{Q}$, such that the sequence of answers $(q_1(x), q_2(x), \ldots, q_k(x))$ uniquely identifies the object $x \in \mathcal{M}$. A method for choosing queries $(q_1, q_2, \ldots, q_k)$ is called a *(combinatorial) search algorithm*. The complexity measure of a search algorithm is the maximal number k of required queries over all $x \in \mathcal{M}$. This implies that we are concerned with *query complexity* only. Precise complexity bounds to combinatorial search problems can be rarely obtained. Instead, one is usually interested in the asymptotic complexity, when $|\mathcal{M}|$ tends to infinity.

Monographs [6,2] present detailed accounts of numerous results on Combinatorial Search problems. Variants of these problems abound in different application domains. For example, paper [10] deals with a problem motivated by genome analysis. Note that Combinatorial Search is closely related to Learning

Theory, where the general framework is similar, except possibly that there is usually an infinity of objects and one is looking not necessarily for the object itself but for its approximation according to a given distance function.

In general, the choice of q_i in the sequence $(q_1, q_2, \ldots, q_n)$ depends on the answers $(q_1(x), \ldots, q_{i-1}(x))$ obtained "so far". If this dependence exists, the algorithm is called *adaptive* (or sequential). Otherwise, when all the queries can be given before any answer is known, the algorithm is called non-adaptive (or predetermined). In this paper, we deal with non-adaptive algorithms. Although they are obviously less powerful in general, non-adaptive algorithms usually admit "nicer" mathematical formulations that allow to use more powerful mathematical methods. Besides, in many cases (including those considered in this paper) non-adaptive algorithms achieve the power of adaptiveness, that is reach the lower bound. Note also that in non-adaptive algorithms all queries can be made in parallel, which is useful in many applications.

For the non-adaptive case, we reformulate the combinatorial search problem as follows: find a minimal number of queries $q_1, q_2, \ldots, q_n \in \mathcal{Q}$ such that for every $x, y \in \mathcal{M}$, there is q_i, $1 \leq i \leq n$, such that $q_i(x) \neq q_i(y)$.

In contrast to the coin weighing problem where objects of $\mathcal{M}$ are just elements or subsets of elements of a given set, the objects may be of a more complex nature, such as graphs or partially ordered sets (see [1,2]). In case of graphs, different combinatorial search problems can be raised. One may look for an unknown edge in a *given* graph by asking, for a subset of vertices, whether one of the edge's endpoints (or both) belongs to the subset. A more general problem, considered in this paper, consists of determining an unknown graph of a given class. Here again, subsets of vertices are queried, but the answer returned characterizes some property of the subgraph induced by the subset. Finally, the third type of problem is to check whether an unknown graph belongs to a given class without actually determining the graph. This problem, known as *property testing*, received much attention in connection with the study of *evasiveness* property (see [18]). Another approach to property testing, in the framework of probabilistic algorithms and approximation, was recently introduced in [8,9].

It is clear that for the same object domain $\mathcal{M}$, different classes of queries $\mathcal{Q}$ lead to combinatorial search problems of different type and different complexity. Under the *additive model*, the domain of answers is the ring of integers $\mathbb{Z}$. This model is also called *quantitative*, as the queries $\mathcal{Q}$ are usually about some quantitative property of the object. A typical example is to identify the subset of counterfeit coins using a spring scale under the knowledge of the difference in weight between a counterfeit and authentic coin (which allows to determine the number of counterfeit coins in a subset by weighing this subset). We will come back to this example in Sect. 2. Some additive models of combinatorial search are studied in [13,7,12].

In this paper we consider the problem of *searching for a graph under the additive model* defined as follows. The domain of objects, denoted $\mathcal{G}_n$ is a class of simple graphs with n vertices labelled by natural numbers $1, 2, \ldots, n$. (A graph is simple if it does not contain loops and multiple edges.) The queries that we are allowed to make about $G \in \mathcal{G}_n$ are of the following form: For a subset $V \subseteq \{1, \ldots, n\}$ of vertices, how many edges are there in G between vertices of V? More formally, how many edges occur in the intersection $G \cap K_V$, where K_V is the complete graph with the set of vertices V?

In this paper we develop new techniques of non-adaptive additive search for a graph. Our main result concerns the search for *bounded degree* graphs, i.e. graphs with the degree of all vertices bounded by a constant d. We prove that such a graph can be reconstructed within $O(dn)$ non-adaptive queries, which matches the information-theoretic lower bound. The key intermediate result shows that a bipartite graph can be reconstructed in $O(dn)$ non-adaptive queries provided that the degree of vertices *on one side* is bounded by d while no restriction on the other part is made (we call such graphs one-sided bounded degree bipartite graphs). We also show how an optimal non-adaptive solution of $O(n^2 / \log n)$ queries for general graphs can be obtained.

The results show the power of the considered model, gained by the possibility of testing a *set* of vertices and *counting* the number of edges between them. For comparison, if we are allowed to query only two vertices (that is, test one edge at a time), $\Omega(n^2)$ queries are needed for many natural classes of graphs, such as trees, matchings, hamiltonian cycles and paths, and others (see [2]).

The paper is organized as follows. Sect. 2 is devoted to separating matrices – the main tool for constructing non-adaptive additive algorithms. We consider two classes of separating matrices – so called d-separating and d-detecting matrices. Our main contribution here is a probabilistic proof of a new upper bound for d-separating matrices which is only two times more than the best known lower bound. In Sect. 3 we turn to our main subject of interest – searching for graphs under the additive model. We consider bounded degree graphs and prove that such graphs can be reconstructed in $O(dn)$ queries. Finally, in Sect. 4 we consider several particular classes of graphs. For some of them, that are not subclasses of bounded degree graphs, we show that our technique still applies. For others, we show that the constant factor can be improved.

2 Separating Matrices

Consider the following setting. Assume we have a set of *items* and each of them is assigned an integer value. Assume that we want to reconstruct the values by making queries about subsets of items. As noted in the introduction, this type of search problems is very common and is called *combinatorial group testing*. Note that each query can be associated to a $(0, 1)$-vector q which is the incidence vector of the corresponding subset. Assume further that the result of a query is the sum of item values of the corresponding subset. This assumption typically corresponds to the *additive model* discussed in the introduction. It implies that if v is the vector of item values, the query result is the scalar product $< q, v >$. Let us now restrict ourselves to non-adaptive algorithms. Then the whole algorithm can be represented by a $(0, 1)$-matrix where each row is a query vector and each column corresponds to an item. This leads us to the following notion.

Definition 1. A $k \times n$ $(0,1)$-matrix M is called *separating* for a finite set of integer vectors $V \subseteq \mathbb{Z}^n$ iff for every $v_1, v_2 \in V$, $Mv_1 \neq Mv_2$ provided that $v_1 \neq v_2$.

In this section we study two important subclasses of separating matrices.

2.1 Optimal d-Separating Matrices

Definition 2. For a constant $d \in \mathbb{N}$, a *d-separating matrix* is a separating matrix for the set of $(0,1)$-vectors containing at most d entries equal to 1.

Equivalently, for a d-separating matrix, all the sums of any up to d columns are distinct. If a matrix has n columns, there will be $\sum_{i=0}^{d} \binom{n}{i}$ different sums of at most d columns. Since each entry of such a sum is at most d, a d-separating matrix has at least $\log_{d+1} \binom{n}{d} = (1 + o(1))d \log_d(n)$ rows (recall that d is a constant). A better lower bound was proved by Noga Alon [3]. Using the second moment method (cf. [4]; see also Appendix A), it was shown in [3] that there exists an absolute constant c such that for every $n > d$ any d-separating matrix has at least $\frac{2d}{\log d + c} \log (n/d)$ rows.

By definition, a matrix is 1-separating iff all its columns are different. Clearly, a 1-separating matrix with n columns and $\lceil \log_2 n \rceil$ rows can be easily constructed by setting the columns to be the binary representations of numbers $1, 2, \ldots, n$. This matrix corresponds to the *non-adaptive binary search*, as it provides a non-adaptive analogue to the binary search procedure. For an arbitrary constant d, it is known that a d-separating matrix with asymptotically $d \log_2 n$ rows can be constructed (see [17] and [2, exercise 2.3.5]). For $d = 2$, the lower bound $(5/3) \log_2 n$ has been proved by Lindström [16], while no better upper bound than $2 \log_2 n$ is known. This suggests that settling the multiplicative factor for the case of arbitrary constant d is difficult.

In this section we give a probabilistic proof that there exists a d-separating matrix which asymptotically meets the lower bound $\Omega(d \log_d n)$ up to a multiplicative constant independent of d. More precisely, we obtain the upper bound which is within the factor two of the lower bound $(2 + o(1))d \log_d n$ from [3]. This answers the question whether the upper bound $d \log_2 n$ can be improved, posed by Lindström in [17].

Before proceeding to the proof, we note a straightforward connection between d-separating matrices and a classical problem of *counterfeit coins*. A d-separating matrix with n columns solves the following problem. *Suppose we have n coins of which at most d are counterfeit. We are allowed to ask how many counterfeit coins occur in a subset. Find an optimal non-adaptive algorithm that determines all counterfeit coins.*

We now prove the main result of this section.

Theorem 3 (d-separating matrix). *For fixed d, there exists a d-separating matrix with n columns and asymptotically $4d \log_d n$ rows.*

Proof: We show that a random matrix M with asymptotically $4d \log_d n$ rows is d-separating with a positive probability. This will imply that such a matrix exists [4].

Consider a $(0, 1)$-matrix M. Let A and B be two different subsets of columns, each of size at most d. We say that (A, B) is a conflicting pair for M iff the sum of columns A in M is equal to the sum of columns B. Without loss of generality, we assume (A, B) to have two additional properties:

1. $A \cap B = \emptyset$. Indeed, if (A, B) is a conflicting pair, then $(A \setminus B, B \setminus A)$ is a conflicting pair too.
2. A and B have the same size $0 < |A| = |B| \leq d$. This can be insured by adding to M an additional row with all entries equal to 1. This row will not be subject to the random choice of matrix' entries. Obviously, adding one row does not affect the asymptotic bound.

We are going to estimate the expected number of conflicting pairs in a random $k \times n$ matrix (k will be chosen later). We define

$$\chi(A, B, M) = \begin{cases} 1 \text{ if } (A, B) \text{ is a conflicting pair for } M \\ 0 \text{ otherwise} \end{cases}$$

We assume the uniform probabilistic distribution over the $n \times k$ $(0, 1)$-matrices. This implies that each entry of M is chosen independently to be 0 or 1 with probability $1/2$. For each fixed i, $1 \leq i \leq d$, we consider all possible partitions σ of the set of $2i$ columns into two equal parts of i columns. There are clearly $\frac{1}{2}\binom{2i}{i}$ different partitions. For each partition σ, we consider all possible $\binom{n}{2i}$ subsets of $2i$ columns. Each subset R of $2i$ columns is thought to be split according to the partition σ above into two non-intersecting sets R_σ^+ and R_σ^- of size i. Compute now the expectation $\mathbb{E}(\chi(R_\sigma^+, R_\sigma^-, M))$ for fixed non-intersecting R_σ^+ and R_σ^-, both of cardinality $i \leq d$. Clearly, the events "two sums of entries in a row are equal" are independent for different rows. Furthermore, since $A \cap B = \emptyset$, the probability for one row is $\sum_{j=0}^{i}(\binom{i}{j}2^{-i})^2 = \binom{2i}{i}2^{-2i} = \frac{1}{\sqrt{\pi i}} + O(\frac{1}{i})$. However, we will need an upper bound *for all i*, and we use that $\binom{2i}{i}2^{-2i} \leq \frac{2}{3\sqrt{i}}$ for *all $i \geq 1$*. A way to show this is to prove by induction over $i \geq 1$ the stronger inequality $\binom{2i}{i}2^{-2i} \leq \frac{2}{3}\frac{1}{e^{\frac{1}{8i}}\sqrt{i}}$ (consider the ratio of consecutive elements and use the fact that $e^{\frac{1}{8i}} \geq 1 + \frac{1}{8i}$). Using this, we have $\mathbb{E}(\chi(A, B, M)) = (\binom{2i}{i}2^{-2i})^k \leq (\frac{2}{3})^k(\frac{1}{i})^{k/2}$.

Now the expected number of conflicting pairs can then be estimated as

$$\mathbb{E}\left(\sum_{i=1}^{d}\sum_{\sigma}\sum_{R}\chi(R_\sigma^+, R_\sigma^-, M)\right) = \sum_{i=1}^{d}\binom{2i}{i}\binom{n}{2i}\left(\binom{2i}{i}2^{-2i}\right)^k \leq$$

$$\leq \sum_{i=1}^{d}2^{2i}\left(\frac{en}{2i}\right)^{2i}\left(\frac{2}{3}\right)^k\left(\frac{1}{i}\right)^{k/2} = \sum_{i=1}^{d}\left(\frac{2}{3}\right)^k\frac{(en)^{2i}}{i^{2i+k/2}}$$

An elementary analysis shows that when $k = 4d \log_d n$, every summand is $o(1)$ and the whole sum can be made smaller than 1. Since the expected value of the number of conflicting pairs is smaller than 1, there exists at least one matrix M without conflicting pairs and therefore d-separating. □

2.2 d-Detecting Matrices

In this section we consider another important class of separating matrices.

Definition 4. Let d be a constant. A $k \times n$ $(0, 1)$-matrix, with n columns, is called d-*detecting* iff it is separating for the set of n-vectors $\{0, \ldots, d - 1\}^n$.

Let $v_1, v_2, \ldots, v_n$ be the columns of a $(0, 1)$-matrix. Then this matrix is d-detecting iff all the sums $\sum_{i=1}^n \epsilon_i v_i$ $(\epsilon_i = 0, 1, \ldots, d - 1)$ are different. Such a set of vectors is called *detecting* in [15], hence our terminology.

Given n and d, we are interested in d-detecting matrices with minimal number of rows. Let k be the number of rows. An information-theoretic reasoning gives the inequality $d^n \leq (dn)^k$, and the lower bound $\Omega(n/(1 + \log_d n))$ for k. The problem has been studied by several authors, and particularly by Bernt Lindström in a series of papers [14,15,17]. In [15] Lindström presents a construction of a detecting matrix, using the theory of Möbius functions. This construction gives a solution of order $2n/\log_d n$, although this was not explicitly pointed out by the author. Moreover, this bound turns out to be optimal, that will be stated in Lemma 6 below.

In [17] Lindström concentrates on the case $d = 2$ for which he proposes a construction of a detecting matrix based on elementary methods (the construction is also described in [2,6]). The matrix has asymptotically $2n/\log_2 n$ rows. Lindström also proves that the construction is optimal, that is the bound $2n/\log_2 n$ is the asymptotic lower bound. Further references for the case $d = 2$ can be found in [17].

We refer to the full version of this paper [11] for a summary of main Lindström's results from [14,15], and their application to the construction of an optimal d-detecting matrix. Here we summarize this construction in the following theorem.

Theorem 5. *For fixed d, a d-detecting matrix can be effectively constructed with n columns and asymptotically $2n/\log_d n$ rows.*

It is interesting that this explicit construction matches the asymptotic lower bound for a d-detecting set.

Lemma 6. *For a fixed d, any d-detecting matrix with n columns has at least $2n/\log_d n$ rows asymptotically.*

The proof is given in Appendix A.

Similarly to d-separating matrices, d-detecting matrices have a natural interpretation in terms of "generalized counterfeit coins problem". *Assume we have n coins and an unknown arbitrary number of them are false. Assume further that* *we know the weight α of an authentic coin, and that the weight of each false coin takes one of the values $\alpha + \delta i$ for $i = 1, \ldots, d - 1$. One can think of i (the over-weight of a coin) as the "measure of falsity". We are allowed to weigh subsets of coins and thus measure the overall overweight of a subset. Determine the false coins and their falsity by possibly minimal number of weighing.* It is easily seen that finding a non-adaptive solution of the generalized counterfeit coins problem is directly translated to constructing a d-detecting matrix with minimal number of rows. Note that for $d = 2$ we get the counterfeit coins problem described in Sect. 2.1 but with an arbitrary non-fixed number of false coins.

3   Reconstructing Bounded Degree Graphs

Now we turn to our main subject of interest – the problem of graph reconstruction under the additive model. Let $\mathcal{G}_n$ be a class of labelled undirected graphs with n vertices labelled by $\{1, 2, \ldots, n\}$. We consider simple graphs, that is graphs without loops or multiple edges.

We address the following problem. Reconstruct an unknown graph $G \in \mathcal{G}_n$ by means of queries of the following type: For a subset $V \subseteq \{1, \ldots, n\}$ of vertices, how many edges are there in the intersection $G \cap K_V$, where K_V is the complete graph with the set of vertices V? Clearly, each query is simply associated with a subset of $\{1, \ldots, n\}$.

Using the results on separating matrices presented in Sect. 2, we solve this problem for an important subclass of graphs, namely the *bounded degree graphs*. These are graphs with the degree of vertices bounded by some constant d. Bounded degree graphs are quite common objects and cover such classes as matchings, circles and paths, trees with bounded branching degree, etc. Property testing for bounded degree graphs was considered in [9]. In this section we propose an asymptotically optimal (modulo a constant factor) predetermined search algorithm for this class.

We first prove an auxiliary result. Using the results of Sect. 2.1 and 2.2, we prove the existence of an optimal predetermined algorithm for the class of *one-sided bounded degree bipartite graphs*. Consider a bipartite graph $G = (V, W, E)$, where $V \cup W$ is the set of vertices, $V \cap W = \emptyset$, and $E \subseteq V \times W$. For a constant d, G is called a one-sided (d-)bounded degree graph if $deg(v) \leq d$ for every vertex $v \in V$.

Assume that $|V| = |W| = n$. By assuming that every node of V has degree d, it is easy to estimate the number of such graphs from below as $\binom{n}{d}^n = \Omega((n/d)^{dn})$. Since the answer to a query has $nd + 1$ potential values, any search algorithm requires at least $\log_{dn+1}(n/d)^{dn} = dn(1 + o(1))$ queries. We now prove that this lower bound can be met, modulo a constant factor, by a non-adaptive algorithm.

Theorem 7. *For a constant d, there exists a non-adaptive search algorithm for the class of one-sided d-bounded degree bipartite graphs with n vertices on each side, that requires $8dn$ queries asymptotically.*

Proof: Consider a bipartite graph $G = (V, W, E)$, where $V \cup W$ is the set of vertices, $V \cap W = \emptyset$, $|V| = |W| = n$, and $E \subseteq V \times W$ is the set of edges. Assume that $deg(v) \leq d$ for all $v \in V$. By definition, each query is associated with a couple (V', W'), $V' \subset V$, $W' \subset W$, and has the form: how many edges of G are between vertices of V' and W'? (what is $|E \cap (V' \times W')|$?)

For a vertex $v \in V$ and a subset $W' \subseteq W$, denote by $deg_{W'}(v)$ the number of vertices of W' adjacent to v. Note that $deg_{W'}(v) \leq d$ and can be determined by one query.

Fix a vertex $v \in V$. According to Theorem 3, we can find all its adjacent vertices in W using $4d\frac{\log n}{\log d}$ queries (think of adjacent vertices as being "counterfeit", all other vertices in W being "authentic"). Each query asks about $deg_{W'}(v)$ for some subset $W' \subseteq W$. Let $W_1, W_2, \ldots, W_k \subseteq W$ (k asymptotically to $4d\frac{\log n}{\log d}$) be these subsets. Since the algorithm is predetermined, the subsets $W_1, W_2, \ldots, W_k$ are independent of v. In other words, if for some $v \in V$ we know $deg_{W_i}(v)$ for every W_i, we can reconstruct all the adjacent vertices of v in W.

Now fix some W_i. By Theorem 5, we can find a sequence of subsets $V_1, \ldots, V_l$, with l asymptotically to $2n\frac{\log d}{\log n}$, such that the queries $< (W_i, V_j), j = 1, \ldots, l >$ allow to reconstruct $deg_{W_i}(v)$ for every $v \in V$ (think of $deg_{W_i}(v)$ as "the degree of falsity" of v).

Repeating this algorithm for each W_i, we can determine $deg_{W_i}(v)$ for all $v \in V$ and all W_i via $4d\frac{\log n}{\log d} \cdot 2n\frac{\log d}{\log n} = 8dn$ queries asymptotically. This allows us to reconstruct the adjacent vertices in W of each $v \in V$, that is to reconstruct the whole graph. $\square$

The key argument of the proof is that the algorithm implied by Theorem 3 is non-adaptive, i.e. the sets W_i don't depend on vertices of V. The fact that the algorithm implied by Theorem 5 is also non-adaptive does not affect the complexity bound but insures that the resulting algorithm is completely predetermined too. Specifically, it insures that all the sets V_j are predetermined, and therefore all the queries (W_i, V_j) are mutually independent and can be made in any order.

We now use Theorem 7 to construct a separating set of queries for general bounded degree graphs. We start with computing the information-theoretic lower bound and for that we estimate from below the number of graphs with bounded degree. Instead of counting all such graphs, we will count a subclass of them, and show that their number is already sufficiently big.

Denote by $D(n, d)$ the set of labelled bipartite graphs with n vertices on each side with the degree of each vertex equal to d (d constant). This d-regular graph is a union of d disjoint matchings. Clearly, $|D(n, 0)| = 1$ and $|D(n, 1)| = n!$.

Consider a graph $G \in D(n, d)$. From G we can obtain a graph in $D(n, d+1)$ by adding a matching which doesn't intersect with G. To estimate the number of possible extensions, consider the complement bipartite graph $\bar{G}$ (an edge connecting the sides belongs to $\bar{G}$ iff it does not belong to G). It is an $(n - d)$-regular graph. Since the number of matching in a bipartite graphs is equal to the permanent of the adjacency matrix, from the Van der Waerden conjecture, proved by Egorychev and Falikman(see [5]), it follows that this graph has at least $(n-d)^n \frac{n!}{n^n}$ matchings. Obviously, none of them intersects with G.

On the other hand, consider a graph $G' \in D(n, d+1)$. The number of matchings it contains is bounded from above by $(d+1)^n$ (a better estimation is not important for our purposes). Thus, $|D(n, 0)| = 1$ and $|D(n, d)|(n - d)^n \frac{n!}{n^n} \leq |D(n, d+1)|(d+1)^n$. From this recurrence, $|D(n, d)| \geq \binom{n}{d}^n (\frac{n!}{n^n})^d \geq (n/d)^{dn}(e^{-nd}) = (\frac{n}{ed})^{nd}$. We obtain $\log_{nd+1} |D(n, d)| \geq nd(1 + o(1))$, and thus any search algorithm for $D(n, d)$ requires $\Omega(nd)$ queries. As $D(n/2, d)$ is a subclass of d-bounded degree graphs with n vertices, we conclude that at least $\Omega(\frac{nd}{2})$ queries are needed for this class.

Using theorem 7 we now show that this lower bound can be achieved, modulo a constant factor, by a predetermined algorithm.

Theorem 8. *For a constant d, there exists a predetermined search algorithm for the class of d-bounded degree graphs with n vertices, that requires $24dn$ queries asymptotically.*

Proof: Consider a graph $G = (V, E)$ with $deg(v) \leq d$ for all $v \in V$, and $|V| = n$. Let μ be the query function, that is $\mu(W) = |E \cap (W \times W)|$ for $W \subseteq V$. Recall that the nodes V are identified with the set of labels $\{1, \ldots, n\}$. We associate to G a bipartite graph $G' = (V', V'', E')$, where $V' = V'' = \{1, \ldots, n\}$, and $(v_1, v_2) \in E'$ iff $(v_1, v_2) \in E$. Note that $deg(v) \leq d$ for every $v \in V' \cup V''$. We want to reconstruct graph G by applying Theorem 7 to graph G'. The query function for graph G' is $\mu'(W', W'') = |E' \cap (W' \times W'')|$ for $W' \subseteq V'$, $W'' \subseteq V''$. We now show that a query μ' can be simulated by a constant number of queries μ.

Observe the following properties of μ':

1. $\mu'(W', W'') = \mu'(W'', W')$
2. $\mu'(W, W) = 2\mu(W)$
3. If $W' \cap W'' = \emptyset$ then $\mu'(W', W'') = \mu(W' \cup W'') - \mu(W') - \mu(W'')$
4. If $W_1 \cap W_2 = \emptyset$ then $\mu'(W_1 \cup W_2, W) = \mu'(W_1, W) + \mu'(W_2, W)$ for any $W \in V''$.

For arbitrary $W_1, W_2 \subseteq V$, let $W = W_1 \cap W_2$. Then

$$\mu'(W_1, W_2) = \mu'((W_1 \setminus W_2) \cup W, (W_2 \setminus W_1) \cup W) =$$
$$\mu'(W_1 \setminus W_2, W_2 \setminus W_1) + \mu'(W_1 \setminus W_2, W) + \mu'(W_2 \setminus W_1, W) + \mu'(W, W).$$

Using properties 1-4, we obtain

$$\mu'(W_1, W_2) =$$
$$\mu((W_1 \setminus W_2) \cup (W_2 \setminus W_1)) - 2\mu(W_1 \setminus W_2) - 2\mu(W_2 \setminus W_1) + \mu(W_1) + \mu(W_2).$$

Thus, one query μ' can simulated by five queries μ. By Theorem 7, graph G', and therefore G, can be reconstructed through $8nd$ queries $\mu'(W_i', W_j'')$. The number of corresponding queries μ can be optimized, if queries $\mu(W_i)$, $\mu(W_j)$ are computed once. This gives us $24nd + 4d\log_d n + 2n/\log_d n = 24nd(1 + o(1))$ queries μ. $\qquad \square$

4 Case Studies

Here we consider some particular classes of graphs. For some of them, which are not subclasses of degree bounded graphs, we show that our technique still applies. For others, we show that the multiplicative factor can be improved.

4.1 General and c-Colorable Graphs

The results above assume some knowledge about the class that the unknown graph is drawn from. What can be said when no prior information about the structure of the graph is given, i.e. all graphs are possible? The information-theoretic lower bound for this case is immediate. There are $2^{\frac{n(n-1)}{2}}$ labelled graphs with n vertices and each query can yield up to $1 + n(n-1)/2$ answers. Therefore, any algorithm should make at least $\log_{(1+n(n-1)/2)} 2^{\frac{n(n-1)}{2}} = \Omega(\frac{n^2}{4\log_2 n})$ queries. Note that since graphs can be represented by $(0,1)$-vectors of length $n(n-1)/2$, any non-adaptive algorithm for reconstructing general graphs gives a 2-detecting matrix with $n(n-1)/2$ columns. By Lemma 6, we can then obtain a better lower bound of $2 \cdot \frac{n(n-1)}{2} / \log_2 \frac{n(n-1)}{2} = \Omega(\frac{n^2}{2\log_2 n})$ for non-adaptive algorithms for reconstructing general graphs. This lower bound can be achieved, within a factor of 4, using the technique of the proof of Theorem 8. Represent $G = (V, E)$ as a bipartite graph $G' = (V_1, V_2, E')$ where each side consists of a copy of vertices of V ($V_1 = V_2 = V$) and $(v_1, v_2) \in E'$ iff $(v_1, v_2) \in E$. For each vertex $v_1 \in V_1$, we can find all its adjacent vertices in V_2 (and then in G) through $\frac{2n}{\log_2 n}$ queries of the form "How many adjacent vertices does v_1 have in a subset $W \subseteq V_2$?". Every such query can be simulated by two queries to the initial graph G. Similar to the proof of Theorem 8, let μ (resp. μ') denote the query function for graph G (resp. G'). Then $\mu'(\{v_1\}, W) = \mu(W \cup \{v_1\}) - \mu(W \setminus \{v_1\})$. To reconstruct the graph, we find for every vertex i the adjacent vertices among $1, \ldots, i-1$. Then the overall complexity is $2 \cdot \sum_{i=2}^{n} \frac{2i}{\log_2 i} = \frac{2n^2}{\log_2 n}(1 + o(1))$ which is 4 times the lower bound. Does the knowledge of the graph's chromatic number $c = \chi(G)$ help? Not much, as there are at least $2^{\frac{c-1}{2c}n^2}$ such graphs (divide n vertices into c parts evenly and consider all possible edge combinations between different parts). The information-theoretic lower bound is then $\Omega(\frac{c-1}{4c} \cdot \frac{n^2}{\log n})$, and the algorithm above for the general case is again optimal up to a constant factor.

4.2 h-Edge Colorable Graphs

If the graph is known to be h-edge-colorable, the degree of vertices is bounded by h and by Theorem 8, it can be reconstructed within $O(hn)$ non-adaptive queries. Note that this is asymptotically best possible, as by Vizing's theorem (see [5]), the graphs with the edge chromatic number less than or equal to h contain all the $(h-1)$-bounded degree graphs.

4.3 Matchings in Bipartite Graphs

Matchings occur in numerous applications and we consider important to present a refinement of the general technique that can be obtained for this class. This refinement is valid for a more general class, namely the 1-bounded one-sided bipartite graphs (see Sect. 3). Let $G = (V, W, E)$ be a bipartite graph, where $|V| = n$, $|W| = m$ and all vertices in V have degree at most 1. According to the proof of Theorem 7, to reconstruct G we need a 1-separating matrix for m objects and 2-detecting matrix for n objects. A 1-separating matrix with m columns has $\log m$ rows (see Sect. 2.1). As for 2-detecting matrix with n columns, $\frac{2n}{\log_2 n}(1+o(1))$ rows are necessary and sufficient, as it was mentioned in Sect. 2.2. Putting together, G can be reconstructed by $\log_2(m)\frac{2n}{\log_2 n}(1+o(1))$ non-adaptive queries. Note that the probabilistic proof of Theorem 3 is not involved here, and the queries can be constructed effectively.

In the case of matchings in bipartite graphs we have $n = m$ which gives a non-adaptive algorithm to reconstruct a matching within $2n(1 + o(1))$ queries. This bound is asymptotically optimal within a factor of 2.

4.4 Hamiltonian Cycles and Paths

Let us consider the 2-bounded degree graphs with n vertices. Any such graph is a collection of paths and cycles without common vertices. In particular, this class contains the Hamiltonian cycles and the Hamiltonian paths. The problem of reconstructing Hamiltonian cycles under different models was considered in [10] in connection with its application to genome physical mapping.

Theorem 8 suggests a non-adaptive solution that requires $48n$ queries asymptotically. A better performance can be obtained if we sacrifice the requirement for the algorithm to be fully non-adaptive. Instead, we propose a two-stage algorithm – the first stage does an adaptive "pre-processing" and the second stage reconstructs the graph non-adaptively.

At the first stage, we sort out the vertices into three disjoint independent subsets, that is without adjacent pairs in each subset. As each vertex has at most two neighbours, this sorting can be easily done in at most $2n$ queries by processing the vertices consecutively and testing each vertex against at most two of the already formed subsets.

At the second stage, we reconstruct separately each of the three bipartite 2-degree bounded graphs resulting from the first stage. Again, applying the proof of Theorem 8, we need a 2-separating and a 3-detecting matrices. As noted in Sect. 2.1, a 2-separating matrix with n columns and $2\log_2 n$ rows can be effectively constructed. On the other hand, by adapting the proof of Theorem 5 to the case $d = 3$, it can be shown that there exists a 3-detecting matrix with n columns and $\frac{4n}{\log n}$ rows. By applying the algorithm of reconstructing bipartite graphs in such a way that the detecting matrix always acts on a smaller part (see proof of Theorem 7), we can reconstruct each bipartite graph in $2\log_2 n \cdot \frac{4n/2}{\log n/2} = 4n$ queries. Putting two stages together, this gives an algorithm with $2n + 3 \cdot 4n = 14n$ queries.

5 Remarks

An interesting open question is to give an explicit construction of d-separating matrices with $4 \log_d n$ rows, the existence of which has been proved in Theorem 3 by a probabilistic method. An explicit construction could open a way to the study of computational complexity of reconstructing the unknown graph given a vector of answers. This question is another direction for future research.

References

1. Martin Aigner. Search problems on graphs. *Discrete Applied Mathematics*, 14:215–230, 1986.
2. Martin Aigner. *Combinatorial Search*. John Wiley & Sons, 1988.
3. Noga Alon. Separating matrices. private communication, May 1997.
4. Noga Alon and Joel Spencer. *The Probabilistic Method*. Wiley Interscience, New York, 1992.
5. Peter J. Cameron. *Combinatorics: Topics, Techniques, Algorithms*. Cambridge University Press, 1994.
6. Ding-Zhu Du and Frank K. Hwang. *Combinatorial Group Testing and its applications*, volume 3 of *Series on applied mathematics*. World Scientific, 1993.
7. L. Gargano, V. Montuori, G. Setaro, and U. Vaccaro. An improved algorithm for quantitive group testing. *Discrete Applied Mathematics*, 36:299–306, 1992.
8. Oded Goldreich, Shafi Goldwasser, and Dana Ron. Property testing and its connection to learning and approximation. In *Proceedings of the 37th Annual Symposium on Founations of Computer Scince*, pages 339–348, 1996.
9. Oded Goldreich and Dana Ron. Property testing in bounded degree graphs. In *The 29th Annual ACM Symposium on Theory of Computing*, 1997. to appear. Full version available at http://theory.lcs.mit.edu/~danar/papers.html.
10. Vladimir Grebinski and Gregory Kucherov. Optimal query bounds for reconstructing a hamiltonian cycle in complete graphs. In *Proceedings of the 5th Israeli Symposium on Theory of Computing and Systems*. IEEE Press, June 1997. to appear.
11. Vladimir Grebinski and Gregory Kucherov. Optimal reconstruction of graphs under the additive model. Rapport de Recherche 3171, INRIA, Mai 1997.
12. Fred H. Hao. The optimal procedures for quantitive group testing. *Discrete Applied Mathematics*, 26:79–86, 1990.
13. V. Koubek and J. Rajlich. Combinatorics of separation by binary matrices. *Discrete Mathematics*, 57:203–208, 1985.
14. Bernt Lindström. On a combinatorial problem in number theory. *Canad. Math. Bull.*, 8:477–490, 1965.
15. Bernt Lindström. On Möbius functions and a problem in combinatorial number theory. *Canad. Math. Bull.*, 14(4):513–516, 1971.
16. Bernt Lindström. On B_2-sequences of vectors. *Journal of Number Theory*, 4:261–265, 1972.
17. Bernt Lindström. Determining subsets by unramified experiments. In J.N. Srivastava, editor, *A Survey of Statistical Designs and Linear Models*, pages 407–418. North Holland, Amsterdam, 1975.
18. R.L. Rivest and J. Vuillemin. On reconstructing graph properties from adjacency matrices. *Theoretical Computer Science*, 3:371–384, 1976.

Appendix A

Here we give a proof of Lemma 6 that an optimal d-detecting matrix has at least $2\log d\frac{n}{\log n}$ rows asymptotically. The proof is by the second moment method attributed to L.Moser (see, e.g., [17, p. 415]).

Let the $\{v_1, v_2, \ldots, v_n\}$ be the columns of a d-detecting matrix. We show that these vectors must have dimension at least $2\log d\frac{n}{\log n}$ asymptotically. Let m be the dimension of v_i's. The idea of the proof is to show that at least half of the vectors of the set $\{\epsilon_1 v_1 + \epsilon_2 v_2 + \cdots + \epsilon_n v_n | \epsilon_i = 0, \ldots, d-1\}$ belongs to an m-dimensional sphere of a "small" radius. This gives an estimation for m.

Consider the uniform probabilistic space $\{(\epsilon_1, \epsilon_2, \ldots, \epsilon_n) | \epsilon_i = 0, \ldots, d-1\} \cong [0, \ldots, d-1]^n$. The random variable $\xi = \epsilon_1$ has the expectation $E(\xi) = (d-1)/2$ and the variance $\sigma^2 = Var(\xi) = \frac{(d+2)d}{12}$. Denote by v_i^j the value of j-th coordinate of v_i. Then the random variable $\varsigma_k = \epsilon_1 v_1^k + \cdots + \epsilon_n v_n^k$ $(k = 1 \ldots m)$ is a sum of independent random variables ϵ_i with coefficients $0, 1$. It means that

$$Var(\varsigma_k) \leq \sum_{i=1}^{n} Var(\epsilon_i) = n\sigma^2.$$

By definition of variance and the linearity of expectation, we have:

$$E\left(\sum_{k=1}^{m}(\varsigma_k - \bar{\varsigma}_k)^2\right) = \sum_{k=1}^{m} E(\varsigma_k - \bar{\varsigma}_k)^2 \leq m * n\sigma^2.$$

It follows from the Chebyshev inequality,

$$Prob\left(\sum_{k=1}^{m}(\varsigma_k - \bar{\varsigma}_k)^2 \leq 2mn\sigma^2\right) \geq 1/2.$$

Since $(\epsilon_1, \ldots, \epsilon_n) \to \epsilon_1 v_1 + \cdots + \epsilon_n v_n = (\varsigma_1, \ldots, \varsigma_m)^T$ is a bijection, at least a half of these sums belong to a sphere with center $(\bar{\varsigma}_1, \ldots, \bar{\varsigma}_m)$ and radius $r = (2mn\sigma^2)^{1/2}$. By the volume argument, the number of integer-valued points in a sphere with radius r is less than $(c/m)^{m/2} r^m$ for a constant c. Therefore, $1/2 \cdot d^n \leq (c/m \cdot r^2)^{m/2} = (c/m \cdot 2mn\sigma^2)^{m/2} = (2cn\sigma^2)^{m/2}$. It follows that $m \geq 2 \cdot \frac{n\log d - \log(2)}{\log n + \log(2 \cdot c \cdot (d+2) \cdot d/12)} = 2\log d\frac{n}{\log(n)} + o(\frac{n}{\log n})$. $\qquad\square$

Fixing Variables in Semidefinite Relaxations

Christoph Helmberg

Konrad-Zuse-Zentrum für Informationstechnik Berlin,
Takustraße 7, D–14195 Berlin, Germany.
e-mail: helmberg@zib.de,
URL: http://www.zib.de/helmberg.

Abstract. The standard technique of reduced cost fixing from linear programming is not trivially extensible to semidefinite relaxations as the corresponding Lagrange multipliers are usually not available. We propose a general technique for computing reasonable Lagrange multipliers to constraints which are not part of the problem description. Its specialization to the semidefinite $\{-1, 1\}$ relaxation of quadratic 0-1 programming yields an efficient routine for fixing variables. The routine offers the possibility to exploit problem structure. We extend the traditional bijective map between $\{0, 1\}$ and $\{-1, 1\}$ formulations to the constraints such that the dual variables remain the same and structural properties are preserved. In consequence the fixing routine can efficiently be applied to optimal solutions of the semidefinite $\{0, 1\}$ relaxation of constrained quadratic 0-1 programming, as well. We provide numerical results showing the efficacy of the approach.

Key words. semidefinite programming, semidefinite relaxations, quadratic 0-1 programming, reduced cost fixing

AMS subject classifications. 90C31, 90C25, 65K10, 49K40

1 Introduction

Semidefinite programming is well known to offer good possibilities for desigining powerful relaxations for many combinatorial problems [11, 6, 12, 4]. However, only few papers presenting computational experience with semidefinite relaxations are published so far [8, 9, 17, 16]. Although the bounds prove to be of good quality in practice, implementations suffer from the high computational cost involved in solving semidefinite programs. Within a branch and bound (or branch and cut) framework the efficiency of an expensive bound hinges on the tradeoff between the number of branch and bound nodes and the computation time needed for each node. In linear programming the possibility of fixing variables by reduced costs often justifies the use of more expensive relaxations. In semidefinite relaxations the reduced costs are usually burried inside a positive semidefinite dual slack matrix and so far it was not known how to extract the necessary information efficiently.

In this paper we show that the dual slack matrix can be interpreted as a variable subsuming all dual costs associated with the active constraints enforcing the primal semidefinite constraint. Based on this insight we provide a practically efficient routine for fixing variables in quadratic 0-1 programming. The use of this routine in the branch and cut code of [8] results in serious speedups of the algorithm.

There are two standard models for quadratic 0-1 programming, one formulated in $\{0,1\}$ variables, the other in $\{-1,1\}$ variables. Both lead, in a canonical way, to semidefinite relaxations that are slightly different in appearance. In particular the routine for fixing variables is based on the dual of the $\{-1,1\}$ relaxation. It is well known that both problems and their primal relaxations are equivalent [3, 7, 10]. The dual variables, however, will differ for varying representations of the same primal set. We present a canonical transformation between the constraints of both formulations such that the dual variables are the same for both, and most of the structure of the constraints is preserved. If the $n+1$ fundamental constraints of the $\{0,1\}$ relaxation are modeled correctly this enables us to use the fixing procedure of the $\{-1,1\}$ formulation even for optimal solutions computed in the $\{0,1\}$ setting.

In Section 2 we introduce the semidefinite relaxation of quadratic 0-1 programming in $\{-1,1\}$ variables which motivated the considerations to follow. Section 3 provides the theoretical framework for extracting information from the dual in the general setting of semidefinite programming. Section 4 explains the practical difficulties in implementing the theoretic approach and presents an efficient alternative within the $\{-1,1\}$ setting. In Section 5 the equivalence transformation between $\{0,1\}$ and $\{-1,1\}$ formulations is extended to the constraints such that dual variables and structural properties are preserved. Section 6 presents numerical results underlining the efficacy of the fixing routine.

Notation

$\mathbb{R}$ ($\mathbb{R}^n$) is the set of real numbers (real column vector of dimension n). $M_{m,n}$ (M_n) is the linear space of $m \times n$ ($n \times n$) matrices of real numbers. S_n denotes the set of $n \times n$ symmetric real matrices. For $A, B \in S_n$, $A \succeq B$ refers to the Löwner partial order, i.e. $A \succeq B$ ($A \succ B$) if $A - B$ is a positive semidefinite (definite) matrix. I (I_n) is the identity of appropriate size (of size n). e_i is the i-th column of I and e the vector of all ones of appropriate dimension. $\lambda_i(A)$ denotes the i-th eigenvalue of $A \in M_n$, usually $\lambda_1 \geq \lambda_2 \geq \ldots \geq \lambda_n$. $\lambda_{\min}(A)$ and $\lambda_{\max}(A)$ are the minimal and maximal eigenvalue of A. Λ_A is the spectrum of A, a diagonal matrix with $(\Lambda_A)_{ii} = \lambda_i(A)$. $\operatorname{tr}(A)$ is the trace of $A \in M_n$, $\operatorname{tr}(A) = \sum_{i=1}^{n} a_{ii} = \sum_{i=1}^{n} \lambda_i(A)$. The inner product in the linear space of matrices is $\langle A, B \rangle = \operatorname{tr}(B^T A)$ for $A, B \in M_{m,n}$. The rank of a matrix is denoted by $\operatorname{rank}(A)$. $\operatorname{diag}(A)$ is the vector of diagonal elements of A, $\operatorname{diag}(A) = [a_{11}, \ldots, a_{nn}]^T$, and, for $v \in \mathbb{R}^n$, $\operatorname{Diag}(v)$ is the diagonal matrix with diagonal v. Unless explicitly stated otherwise all matrices considered are symmetric and vectors are columns.

2 Quadratic 0-1 Programming in $\{-1, 1\}$ Variables

Two canonical formulations of quadratic 0-1 programming appear in the literature, one in terms of $\{0, 1\}$ variables and one in $\{-1, 1\}$ variables. For our purposes the semidefinite relaxation of the $\{-1, 1\}$ formulation (which is better known as the semidefinite relaxation of max-cut) is more convenient. We will return to the $\{0, 1\}$ formulation and the equivalence of both in Section 5.

The combinatorial problem to be investigated reads

$$\text{(MC)} \qquad \max \ x^T C x \ \text{ s.t. } \ x \in \{-1, 1\}^n$$

The standard semidefinite relaxation is derived by observing that $x^T C x = \langle C, xx^T \rangle$. For all $\{-1, 1\}^n$ vectors, xx^T is a positive semidefinite matrix with all diagonal elements equal to one. We relax xx^T to $X \succeq 0$ and $\text{diag}(X) = e$ and obtain the following primal dual pair of semidefinite programs,

$$\text{(PMC)} \quad \begin{array}{c} \max \ \langle C, X \rangle \\ \text{s.t. } \text{diag}(X) = e \\ X \succeq 0 \end{array} \qquad \text{(DMC)} \quad \begin{array}{c} \min \ e^T u \\ \text{s.t. } C + Z - \text{Diag}(u) = 0 \\ Z \succeq 0. \end{array}$$

The relaxation can be strengthened by adding a few of the so called triangle inequalities,

$$x_{ij} + x_{ik} + x_{jk} \geq -1$$
$$x_{ij} - x_{ik} - x_{jk} \geq -1$$
$$-x_{ij} + x_{ik} - x_{jk} \geq -1$$
$$-x_{ij} - x_{ik} + x_{jk} \geq -1$$

for $i < j < k$ from $\{1, \ldots, n\}$.

A bound of this kind is used in [8] in a branch and cut scheme. We branch by setting one x_{ij} to either 1 or -1. Each subproblem can be expressed as a quadratic $\{-1, 1\}^{n-1}$ problem.

If the bound is not good enough to fathom a node, but $x_{ij} = 1$ ($x_{ij} = -1$) for some $i \neq j$ in the optimal solution of the relaxation we can expect that indeed i and j belong together (apart). If we force the opposite a drop in the bound is to be expected. Can we prove that this drop will be large enough without recomputing the bound for this case?

In a linear cutting plane algorithm for the $\{-1, 1\}$ model the constraints $-1 \leq x_{ij} \leq 1$ are typically included in the initial relaxation. If the optimal solution of the linear relaxation exhibits $|x_{ij}| = 1$ then the dual variable of the corresponding active constraint yields a lower bound on the change of the objective value that would result from forcing x_{ij} to the opposite sign. This bound may suffice to prove that the current value of x_{ij} is correct for all optimal solutions of (MC).

In the semidefinite relaxation (PMC) the constraints $-1 \leq x_{ij} \leq 1$ are already implied by the diagonal constraints and the semidefiniteness of X. Therefore they are not included in the semidefinite relaxation and the corresponding dual variables are not available.

However, we can associate with each active constraint $x_{ij} \geq -1$ or $x_{ij} \leq 1$ an active constraint $v^T X v \geq 0$ from the set of constraints ensuring the positive semidefiniteness of X as follows. Let $|x_{ij}| = 1$ for some $i \neq j$ in the optimal solution (X^*, u^*, Z^*) of the current relaxation. Then the vector $v \in \mathbb{R}^n$ with

$$v_k = \begin{cases} 1 & k = i \\ -\text{sgn}(x_{ij}) & k = j \\ 0 & \text{otherwise} \end{cases}, \qquad (1)$$

is in the null space of X^*. Although this does not yet yield the Lagrange multiplier corresponding to the constraint $\langle vv^T, X \rangle \geq 0$ it suggests to look for it in the dual slack matrix Z^*. We will do so in the general setting of semidefinite programming.

3 The Theoretical Framework

Consider a standard primal dual pair of semidefinite programs,

$$\begin{array}{ll} & \min \langle C, X \rangle \\ \text{(P)} & \text{s.t. } \mathcal{A}(X) = b \\ & X \succeq 0 \end{array} \qquad \begin{array}{ll} & \max \langle b, u \rangle \\ \text{(D)} & \text{s.t. } \mathcal{A}^T(u) + Z = C \\ & Z \succeq 0. \end{array}$$

$\mathcal{A} : S_n \to \mathbb{R}^m$ is a linear operator and $\mathcal{A}^T : \mathbb{R}^m \to S_n$ is its adjoint operator, i.e. it satisfies $\langle \mathcal{A}(X), u \rangle = \langle X, \mathcal{A}^T(u) \rangle$ for all $X \in S_n$ and $u \in \mathbb{R}^m$. They are of the form

$$\mathcal{A}(X) = \begin{bmatrix} \langle A_1, X \rangle \\ \vdots \\ \langle A_m, X \rangle \end{bmatrix} \qquad \text{and} \qquad \mathcal{A}^T(u) = \sum_{i=1}^{m} u_i A_i$$

with $A_i \in S_n$, $i = 1, \ldots, m$.

We examine possibilities to extract duality information for equality or inequality constraints that are not explicitly given in the problem description. Assume that optimal solutions X^* of (P) and (u^*, Z^*) of (D) are given. How much does the optimal value of (D) decrease if an additional constraint $\langle A_0, X \rangle = b_0$ is added to the problem?

Let u_0 denote the new dual variable associated with the new constraint. The corresponding primal dual pair reads

$$\begin{array}{ll} & \min \langle C, X \rangle \\ & \text{s.t. } \langle A_0, X \rangle = b_0 \\ \text{(P}_0\text{)} & \mathcal{A}(X) = b \\ & X \succeq 0 \end{array} \qquad \begin{array}{ll} & \max b_0 u_0 + \langle b, u \rangle \\ \text{(D}_0\text{)} & \text{s.t. } u_0 A_0 + \mathcal{A}^T(u) + Z = C \\ & Z \succeq 0. \end{array}$$

Computing the optimal solution is as hard as solving the original problem. Therefore we try to improve the "good" dual feasible solution for (D$_0$), $(u_0 = 0, u^*, Z^*)$, along a dual ascent direction $(\Delta u_0, \Delta u, \Delta Z)$ with

$$b_0 \Delta u_0 + \langle b, \Delta u \rangle > 0$$
$$\Delta u_0 A_0 + \mathcal{A}^T(\Delta u) + \Delta Z = 0$$
$$Z + t\Delta Z \succeq 0 \qquad \text{for some } t \geq 0.$$

To determine the best search direction is again as difficult as the problem itself. The choice of a good direction will depend on our understanding of the problem at hand.

Having fixed an ascent direction $(\Delta u_0, \Delta u, \Delta Z)$ it remains to compute the maximal step size t such that $Z + t\Delta Z$ is still positive semidefinite, because the objective function is linear. With $S = C - \mathcal{A}^T(u^*) = Z^* \succeq 0$ and $B = \Delta u_0 A_0 + \mathcal{A}^T(\Delta u)$ the problem reduces to

$$(LS) \quad \max t \text{ s.t. } S - tB \succeq 0.$$

Problems of this form appear as matrix pencils in the literature (see e.g. [5], Chapters 7.7, 8.7, and references therein). Indeed, the optimal t can be computed explicitly.

Let $P\Lambda_S P^T = S$ denote an eigenvalue decomposition of S with P an orthonormal matrix and Λ_S a diagonal matrix having the eigenvalues $\lambda_1(S) \geq \ldots \geq \lambda_n(S)$ on its diagonal in this order. Then $S - tB \succeq 0$ is equivalent to $\Lambda_S - tP^T BP \succeq 0$. If the rank of S is k then $\lambda_i(S) = 0$ for $i = k+1, \ldots, n$. Multiplying by $D = \text{Diag}(\lambda_1(S)^{-\frac{1}{2}}, \ldots, \lambda_k(S)^{-\frac{1}{2}}, 1, \ldots, 1)$ from left and right and dividing by t (for the moment assume $t > 0$) we obtain

$$\begin{bmatrix} \frac{1}{t}I_k & 0 \\ 0 & 0 \end{bmatrix} - \begin{bmatrix} B_{11} & B_{12} \\ B_{12}^T & B_{22} \end{bmatrix} \succeq 0 \quad \text{with} \quad \begin{bmatrix} B_{11} & B_{12} \\ B_{12}^T & B_{22} \end{bmatrix} = DP^T BPD,$$

$B_{11} \in M_k$, $B_{22} \in M_{n-k}$, and $B_{12} \in M_{k,n-k}$. In case B_{12} and B_{22} are both zero, $\frac{1}{t} \geq \lambda_{\max}(B_{11})$ is the best choice. Note, that for $\lambda_{\max}(B_{11}) \leq 0$ the problem is unbounded. If $-B_{22}$ is non-zero it must be positive semidefinite, otherwise $t = 0$ is the only feasible solution. If $-B_{22}$ is positive semidefinite with rank h we can apply a similar sequence of steps to obtain a condition

$$\begin{bmatrix} \frac{1}{t}I_k - B_{11} & \bar{B}_{12} & \bar{B}_{13} \\ \bar{B}_{12}^T & I_h & 0 \\ \bar{B}_{13}^T & 0 & 0 \end{bmatrix} \succeq 0.$$

If $\bar{B}_{13}$ is non-zero then again t must be zero. Otherwise we can apply the Schur complement Theorem to obtain the condition $\frac{1}{t}I_k - B_{11} \succeq \bar{B}_{12}\bar{B}_{12}^T$. This yields $\frac{1}{t} \geq \lambda_{\max}(B_{11} + \bar{B}_{12}\bar{B}_{12}^T)$.

We specialize this general procedure to a case of particular importance in semidefinite programming. For the purpose of explanation assume that X^* and (u^*, Z^*) are a strictly complementary pair of optimal solutions, i.e. , $\text{rank}(X^*) + \text{rank}(Z^*) = n$ (these do not necessarily exist, see e.g. [1]). Furthermore let A_0 be a dyadic product vv^T for some $v \in \mathbb{R}^n$ with $\langle vv^T, X^* \rangle = 0$, i.e. , v is in the null space of X^*. vv^T may be interpreted as one of the active constraints ensuring the positive definiteness of X. The right hand side b_0 of the new constraint must be greater than zero, otherwise there is certainly no feasible primal solution for the new problem. As ascent direction we choose $\Delta u_0 = 1$ and $\Delta u = 0$. This yields the following line search problem,

$$\max t \text{ s.t. } Z^* - tvv^T \succeq 0.$$

Because X^* and Z^* are strictly complementary solutions and v is in the null space of X we conclude that v lies in the span of the eigenvectors to non-zero eigenvalues of Z^*. Assume that $\text{rank}(Z^*) = k$ and let $P\Lambda_{Z^*}P = Z^*$ denote the eigenvalue decomposition of Z^* with $P \in M_{n,k}$, $P^T P = I_k$, and the spectrum of non-zero eigenvalues $\Lambda_{Z^*} \in S_k$. Then the maximal t is given by

$$t^* = \frac{1}{v^T P \Lambda_{Z^*}^{-1} P^T v}. \tag{2}$$

If in particular v happens to be an eigenvector of Z^* then t^* is the corresponding eigenvalue of Z^*. Relating this to linear programming we might formulate, the dual slack matrix Z^* subsumes the dual variables to the constraints generating the primal cone $X \succeq 0$.

This interpretation can be extended to the case that X^* and Z^* are not strictly complementary. For any vector v in the null space of X^* but not in the span of the non-zero eigenvectors of Z^* the optimal t is zero.

4 A Practical Algorithm

With respect to the semidefinite relaxation (PMC) the previous section suggests a convenient procedure for constructing Lagrange multipliers for the constraints of form (1). Assuming that the eigenvalue decomposition of Z^* into $P\Lambda_{Z^*}P^T$ is available ($k = \text{rank}(Z^*)$, $P \in M_{n,k}$, $P^T P = I_k$), it is easy to check whether v is in the span of the eigenvectors P. If it is not, then $t^* = 0$, otherwise $t^* = -1/(v^T P \Lambda_{Z^*}^{-1} P^T v)$ is the best Lagrange multiplier for u fixed to u^*. The bound corresponding to forcing i and j into opposite sets can be modeled by changing the right hand side of the (currently active) constraint $v^T X v = 0$ to $v^T X v = 4$ in the current relaxation. Therefore the bound obtained from the relaxation with i and j in opposite sets is less than or equal to $e^T u^* + 4t^*$.

In implementing this approach several difficulties are encountered. Indeed, real world algorithms do not deliver the true optimal solution (X^*, u^*, Z^*) of (PMC). In a computed solution $(\hat{X}, \hat{u}, \hat{Z})$ both, $\hat{X}$ and $\hat{Z}$, will be (rather ill conditioned) full rank matrices. Even in case the gap $\langle \hat{X}, \hat{Z} \rangle$ between primal and dual solution is almost zero, it is difficult to decide which of the eigenvalues of $\hat{X}$ and $\hat{Z}$ will eventually converge to zero. The space spanned by the eigenvectors corresponding to the "non-zero" eigenvalues of $\hat{X}$ and $\hat{Z}$ may still differ substantially from the true eigenspaces of X^* and Z^*. The vectors v of (1) will neither be contained in the null space of $\hat{X}$ nor in the space spanned by the "non-zero" eigenvectors of $\hat{Z}$ because no $|x_{ij}|$ will be strictly one. In consequence the line search will allow for a very short step only and the approach fails.

However, in the case of (PMC) the diagonal variables may be used to compensate for numerical instabilities. We mention, that the following framework can be applied in the presence of additional primal constraints, as well, but as these have no influence on the considerations to follow, we ignore them here.

Within the branch and bound scenario let $(\hat{X}, \hat{u}, \hat{Z})$ be the solution computed for the relaxation of the current branch and bound node yielding the upper bound

$e^T \hat{u}$ and let c^* denote the lower bound. Let i, j and v as in (1) with $v^T \hat{X} v$ almost zero. How much does the bound improve if we add the constraint $\langle A_0, \hat{X} \rangle = b_0$ ($b_0 > 0$ w.l.o.g.) to the current relaxation? We denote the Lagrange multiplier for the new constraint by u_0. We would like to compute an upper bound, ideally smaller than c^*, for the problem

$$\min\ b_0 u_0 + e^T u$$
$$\text{s.t.}\ Z = u_0 A_0 + \text{Diag}(u) - C$$
$$Z \succeq 0.$$

Consider the situation of setting u_0 to some (negative) value required for achieving $b_0 u_0 + e^T \hat{u} < c^*$. If $\hat{Z} + u_0 A_0$ is still positive semidefinite then we are done . If not, we add $-\lambda_{\min}(\hat{Z} + u_0 A_0)e$ to $\hat{u}$, $u = \hat{u} - \lambda_{\min}(\hat{Z} + u_0 A_0)e$. This worsens the original bound of $e^T \hat{u}$ by $-n\lambda_{\min}$ but the new Z is feasible again. Thus we are looking for an u_0 such that $b_0 u_0 + e^T \hat{u} - n\lambda_{\min}(\hat{Z} + u_0 a_0) < c^*$.

Summing up we specialize the semidefinite program above to

$$\min_{u_0 \in \mathbb{R}}\ b_0 u_0 + e^T \hat{u} - n\lambda_{\min}(u_0 A_0 + \hat{Z}). \tag{3}$$

The minimal eigenvalue is a concave function, so (3) is a convex line search problem. Each evaluation requires the computation of the minimal eigenvalue of $u_0 A_0 + \hat{Z}$. Extremal eigenvalues and eigenvectors are best determined via iterative methods such as the Lanczos method, which can exploit problem structure (see e.g. [5]). In particular these methods are very fast if a good starting vector is known. For the first computation we suggest the eigenvector v corresponding to $\lambda_{\max}(A_0)$, for all further iterations the last eigenvector computed is the natural choice. The function is differentiable if and only if the minimal eigenvalue has multiplicity one. In this case the gradient is determined by

$$\nabla_{u_0}(b_0 u_0 - n\lambda_{\min}(u_0 A_0 + \hat{Z})) = b_0 - n\,q(u_0)^T A_0 q(u_0)$$

with $q(u_0)$ denoting the (normalized) eigenvector to the minimal eigenvalue of $u_0 A_0 + \hat{Z}$. As explained above, it can be expected that $\hat{Z}$ has eigenvalue zero with high multiplicity. Therefore the function is not differentiable for $u_0 = 0$. It seems appropriate to choose the starting value u_0 with respect to the gap $c^* - e^T \hat{u}$, e.g. $u_0 = 1.2(c^* - e^T \hat{u})$. For reasonably large $|u_0|$ the minimal eigenvalue will be well separated and the gradient can be used to speed up the line search process. We expect that this method is efficiently applicable even in case approximate solutions of rather large sparse problems are given.

For testing whether variable x_{ij} can be fixed to 1 (w.l.o.g) we use $A_0 = E_{ij}$ (the matrix E_{ij} has a 1 in positions ij and ji and is zero otherwise) and $b_0 = -2$. In contrast to the original representation (1) this representation has no common support with the diagonal constraints. The fact that E_{ij} is a rank two matrix can be exploited to avoid repeated eigenvalue computations in the line search[1]. Let

---

[1] We are grateful to Kurt Anstreicher for pointing this out to us.

$E_{ij} = vv^T - ww^T$ for appropriate $v, w, \in \mathbb{R}^n$ and let $\hat{Z} = P\Lambda P^T$ with $P^T P = I_n$ be an eigenvalue decomposition of $\hat{Z}$. Then (3) specializes to

$$\min_{u_0 \geq 0, \lambda \geq 0} b_0 u_0 + n\lambda \quad \text{s.t.} \quad u_0 P^T (vv^T - ww^T) P + \lambda I + \Lambda \succeq 0.$$

For a given $\lambda > 0$ we obtain from (2) together with $\bar{v} = P^T v$ and $\bar{w} = P^T w$

$$u_0 \leq \frac{1}{\bar{w}^T (\Lambda + \lambda I + u_0 \bar{v} \bar{v}^T)^{-1} \bar{w}}.$$

Using a low rank update formula for the inverse of a matrix this leads to

$$\left(\sum_{i=1}^n \frac{\bar{v}_i \bar{w}_i}{\lambda_i(\hat{Z}) + \lambda} \right)^2 \geq \left(\sum_{i=1}^n \frac{\bar{v}_i^2}{\lambda_i(\hat{Z}) + \lambda} + \frac{1}{u_0} \right) \left(\sum_{i=1}^n \frac{\bar{w}_i^2}{\lambda_i(\hat{Z}) + \lambda} - \frac{1}{u_0} \right).$$

From this quadratic relation the best u_0 can be computed explicitely for any given $\lambda > 0$ in $O(n)$ arithmetic operations. This speeds up the line search, which is now formulated in λ, considerably. The factorization of $\hat{Z}$ has to be computed only once for all x_{ij} that are considered for fixing. In Section 6 we will present some experimental results indicating the efficacy of this approach.

5 Quadratic 0-1 Programming

It is well known that quadratic 0-1 programming in n variables is equivalent to quadratic $\{-1, 1\}$ programming in $n+1$ variables via a linear transformation [3]. This equivalence also extends to the canonical semidefinite relaxations [7, 10]. Here we show, that, formulated correctly, the same transformation preserves dual variables and problem structure.

Quadratic 0-1 programming in $\{0, 1\}$ variables asks for the optimal solution of

$$\text{(QP)} \qquad \max y^T B y \quad \text{s.t. } y \in \{0, 1\}^n$$

The canonical semidefinite relaxation for quadratic 0-1 programming is derived by adding an additional component 1 (with index 0) to the vector y and by forming the dyadic product $\begin{bmatrix} 1 & y^T \end{bmatrix}^T \begin{bmatrix} 1 & y^T \end{bmatrix}$ of this extended vector. The latter matrix is positive semidefinite and its diagonal is equal to the first column and the first row for all $y \in \{0, 1\}^n$. An intuitive way to write the semidefinite relaxation is

$$\text{(PQ)} \qquad \begin{aligned} &\max \langle B, Y \rangle \\ &\text{s.t. } \bar{Y} = \begin{bmatrix} 1 & \text{diag}(Y)^T \\ \text{diag}(Y) & Y \end{bmatrix} \succeq 0. \end{aligned}$$

There are several possibilities to model linear constraints ensuring the diagonal property of $\bar{Y}$. We will construct a representation ensuring that the dual variables are the same as those of the equivalent problem (PMC) in $n + 1$ variables.

To this end we present some well known facts about transformations of the type $W = QXQ^T$ for nonsingular $Q \in M_n$ in the general setting of the primal dual pair (P) and (D). These transformations belong to the automorphism group of the semidefinite cone (the set of all bijective linear maps leaving the semidefinite cone invariant) and appear several times in the interior point literature in connection with scaling issues (see e.g. [13, 14]). Clearly, $W = QXQ^T$ is positive semidefinite if and only if X is. How do we have to change the constraints of (P) such that we get the same semidefinite program in terms of W? Since $X = Q^{-1}WQ^{-T}$ and, for arbitrary $A \in S_n$,

$$\langle A, X \rangle = \langle A, Q^{-1}WQ^{-T} \rangle = \langle Q^{-T}AQ^{-1}, W \rangle,$$

the correct transformation of a coefficient matrix A is $Q^{-T}AQ^{-1}$. Note, that this is the adjoint to the inverse transformation of QXQ^T. With

$$\bar{C} = Q^{-T}CQ^{-1}, \quad \bar{A}_i = Q^{-T}A_iQ^{-1} \quad i = 1, \ldots, m,$$

and the linear operators $\bar{A}$ and $\bar{A}^T$ formed by the $\bar{A}_i$, we obtain the transformed primal dual pair

$$
\begin{array}{ll}
\text{(P}_Q\text{)} & \begin{aligned} \min &\ \langle \bar{C}, W \rangle \\ \text{s.t.} &\ \bar{A}(W) = b \\ &\ W \succeq 0 \end{aligned} \qquad\qquad
\text{(D}_Q\text{)} & \begin{aligned} \max &\ \langle b, \bar{u} \rangle \\ \text{s.t.} &\ \bar{A}^T(\bar{u}) + \bar{Z} = \bar{C} \\ &\ \bar{Z} \succeq 0. \end{aligned}
\end{array}
$$

Proposition 1. *X is a feasible solution of (P) if and only if the associated $W = QXQ^T$ is a feasible solution of (P$_Q$). Furthermore X and W satisfy $\langle C, X \rangle = \langle \bar{C}, W \rangle$. (u, Z) is a feasible solution of (D) if and only if the associated $(\bar{u}, \bar{Z}) = (u, Q^{-T}ZQ^{-1})$ is a feasible solution of (D$_Q$). Trivially, $\langle b, u \rangle = \langle b, \bar{u} \rangle$.*

Proof. Clear by construction.

In particular this implies that given an optimal primal dual solution for one of the problems we can construct an optimal primal dual solution for the other.

We apply this approach to the transformation between $\{-1, 1\}$ and $\{0, 1\}$ representation of 0-1 quadratic programming

Proposition 2. *Let $Q \in M_{n+1}$ be the matrix*

$$Q = \begin{bmatrix} 1 & 0 \\ \frac{1}{2}e & \frac{1}{2}I_n \end{bmatrix},$$

then $\varphi : S_{n+1} \to S_{n+1}, X \mapsto \bar{Y} = QXQ^T$ bijectively maps feasible solutions of (PMC) (for $n + 1$ variables) to feasible solutions of (PQ).

Proof. Q is nonsingular, therefore X is positive definite if and only if $\varphi(X)$ is. The properties concerning the diagonals are verified by direct computation.

This is a slight simplification with respect to earlier proofs of this fact ([7, 10]). However, the advantage of this approach is that by Proposition 1 we know how to formulate the constraints such that we can go back and forth between both models without changing the dual variables. In particular the correct transformation of a coefficient matrix A in the $\{-1, 1\}$ formulation to a coefficient matrix

B in the $\{0,1\}$ formulation is achieved by the adjoint operator to φ^{-1} (with φ as in Proposition 2),

$$(\varphi^{-1})^* : S_{n+1} \to S_{n+1}, \bar{A} \mapsto \bar{B} = Q^{-T}\bar{A}Q^{-1} \quad \text{with} \quad Q^{-1} = \begin{bmatrix} 1 & 0 \\ -e & 2I_n \end{bmatrix}.$$

It is easy to show that this transformation preserves sparsity (except for the first row and column) and low rank representations.

Returning to the fixing procedure for (PMC) we mention that in the $\{0,1\}$ model it is not obvious how to guarantee the positive semidefiniteness of S in the corresponding dual to (PQ) by a similar approach. However, for given optimal solutions of the $\{0,1\}$ model we can switch to the $\{-1,1\}$ setting without changing the dual variables and compute appropriate Lagrange multipliers. These are also correct multipliers in the $\{0,1\}$ model.

Remark. In theory, the fixing routine can be applied to any problem (D) for which there exists a $\hat{u}$ with $\mathcal{A}^T(\hat{u}) \succ 0$. By scaling the problem with $Q = \mathcal{A}^T(\hat{u})^{\frac{1}{2}}$ we obtain $Q^{-T}\mathcal{A}^T(\hat{u})Q^{-1} = I$. Therefore it is possible to shift the eigenvalues by adding or subtracting scalar multiples of $\hat{u}$; this is the basic property required by the fixing routine. In practice such a $\hat{u}$ may not exist or is not known in advance and an associated Q is likely to destroy problem structure.

6 Implementation

We have implemented the algorithm for fixing variables of Section 4 within our branch and cut code for solving (MC) as described in [8]. Here, we improve the semidefinite relaxation (PMC) with triangle inequalities only. Eigenvalues and eigenvectors are computed by the EISPACK routines tred2 and imtql2 (translated from Fortran to C with f2c). The fixing procedure is applied whenever a variable x_{ij} of the current optimal solution satisfies $|x_{ij}| > .98$. This leads to literally no additional cost for problems in which no variables satisfy this bound. Whenever variables of this size appeared then usually some of them could be fixed. We tested the code on the same classes of problems as in [8], $G_{.5}$, $G_{-1/0/1}$, Q_{100}, and $Q_{100,.2}$.

$G_{.5}$ consists of unweighted graphs with edge probability $1/2$, $G_{-1/0/1}$ of weighted (complete) graphs with edge weights chosen uniformly from $\{-1,0,1\}$. Q_{100} and $Q_{100,.2}$ were used in [15, 2]. Formulating Q_{100} with respect to (QP) the lower triangle of B is set to zero, the upper triangle (including the diagonal) is chosen uniformly from $\{-100,\ldots,100\}$. The diagonal takes the role of the linear term. $Q_{100,.2}$ represents instances with a density of 20%.

It was observed in [8] that in practice $G_{.5}$ and $G_{-1/0/1}$ are substantially more difficult to solve than Q_{100} and $Q_{100,.2}$. Indeed, for these classes the fixing routine was hardly ever called, because no variables satisfied $|x_{ij}| > .98$. Accordingly the additional cost of the routine was neglectable. However, for the "easy" classes of problems Q_{100} and $Q_{100,.2}$ the fixing routine was very successful and we present the results in Table 1.

Table 1. Average branch and bound results

		no fixing		with fixing	
n	nr	h:mm:ss	nodes	h:mm:ss	nodes
Q_{100}					
41	10	1:42	23	55	10
51	10	5:24	39	2:30	14
61	5	10:40	51	5:10	16
71	3	26:52	71	16:15	35
81	2	49:33	86	39:43	45
91	1	1:17:44	79	5:26	3
101	1	2:00:31	137	1:56:29	104
$Q_{100,.2}$					
41	10	9	1	9	1
51	10	32	1	24	1
61	5	4:11	15	2:59	9
71	3	32:57	71	16:28	32
81	2	32:10	33	14:02	11
91	1	1:15:16	73	11:42	7
101	1	2:05:00	113	17:17	7

Column n gives the dimension of the problem within the $\{-1,1\}$ setting (the additional one is due to the transformation) and nr refers to the number of instances solved. The average computation time[2] and number of branch and bound nodes follow. Clearly, fixing variables pays off in all cases and leads to considerable savings for most of the problems.

In future work it will be important to narrow the number of candidates by analyzing the relation of the respective vectors v to the spectrum and the eigenvectors of Z^*. Finally it remains to investigate other branching schemes, e.g. branching with respect to triangle inequalities.

I would like to thank Kurt Anstreicher for drawing my attention to the possibility of exploiting the low rank structure of the constraint matrix and Stefan E. Karisch and Franz Rendl for pointing out some missing references and for their constructive criticism with respect to the presentation.

References

1. F. Alizadeh, J.-P. A. Haeberly, and M. L. Overton. Coplementarity and nondegeneracy in semidefinite programming. *Mathematical Programming*, 77(2):111–128, 1997.
2. F. Barahona, M. Jünger, and G. Reinelt. Experiments in quadratic 0-1 programming. *Mathematical Programming*, 44:127–137, 1989.

---

[2] All times were computed on a Sun SPARCstation-4 with a 110 MHz microSPARC II CPU.

3. C. De Simone. The cut polytope and the boolean quadric polytope. *Discrete Applied Mathematics*, 79:71-75, 1989.
4. M. X. Goemans and D. P. Williamson. Improved approxiamtion algorithms for maximum cut and satisfiability problems using semidefinite programming. *J. ACM*, 42:1115–1145, 1995.
5. G. H. Golub and C. F. van Loan. *Matrix Computations*. The Johns Hopkins University Press, 2nd edition, 1989.
6. M. Grötschel, L. Lovász, and A. Schrijver. *Geometric Algorithms and Combinatorial Optimization*, volume 2 of *Algorithms and Combinatorics*. Springer, 2nd edition, 1988.
7. C. Helmberg, S. Poljak, F. Rendl, and H. Wolkowicz. Combining semidefinite and polyhedral relaxations for integer programs. In E. Balas and J. Clausen, editors, *Integer Programming and Combinatorial Optimization*, volume 920 of *Lecture Notes in Computer Science*, pages 124–134. Springer, May 1995.
8. C. Helmberg and F. Rendl. Solving quadratic (0,1)-problems by semidefinite programs and cutting planes. ZIB Preprint SC-95-35, Konrad Zuse Zentrum für Informationstechnik Berlin, Takustraße 7, D-14195 Dahlem, Germany, Nov. 1995.
9. S. E. Karisch and F. Rendl. Semidefinite programming and graph equipartition. Technical Report 302, Department of Mathematics, Graz University of Technology, Graz, Austria, Dec. 1995.
10. M. Laurent, S. Poljak, and F. Rendl. Connections between semidefinite relaxations of the max-cut and stable set problems. *Mathematical Programming*, 77(2):225–246, 1997.
11. L. Lovász. On the Shannon capacity of a graph. *IEEE Transactions on Information Theory*, IT-25(1):1–7, Jan. 1979.
12. L. Lovász and A. Schrijver. Cones of matrices and set-functions and 0-1 optimization. *SIAM J. Optimization*, 1(2):166–190, May 1991.
13. Y. Nesterov and M. J. Todd. Self-scaled barriers and interior-point methods for convex programming. Technical Report TR 1091, School of Operations Research and Industrial Engineering, Cornell University, Ithaca, New York 14853, Apr. 1994. Revised June 1995, to appear in Mathematics of Operations Research.
14. L. Tunçel. Primal-dual symmetry and scale invariance of interior-point algorithms for convex optimization. CORR Report 96–18, Department of Combinatorics and Optimization, Univeristy of Waterloo, Ontario, Canada, Nov. 1996.
15. A.C. Williams. *Quadratic 0-1 programming using the roof dual with computational results*. RUTCOR Research Report 8-85, Rutgers Unversity, 1985.
16. H. Wolkowicz and Q. Zhao. Semidefinite programming relaxations for the graph partitioning problem. Corr report, University of Waterloo, Ontario, Canada, Oct. 1996.
17. Q. Zhao, S. E. Karisch, F. Rendl, and H. Wolkowicz. Semidefinite programming relaxations for the quadratic assignment problem. CORR Report 95/27, University of Waterloo, Ontario, Canada, Sept. 1996.

Test Sets of the Knapsack Problem and Simultaneous Diophantine Approximation

Martin Henk and Robert Weismantel

Konrad-Zuse-Zentrum für Informationstechnik Berlin, Takustr. 7,
14195 Berlin, Germany.

Abstract. This paper deals with the study of test sets of the knapsack problem and simultaneous diophantine approximation. The Graver test set of the knapsack problem can be derived from minimal integral solutions of linear diophantine equations. We present best possible inequalities that must be satisfied by all minimal integral solutions of a linear diophantine equation and prove that for the corresponding cone the integer analogue of Caratheodory's theorem applies when the numbers are divisible.

We show that the elements of the minimal Hilbert basis of the dual cone of all minimal integral solutions of a linear diophantine equation yield best approximations of a rational vector "from above". A recursive algorithm for computing this Hilbert basis is discussed. We also outline an algorithm for determining a Hilbert basis of a family of cones associated with the knapsack problem.

Keywords: knapsack problem, simultaneous diophantine approximation, diophantine equation, Hilbert basis, test sets.

1 Introduction

This paper deals with the study of test sets of the knapsack problem and simultaneous diophantine approximation. Both topics play a role in various branches of mathematics such as number theory, geometry of numbers and integer programming.

From the viewpoint of integer programming, minimal integral solutions of a linear diophantine equation allow to devise an exact primal algorithm for solving knapsack problems in non-negative integral variables,

$$\max c^T x : \; \alpha^T x = \beta, \; x \in \mathbb{N}^n, \tag{1}$$

where $c \in \mathbb{Z}^n$, $\alpha \in (\mathbb{N} \setminus \{0\})^n$ and $\beta \in \mathbb{N}$. More precisely, the primal methods that we consider here are augmentation algorithms, and the question we address is to describe the set of all possible augmentation vectors. This leads us to *test sets*.

A test set is a collection of all augmenting directions that one needs in order to guarantee that every non-optimal feasible point of a linear integer program can be improved by one member in the test set. There are various possible ways of defining test sets depending on the view that one takes: the *Graver test set* is naturally derived from a study of the integral vectors in cones [G75]; the *neighbors of the origin* are strongly connected to the study of lattice point free convex bodies [S86]; the so-called *reduced Gröbner basis* of an integer program is obtained from generators of polynomial ideals that is a classical field of algebra, [CT91]. We refrain within this paper from introducing all these three kinds of test sets, but concentrate on the Graver test set, only. In order to introduce the *Graver test set* for the family of knapsack problems with varying $c \in \mathbb{Z}^n$ and $b \in \mathbb{N}$, the notion of a rational polyhedral cone and its Hilbert basis is needed.

Definition 1. For $z^1, \ldots, z^m \in \mathbb{Z}^n$, the set

$$C := \mathrm{pos}\,\{z^1, \ldots, z^m\} = \left\{ \sum_{i=1}^{m} \lambda_i z^i \,:\, \lambda \in \mathbb{R}_{\geq 0}^m \right\}$$

is called a *rational polyhedral cone*. It is called *pointed* if there exists a hyperplane $\{x \in \mathbb{R}^n : a^T x = 0\}$ such that $\{0\} = \{x \in C : a^T x \leq 0\}$.

Definition 2. Let $C \subseteq \mathbb{R}^n$ be a rational polyhedral cone. A finite subset $H = \{h^1, \ldots, h^t\} \subseteq C \cap \mathbb{Z}^n$ is a *Hilbert basis* of C if every $z \in C \cap \mathbb{Z}^n$ has a representation of the form

$$z = \sum_{i=1}^{t} n_i h^i,$$

with non-negative integral multipliers $n_1, \ldots, n_t$.

The name Hilbert basis was introduced by Giles and Pulleyblank [GP79] in the context of totally dual integral systems. Essential is (see [G1873], [C31])

Theorem 3. *Every rational polyhedral cone has a Hilbert basis. If it is pointed, then there exists a unique Hilbert basis that is minimal w.r.t. inclusion.*

In the following by a cone we always mean a rational polyhedral cone.

Let O_j denote the j-th orthant in $\mathbb{R}^n$. For $A \in \mathbb{Z}^{m \times n}$, the set $C_j := \{x \in O_j : Ax = 0\}$ is a pointed cone in $\mathbb{R}^n$. Denoting by H_j the minimal Hilbert basis of C_j, Graver proved the following: The set $H := \bigcup_j H_j$ is a test set for the family of integer programs of the form $\max c^T x : Ax = b, \ x \in \mathbb{N}^n$ for a fixed matrix $A \in \mathbb{Z}^{m \times n}$ and varying $c \in \mathbb{Z}^n$ and $b \in \mathbb{Z}^m$.

This result is the starting point for our discussions. Namely, in order to devise an exact primal algorithm for solving a knapsack problem of the form (1), we need to determine, for every orthant O_j in $\mathbb{R}^n$, a Hilbert basis H_j of the so-called *knapsack cone* $C_j = \{x \in O_j : \alpha^T x = 0\}$.

We present in this paper best possible inequalities that must be satisfied by all the elements of the minimal Hilbert basis of C_j and prove that for C_j the integer analogue of Caratheodory's theorem applies when the numbers $\{\alpha_1, \ldots, \alpha_n\}$ are pairwise divisible. We also show that the elements of the minimal Hilbert basis of the dual of C_j yield best approximations of a rational vector "from above". A recursive algorithm for computing this Hilbert basis is discussed. A similar type of procedure applies to the cone C_j. Therefore this method can also be used to find a test set of the knapsack problem.

2 The knapsack cone

Up to a permutation of the coordinates, a knapsack cone C_j can be identified with the set $K_{n,m}$ of all non-negative solutions of a linear diophantine equation, i.e.,

$$
K_{n,m} = \left\{ (x,y)^T \in \mathbb{R}^n_{\geq 0} \times \mathbb{R}^m_{\geq 0} : \sum_{i=1}^{n} a_i x_i = \sum_{j=1}^{m} b_j y_j \right\},
$$

where we always assume that $a = (a_1, \ldots, a_n)^T \in \mathbb{N}^n$, $b = (b_1, \ldots, b_m)^T \in \mathbb{N}^m$, $n \geq m \geq 1$ and $a_1 \leq a_2 \leq \cdots \leq a_n$, $b_1 \leq b_2 \leq \cdots \leq b_m$. It is easy to see that

$$
K_{n,m} = \operatorname{pos} \left\{ b_j e^i + a_i e^{n+j} : 1 \leq i \leq n,\, 1 \leq j \leq m \right\}, \tag{2}
$$

where $e^i \in \mathbb{R}^{n+m}$ denotes the i-th unit vector. The minimal Hilbert basis of $K_{n,m}$ is denoted by $\mathcal{H}_{n,m}$.

One of the major results of this paper is to show that every element in $\mathcal{H}_{n,m}$ satisfies $n + m$ special inequalities that generalize the two inequalities

$$
\sum_{i=1}^{n} x_i \leq b_m, \text{ and } \sum_{j=1}^{m} y_j \leq a_n \tag{3}
$$

proved by Lambert ([L87]) and independently by Diaconis, Graham & Sturmfels [DGS94].

Theorem 4. *Every* $(x,y)^T \in \mathcal{H}_{n,m}$ *satisfies the inequalities*

$$
[j_l]: \quad \sum_{i=1}^{n} x_i + \sum_{j=1}^{l-1} \left\lfloor \frac{b_l - b_j}{a_n} \right\rfloor y_j \leq b_l + \sum_{j=l+1}^{m} \left\lceil \frac{b_j - b_l}{a_1} \right\rceil y_j, \quad l = 1, \ldots, m,
$$

$$
[i_k]: \quad \sum_{j=1}^{m} y_j + \sum_{i=1}^{k-1} \left\lfloor \frac{a_k - a_i}{b_m} \right\rfloor x_i \leq a_k + \sum_{i=k+1}^{n} \left\lceil \frac{a_i - a_k}{b_1} \right\rceil x_i, \quad k = 1, \ldots, n.
$$

From an algorithmic point of view Theorem 4 allows to assert that an integral point in $K_{n,m}$ does not belong to a minimal Hilbert basis of this cone. This problem is in general $\mathcal{NP}$-complete, see Sebö [S90].

Theorem 5 (The Decomposition Problem). *For the pointed cone* $K_{n,m}$, *and a vector* $(x,y)^T \in K_{n,m} \cap \mathbb{Z}^{n+m}$ *it is co-$\mathcal{NP}$-complete to decide whether* $(x,y)^T$ *is contained in* $\mathcal{H}_{n,m}$.

Theorem 5 asserts the difficulty of testing for non-membership in $\mathcal{H}_{n,m}$. On the other hand, every integral vector in this cone can be decomposed by vectors of the basis. In fact we can write every integral vector in any pointed cone of dimension n as the non-negative integral combination of at most $2n - 2$ vectors from the basis. This was shown by Sebö [S90] and gives currently the best bound in general; it improves the bound given by Cook, Fonlupt & Schrijver [CFS86] by 1, yet is still quite far from what many researchers conjecture to be true, namely: every integral vector in a pointed cone is the non-negative integral combination of at most n vectors of the Hilbert basis. We now prove that this *integer Version of Caratheodory's Theorem* holds for the knapsack cone when the numbers are divisible.

Theorem 6. *Let positive integers $a_1, \ldots, a_n$ and $b_1, \ldots, b_m$ be given such that there exist $p_i, q_j \in \mathbb{N}$ with*

$$a_i = p_i \cdot a_{i-1}, \ i = 2, \ldots, n, \quad b_1 = q_1 \cdot a_n, \ b_j = q_j \cdot b_{j-1}, \ j = 2, \ldots, m.$$

Every integral point in $K_{n,m}$ can be written as the non-negative integral combination of at most $n + m - 1 = \dim(K_{n,m})$ elements of $\mathcal{H}_{n,m}$.

Proof. Let $(\widetilde{x}, \widetilde{y})^T \in K_{n,m}$. We have to show that there exist $(x^i, y^i)^T \in \mathcal{H}_{n,m}$, $n_i \in \mathbb{N}$, $1 \le i \le n + m - 1$ such that $(\widetilde{x}, \widetilde{y})^T = \sum_{i=1}^{n+m-1} n_i (x^i, y^i)^T$.

In order to show this statement we use induction w.r.t. n. W.l.o.g. we may assume that $a_1 = 1$. If $n = 1$ then we see by (3) that $b_i e^1 + e^{n+i}$, $i = 1, \ldots, m$, is the Hilbert basis of $K_{n,m}$ and we are done. So let $n \ge 2$, $a_i^* = a_i / a_2$, $2 \le i \le n$, $b_j^* = b_j / a_2$, $1 \le j \le m$, and let

$$K_{n,m}^* = \left\{ (x,y)^T \in \mathbb{R}_{\ge 0}^n \times \mathbb{R}_{\ge 0}^m : x_1 + x_2 + \sum_{i=3}^n a_i^* x_i = \sum_{j=1}^m b_j^* y_j \right\}.$$

Let $f : \mathbb{R}^{n+m} \to \mathbb{R}^{n+m}$ be the map given by $f((x_1, \ldots, x_n, y_1, \ldots, y_m)^T) = (x_1/a_2, x_2, \ldots, x_n, y_1, \ldots, y_m)^T$. It is easy to see that the linear map f induces a bijection between $K_{n,m} \cap \mathbb{Z}^{n+m}$ and $K_{n,m}^* \cap \mathbb{Z}^{n+m}$. Therefore it suffices to prove that $f((\widetilde{x}, \widetilde{y})^T)$ can be written as integral combination of at most $n + m - 1$ elements of the Hilbert basis of $K_{n,m}^*$, which is denoted by $\mathcal{H}_{n,m}^*$.

For abbreviation we set $(x^*, y^*) = f((\widetilde{x}, \widetilde{y})^T)$. Now let

$$K_{n-1,m} = \left\{ (x,y)^T \in \mathbb{R}_{\ge 0}^{n-1} \times \mathbb{R}_{\ge 0}^m : x_1 + \sum_{i=2}^{n-1} a_{i+1}^* x_i = \sum_{j=1}^m b_j^* y_j \right\}$$

and let $g : \mathbb{R}^{n+m} \to \mathbb{R}^{n-1+m}$ defined by $g((x_1, \ldots, x_n, y_1, \ldots, y_m)^T) = (x_1 + x_2, x_3, \ldots, x_n, y_1, \ldots, y_m)^T$. Next we make use of our induction hypothesis w.r.t. to the point $g((x^*, y^*)^T) \in K_{n-1,m}$, i.e., there exist $n_i \in \mathbb{N}$, $(\bar{x}^i, \bar{y}^i)^T \in \mathcal{H}_{n-1,m}$, $1 \le i \le n + m - 2$, such that

$$g((x^*, y^*)^T) = \sum_{i=1}^{n+m-2} n_i (\bar{x}^i, \bar{y}^i)^T,$$

where $\mathcal{H}_{n-1,m}$ denotes the Hilbert basis of $K_{n-1,m}$ w.r.t. $\mathbb{Z}^{n-1+m}$. In particular, we have

$$x_1^* + x_2^* = \sum_{i=1}^{n+m-2} = n_i \, \bar{x}_1^i. \tag{4}$$

W.l.o.g. let $n_1\bar{x}_1^1 \geq n_2\bar{x}_1^2 \geq \cdots \geq n_{n+m-2}\bar{x}_1^{n+m-2}$. From the identity (4) we see that we can find $\epsilon_i \in \{0,1\}$, $2 \leq i \leq n + m - 2$ such that

$$\begin{pmatrix} v_1 \\ v_2 \end{pmatrix} = \begin{pmatrix} x_1^* \\ x_2^* \end{pmatrix} - \sum_{i=2}^{n+m-2} n_i \left(\epsilon_i \begin{pmatrix} \bar{x}_1^i \\ 0 \end{pmatrix} + (1-\epsilon_i) \begin{pmatrix} 0 \\ \bar{x}_1^i \end{pmatrix} \right) \geq_{\text{lex}} 0. \tag{5}$$

Obviously, we have $v_1 + v_2 = n_1 \bar{x}_1^1$. Now if we consider the 2-dimensional knapsack cone $C = \{(x,y) \in \mathbb{R}_{\geq 0}^2 \times \mathbb{R}_{\geq 0} : x_1 + x_2 = \bar{x}_1^1 y_1\}$ then all elements of the Hilbert basis have the form $(\alpha_1, \alpha_2, 1)^T$ (cf. (3)). Further, for a 2-dimensional cone it is well known that each element can be written as an integral combination of at most two elements of the Hilbert basis. Therefore, since $(v_1, v_2)^T \in C$, there exist $n_0^*, n_1^* \in \mathbb{N}$, $w^0, w^1 \in \mathbb{N}^2$, such that

$$\begin{pmatrix} v_1 \\ v_2 \end{pmatrix} = n_0^* w^0 + n_1^* w^1, \quad n_0^* + n_1^* = n_1, \quad w_1^0 + w_2^0 = w_1^1 + w_2^1 = \bar{x}_1^1. \tag{6}$$

Finally, for $i = 2, \ldots, n + m - 2$ let $(x^i, y^i)^T \in \mathbb{R}^n \times \mathbb{R}^m$ be defined by

$$(x^i, y^i) = \left(\epsilon_i \bar{x}_1^i, (1-\epsilon_i)\bar{x}_1^i, \bar{x}_2^i, \ldots, \bar{x}_{n-1}^i, \bar{y}_1^i, \ldots, \bar{y}_m^i \right).$$

and for $i = 0, 1$ let

$$(x^i, y^i) = \left(w_1^i, w_2^i, \bar{x}_2^1, \ldots, \bar{x}_{n-1}^1, \bar{y}_1^1, \ldots, \bar{y}_m^1 \right).$$

By definition and (6) we have $(x^i, y^i)^T \in K_{n,m}$ and indeed it is easy to check that $(x^i, y^i)^T \in \mathcal{H}_{n,m}$, $0 \leq i \leq n + m - 2$. By (5) and (6) we get

$$(x^*, y^*)^T = n_0^*(x^0, y^0)^T + n_1^*(x^1, y^1)^T + \sum_{i=2}^{n+m-2} n_i(x^i, y^i)^T.$$

$\square$

Let us point out that, although Theorem 4 gives the best inequalities known so far to assert that an integral point in $K_{n,m}$ does not belong to the minimal Hilbert basis, we believe that a much stronger and more general statement is true: every element in the minimal Hilbert basis of $K_{n,m}$ is a convex combination of 0 and the generators $b_j e^i + a_i e^{n+j}$ of $K_{n,m}$. More formally, let

$$P_{n,m} = \text{conv}\left\{ 0, b_j e^i + a_i e^{n+j} : 1 \leq i \leq n, \, 1 \leq j \leq m \right\}.$$

One might conjecture that

Conjecture 21. $\mathcal{H}_{n,m} \subset P_{n,m}$.[1]

For $m = 1$ Theorem 4 implies the inclusion $\mathcal{H}_{n,1} \subset P_{n,1}$. This can easily be read off from the representation

$$P_{n,1} = \left\{ (x,y)^T \in \mathbb{R}^n \times \mathbb{R} : a^T x = b_1 y; x, y \geq 0, \sum_{i=1}^{n} x_i \leq b_1 \right\}.$$

One way of verifying the correctness of the conjecture could be to find all facets defining inequalities of $P_{n,m}$ and to check that these inequalities are satisfied by the elements of $H_{n,m}$. A subset of the facets defining inequalities is given by

Proposition 7. *For $l = 1, \ldots, m$ let*

$$J_l = \left\{ (x,y) \in \mathbb{R}^n \times \mathbb{R}^m : \sum_{i=1}^{n} x_i + \sum_{j=1}^{l-1} \frac{b_l - b_j}{a_n} y_j \leq b_l + \sum_{j=l+1}^{m} \frac{b_j - b_l}{a_1} y_j \right\}$$

and for $k = 1, \ldots, n$ let

$$I_k = \left\{ (x,y) \in \mathbb{R}^n \times \mathbb{R}^m : \sum_{j=1}^{m} y_j + \sum_{i=1}^{k-1} \frac{a_k - a_i}{b_m} x_i \leq a_k + \sum_{i=k+1}^{n} \frac{a_i - a_k}{b_1} x_i \right\}.$$

The inequalities defining the halfspaces J_l and I_k define facets of $P_{n,m}$, $1 \leq l \leq m$, $1 \leq k \leq n$.

Proof. It is quite easy to check that all vectors $b_j e^i + a_i e^{n+j}$, $1 \leq i \leq n$, $1 \leq j \leq m$, are contained in J_l, $l = 1, \ldots, m$. Moreover, the inequality corresponding to J_l is satisfied with equality by the $n+m-1$ linearly independent points $b^l e_i + a_i e^{n+l}$, $1 \leq i \leq n$, $b_j e^n + a_n e^{n+j}$, $1 \leq j \leq l-1$, $b_j e^1 + a_1 e^{n+j}$, $l+1 \leq j \leq m$. The halfspaces I_k can be treated in the same way.

Remark 8. Since $P_{n,2} = \{(x,y)^T \in \mathbb{R}^n \times \mathbb{R}^2 : a^T x = b^T y; x, y \geq 0, (x,y)^T \in I_k, 1 \leq k \leq n\}$, Theorem 4 shows that the conjecture is "almost true" when $m = 2$.

3   Best approximations "from above"

In this section we deal with a cone that on the first view seems to be not related to the knapsack cone investigated before.

Let $e^1, \ldots, e^n$ denote the n unit vectors in $\mathbb{R}^{n+1}$ having a 1 in coordinate $1, \ldots, n$, respectively. For $p \in \mathbb{Z}^{n+1}$ such that $\gcd(p_1, \ldots, p_{n+1}) = 1$, $p_1, \ldots, p_k > 0$, $p_{k+1}, \ldots, p_n < 0$ and $p_{n+1} > 0$, let

$$C(p) = \text{pos}\{e^1, \ldots, e^n, p\}. \tag{7}$$

---

[1] This conjecture was independently made by Hosten and Sturmfels, private communication

It turns out that the dual cone $C(p)^*$ of $C(p)$ is "essentially" the knapsack cone. This result builds the bridge towards the previous section. By definition, $C(p)^*$ can be written as $C(p)^* = \{v \in \mathbb{R}^{n+1} : v^T x \geq 0, \ \forall \ x \in C(p)\}$. Since the generators of $C(p)$ consist of the unit vectors $e^1, \ldots, e^n$ plus the vector $p \in \mathbb{Z}^{n+1}$, we obtain

$$C(p)^* = \left\{ v \in \mathbb{R}^n_{\geq 0} \times \mathbb{R} : \sum_{i=1}^{k} v_i \cdot p_i \geq \sum_{i=k+1}^{n} v_i \cdot (-p_i) - v_{n+1} p_{n+1} \right\}.$$

Depending on the sign of v_{n+1}, we partition $C(p)^*$ into the following two cones

$$C(p)^*_{\geq} = \left\{ v \in \mathbb{R}^{n+1}_{\geq 0} : \sum_{i=1}^{k} v_i p_i + v_{n+1} p_{n+1} \geq \sum_{i=k+1}^{n} v_i \cdot (-p_i) \right\},$$

$$C(p)^*_{\leq} = \left\{ v \in \mathbb{R}^n_{\geq 0} \times \mathbb{R}_{\leq 0} : \sum_{i=1}^{k} v_i p_i \geq \sum_{i=k+1}^{n} v_i \cdot (-p_i) + (-v_{n+1}) p_{n+1} \right\}.$$

Both cones, $C(p)^*_{\geq}$ and $C(p)^*_{\leq}$ may be regarded as "$\geq$-knapsack cones", or, the facet of the cone $C(p)^*_{\geq}$ ($C(\overline{p})^*_{\leq}$) induced by the non-trivial inequality is a knapsack cone of the form $K_{k+1,n-k}$ ($K_{k,n-k+1}$) that we studied in Section 2.

In the remainder of this section we study the minimal Hilbert basis of $C(p)$. It turns out that this basis is closely related to the problem of simultaneous diophantine approximation of rational numbers by other rational numbers with an upper bound on the denominator. More precisely, we consider the following approximation problem:

Simultaneous Diophantine Approximation "from above":

Let $p_1, \ldots, p_{n+1} \in \mathbb{Z}$, $p_{n+1} > 0$, and $N \in \mathbb{Z}$, $N > 0$.
Find integers $q_1, \ldots, q_{n+1}$, $N \geq q_{n+1} > 0$ such that $q_i/q_{n+1} \geq p_i/p_{n+1}$, $i = 1, \ldots, n$, and $\sum_{i=1}^{n} \left(\frac{q_i}{q_{n+1}} - \frac{p_i}{p_{n+1}} \right)$ is as small as possible.
The vector $q' = \left(\frac{q_1}{q_{n+1}}, \ldots, \frac{q_n}{q_{n+1}} \right)$ is called a best approximation of $p' = \left(\frac{p_1}{p_{n+1}}, \ldots, \frac{p_n}{p_{n+1}} \right)$ from above with respect to N.

It is clear that if $N \geq p_{n+1}$, then $p' = \left(\frac{p_1}{p_{n+1}}, \ldots, \frac{p_n}{p_{n+1}} \right)$ itself is its best approximation from above. It is, however, not clear how one can characterize a best approximation of p' from above when $N < p_{n+1}$. We show that a best approximation of p' from above can be read off from the minimal Hilbert basis of $C(p)$.

Theorem 9. *Let $p_1, \ldots, p_{n+1} \in \mathbb{Z}$, $p_{n+1} > 0$, and $N \in \mathbb{Z}$, $N > 0$. There exists an element $(q_1, \ldots, q_{n+1})$ of the minimal Hilbert basis of $C(p)$ such that $q' = \left(\frac{q_1}{q_{n+1}}, \ldots, \frac{q_n}{q_{n+1}} \right)$ is a best approximation of $p' = \left(\frac{p_1}{p_{n+1}}, \ldots, \frac{p_n}{p_{n+1}} \right)$ from above with respect to N. Moreover, among all such best approximations of p', q' is the unique one with smallest denominator q_{n+1}.*

Proof. Let $x' = \left(\frac{x_1}{x_{n+1}}, \ldots, \frac{x_n}{x_{n+1}} \right)$ be the best approximation of p' from above with $0 < x_{n+1}$ as small as possible. Then $(x_1, \ldots, x_{n+1}) \in C(p) \cap \mathbb{Z}^{n+1}$, because

$\frac{x_i}{x_{n+1}} \geq \frac{p_i}{p_{n+1}}$ for all $i = 1, \ldots, n$. If $x \notin \mathcal{H}$, then there exist $v, w \in C(p) \cap \mathbb{Z}^{n+1}$ such that $x = v + w$. For abbreviation we define $R(z) := \sum_{i=1}^{n} (\frac{z_i}{z_{n+1}} - \frac{p_i}{p_{n+1}})$ for $z \in C(p) \cap \mathbb{Z}^{n+1}$. Since

$$R(x) = \frac{\sum_{i=1}^{n}(x_i p_{n+1} - p_i x_{n+1})}{x_{n+1} \cdot p_{n+1}} = \frac{\sum_{i=1}^{n}(v_i \cdot p_{n+1} - p_i v_{n+1} + w_i p_{n+1} - p_i w_{n+1})}{(v_{n+1} + w_{n+1})p_{n+1}}$$

$$= \frac{\sum_{i=1}^{n}[v_i - \frac{p_i v_{n+1}}{p_{n+1}} + w_i - \frac{p_i w_{n+1}}{p_{n+1}}]}{(v_{n+1} + w_{n+1})} = \frac{v_{n+1}R(v) + w_{n+1}R(w)}{v_{n+1} + w_{n+1}}$$

we have $R(v) = R(w) = R(x)$. However, this contradicts the choice of x as a best approximation of p' from above for which x_{n+1} is as small as possible. $\square$

Instead of restricting our attention to approximations of a rational vector p' from above, one could ask for approximations where, for any of the components of p', one would specify a-priori, whether the approximation should lie below or above the corresponding value of p'. Theorem 9 can be extended to this situation.

Theorem 10. *Let $\sigma \in \{-1, +1\}^n$ be the sign pattern associated with one orthant of $\mathbb{R}^n$. Let $p_1, \ldots, p_{n+1} \in \mathbb{Z}$, $p_{n+1} > 0$, and $N \in \mathbb{Z}$, $N > 0$. There exists an element $(q_1, \ldots, q_{n+1})$ of the minimal Hilbert basis of $\mathrm{pos}\,\{\sigma_1 e^1, \ldots, \sigma_n e^n, p\}$ such that*

$$\sum_{i=1}^{n}\left|\frac{q_i}{q_{n+1}} - \frac{p_i}{p_{n+1}}\right| = \min\left\{\sum_{i=1}^{n}\left|\frac{x_i}{x_{n+1}} - \frac{p_i}{p_{n+1}}\right| : x_1, \ldots, x_{n+1} \in \mathbb{Z},\right.$$

$$\left. N \geq x_{n+1} > 0,\ \sigma_i\left(\frac{x_i}{x_{n+1}} - \frac{p_i}{p_{n+1}}\right) \geq 0\right\}.$$

Among all solutions of this diophantine approximation problem, $(\frac{q_1}{q_{n+1}}, \ldots, \frac{q_n}{q_{n+1}})$ is the unique one with smallest denominator.

4 A recursive algorithm for the Hilbert basis of $C(p)$ and the knapsack cone

We have motivated in the previous sections why the Hilbert basis of the knapsack cone and the cone of best approximations from above is of particular interest. In this section we treat algorithmic questions related to these bases. We first deal with the cone $C(p) = \mathrm{pos}\,\{e^1, \ldots, e^n, p\} \subseteq \mathbb{R}^{n+1}$ related to the best approximations from above. Applying a unimodular transformation we may assume that $p = (p_1, \ldots, p_{n+1}) \in \mathbb{N}^{n+1}$.

We remark that it is trivial to find a Hilbert basis of $C(p)$, because it is well known that $\{e^1, \ldots, e^n, p\} \cup \{z \in \mathbb{Z}^{n+1} : z = \sum_{i=1}^{n} \lambda_i e^i + \lambda_{n+1}p, 0 \leq \lambda_i < 1\}$ actually is a Hilbert basis of $C(p)$ (cf. [C31]). All we are left with is to enumerate these integral points. However, in general, the size of this Hilbert basis is exponentially larger than the size of the minimal Hilbert basis, and, of course, we are interested in computing a "small" one.

We proceed in an inductive fashion to compute the basis of $C(p)$: Let H_2 be the minimal Hilbert basis of the 2-dimensional cone

$$C_2 := \text{pos}\,\{e^1, (p_n, p_{n+1})^T\}.$$

It is clear that $e^1 \in H_2$. Let $(w_1, h_1) < \ldots < (w_m, h_m) \in H_2 \setminus \{e^1\}$ be the remaining elements in H_2. It follows that, for every $x \in C(p) \cap \mathbb{Z}^{n+1}$ with $x_{n+1} > 0$, the coordinate x_{n+1} has a representation of the form $x_{n+1} = \sum_{v=1}^{m} \mu_v h_v$ with $\mu_1, \ldots, \mu_m \in \mathbb{N}$.

Definition 11. For $i, j \in \{1, \ldots, m-1\}$, $i < j$, let

$$C^{i,j} := \left\{ x \in C(p) \cap \mathbb{Z}^{n+1} : \exists\, \mu_i, \ldots, \mu_j \in \mathbb{N},\ 0 < x_{n+1} = \sum_{v=i}^{j} \mu_v h_v < h_{j+1} \right\}.$$

We say that $\{g^1, \ldots, g^t\} \subseteq L$ are generators of a set $L \subseteq \mathbb{Z}^n$ if, for every $x \in L$, there exist $\sigma_1, \ldots, \sigma_t \in \mathbb{N}$ such that $x = \sum_{v=1}^{t} \sigma_v g^v$.

Noting that $h_m = p_{n+1}$, the following statement is immediate.

Lemma 12. *The union of $\{e^1, \ldots, e^n, p\}$ and a set of generators of $C^{1,m-1}$ defines a Hilbert basis of $C(p)$.*

In fact, an inclusionwise ordering of all the subsets $C^{i,j}$ is possible.

Lemma 13. *For every $i' \leq i \leq j \leq j' \in \{1, \ldots, m-1\}$, a minimal set of generators of $C^{i,j}$ (w.r.t. inclusion) defines a subset of any set of generators of $C^{i',j'}$.*

On account of Lemma 12 it suffices to find generators of $C^{1,m-1}$ in order to determine a Hilbert basis of $C(p)$. A set of generators of $C^{1,m-1}$ can be computed in a recursive manner.

Algorithm 14. Recursion formula to find the generators of $C^{1,m-1}$.
   (1)   For $i = 1, \ldots, m-1$ determine generators of $C^{i,i}$.
   (2)   For $i = m-2, \ldots, 1$ determine generators of $C^{i,m-1}$.

The reason why this recursion makes sense is that the task of finding a set of generators of $C^{i,i}$ can be solved with a procedure to determine the Hilbert basis of a cone similar to $C(p)$, but in one dimension less. Secondly, if one has, for $i = m-2, \ldots, 1$ as input a set of generators of $C^{i,i}$ and $C^{i+1,m-1}$, then one can devise a procedure that returns generators of $C^{i,m-1}$. This is the idea behind the recursion.

Lemma 15. *Let $\tilde{p} = (p_1, \ldots, p_{n-1}, h_i p_{n+1}) \in \mathbb{N}^n$ and let $e^1, \ldots, e^{n-1}$ denote the first $n - 1$ unit vectors of $\mathbb{R}^n$. For $i \in \{1, \ldots, m-1\}$, let H^i be a Hilbert basis of the n-dimensional cone $C(\tilde{p}) = \mathrm{pos}\,\{e^1, \ldots, e^{n-1}, \tilde{p}\}$. The set $\{(x_1, \ldots, x_{n-1}, zw_i, zh_i) : (x_1, \ldots, x_{n-1}, z) \in H^i, \, 0 < zh_i < h_{i+1}\}$ is a set of generators of $C^{i,i}$.*

Lemma 15 shows that a set of generators of $C^{i,i}$ can easily be reconstructed from a Hilbert basis of the cone $C(\tilde{p})$. This is where the inductive step comes into play. In order to solve Step (2) of Algorithm 14 one must be able to turn a set of generators of $C^{i,i}$ and $C^{i+1,m-1}$ into a set of generators of $C^{i,m-1}$. This task may again be solved with a recursive algorithm that would read as follows.

Algorithm 16. (Recursion to find $C^{i,m-1}$)

Input: *An ordered set $\{g^1, \ldots, g^t\}$ of generators of $C^{i+1,m-1}$ with $g^1_{n+1} < \ldots < g^t_{n+1}$; a set of generators of $C^{i,i}$.*
Output: *A set of generators of $C^{i,m-1}$;*
For every $v \in \{1, \ldots, t\}$ determine a set of generators of the set $G_v := \{y \in C^{i,m-1} : y_{n+1} < g^v_{n+1}\}$.

Lemma 17. *$G_1 = C^{i,i}$ and $G_{t+1} = C^{i,m-1}$.*

Lemma 17 shows that when we enter the for-loop of Algorithm 16, a set of generators of $C^{i,i}$ is also a set of generators of G_1. Then we proceed through all values of v and determine a set of generators of G_v using a set of generators of G_{v-1}. When v equals $t + 1$, we terminate with a set of generators of G_{t+1} that corresponds to a set of generators of $C^{i,m-1}$.

The key of Algorithm 16 is a subroutine for returning a set of generators of G_{v+1} whose input consists of a set of generators of G_v. It is not difficult to see that G_{v+1} is equal to the set

$$S = \{x \in C^{i,m-1} : \exists\, \lambda \in \{0, \ldots, \lambda_v\}, y \in G_v \text{ s.t.} \atop x_{n+1} = \lambda g^v_{n+1} + y_{n+1} < g^{v+1}_{n+1}\}, \tag{8}$$

where $\lambda_v := \max\{\lambda \in \mathbb{N} : \lambda g^v_{n+1} < g^{v+1}_{n+1}\}$ and $g^{t+1}_{n+1} := h_m$. The generators of G_{v+1} are points of the form

$$(x_1, \ldots, x_{n+1}) \in S : x_i = \lceil \frac{p_i x_{n+1}}{p_{n+1}} \rceil, \, i = 1, \ldots, n. \tag{9}$$

To each point x of the form (9) there corresponds a n-dimensional vector of residua of components $x_i p_{n+1} - x_{n+1} p_i$, $i = 1, \ldots, n$. In fact, one can show that the minimal set of generators of G_{v+1} can be characterized as follows: we order the integral points of the form (9) w.r.t. increasing last coordinate; the vector of residua of a point that appears later in this sequence is incomparable with the vectors of residua of all points that occur earlier in this sequence. Resorting to appropriate data structures that contain information about the vector of residua for every element in G_v one can determine a set of generators of G_{v+1} without testing every integral point in S.

For precisely this reason, Algorithm 14 yields a much more sophisticated algorithm for determining the Hilbert basis of $C(p)$ than the trivial method discussed at the beginning of the section.

We now illustrate the essential steps of Algorithm 14 on an example.

Example. Let $n = 2$ and $p = (30, 29, 17)$. The elements of the Hilbert basis of $C(p) = \text{pos}\{e^1, e^2, p\}$ that we are interested in consists of integral points of the form $(x_1, x_2, x_3) \in \mathbb{N}^3$ with $x_3 \leq p_{n+1} = 17$ and $x_i = \lceil \frac{p_i x_3}{p_3} \rceil$ for $i = 1, 2$. To each such point x there corresponds the 2-dimensional vector of residua $(x_1 p_3 - x_3 p_1, x_2 p_3 - x_3 p_2)$. Table 4 includes this information.

By $(w_1, h_1), \ldots, (w_m, h_m)$ we denote all elements in the Hilbert basis of the 2-dimensional subcone $C_2 := \{e^1, (29, 17)^T\}$, except for e^1. In our example we have that $m = 4$ and $(w_1, h_1) = (2, 1)$, $(w_2, h_2) = (7, 4)$, $(w_3, h_3) = (12, 7)$ and $(w_4, h_4) = (29, 17)$.

Following Algorithm 14 we execute Step (1) to find generators of the sets $C^{1,1}, C^{2,2}$ and $C^{3,3}$. Lemma 15 implies that a generator of $C^{1,1}$ is the element $(2, 2, 1)$. Accordingly, we see that a generator of $C^{2,2}$ is the element $(8, 7, 4)$ and that the vectors $(13, 12, 7)$ and $(25, 24, 14)$ define generators of $C^{3,3}$.

x_3	vector	residuum	x_3	vector	residuum
1	**(2,2,1)**	$(4, 5)$	2	$(4, 4, 2)$	$(8, 10)$
3	$(6, 6, 3)$	$(12, 15)$	4	**(8,7,4)**	$(16, 3)$
5	**(9,9,5)**	$(3, 8)$	6	$(11, 11, 6)$	$(7, 13)$
7	**(13,12,7)**	$(11, 1)$	8	$(15, 14, 8)$	$(15, 6)$
9	**(16,16,9)**	$(2, 11)$	10	$(18, 18, 10)$	$(6, 16)$
11	**(20,19,11)**	$(10, 4)$	12	$(22, 21, 12)$	$(14, 9)$
13	**(23,23,13)**	$(1, 14)$	14	**(25,24,14)**	$(5, 2)$
15	$(27, 26, 15)$	$(9, 7)$	16	$(29, 28, 16)$	$(13, 12)$
17	**(30,29,17)**	$(0, 0)$			

Table 1

All vectors that are written in bold together with the unit vectors e^1, e^2 define the minimal Hilbert basis of $C(p)$.

With this information we can start Step (2) of Algorithm 14. There we first determine generators of $C^{2,3} = \{x \in C \cap \mathbb{Z}^3 : \exists \, \mu_2, \mu_3 \in \mathbb{N} \text{ with } p_3 > x_{n+1} = 4\mu_2 + 7\mu_3\}$. The generators of this set coincide with the union of the generators of $C^{2,2}$ and $C^{3,3}$. It remains to find the generators of $C^{1,3}$. To find these, we inspect the generators of $C^{2,3}$ in the following order: first $g^1 = (8, 7, 4)$, then $g^2 = (13, 12, 7)$ and finally $g^3 = (23, 23, 13)$.

For every such element we determine the maximal natural number λ_v such that $\lambda_v g_3^v < g_3^{v+1}$. The corresponding numbers in this case are $\lambda_1 = \lambda_2 = \lambda_3 = 1$.

For $g^1 = (8, 7, 4)$, we determine the minimal natural number μ such that the residuum of the vector $g^1 + \mu(2, 2, 1)$ exceeds the value of $p_3 = 17$ in one component. This gives $\mu = 1$, and the corresponding vector is $(9,9,5)$ that we add to the generators of $C^{1,3}$. For all the other vectors whose 3rd. coordinate is of the form $g_3^1 + \mu < g_3^2$, the associated vectors of residua are greater than the residuum of the point $(9,9,5)$.

Next we proceed to g^2. We know from the previous iteration the generators of $G_2 := \{y \in C^{1,3} : y_3 < g_3^2 = 7\}$. This was the set $\{(2,2,1),(8,7,4),(9,9,5)\}$. On account of (8), G_3 is of the form $G_3 = \{x \in C^{1,3} : \exists \lambda \in \{0,\dots,\lambda_2 = 1\}$ and $y \in G$ such that $x_3 = \lambda g_3^2 + y_3 < g_3^3\}$. In order to find generators of G_3 we have to examine the vector of residua of the points $x \in G_3$. There are precisely two additional vectors for which the vector of residua is incomparable with every vector of residua of the generators of G_2: $(16,16,9)$ and $(20,19,11)$. A set of generators of G_3 is

$$\{(2,2,1),(8,7,4),(9,9,5),(13,12,7),(16,16,9),(20,19,11)\}.$$

Next we proceed to g^3. Because $\lambda_3 = 1$ we need to find the set of all points $x \in G_4 = \{x \in C^{1,3} : \exists \lambda \in \{0,1\}$ and $y \in G_3$ such that $x_3 = \lambda g_3^3 + y_3 < 17\}$ and the vector of residua is incomparable with every vector of residua associated with the generators of G_3. This yields the vector $(25,24,14)$. On account of lemma 17, the sets $C^{1,3}$ and G_4 coincide. The generating set of $C^{1,3}$ consists of the following vectors: $(2,2,1)$, $(8,7,4)$, $(9,9,5)$, $(13,12,7)$, $(16,16,9)$, $(20,19,11)$, $(23,23,13)$, $(25,24,14)$. These vectors plus the vectors e^1, e^2 and p define a Hilbert basis of $C(p)$ in this example.

We want to remark that when $p \in \mathbb{N}^{n+1}$, then $C(p)^*$ partitions into the two cones $\mathbb{R}_{\geq 0}^{n+1}$ and a $\geq$-knapsack cone of the form (cf. Section 2)

$$C(p)_{\geq}^* = \left\{ x \in \mathbb{R}_{\geq 0}^{n+1} : \sum_{i=1}^{n} p_i x_i \geq p_{n+1} x_{n+1} \right\}.$$

There is a similar recursive way of computing a "small" Hilbert basis of $C(p)_{\geq}^*$. Let H_2 be the minimal Hilbert basis of the 2-dimensional cone $C_2 := \{(y,z) \in \mathbb{R}_{\geq 0}^2 : p_n y \geq p_{n+1} z\}$. Then $e^1 \in H$. Let $(h_1, w_1) < \dots < (h_m, w_m)$ be all the elements of $H \setminus \{e^1\}$ ordered in this way. For $i \in \{1,\dots,m-1\}$ we introduce a parameter λ_i to denote the maximal natural number such that $\lambda_i h_i < h_{i+1}$. Then we know that for every $x \in C(p)_{\geq}^* \cap \mathbb{Z}^{n+1}$ with $x_n \neq 0$, x_n can be written as $x_n = \sum_{v=1}^{m} \mu_v h_v$ with $\mu_1, \dots, \mu_m \in \mathbb{N}$. We define, for every $i \in \{1,\dots,m-1\}$, the set $C^i := \{x \in C(p)_{\geq}^* \cap \mathbb{Z}^{n+1} : h_{i+1} > x_n = \lambda h_i + y_n > 0, \ y \in \bigcup_{j=1}^{i-1} C^j, \ \lambda \in \{0,\dots,\lambda_i\}\}$. Realizing that a Hilbert basis of $C(p)_{\geq}^*$ consists of the union of the element $h_m e^n - w_m e^{n+1}$, the set $\{e^1,\dots,e^{n+1}\}$ and a generating set of C^{m-1}, the following recursive procedure to determine the Hilbert basis of $C(p)_{\geq}^*$ becomes obvious.

Algorithm 18. *For $i = 1,\dots,m-1$ determine a set of generators of C^i.*

Finally, we remark that a similar recursion can be formulated to determine the Hilbert basis of a knapsack cone $K_{n,m}$, see Section 2.

Acknowledgement The first author is supported by a "Leibniz Preis" of the German Science Foundation (DFG) awarded to M. Grötschel. The second author is supported by a "Gerhard-Hess-Forschungsförderpreis" of the German Science Foundation (DFG).

References

[CT91]   P. Conti, C. Traverso, *Buchberger algorithm and integer programming*, Proceedings AAECC-9 (New Orleans), Springer LNCS 539, 130 - 139 (1991).

[CFS86]   W. Cook, J. Fonlupt, and A. Schrijver, *An integer analogue of Caratheodory's theorem*, J. Comb. Theory (B) **40**, 1986, 63–70.

[C31]   J.G. van der Corput, *Über Systeme von linear-homogenen Gleichungen und Ungleichungen*, Proceedings Koninklijke Akademie van Wetenschappen te Amsterdam 34, 368 - 371 (1931).

[DGS94]   P. Diaconis, R. Graham, and B. Sturmfels, *Primitive partition identities*, Paul Erdös is 80, Vol. II, Janos Bolyai Society, Budapest, 1-20 (1995).

[GP79]   F.R. Giles and W.R. Pulleyblank, *Total dual integrality and integer polyhedra*, Lineare Algebra Appl. 25, 191 - 196 (1979).

[G1873]   P. Gordan, *Über die Auflösung linearer Gleichungen mit reellen Coefficienten*, Math. Ann. 6, 23 - 28 (1873).

[G75]   J. E. Graver, *On the foundations of linear and integer programming I*, Mathematical Programming 8, 207 - 226 (1975).

[L87]   J.L. Lambert, *Une borne pour les générateurs des solutions entiéres postives d'une équation diophnatienne linéaire*, C.R. Acad. Sci. Paris **305**, Série I, 1987, 39–40.

[S86]   H. E. Scarf, *Neighborhood systems for production sets with indivisibilities*, Econometrica 54, 507 - 532 (1986).

[S90]   A. Sebö, *Hilbert bases, Caratheodory's Theorem and combinatorial optimization*, in Proc. of the IPCO conference, Waterloo, Canada, 1990, 431–455.

Three-Dimensional Meshes are Less Powerful than Two-Dimensional Ones in Oblivious Routing

Kazuo Iwama[*] and Eiji Miyano

Department of Computer Science and Communication Engineering,
Kyushu University, Hakozaki, Higashi-ku, Fukuoka 812-81, JAPAN
E-mail: {iwama, miyano}@csce.kyushu-u.ac.jp

Abstract. This paper shows an important exception against the common perception that three-dimensional meshes are more powerful than two-dimensional ones: Let N be the total number of processors. Then permutation routing over three-dimensional mesh computers needs $\Theta(N^{2/3})$ steps while it takes $\Theta(N^{1/2})$ steps over two-dimensional ones under the following condition: (1) The path of each packet must be determined solely by its initial position and destination, i.e., the algorithm must be oblivious. (2) Each path must be "elementary," i.e., it must be shortest and as straight as possible. Thus the conditions are quite reasonable in practice, under which, little surprisingly, three-dimensional meshes are significantly less powerful than two-dimensional ones for the fundamental network operation.

1 Introduction

It is a common perception that three-dimensional (3D) meshes are more powerful than two-dimensional (2D) ones in the parallel architecture [Lei92]. There is a lot of evidence: If the total number of processors is N, the diameter of the network decreases from $2N^{1/2}$ to $3N^{1/3}$, its bisection width increases from $N^{1/2}$ to $N^{2/3}$, and the degree of communication links in each processor increases from four to six. In fact, we can develop faster algorithms on 3D meshes: Sorting takes $\Theta(N^{1/2})$ steps over 2D meshes while it takes $\Theta(N^{1/3})$ steps over 3D ones [KK92, Kun87, SS86, Sue94]. Matrix multiplication also becomes faster using 3D meshes without losing the work efficiency [Lei92].

In this paper, we show that there is an important exception against this common perspective: Under a practical setting described later, permutation routing needs $\Theta(N^{2/3})$ steps over 3D meshes against $\Theta(N^{1/2})$ steps over 2D meshes. Thus, little surprisingly, 3D meshes are significantly worse than 2D meshes. Our "practical setting" means: (1) The path of each packet must be determined solely by its initial position and destination. (2) The path must be "elementary," namely, it must be shortest and be as straight as possible. In other words,

---

[*] New address: Department of Information Science, Kyoto University, Kyoto 606-01, JAPAN. Email: iwama@kuis.kyoto-u.ac.jp.

each packet can change its direction at most twice in the 3D case. (See Figure 1 for the 2D case. Packet A changes its direction only once but B changes four times until arriving at their destinations.) The first condition (1), called the *oblivious* condition, is very common [e.g., BH85, KKT91]. The second condition is also quite reasonable in practice. Even more rigid routing algorithms are often considered to be reasonable in real-world application, for example, the *dimension-order* routing [BH85, Nes95, IM97] requires all the packets to move in the same direction in each of its three phases, e.g., along x-dimension in the first phase, then along y-dimension in the second phase and finally along z-dimension in the third phase.

Routing on the mesh architecture is one of the most extensively studied topics in parallel computation. There is a huge amount of literature even if we focus our attention on deterministic, permutation routing. First of all, there is a very general lower bound for oblivious routing, i.e., $N^{1/2}/d$ for degree-d networks of any type [KKT91]. This is tight for the hypercube, namely, $N^{1/2}/\log N$ is both upper and lower bounds for oblivious routing over the hypercube [KKT91]. This is also tight within constant factor for the 2D mesh, where $2N^{1/2} - 2$ steps is an upper bound under the most rigid, dimension-order condition [Tom94] and also is a lower bound without any condition [BRST93, Lei92, RT92]. That means we have little room for further research (but that is not the case if the size of buffer in each processor is restricted [e.g., GG95, LMT95, RO92]).

Thus, tight bounds are known for two extremes, namely, $2N^{1/2} - 2$ steps for the 2D mesh and $N^{1/2}/\log N$ steps for the $\log N$-dimensional mesh (= the hypercube). This fact, combined with the common perspective mentioned at the beginning, could lead us to conjecture some $cN^{1/2}$ ($c < 2$) bound for the 3D mesh. The true bound proved in this paper is surprisingly higher, although our proof needs the elementary-path restriction. In many cases of this type of lower bound proof, one has only to consider a single instance, i.e., a set of N packets and their destinations, that includes a lot of different (and usually long) movements [CL93, LMT95, LS91, LS94]. This "hard" instance can produce a sufficient amount of conditions to derive the lower bounds. In our present case, a single instance is not enough to obtain the sufficient amount of conditions. Note that if we use a single instance, it then means that a single packet which has a specific destination is originally placed at a single position. In this paper, each single packet is placed at $N^{1/3}$ different original places. Then one can generate a great many different instances compared to a relatively small number of all the different paths for those packets that is $N^{4/3}$. This is the basic idea of the proof, whose implementation is quite tricky.

There are two similar models. One is called the mesh of buses (MBUS) which is equipped with global busses instead of local connections (See Figure 2; (a) is a 2D mesh computer and (b) is a 2D MBUS.) For this model, we can obtain a tight $\lceil 1.5N^{1/2} \rceil$ bound for permutation routing if we impose so-called "strongly oblivious" condition and a $(1-\varepsilon)N^{1/2}$ lower bound without any condition [IMK96]. Under the same condition we can also prove $\Omega(N^{2/3})$ lower bound for 3D MBUSs. The strong oblivious condition seems to be too strong for higher dimensions but this lower bound was a strong motivation towards the work in this paper. The other model is the one equipped with both local connections and buses, for which $(1 + \varepsilon)N^{1/2} + o(N^{1/2})$ upper and $0.69N^{1/2}$ lower bounds are know for the 2D model and $(2.33 + \varepsilon)N^{1/3} + o(N^{1/3})$ upper and $0.72N^{1/3}$ lower bounds for the 3D model [CL93, LS94, SKR93].

Since routing is the most fundamental operation, its somewhat large inefficiency that is proved here for 3D meshes might cancel their many merits over 2D meshes. To avoid this inefficiency does not appear to be so easy: For example, if we use sorting-type routing algorithms, then we could enjoy the same performance as the sorting (i.e., $O(N^{1/3})$ for 3D meshes). However, it is clearly not accepted as a practical answer (and that is why so many studies have been done for oblivious routing). Another possibility is to add a small amount of adaptiveness into the oblivious condition, like the dimension-exchange approach [BRST93, CLT94]. Unfortunately, we cannot hope too much either since a quite high lower bound has been proved for 2D meshes [CLT94]. A slight possibility exists in exploiting non-elementary moves of packets, but our conjecture is again negative.

2   Models, Problems and Assumptions

For simplicity, the total number of processors in 3D meshes is hereafter denoted by not N but n^3. See Figure 3. The three dimensions are denoted by *x-dimension*, *y-dimension* and *z-dimension*. Let $1 \leq i \leq n$, $1 \leq j \leq n$ and $1 \leq k \leq n$. A *position* is denoted by (i, j, k) (i.e., $x = i$, $y = j$ and $z = k$) and a processor whose position is (i, j, k) is denoted by $P_{i,j,k}$. A connection between the neighboring processors is called a *(communication) link*.

A *packet* is denoted by $[i, j, k]$, $1 \leq i \leq n$, $1 \leq j \leq n$ and $1 \leq k \leq n$, which shows that the destination of the packet is (i, j, k). (A real packet includes more information besides its destination such as its original position and body data, but they are not important within this paper and are omitted.) So we have n^3 different packets in total. An *instance* of *(permutation) routing* consists of a sequence of the n^3 packets $\sigma_1, \sigma_2, \cdots, \sigma_{n^3}$ where σ_1 is originally placed in $P_{1,1,1}$, σ_2 in $P_{2,1,1}$ and so on. Suppose that a packet $\sigma = [i, j, k]$ is originally placed at (l, m, h) such that $(i, j, k) \neq (l, m, h)$. Then a *path* Q of σ is denoted by a sequence of positions $(x_1, y_1, z_1), (x_2, y_2, z_2), \cdots, (x_t, y_t, z_t)$ such that $(x_1, y_1, z_1) = (l, m, h)$, $(x_t, y_t, z_t) = (i, j, k)$ and each consecutive two positions differ in exactly one coordinate (for example, $x_1 = x_2, y_1 = y_2$ but $z_1 \neq z_2$). Q is said to be *elementary* if Q consists of b positions where $b - 1$ ($2 \leq b \leq 4$) is the number of different coordinates between its original position and destination. One can see that if σ moves along path Q and Q is elementary, then σ moves along its shortest route and the number of "turns" is also minimum.

In this paper, we do not discuss constant factors, so some details of the models and algorithms are not important. The one-step computation of each processor consists of (a) receiving a packet from each communication link, (b) executing arbitrarily complicated instructions and (c) if necessary, writing a packet to each link. It is not allowed that two or more packets are written into a single link at the same time. The buffer size is not bounded, i.e., an arbitrary number of packets can stay on a single processor temporarily.

Simply speaking, a routing algorithm, A, determines a path of each packet. In general, A allows each processor P to use every information P has obtained so far to determine the next move of each packet now staying in P. In practice, however, such a general algorithm is not so popular. The most common regulation is called *oblivious*. A is said to be *oblivious* if the path of each packet σ is completely determined by σ's initial position and destination. Also A is said to be *elementary-path* if the path of every packet is elementary.

3 Lower Bounds for Three Dimensional Routing

Theorem 1. *An oblivious and elementary-path routing algorithm on the 3D mesh needs $\Omega(n^2)$ steps.*

Proof. In this proof, we frequently use the $4 \times 4 \times 4$ mesh shown in Figure 3, which will be simply denoted by 4MC. Also, x_i-plane $(1 \leq i \leq n)$ means the two-dimensional plane determined by $x = i$. Similarly for y_j-plane and z_k-plane. For example, the y_1-plane on 4MC is the top, horizontal plane in Figure 3, which includes 16 (n^2 in general) positions $(1, 1, 1), \cdots, (1, 1, 4), (2, 1, 1), \cdots, (4, 1, 4)$.

Recall that our algorithm is oblivious and elementary-path. Suppose, for example, that packet $\sigma = [1, 1, 2]$ (recall that its destination is $(1, 1, 2)$) is originally placed in position $(4, 3, 1)$. Then, by the elementary-path condition, σ must move along a shortest path, and furthermore, the path consists of three sub-paths, one along each dimension. (In this proof, the original positions of almost all packets we use are different from those destinations in all three coordinates, so the packets always behave like this.) For example, σ first moves along y-dimension, i.e., moves from $(4, 3, 1)$ to $(4, 2, 1)$ and then to $(4, 1, 1)$. Similarly, σ goes to $(1, 1, 1)$ along x-dimension and finally arrives at its destination $(1, 1, 2)$ along z-dimension. Thus, the path is determined by the order of dimensions, for example, yxz in this example. In other words, there are six different choices as σ's path, xyz, xzy, yxz, yzx, zxy and zyx, one of which must be determined only by the initial position and destination (the oblivious condition).

Now suppose for contradiction that there exists a routing algorithm, say A, which runs in $f(n)$ steps for any instance, where $f(n)$ satisfies the condition that for any constant $c > 0$, there exists n_0 such that $f(n) < cn^2$ if $n \geq n_0$. (For simplicity we will often say that "A runs in $o(n^2)$ steps.") In the following, we shall construct a set, Σ, of 4^4 (n^4 in general) elements. Each element is a pair of a packet and its original position, denoted by (σ, τ). Therefore, the set Σ can also be regarded as the set of n^4 paths in general. Recall that there are n^3 different packets in total. For each packet σ, Σ includes n different pairs $(\sigma, \tau_1), \cdots, (\sigma, \tau_n)$.

Let Σ_1 be a subset of Σ. Then Σ_1 is said to be *disjoint* if for any two pairs, (σ_1, τ_1) and (σ_2, τ_2), in Σ_1, both $\sigma_1 \neq \sigma_2$ and $\tau_1 \neq \tau_2$ are met. Namely, if Σ_1

is disjoint, it can be regarded as a portion of some single instance. Algorithm A determines $|\Sigma_1|$ paths, which are denoted by $A(\Sigma_1)$ for Σ_1. Now let $N(\Sigma_1, L)$ be the number of paths in $A(\Sigma_1)$ which goes through the communication link L. An apparent fact is that for any disjoint Σ_1 and any link L, $N(\Sigma_1, L)$ must be $o(n^2)$ since at most one packet can flow on a single link each step. Roughly speaking, our argument proceeds as follow: We will construct the set Σ of pairs step by step. Then we consider various disjoint subsets $\Sigma_1 \subseteq \Sigma$ and apply the above fact. This means we can get several conditions that must be met by the set of paths $A(\Sigma)$. Analyzing these conditions, we can finally conclude that $A(\Sigma)$ can contain only $o(n^4)$ different paths, which is a contradiction.

Now let us construct Σ step by step. We will use figures for better exposition using 4MC.

Step 1. See Figure 4(a), which shows first n^2 pairs among the n^4 ones in Σ. Namely, those n^2 packets are placed on the z_1-plane as illustrated in the figure. Note that the destinations of those n^2 packets are all on the y_1-plane, i.e., those n^2 packets placed on z_1-plane should move to y_1-plane. Also one can see that those n^2 pairs are disjoint. Now let $\mathcal{A}_1$ be the number of the packets among those n^2 ones whose first moves are along y-dimension, i.e., which go "up" in the figure. Then as the first condition that should be satisfied by the set $A(\Sigma)$ of paths, we can obtain the following:

Remark 1. Suppose that $\mathcal{A}_1 = cn^2$ for some constant $0 < c \leq 1$. Then by the elementary-path assumption, those cn^2 packets arrive at the n processors whose y and z coordinates are both one, i.e., $P_{1,1,1}$ through $P_{n,1,1}$. Then each of those cn^2 packets must move along x-dimension or z-dimension. However, we can claim that the number of packets which go along x-dimension must be $o(n^2)$. The reason is as follows: Suppose for contradiction that $\Omega(n^2)$ packets go along x-dimension. Notice the important fact that those n^2 pairs shown in Figure 4(a) satisfies the condition that most of them have to move $\Omega(n)$ positions along x-dimension. Hence there are $\Omega(n^2) \times \Omega(n) = \Omega(n^3)$ movements in total along x-dimension. However, the $n - 1$ communication links among the n processors $P_{1,1,1}$ through $P_{n,1,1}$ can handle at most $n-1$ movements each step. That means there is at least one link that has to carry $\Omega(n^2)$ packets, which is a contradiction. Thus we can conclude that $\Omega(n^2)$ packets must go along z-dimension.

Step 2. See Figures 4(b), (c) and (d). n^2 packets as Figure 4(a) are now placed on z_2-plane, z_3-plane and z_4-plane, respectively (z_2-plane through z_n-plane in general). Thus we have so far constructed n^3 pairs among the n^4 pairs in Σ (although they include only n^2 different packets). Here, the difference between Figures 4(a) and (b) (z_1-plane and z_2-plane) should be carefully noticed: Although the whole set of the n^2 packets is the same both in (a) and (b), each position is shifted (cyclically) to the right-down direction. This is the same for between z_2-plane and z_3-plane, etc.

Now we can make the same argument for z_i-plane ($i \geq 2$) as we did for z_1-plane in Remark 1. Let $\mathcal{A}_i$, $2 \leq i \leq n$, be the number of the packets on z_i-plane that move along y-dimension (i.e., up) first. Then, if $A_i = cn^2$ for some constant $c > 0$, then "most" (i.e., $\Omega(n^2)$) of those A_i packets then move not along x-dimension but along z-dimension. Let us take a close look at Figures 4(a)–(d) again. One can see that the four rightmost columns of (a)–(d) constitute x_1-plane, and more importantly, those n^2 packets placed on this x_1-plane are all different and similarly for x_2-plane, etc. Thus, we can make a similar argument like Remark 1 for this set of n^2 disjoint pairs on x_1-plane, for x_2-plane, and so on.

Remark 2. Let B_i $(1 \leq i \leq n)$ be the number of packets on x_i-plane whose first movements are along y-dimension (i.e., up). Then if $B_i = cn^2$ for constant $c > 0$, then again "most" (i.e., $\Omega(n^2)$) of them next move not along z-dimension but along x-dimension. (Be careful: z-dimension and x-dimension here are switched compared to before.)

Now we can prove the following key lemma:

Lemma 1. $\Sigma_i A_i = \Sigma_i B_i = o(n^3)$

Proof. Consider the n^3 pairs so far constructed and the n^3 paths associated with them. Now let $\mathcal{A}$ be the number of the paths among those n^3 ones that go up first. Then one can see that $\Sigma_i A_i = \mathcal{A}$ and $\Sigma_i B_i = \mathcal{A}$. Hence $\Sigma_i A_i = \Sigma_i B_i$. Each of those $\mathcal{A}$ paths goes along x-dimension or z-dimension after it gets to the y_1-plane. The total number of paths that go along x-dimension is bounded by $n \times o(n^2) = o(n^3)$ by Remark 1. The total number of paths that go along z-dimension is also bounded by $n \times o(n^2) = o(n^3)$ by Remark 2. Thus $\mathcal{A} = o(n^3) + o(n^3) = o(n^3)$. $\qquad\square$

Step 3. Recall that we have already constructed only n^3 pairs out of the n^4 pairs in Σ. See Figures 5(a)–(d), which show z_1-plane through z_4-plane of 4MC, respectively, as Figures 4(a)–(d). However, this time the n^2 packets in each of those planes should go to y_2-plane (the second-top, horizontal plane). Thus we have constructed another n^3 pairs.

Step 4. See Figures 6 and 7. The packets of Figure 6 go to y_3-plane and those of Figure 7 go to y_4-plane. Thus we have constructed all the n^4 pairs.

Now we can see what n^4 paths associated with the n^4 pairs in Σ look like. Let Δ^y be the number of those paths that go along y-dimension (i.e., up or down) first.

Lemma 2. $\Delta^y = o(n^4)$.

Proof. Δ^y can be written as $\Delta_1^y + \Delta_2^y + \cdots + \Delta_n^y$ where Δ_1^y is the number of such (first going up) paths that get to y_1-plane. By Lemma 1, $\Delta_1^y = o(n^3)$. Obviously we can obtain similar lemmas for all Δ_i^y's. Hence $\Delta^y = n \times o(n^3) = o(n^4)$. $\qquad\square$

At this moment, we have shown that there are only $o(n^4)$ paths among the n^4 ones in Σ that first go along y-dimension. We next count the number of paths that go along x-direction first. To do so, we shall consider new disjoint subsets of Σ. (One should avoid confusion. This Σ is not changed at all.) See Figures 4(a), 5(a), 6(a) and 7(a). Let Σ_2 be 16 (n^2 in general) pairs which include the leftmost column of Figure 4(a), the second column from the left of Figure 5(a), the third column of Figure 6(a) and the fourth column of Figure 7(a). One can see that the n^2 packets in this Σ_2 are placed on z_1-plane and should go to x_1-plane. Thus we can make a very similar argument as Remark 1, i.e., if those n^2 packets first move along x-dimension (to the left in the figure), then most packets then move to not y-dimension but z-dimension.

We next consider similar disjoint subsets of Σ whose original positions are z_2-, z_3- and z_4-planes and their destinations are x_1-plane. This can be done by taking similar columns from Figures 4(b), 5(b), 6(b) and 7(b) for z_2-plane (but the second column from 4(b), the third column from 5(b), the fourth column from 6(b) and the first column from 7(b)) and so on. Now, let us take the top row from each of those z_1-plane through z_4-plane packets. Then these four top rows constitute y_1-plane and the n^2 packets on that plane should move to x_1-plane. Thus we can make a similar argument as Remark 2. Namely, if those n^2 packets on y_1-plane move along x-dimension first, then the most of them next move along y-dimension.

This consideration is further extended to more disjoint subsets whose destinations are x_2-plane, x_3-plane and so on. (For example, for the n^2 packets moving from z_2-plane to x_3-plane, we can take the following four columns: The rightmost column of Figure 4(b), the leftmost column of Figure 5(b), the second column from left of Figure 6(b) and the third column Figure 7(b).) Now the following lemma is proved in exactly the same manner as before. Let Δ^x be the number of paths among the n^4 ones associated with Σ that go along x-dimension first.

Lemma 3. $\Delta^x = o(n^4)$.

Finally we compute the number of packets that move along z-dimension first: We repeat exactly the same argument once more to obtain n^2 disjoint subsets of Σ each of which now moves from y_i-plane to z_j-plane. Details may be omitted but, for example, the 16 pairs moving from y_1-plane to z_4-plane can be obtained by gathering all the top rows in Figure 4(a), 5(b), 6(c) and 7(d). Let Δ^z be the number of paths among the n^4 ones associated with Σ that go along z-dimension first.

Lemma 4. $\Delta^z = o(n^4)$.

By the assumption $\Delta^x + \Delta^y + \Delta^z = |\Sigma|$. However, Lemmas 2, 3 and 4 imply that $\Delta^x + \Delta^y + \Delta^z = o(n^4)$, which contradicts the fact that $|\Sigma| = n^4$. This concludes the proof of Theorem 1. $\qquad\square$

4 Concluding Remarks

An apparently interesting question is whether or not we are able to remove the elementary-path condition. Our conjecture is yes; here is some intuition: (1) Suppose that the elementary-path condition is a little bit relaxed, i.e., a packet can go off the elementary path within a constant radius $d/2$ (see Figure 8). Then we can make almost the same argument and can obtain the same lower bound. (2) That means each packet has to move along more direct path as shown by the dotted line in Figure 8. However, this seems to create a congestion at processors around the center position if we are given an instance where each packet is to move to its symmetrical position against the center.

One intermediate goal might be to remove only the shortest path condition. Namely we allow one or more intermediate destinations for each packet, but has to follow the elementary path conditions when moving to and from the intermediate destinations.

References

[BRST93] A. Bar-Noy, P. Raghavan, B. Schieber and H. Tamaki, "Fast deflection routing for packets and worms," In *Proc. ACM Symposium on Principles of Distributed Computing*, 75-86 (1993).

[BH85] A. Borodin and J.E. Hopcroft, "Routing, merging, and sorting on parallel models of computation," *J. Computer and System Sciences* 30 (1985) 130-145.

[CL93] S. Cheung and F.C.M. Lau, "A lower bound for permutation routing on two-dimensional bused meshes," *Information Processing Letters* 45 (1993) 225-228.

[CLT94] D.D. Chinn, T. Leighton and M. Tompa, "Minimal adaptive routing on the mesh with bounded queue size," In *Proc. 1994 ACM Symp. on Parallel Algorithms and Architectures* (1994) 354-363.

[GG95] Q.P. Gu and J. Gu, "Two packet routing algorithms on a mesh-connected computer," *IEEE Trans. on Parallel and Distributed Systems*, Vol. 6, No. 4 (1995) 436-440.

[IMK96] K. Iwama, E. Miyano and Y. Kambayashi, "Routing problems on the mesh of buses," *J. Algorithms* 20 (1996) 613-631.

[IM97] K. Iwama and E. Miyano, "Oblivious routing algorithms on the mesh of buses," In *Proc. 11th International Parallel Processing Symposium*, (1997) 721-727.

[KKT91] C. Kaklamanis, D. Krizanc and A. Tsantilas, "Tight bounds for oblivious routing in the hypercube," *Math. Systems Theory 24* (1991) 223-232.

[KK92] C. Kaklamanis and D. Krizanc, "Optimal sorting on mesh-connected processor arrays," In *Proc. 1991 ACM Symposium on Parallel Algorithms and Architectures*, (1991) 17-28.

[Kun87] M. Kunde, "Lower bounds for sorting on mesh-connected architectures," *Acta Informatica*, 24 (1987) 121-130.

[Lei92] F.T. Leighton, *Introduction to Parallel Algorithms and Architectures: Arrays, Trees, Hypercubes*, Morgan Kaufmann (1992).

[LMT95]   F.T. Leighton, F. Makedon and I. Tollis, "A $2n-2$ step algorithm for routing in an $n \times n$ array with constant queue sizes," *Algorithmica* 14 (1995) 291-304.

[LS91]   L.Y.T. Leung and S.M. Shende, "Packet routing on square meshes with row and column buses," In *Proc. 1989 ACM Symp. on Parallel Algorithms and Architectures* (1989) 328-335.

[LS94]   L.Y.T. Leung and S.M. Shende, "On multidimensional packet routing for meshes with buses," *J. Parallel and Distributed Computing 20* (1994) 187-197.

[Nes95]   T. Nesson, *Randomized, oblivious, minimal routing algorithms for multicomputers,* Ph.D Thesis, Harvard University (1995).

[SS86]   C.P. Schnorr and A. Shamir, "An optimal sorting algorithm for mesh-connected computers," In *Proc. ACM Symp. on Theory of Computing,* (1986) 255-263.

[SKR93]   J.F. Sibeyn, M. Kaufmann and R. Raman, "Randomized routing on meshes with buses," In *Proc. European Symposium on Algorithms,* (1993) 333-344.

[Sue94]   T. Suel, "Routing and sorting on fixed topologies," Ph.D Thesis, The University of Texas at Austin (1994).

[RO92]   S. Rajasekaran and R. Overholt, "Constant queue routing on a mesh," *J. Parallel and Distributed Comput.,* 15 (1992) 160-166.

[RT92]   S. Rajasekaran and T. Tsantilas, "Optimal routing algorithms for mesh-connected processor arrays," *Algorithmica* 8 (1992) 21-38.

[Tom94]   M. Tompa, *Lecture notes on message routing in parallel machines,* Technical Report # 94-06-05, Department of Computer Science and Engineering, University of Washington (1994).

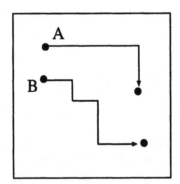

Fig. 1. A: Elementary path   B: Non-elementary path

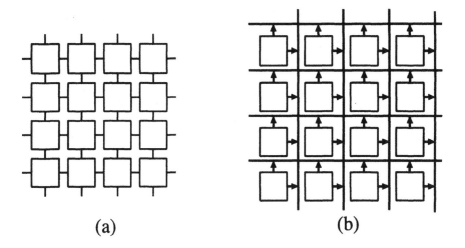

(a)   (b)

Fig. 2. (a) 2D mesh computer   (b) 2D MBUS

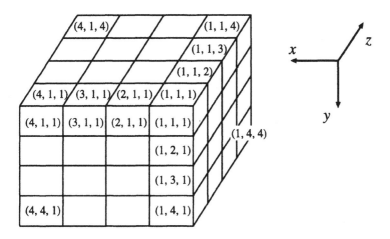

Fig. 3. $4 \times 4 \times 4$ - 3D mesh

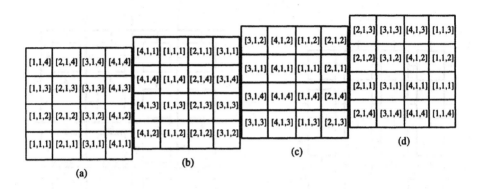

Fig. 4. (a) z_1-plane (b) z_2-plane (c) z_3-plane (d) z_4-plane

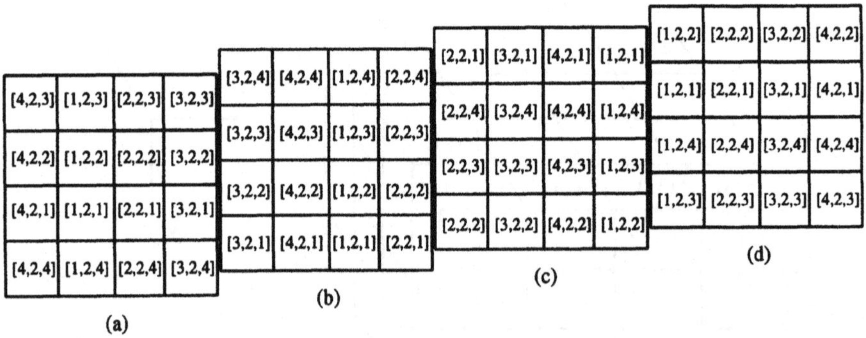

Fig. 5. (a) z_1-plane (b) z_2-plane (c) z_3-plane (d) z_4-plane

Fig. 6. (a) z_1-plane (b) z_2-plane (c) z_3-plane (d) z_4-plane

Fig. 6 (a) z_1-plane:

[3,3,2]	[4,3,2]	[1,3,2]	[2,3,2]
[3,3,1]	[4,3,1]	[1,3,1]	[2,3,1]
[3,3,4]	[4,3,4]	[1,3,4]	[2,3,4]
[3,3,3]	[4,3,3]	[1,3,3]	[2,3,3]

(b) z_2-plane:

[2,3,3]	[3,3,3]	[4,3,4]	[1,3,3]
[2,3,2]	[3,3,2]	[4,3,2]	[1,3,2]
[2,3,1]	[3,3,1]	[4,3,1]	[1,3,1]
[2,3,4]	[3,3,4]	[4,3,4]	[1,3,4]

(c) z_3-plane:

[1,3,4]	[2,3,4]	[3,3,4]	[4,3,4]
[1,3,3]	[2,3,3]	[3,3,3]	[4,3,3]
[1,3,2]	[2,3,2]	[3,3,2]	[4,3,2]
[1,3,1]	[2,3,1]	[3,3,1]	[4,3,1]

(d) z_4-plane:

[4,3,1]	[1,3,1]	[2,3,1]	[3,3,1]
[4,3,4]	[1,3,4]	[2,3,4]	[3,3,4]
[4,3,3]	[1,3,3]	[2,3,3]	[3,3,3]
[4,3,2]	[1,3,2]	[2,3,2]	[3,3,2]

Fig. 7. (a) z_1-plane (b) z_2-plane (c) z_3-plane (d) z_4-plane

Fig. 7 (a) z_1-plane:

[2,4,1]	[3,4,1]	[4,4,1]	[1,4,1]
[2,4,4]	[3,4,4]	[4,4,4]	[1,4,4]
[2,4,3]	[3,4,3]	[4,4,3]	[1,4,3]
[2,4,2]	[3,4,2]	[4,4,2]	[1,4,2]

(b) z_2-plane:

[1,4,2]	[2,4,2]	[3,4,2]	[4,4,2]
[1,4,1]	[2,4,1]	[3,4,1]	[4,4,1]
[1,4,4]	[2,4,4]	[3,4,4]	[4,4,4]
[1,4,3]	[2,4,3]	[3,4,3]	[4,4,3]

(c) z_3-plane:

[4,4,3]	[1,4,3]	[2,4,3]	[3,4,3]
[4,4,2]	[1,4,2]	[2,4,2]	[3,4,2]
[4,4,1]	[1,4,1]	[2,4,1]	[3,4,1]
[4,4,4]	[1,4,4]	[2,4,4]	[3,4,4]

(d) z_4-plane:

[3,4,4]	[4,4,4]	[1,4,4]	[2,4,4]
[3,4,3]	[4,4,3]	[1,4,3]	[2,4,3]
[3,4,2]	[4,4,2]	[1,4,2]	[2,4,2]
[3,4,1]	[4,4,1]	[1,4,1]	[2,4,1]

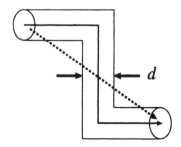

Fig. 8. Relaxed elementary path

Fault-Tolerant Real-Time Scheduling

Bala Kalyanasundaram and Kirk Pruhs

Computer Science Dept., University of Pittsburgh, Pittsburgh, PA 15260, USA.
{kalyan,kirk}@cs.pitt.edu, http://www.cs.pitt.edu/{~kalyan,~kirk}

Abstract. We use competitive analysis to study how to best use redundancy to achieve fault-tolerance in online real-time scheduling. We
show that the optimal way to make use of spatial redundancy depends
on a complex interaction of the benefits, execution times, release times,
and latest start times of the jobs. We give a randomized online al-
gorithm whose competitive ratio is $O(\log \Phi \log \Delta \frac{\log^2 n \log m}{\log \log m})$ for transient faults. Here n is the number of jobs present in the system at
any one time, m is the number of processors, Φ is the ratio of maximum value density of a job to the minimum value density of a job,
and Δ the ratio of the longest possible execution time to the shortest possible execution time. We show that this bound is close to opti-
mal by giving an $\Omega(\log \Delta \Phi(\frac{\log m}{\log \log m \log \log(m \Delta \Phi)})^2)$ lower bound on the
competitive ratio of any randomized algorithm. In the case of perma-
nent faults, there is a randomized online algorithm that has a com-
petitive ratio of $O(\log \Phi \log \Delta \frac{\log m}{\log \log m})$. We also show a lower bound of
$\Omega(\log \Delta \Phi \frac{\log m}{\log \log(m \Delta \Phi)})$ on the competitive ratio for interval scheduling
with permanent faults.

1 Introduction

We use competitive analysis to study how to best utilize redundancy to achieve
fault tolerance in the setting of online real-time scheduling. In an online scheduling environment jobs arrive over time and must be scheduled without knowledge
of future jobs. In real-time scheduling each job has a deadline, and a job must be
finished by its deadline to be of full use to the system. Fault-tolerance is achieved
by having multiple processors performing the same job so that if any one of these
processors does not fault, the job will be completed. The question we examine
is, "How should a fixed number of processors be allocated among the jobs". In
almost all of the fault tolerant settings that we are aware of, the answer to this
question is essentially that each equivaluable job should get an equal number of
processors. We show that with job deadlines and laxity, the optimal online allocation depends on a complex interaction of various characteristics of the jobs,
including release time, length and latest start times.

As just one concrete application we cite the problem of routing messages in
a real-time network. In this setting the jobs are messages, and the processors
are channels between two nodes in a network. A fault would then be a failure
on some link or switch in a channel.

1.1 Problem Statement

We adopt best-effort firm real-time scheduling, a standard models of real-time computation [2, 3, 4, 7, 9, 10, 13], as our model of real-time computation. The setting is a collection $\mathcal{P} = \{P_1, \ldots P_m\}$ of unit speed processors. The scheduler sees over time a collection $\mathcal{J} = \{J_1, \ldots J_n\}$ of jobs. Each job J_i has a *release time* r_i, a processing time x_i, a deadline d_i, and a benefit b_i. The scheduler is unaware of J_i until time r_i, at which time the scheduler additionally learns x_i and b_i. Thus the scheduler may schedule J_i any time after time r_i. If some processor completes J_i by time d_i then the scheduler receives a benefit of b_i, otherwise no benefit is gained from this job. Note that we allow many processors to run a job J_i simultaneously and that these processors need not have begun execution of J_i at the same time. The problem is to maximize the sum of the benefits of the jobs completed by their deadlines.

Interval scheduling is special case of best effort firm real-time scheduling where the *laxity* $l_i = d_i - x_i - r_i$ of each job J_i is equal to zero. The *value density* v_i of J_i is b_i/x_i. We assume that the minimum execution time and minimum value density of any job are 1, that the maximum value density is Φ, and that the maximum execution time is Δ. We say an instance is *static* if all release times are 0. The *latest start time* of a job J_i is $d_i - x_i$.

One gets different models depending on the type of preemption allowed. The three standard options are nonpreemptive, preemptive, and preemptive with restart. A scheduling algorithm is *nonpreemptive* if a processor P_j may not interrupt a job J_i once it has begun execution of J_i, is *preemptive* if it is allowed to at no cost abandon J_i and later restart J_i from the point of suspension, and is *preemptive with restart* if it can abandon J_i but must restart J_i from the beginning if it wants to return to J_i.

In this paper we consider only fail-stop faults. A processor P_j experiences a *fail-stop fault* during a period T of time if P_j is not capable of performing any work during the period T, that is, it is as if the processor was switched off during this period. Thus, we do not consider malicious faults. A *permanent fault* is one from which the processor never recovers, otherwise, the fault is *transient*. A *checkpointing system* may at no cost save the state of a job, which is called a checkpoint. Then after a job experiences a fault it may be restarted from the most recent checkpoint, while a *noncheckpointing system* must restart the job from the beginning. For example, a file transfer may be checkpointing at the system level, but (as ftp users are painfully aware) restart at the users level. Note that the power of the system/scheduler to checkpoint is essentially the same as the power to preempt. So we are left with three reasonable models of a fault-tolerant version of best-effort firm real-time scheduling: 1) preemptive and checkpointing, 2) nonpreemptive and noncheckpointing, and 3) preemptive with restart and noncheckpointing.

The competitive ratio of a randomized algorithm $\mathcal{A}$ is the supremum over all instances $\mathcal{I}$ of the ratio of the optimal off-line benefit for $\mathcal{I}$ over the expected benefit achieved by $\mathcal{A}$ on input $\mathcal{I}$. We assume an oblivious adversary, that is, the adversary may not modify $\mathcal{I}$ in response to the outcome of a random event in

$\mathcal{A}$. We assume that processor faults are not caused by the particular job running on that processor. That is the faults are specified in $\mathcal{I}$ by a list with entries of the form (P_i, t, α), meaning that processor P_i is faulty between time t and time $t + \alpha$.

In appendix A, we show that the competitive ratio of every deterministic algorithm is linear in the number of processors for all three reasonable models. We also show in appendix A, that the competitive ratio of every randomized algorithm for preemptive and checkpointing, and for nonpreemptive and noncheckpointing systems, is linear in the number of jobs. *For the rest of this paper we devote our attention to randomized algorithms for systems where the scheduler is preemptive with restart and noncheckpointing.*

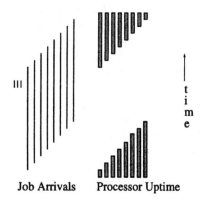

Fig. 1. A Hard Instance for Greedy Algorithms

For preemptive with restart schedulers, figure 1 shows an instance of interval scheduling that is particularly hard for many natural algorithms. Jobs are represented as vertical lines with the length of the line being the execution time and the bottom of the line representing the release date. Processors are represented by vertical columns that are darkened when the processor is faulty. The benefit of every job is 1. From any particular processor's point of view, after recovering from a fault at time s, the processor should intuitively prefer jobs with earliest deadline among those released after s since these jobs are least likely to be affected by future faults. Yet if there are few faults, you don't want every processor picking the early deadline jobs or there would not be enough distinct jobs to be competitive. So the most natural algorithms would be to have some monotone nonincreasing probability distribution on the jobs ranked by increasing deadline. By making the number of short jobs $\Theta(\sqrt{m})$, and the number of long jobs m, this instance shows that such algorithms will have competitive ratio $\Omega(\sqrt{m})$.

1.2 Our Results

Recall that we only consider systems with preemptive with restart and noncheckpointing schedulers that experience only fail-stop faults.

In section 2, we give randomized online algorithm CLASSIFY that has a competitive ratio of $O(\log \Phi \log \Delta \frac{\log^2 n \log m}{\log \log m})$. Here n is really the number of jobs present in the system at any one time, which presumably is generally not too much larger than m, or the offline performance will also be poor. Note that the "n" term in this bound can be replaced by "$m + f$", where f is the maximum number of faults that can occur during a period of time Δ. In section 3, we give an $\Omega(\log \Phi \Delta (\frac{\log m}{\log \log m \log \log (m \Phi \Delta)})^2)$ lower bound on the competitive ratio of any randomized algorithm.

In the case of permanent faults, there is a randomized online algorithm with competitive ratio $O(\log \Phi \log \Delta \frac{\log m}{\log \log m})$. We show that this upper bound is not too far off by giving a lower bound of $\Omega(\log \Phi \Delta \frac{\log m}{\log \log (m \Phi \Delta)})$ on the competitive ratio for interval scheduling with permanent faults.

These results show that the optimal way for a online scheduler in a real-time environment to use redundancy is very nonintuitive. This is in sharp contrast to fault-tolerant scheduler of non-real-time jobs, where it is known that the best way for an online scheduler to use redundancy is to devote an equal number of processors to each job [6]. These results also show that, for real-time scheduling transient, faults are significantly more damaging than permanent faults to the performance of the online scheduler. For scheduling jobs without deadlines, it was known that transient and permanent faults are essentially equally damaging [6].

1.3 Previous Results

We first survey results for online best-effort firm real-time scheduling without faults. All of the following results are for a single processor with preemption unless stated otherwise. Let Λ be the ratio of the largest benefit of a job to the minimum benefit of a job. There are 4-competitive algorithms for the cases $\Phi = 1$ or $\Delta = 1$, and this is optimal [3, 4, 13]. The optimal competitive ratio is $\Theta(\Phi)$ [3, 4, 9]. In [3] a 2-competitive algorithm is given for the case $\Phi = 1$ and $m = 2$. This immediately yields a 2-competitive randomized algorithm for $\Phi = 1$ and $m = 1$. A 3/2 lower bound on the competitive ratio for this case is given in [8]. In [10] a deterministic algorithm for multiple processors is given with competitive ratio $O(\log \Phi)$ for large m, and this is shown to be optimal. In [8] a randomized algorithm is given with competitive ratio $O(\min(\log \Phi, \log \Delta))$, and this is shown to be optimal. In [5] a 1-competitive algorithm is given for the case interval scheduling with $\Lambda = 1$. In [2], a nonconstant deterministic lower bound for the case that $\Lambda = 1$ is given. For nonpreemptive interval scheduling with $\Phi = 1$, [11] gives a randomized algorithm that is $O(\log^{1+\epsilon} \Delta)$-competitive, and give a general randomized lower bound of $\Omega(\log \Delta)$ on the competitive ratio. A related model where benefit is equal to job length can be found in [1].

Faults completely change the nature of these scheduling problems. There seems to be little research that uses competitive analysis to study fault-tolerance.

In [6] fault-tolerant scheduling of jobs without deadlines is studied, and optimal deterministic and randomized algorithms are given for online problems involving minimizing the make-span and minimizing the average flow time. Given λ permanent faults, the optimal deterministic competitive ratio for makespan is $\Theta(\max(\frac{\log m}{\log(\frac{m \log m}{\lambda})}, \frac{m}{\eta}))$ and the optimal randomized competitive ratio is $\Theta(\max(\log^* m - \log^* \frac{m}{\lambda}, \frac{m}{\eta}))$. Here $\eta = m - \lambda$. For transient faults, the optimal deterministic competitive ratio is $\Theta(\max(\frac{\log m}{\log(\frac{m \log m}{\lambda})}, \frac{\lambda}{m}))$, and the optimal randomized competitive ratio is $\Theta(\max(\log^* m - \log^* \frac{m}{\lambda}, \frac{\lambda}{m}))$. For average flow time the optimal (deterministic and randomized) competitive ratios are $\Theta(\frac{m}{\eta})$ for permanent faults, and $\Theta(\max(\frac{\lambda}{m}, 1))$ for transient faults. In all of these problems the optimal way for the online scheduler to use redundancy was essentially to devote an equal number of processors to each job.

2 Upper Bounds

We assume that no pair of events (the release of a job, the deadline a job, or the occurrence of a fault etc.) happen at exactly the same time. All results can be extended to handle these cases, at the expense of slightly complicating the description of the algorithms and the proofs.

2.1 Algorithm Classify

Algorithm CLASSIFY uniformly at random selects an i in the range $1 \ldots \log \Delta$, and a j in the range $1 \ldots \log \Phi$, and only considers jobs with length $\Theta(2^i)$ and value density $\Theta(2^j)$. Hence, from now on we will assume that the execution time x_i of each job J_i satisfies $1 \leq x_i < 2$, and that the benefit of each job is 1. This assumption adds a $\Theta(\log \Phi \log \Delta)$ multiplicative factor to the competitive ratio.

Algorithm CLASSIFY then uniformly at random selects an integer s from $0, 1, 2$. Algorithm CLASSIFY then partitions time into rounds. Round r, $r \geq 0$, occurs between time $3r + s$ and $3(r + 1) + s$. During round r, CLASSIFY considers all unfinished jobs with release time before $3r + s$ and latest start time greater than $3r + s$, and those jobs with release times between $3r + s$ and $3r + s + 1$. During round r, no job will be assigned a processor after time $3r + s + 1$. Hence, all processors will be idle by time $3(r + 1) + s$, when round $r + 1$ starts. Define a time t to be a *selection time* if some processor is assigned a job at time t. A selection time is either at the start of a round or just after a processor recovers from a fault. During each round, CLASSIFY uses an algorithm T1R which we present in the next subsection.

2.2 One Round Problem

We say a job J_i is *delayed* if it is completed by the adversary at some time later than $r_i + x_i$. The online algorithm T1R, with probability $1/2$, runs an algorithm NO-DELAY, which will be competitive with the non-delayed jobs completed by the adversary. And with probability $1/2$, T1R runs an algorithm DELAY, which will be competitive with the delayed jobs.

Before we present algorithms DELAY and NO-DELAY we present an algorithm P1R that will be used as a subroutine in both DELAY and NO-DELAY.

Consider a fixed round. We redefine time 0 to be the start of this round, so that all jobs are released before time 1 and can be completed before time 3. The *earliest completion time* $e_j(s)$ of a job J_j relative to time s is $\max(s, r_j) + x_j$. We call a collection of jobs *agreeable* over a time period $[t_1, t_2]$ if the rankings of the jobs by their earliest completion time relative to t is the same for all t in the interval $[t_1, t_2]$. We now give the algorithm P1R for agreeable jobs.

Algorithm P1R: Let $\mathcal{J}$ be the collection of n jobs and S be the collection of m processors. Let $\mathcal{J}'$ be a maximal subset of $\mathcal{J}$ such that $|\mathcal{J}'| \leq m$ and no job in $\mathcal{J} - \mathcal{J}'$ has an earlier completion time relative to time 0 than any job in $\mathcal{J}'$. Let c be the largest integer such that $c^c \leq m$. Note that $c = \Theta(\log m / \log \log m)$. For $1 \leq j \leq c^c$ let $\delta(j) = 1/(c^k(c+1))$ where k is the unique integer that satisfies $c^{k-1} < j \leq c^k \leq m$. Assume that a processor P_i needs to select a new job at a time t. First, P_i selects an integer γ_i uniformly at random between 0 and c, inclusive, and then selects an integer α_i uniformly at random between 1 and c^{γ_i}. P1R maintains the invariant that P_i has run the job in $\mathcal{J}'$ with the α_ith earliest completion time, relative to time t, exclusively since the latest selection time on P_i. The algorithm guarantees that at all times, the job with the jth earliest completion time relative to time t has been run exclusively since the latest selection time that job on P_i with probability at least $\delta(j)$. ∎

Theorem 1. *The competitive ratio of* P1R *is* $O((\log m / \log \log m))$-*competitive if the jobs are agreeable over some interval covering the collection of selection times.*

Proof. Assume that the adversary completes x (≥ 2) jobs that online has not completed yet. Let g be the largest integer such that $c^g \leq x/2$. Also, let K be the set of c^g jobs with earliest relative completion times. Among the processors that the adversary used to complete x jobs, let D with the set of c^g processors that ran for longest time without fault in this round. Note that each processor in D can complete every job in K. Let $P(i, j)$ be the probability that the online algorithm processed J_j exclusively on P_i since the latest selection time on P_i. For each $P_i \in D$ and each $J_j \in K$, $P(i, j) \geq 1/(c^g(c+1))$, and hence for each $J_j \in K$, $\sum_{P_i \in D} P(i, j) \geq x/(2c^g(c+1))$. Note that by the choice of g, $x/(2c^g(c+1)) \leq 1$. By summing up over all the c^g jobs in K we determine that the benefit gained by the online algorithm is at least $x/2(c+1)$. ∎

Note that this gives an $O(\log m / \log \log m)$-competitive algorithm for permanent faults since the only selection time in the case of permanent faults is at the start of a round.

Before proceeding, we need to introduce the following definitions. Let $S = \{s(1), \ldots, s(\mu)\}$ be a some arbitrary collection of points with $s(0) = 0 < s(1) < s(2) < \ldots < s(\mu) < s(\mu+1) = 3$. Without loss of generality, we assume $\mu+1$ is a perfect power of 2. We define $E_S(a, b)$, $0 \le b \le \log_2(\mu+1)$ and $0 \le a \le (\mu+1)/2^b$, to be the time period from $s(a2^b)$ to $s((a+1)2^b)$. One can think of $E_S(a, b)$ as being the ath node of height b in a complete binary tree that contains $\mu+1$ time periods of the form $[s(k), s(k+1)]$, $0 \le k \le \mu$, at the leaves, and where each interior node is the union of time periods of the children of that node. Define the *time tree* of S, denoted $time(S)$ to be the collection of all $E_S(a, b)$'s.

We define the *forward time cover* of an $s(i)$, denoted by $FTC_S(i)$, to be a minimum cardinality subset of $time(S)$ whose union is $[s(i), 3]$. Similarly, we define the *backward time cover* of an $s(i)$, denoted by $BTC_S(i)$, to be a minimum cardinality subset of $time(S)$ whose union is $[0, s(i)]$. Note that each forward time cover can be computed efficiently by iteratively adding the longest interval beginning at the first uncovered moment in time. A backward time cover can be computed similarly by going backward in time. Figure 2 shows an example of a time tree. In this figure $FTC_S(3) = \{E(3, 0), E(1, 2)\}$, and $BTC_S(3) = \{E(2, 0), E(0, 1)\}$.

We need the following facts about these time covers. The cardinality of each forward time cover and each backward time cover is at most $O(\log \mu)$. For each time t, there exist at most $O(\log \mu)$ different $E_S(a, b) \in time(S)$ such that $t \in E_S(a, b)$.

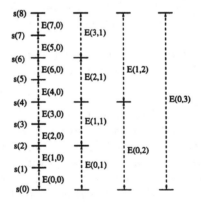

Fig. 2. A Time Tree

We are now ready to present the algorithm NO-DELAY and DELAY.

Algorithm NO-DELAY: Assume that this algorithm is being run on a collection $\mathcal{J}$ of n jobs. Let R be the set of all the release times, and renumber the jobs in $\mathcal{J}$ so the $r_i < r_j$ for $i < j$. A processor needs to select a job to run at time 0 and after each time that it recovers from a fault. Assume that a processor P_i needs to select a job at time t satisfying $r_k < t < r_{k+1}$. P_i uniformly at random selects an $E_R(a, b) \in FTC_R(k+1)$. NO-DELAY then runs the algorithm P1R on jobs with release times interval $E_R(a, b)$. ∎

Note the following: 1) One can determine online when $E_R(a, b)$ begins and ends by counting release times and latest start times. 2) The jobs released during the period $E_R(a, b)$ are agreeable over the period of selection times of the processors that select $E_R(a, b)$.

Algorithm DELAY: Assume that this algorithm is being run on a collection $\mathcal{J}$ of n jobs. Let R be the set of all the release times, and renumber the jobs in $\mathcal{J}$ so the $r_i < r_j$ for $i < j$. A processor needs to select a job to run at time 0 and after each time that it recovers from a fault. Assume that a processor P_i needs to select a job at time t satisfying $r_k \leq t < r_{k+1}$. P_i uniformly at random selects an $E_R(a, b) \in BTC_R(k)$. Let L denote the set of latest start times for jobs released during the period $E_R(a, b)$. Suppose k' be the number of jobs with latest start times strictly less than t. P_i uniformly at random selects an $E_L(c, d) \in FTC_L(k' + 1)$. DELAY then runs the algorithm P1R on jobs with release times interval $E_R(a, b)$, and all jobs with latest start dates in $E_L(c, d)$. Note that the set of jobs with latest start times in $FTC_L(k' + 1)$ are the jobs that P_i might be able to finish if P_i doesn't fault too soon after time t. ∎

Note that the jobs released during $E_R(a, b)$, with latest starts times in a particular $E_L(c, d)$ are agreeable over the period of selection times of processors that pick $E_R(a, b)$ and $E_L(c, d)$.

Theorem 2. *The competitive ratio of* T1R *is* $O(\log^2 n \log m / \log \log m)$ *for a fixed round.*

Proof. Assume that J_j is a delayed job finished by the adversary on processor P_i. With probability $\Omega(1/\log n)$ T1R selected an $E_R(a, b)$ that included r_j. Similarly with probability $\Omega(1/\log^2 n)$ T1R also picked the $E_L(c, d)$ that contained the latest start time of J_j. The result then follows from theorem 1 and the fact that the selected set of jobs and the processors that pick them are agreeable. Note that we are not over-counting the benefit we achieve from any one job since each job will be included in at most $O(\log^2 n)$ different $E_R(a, b)$, $E_L(c, d)$ pairs. A similar argument shows that NO-DELAY is $O(\log n \log m / \log \log m)$ competitive with respect to non-delayed jobs. ∎

3   Lower Bounds

We first consider the following online game that will arise repeatedly in our lower bound constructions. The online player randomly selects an integer a with probability p_a, and the adversary selects an integer b, $1 \leq a, b \leq \ell$. The payoff to the online algorithm is $\sum_{i=1}^{b-1} p_i c^i$. We call a choice of b *weak* if the payoff is no more than $3c^b/c\ell$, and $p_b \leq 4/\ell$.

Lemma 3. *At least $\ell/4$ choices for b are weak.*

3.1 Permanent Fault Case

Lemma 4. *The competitive ratio of every online randomized algorithm is* $\Omega(\log m / \log \log m)$ *even if all jobs have benefit 1.*

Proof. The instance consists of m intervals with the property that no interval contains another and the release times of all jobs precede the deadline of any job. More formally, jobs $J_1, \ldots, J_m$ satisfy $r_i < r_j$ and $d_i < d_j$ for $i < j$, and $d_1 > r_m$. We conceptually partition the jobs into groups $G_1, \ldots G_c$, where $c = \Theta(\log m / \log \log m)$ and G_i contains the c^{i-1} jobs with earliest deadline that aren't in $G_1, \ldots, G_{i-1}$.

The adversary will fault all but c^b processors $\mathcal{P}$ immediately after the release of J_m. The remaining processors will fault after latest deadline of a job in G_b. Now let p_a be the total probability that a processor P_i selects a job in G_a. By lemma 3, at least $c/4$ choices of b will be weak for each processor. Hence, there is at least one group G_b that is weak for at least $m/4$ processors. $\mathcal{P}$ is then selected to be any subset of size c^b of these $m/4$ processors. ∎

Lemma 5. *The competitive ratio of every online randomized algorithm is* $\Omega(\log \Delta \Phi \frac{\log m}{\log \log m \log \log(m \Phi \Delta)})$.

Proof. We consider instances where $\Lambda = \Delta \Phi$. Let $c^c = \Theta(m)$, so $c = \Theta(\frac{\log m}{\log \log m})$. If $\Lambda \geq 2^{\sqrt{m}}$, a lower bound of $\Omega(\log \Lambda)$ follows easily, and if $\log \Lambda \leq c$ the bound follows from lemma 4. So assume neither of these conditions hold. Let $c^\ell = \Lambda$ define ℓ, and hence $\ell = \Theta(\log \Lambda / \log \log m)$. The collection of jobs are of the form $J_{i,b}$ where $0 \leq b < \ell$ and $0 \leq i < m/\ell$. The benefit of a job $J_{i,b}$ is c^b. The deadline (release) of J_{i_1,b_1} occurs before the deadline (release) of J_{i_2,b_2} if $b_1 < b_2$ or if $b_1 = b_2$ and $i_1 < i_2$. Every job is released before the deadline of any other job.

Let G_i to be the set of jobs with benefit c^i. For each processor, at least $\ell/4$ groups are weak. Hence, there exists a group G_b that is weak for at least a collection $\mathcal{P}$ of $m/4$ processors. The adversary faults all processors not in $\mathcal{P}$ immediately after the arrival of the last job. The adversary for G_b and $\mathcal{P}$ is now exactly the same as the adversary for the unit point case. All processors fault before the earliest deadline of a job in G_{b+1}. Assume the adversary completes α job from G_b. Then we know the online algorithm accumulated benefit at most $O(\alpha c^b / c\ell)$ from groups $G_1, \ldots, G_{b-1}$ benefit at most $O(\alpha c^b / c\ell)$ from jobs in G_b. ∎

3.2 Transient Fault Case

Generally speaking, in the interval scheduling problem a processor should prefer jobs with early deadlines and should also prefer jobs with early release dates. The following lemma shows that these two goals are essentially incompatible.

Lemma 6. *The competitive ratio of every randomized algorithm is* $\Omega(\log^2 m / (\log \log m)^2)$ *even if all jobs have benefit 1 and laxity 0.*

Proof. Let $c = \Theta(\log^2 m/(\log\log m)^2)$, and $k = \Theta(\sqrt{c})$. Jobs are partitioned into k groups $G_1, G_2, \ldots, G_k$ where there are exactly c^{i-1} jobs in the group G_i. We imagine that the jobs in the G_i's are nodes of a degree c depth k tree. G_1 is the root of the tree, and more generally, G_i is the collection of jobs/nodes of depth i in the tree. All jobs in G_i have deadlines before the deadline of any every in G_{i+1}. For any two jobs J_a and J_b in a G_i, if $r_a < r_b$ then $d_a < d_b$. Finally, the postorder traversal of the tree gives the release time of jobs. So if J_x is a node in the tree with subtrees $T_1, \ldots, T_c$ the jobs with earliest release time are those in T_1, followed by those in T_2, ..., followed by those in T_c, followed by J_x. See figure 3.

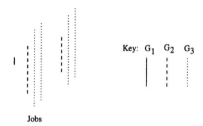

Fig. 3. Lower bound instance for $c = 2$ and $k = 3$

The adversary is going to complete $\Theta(c^b)$ jobs in one group G_b. The adversary will fault all but a collection $\mathcal{P}$ of $\Theta(c^b)$ processors immediately after the arrival of the last job. The processors in $\mathcal{P}$ will have the property that for each $P \in \mathcal{P}$ the expected benefit accrued by P from jobs in $G_1, \ldots, G_{b-a}$ is $c^b/c^{3/2}$. The processors in $\mathcal{P}$ will fault sometime before the earliest deadline of a job in G_{b+1}. Now consider only the jobs in G_b. We renumber the jobs in G_b by release dates. So J_a is now the job in G_b with the ath earliest release date. Denote by P_a the set of $\sqrt{m}$ processors that wake up between r_{a-1} and r_a. We construct $(b-1)$ different strategies for the adversary. In strategy d, $1 \leq d \leq b-1$, each processor P in each P_a either faults so that P can only complete the next c^{d-1} jobs released after r_a (i.e. jobs $J_a, J_{a+1}, \ldots J_{a+c^{d-1}-1}$), or P can complete no jobs. This is done in such a way that the adversary can complete $\Theta(c^b)$ jobs. So on average the processors in each P_a must on average put probability $1/c^d$ on each of the c^{d-1} jobs with earliest deadline that are released after time r_a if the online algorithm is to stay c-competitive. Since on average each processor must spend probability i/c on each G_i, and $\sum_{i=1}^k i/c = \Theta(1)$, one can set the constants so that this exhausts the online algorithm's probability. ∎

The following lemma essentially shows that execution times and latest start dates are incompatible parameters.

Lemma 7. *The competitive ratio of every randomized algorithm is $\Omega(\log^2 m/(\log\log m)^2)$ even if all jobs have benefit 1 and static release dates.*

Lemma 8. *The competitive ratio of every randomized algorithm is $\Omega(\log \Delta\Phi(\frac{\log m}{\log\log m \log\log(m\Phi\Delta)})^2)$.*

References

1. A. Bar-Noy, R. Canetti, S. Kutten, Y. Mansour, and B. Schieber, "Bandwidth Allocation with Preemption", STOC, 1995.
2. S. Baruah, J. Harita, and N. Sharma, "On-line scheduling to maximize task completions", *IEEE Real-time Systems Symposium*, 1994.
3. S. Baruah, G. Koren, D. Mao, B. Mishra, A. Raghunathan, L. Rosier, D. Shasha, and F. Wang, "On the competitiveness of on-line real-time task scheduling", *Journal of Real-Time Systems*, 4, 124–144, 1992.
4. S. Baruah, G. Koren, B. Mishra, A. Raghunathan, L. Rosier, and D. Shasha, "On-line scheduling in the presence of overload", IEEE FOCS, 101–110, 1991.
5. U. Faigle and W. Nawijn, "Note on scheduling intervals online", *Discrete Applied Math.*, **58**, 13–17, 1995.
6. B. Kalyanasundaram, and K. Pruhs, "Fault-tolerant scheduling", STOC, 115–124, 1994.
7. B. Kalyanasundaram and K. Pruhs "Speed is as powerful as clairvoyance", FOCS, 1995.
8. B. Kalyanasundaram and K. Pruhs "Real-time scheduling with fault-tolerance", Technical report, Computer Science Dept. University of Pittsburgh.
9. G. Koren, and D. Shasha, "D^{over}:An optimal on-line scheduling algorithm for overloaded real-time systems", IEEE Real-time Systems Symposium, 290–299, 1992.
10. G. Koren and D. Shasha, "MOCA: A multiprocessor on-line competitive algorithm for real-time systems scheduling", *Theoretical Computer Science*, **128**, 75–97, 1994.
11. R. Lipton, and A. Tomkins, "Online interval scheduling", SODA, 302–311, 1994.
12. J. Vytopil, *Formal techniques in real-time and fault-tolerant systems*, Kluwer Academic Publishers, 1993.
13. G. Woeginger, "On-line scheduling of jobs with fixed start and end time", *Theoretical Computer Science*, **130**, 5–16, 1994..

A   Lower Bounds for Alternate Models

To show that the deterministic competitive ratio is at least linear in the number of processors (for all reasonable models) one need only note that every processor must be running the job with the earliest relative completion time. If not, the online algorithm risks the possibility that it didn't finish any jobs, while the the optimal schedule has positive benefit.

The following instance shows that the competitive ratio of every randomized algorithm in a system with checkpointing and premption is linear in the number of jobs. Job J_i has $r_i = 0$, $x_i = 1 + i/n^2$, $d_i = 1 + 2i/n^2$, $l_i = i/n^2$, and $b_i = 1$, $0 \le i \le n-1$. By having a transient fault from time $1/n^2$ to time $(i+1)/n^2$, then having a permanent fault at time $1 + 2i/n^2$, the adversary can guarantee that only J_i can be finished. The lower bound follows since for any online algorithm, there is a job such that the algorithm runs the job with probability at most $1/n$ in a uniprocessor system.

The following instance shows that the competitive ratio of every randomized algorithm for a system a nonpreemptive and noncheckpointing scheduler is linear in the number of jobs. For $0 \leq i$, the adversary releases a job J_i has $r_i = i/n^2$, $x_i = 1 - 2i/n^2$, $d_i = 1 - i/n^2$, and $b_i = 1$. If the online algorithm runs J_i with probability at most $1/n$ then the adversary releases $n - i - 1$ jobs of length 1 at time $i + 1/n^2$ and deadline $1 + i + 1/n^2$. Finally, the adversary induces a permanent fault at time $1 - i/n^2$. The desired lower bound then follows since the adversary finished job J_i, and no job other than J_i can be finished by any algorithm. Note that by the time nth job is released, the online algorithm must run some job with probability at most $1/n$.

Collecting Garbage Pages in a Distributed Shared Memory with Reduced Memory and Communication Overhead

Dmitry Kogan and Assaf Schuster

Department of Computer Science, Technion
kogand,assaf@cs.technion.ac.il

Abstract. We present a novel algorithm for Garbage Collection (GC) in Distributed Shared Memory systems (DSM). Our algorithm reduces the network traffic overhead (and the memory and computation overheads), essentially eliminating all communication when there is no active collection, and minimizing it when the collection process is turned on. Our algorithm works correctly for asynchronous environments where messages may experience arbitrary delays on the way to their destinations. It also tolerates arbitrary duplication of messages and is thus a suitable "add-on" for fault-tolerant communication protocols. It does not suffer from problems such as weight underflow (which arise in reference counting techniques). In addition, when applied in granularity of pages (which is the most relevant in page-based DSM systems), then the memory overhead is not inflated when the average allocation size is small, and the memory reorganization required due to the GC operations is simplified.

1 Introduction

Manual memory management is programmer-time consuming and error prone. Most programs still contain leaks of memory even after considerable effort by the programmer. Modern API gives an extended set of functions for memory allocation and deallocation. This complicates the possibility of controlling the freeing of the dynamic objects when they become unreachable (i.e., no other object points to them). The problem is even harder in the distributed environment, because a part of code in the thread, which is in charge of deallocating the storage, must verify that no other object still needs that storage. Thus, many modules must cooperate closely. This leads to a tight binding between supposedly independent modules.

Despite the well-developed techniques of uniprocessor garbage collection, distributed techniques still present some difficult problems for researchers. These problems are engendered by several distinctive features of distributed systems:

- To know the status of a remote workstation one must send and receive messages; messages are costly.
- Most distributed systems are asynchronous. A message only tells about the past of the sender. There is usually no single global order of events.

- Failures do occur. Machines crash; processes get into infinite loops. Messages get lost and delivered out-of-order (even with TCP/IP, in the presence of crashes).
- As a consequence of the above, data tends to be inconsistent. If the data being copied is the Garbage Collection (GC) status, then the collection algorithm may decide to reclaim an object based on incorrect information.

As a result of these problems, many of the simple approaches for implementing garbage collection in distributed systems do not work at all or make costly demands on the system, thus reducing efficiency and slowing down the work of other parallel processes.

The easiest GC algorithm to distribute is reference counting. One simple approach is to attach a counter to each object. Duplicating or killing a reference to a global object sends an increment or decrement message to the target object. This is very expensive (one message per pointer operation); more importantly, it does not work. Since there is no guaranteed delivery order for messages, an increment might be overrun by a decrement, and the target object unsafely reclaimed.

A general fix is to provide "casual delivery" of messages, but that's too costly. A simpler fix is Weighted Reference Count (or some variation of it) [3]. When a reference to the global object v is first created, it is given some total weight. Each local and remote reference to the object v has a field containing the partial weight. When a reference is duplicated, its partial weight is divided in two and the partial weight of the duplicated reference is given the new value, which equals half of the previous partial weight. Another half of this partial weight travels to the new reference and the partial weight of this reference is given that value. When deleting a reference, its partial weight is sent to the owner of the object v and is subtracted from the total value. In other words, all the Weighted Reference Counting techniques maintain the following invariant for every object v:

$$total\_weight_v = \Sigma partial\_weight_v$$

When splitting the partial weight of some reference, one may come to partial weight value 1, which cannot be further divided. This problem is known in the literature as weight underflow and is referred to as the most serious problem of the Weighted Reference Counting techniques. There are solutions to this problem [10], but they significantly increase the number of messages sent to update the counters. Weighted Reference Counting is neither resilient to the loss or duplication of messages, nor to hardware failures. Weight underflow, hardware failures and the loss of messages leave a certain number of unreachable objects in the memory, which the collection algorithm is not able to reclaim. The Yu - Cox algorithm [11], for example, avoids costly run-time solutions to these problems in the following way: when a node can not collect the necessary memory space from the garbage, it sends a signal to all the workstations to suspend their computations and runs global collection. Global collection can achieve ideal results because all the nodes are in a consistent state, but it is a very ineffective solution with respect to performance considerations, and is thus used as a last resort. Yu

- Cox and other algorithms based on weighted reference count([5],[6],[11]) also do not solve the problem of messages duplication, as a result of which some global objects may be unsafely reclaimed. This happens when some node observes that the reference to some global object is unreachable and sends it back to the owner along with its weight. If this message is accidentally duplicated, then the owner erroneously counts this returned weight twice. In this case the owner may mark this global object unused before all the references to it become unreachable, which is unacceptable in a loosely coupled distributed system.

Other techniques that try to avoid race conditions on remote counters and are resilient to hardware and message failures use Reference Listing (SSP Chains [7], Garbage Collector for Network Objects [1]). Their most significant shortcoming is the increase in the amount of auxiliary messages sent between nodes.

It should be noted that Reference Counting and Reference Listing techniques have one common drawback. They do not allow collection of cycles of garbage. There are various solutions for collecting these cycles, as well as other unreachable data (Hybrid Garbage Collection [4]), but they all add a considerable amount of messaging to the implementation of their algorithms.

In Distributed Systems, communication causes bottlenecking in computation processes. Thus it is not permissible for a GC algorithm to exchange messages only for the purpose of decreasing the time required for reclaiming an object that has become unreachable. That is why, when we encountered various tradeoffs, we chose to reduce network traffic at the cost of a possible increase in required memory space and local work.

Our solution is based on a variation of Reference Counting as a local part of the algorithm that is executed by every node without sending messages. The counters are allocated for every page in the Distributed Shared Memory (DSM), rather then for every object, as in earlier solutions. This solution is bound up with the widely applicable page-based implementation of DSM. An address or a handle of some object in the page-based DSM can be easily converted to the page table entry number corresponding to the page which contains this object. When the algorithm observes a pointer to some object, it associates this pointer with the proper page. The memory overhead of counters is thus reduced proportionally to the average number of objects in a page. Although a page may contain some unreachable objects, there is no guaranteed time period during which they can be reclaimed, because other objects in this page may still be reachable for an unlimited time, and the unit of reclamation is a whole page. But it is preferable to wait for all objects in the page to become unreachable than to exchange messages in order to receive the collection decision regarding each separate object. Thus, page-based counting considerably reduces the overhead communication for GC as well.

To collect the reachability information about some page, which is distributed between local counters on the nodes of the network, we use the technique first proposed by R.W.Topor [9]. He used this method for Termination Detection, but it scales well enough for collecting garbage information as well. The idea is as follows: one or more waves of tokens go to and from the root node of some spanning tree built for the purpose of garbage collection. These tokens contain the local information created by reference counting. This data is then analyzed at the root, which in turn decides whether or not to collect the page.

It is known that Topor's algorithm doesn't work in the case of asynchronous message passing. We overcome this problem by analyzing the messages being sent and updating the counters according to the number of references the messages contain. So even if some message containing a reference takes a long time to arrive at its destination, the node that sent the message already knows which references were sent. It passes this information via the token to the root. So the root knows that the message, which has not yet arrived, contains a reference, and doesn't allow the corresponding page to be reclaimed.

The algorithm is resilient to hardware and message failures. Duplication of messages is ignored, because the data is sent in the spanning tree and every node is aware of which nodes have already sent their tokens.

But perhaps the most important advantage of our algorithm is that it sends messages that contain auxiliary data for garbage collection only when the chance that some page can be reclaimed is very high. There are no additional messages when there is no attempt to collect pages. The only overhead in the absence of collection is the supplementary expense for maintaining local data structures per page.

Our solution is also tolerant of the consistency protocol and doesn't depend on some global order of operations in the system.

The algorithm proposed in this paper still doesn't collect cycles of garbage. Additional research is required to check the efficiency of using global garbage collection as a last resort, as it was suggested by Yu - Cox [11].

The rest of this paper is as follows. In Section 2, we describe the system model and the basic terminology used in this paper. In Section 3, we show how local counting is carried out. We present two types of counters local and remote. In Section 4, we develop, step by step, an invariant that helps us to describe the global part of the algorithm and to prove its correctness. In Section 5, advantages and drawbacks of the algorithm are shown. In Section 6, we trace how the algorithm can be implemented to avoid the negative effect of message duplication, loss and hardware failure. In Section 7, we use the invariant developed in the Section 4 to prove the safety of the algorithm.

2 System Model

The algorithm is aimed to work on a Distributed Shared Memory, which presents an abstraction of a shared memory for some Parallel Virtual Machine. The real network structure and location of local memory spaces, distributed between the workstations, is totally hidden from the programmer.

We assume here that there are several separate processors using Distributed Shared Memory (DSM). We reference each one of them as a *node*. Each node can have several processes running on it and each process can allocate memory from the DSM. We denote the unit of allocation as an *object*. The DSM consists of pages, containing one or more objects. Every page can be replicated at several nodes.

We say that there is a pointer at node i to some page X if some object in the local memory of node i, that is not in page X itself, is pointing to some object in one of the replicas of page X. We assume that, given a pointer or a handle to some object, it is possible to find which page contains this object.

As a *root set* we denote the data that is immediately available to a program, without following any pointers. Typically this would include local variables from the activation stack, values in machine registers, and global, static, or module variables.

An object becomes *reachable* after allocating memory for that object, receiving a pointer (handle) to that memory and assigning it to one of the previously reachable objects.

An object becomes *garbage* when all reachable objects that were pointing to it are pointing to another object or NULL.

A page becomes garbage when all objects contained in all of the replicas of this page are garbage themselves.

A node i is *clean* for page X when there are no pointers to page X in the local memory of that node. Otherwise, node i is *dirty* for page X.

There are messages traveling between the nodes that can contain reachable objects pointing to pages of DSM. These are the only messages that interest us: when we mention the word *message*, we refer to a message containing pointers to the page that is being checked for garbage.

Each node has in its local memory page tables containing an entry for every page in the DSM.

3 Local Counting

Before introducing the Garbage Collection algorithm, we want to describe some local information that is gathered for each page even when the GC algorithm is not actively reclaiming garbage pages. This information consists of two counters attached to each entry of the page table in the local memory - *local* and *remote* counters.

3.1 Local Counters

For every node i we want to count the number of pointers located at node i to each page allocated in the DSM. When page X is created, local counters for this page at every node are 0. At node i we denote such a counter as $local_i(X)$.

$local_i(X)$ is increased:

- when some reachable object at node i, that is not in page X itself, is assigned a pointer (handle) to some object in page X ($local_i(X)$ is increased by 1);
- when receiving a message containing pages with pointers to page X, $local_i(X)$ is increased by the total number of the pointers to page X in this message.

$local_i(X)$ is decreased:

- when reassigning a pointer at node i, that was previously pointing to page X, to another value($local_i(X)$ is decreased by 1);
- when deleting the object in page X, which was pointed to from node i by another object, which is not in page X itself ($local_i(X)$ is decreased by 1);
- when some page $Y \neq X$, which contains pointers to page X, is discarded from the local memory of node i ($local_i(X)$ is increased by the number of pointers to page X in page Y).

A node with a $local(X)$ counter 0 is called *clean* regarding page X. If a $local(X)$ counter at the node is greater then 0 then the node is denoted as *dirty* with respect to page X. It is obvious that $local_i(X) \geq 0$ is always true.

3.2 Remote Counters

The remote counters are almost exactly the same as the local counters except for one thing. Remote counters are not increased when a message containing pages with pointers to page X is received; rather, they increase when such a message is sent. In other words, when some message containing pointers to page X is sent from node i, $remote_i(X)$ is increased by the total number of these pointers in the message. Initialization of remote counters and other cases of their modification are the same as for local counters. It is thus easy to see that when all the messages containing pointers to page X which were sent have been received then $\sum remote_i(X) = \sum local_i(X)$. This will be proven in Section 7.

4 Determination of a Page as Garbage

We develop an algorithm for determining if a certain page X is garbage using a sequence of approximations to help us find invariant R and condition C, such that

$$R \cap C \Longrightarrow \text{page X is garbage.} \tag{1}$$

Note that all actions here with regard to the collection algorithm concern the same page X. In other words, page X is the default, when no other page is indicated.

Let some node be the root of a fixed spanning tree of the given graph of all the nodes participating in a process of calculation. Let's refer to the root as node 0.

At the start of "garbage scanning", node 0 sends signals to all the leaves of the tree to start sending tokens in the direction of the root through their ancestors. We call this moment the beginning of the wave. When a leaf receives a signal it sends a token that moves through the nodes of the tree to the root. When the root receives tokens from all of its siblings, it detects if page X is garbage. If yes, then it reclaims all the replicas of this page and the scanning ends. Otherwise, next wave begins. It is important to note that signals and tokens regarding some page are sent only when the chance that this page is a garbage is very high. This will be shown in Section 5.1.

The set of nodes holding one or more tokens in the wave at a certain moment and the set of leaves that still haven't received a signal from the beginning of the wave is called S.

Definition 1. A node j is outside of S if j is not in S and the path from j to node 0 (the root) includes nodes from S.

Let the invariant R be set initially to $R0$.

$$R0 \equiv \text{all the nodes outside S are clean.} \tag{2}$$

Initially $R0$ is true when node 0 sends signals to the leaves and there is no leaf which has already received one of these signals, so the set of nodes outside S is empty in the start of the wave.

Tokens then move inward (towards the root) according the following rule:

Rule 1 *A leaf that has a local counter 0 transmits a token to its parent; an internal node with $local(X) = 0$ that has received a token from each of its children transmits a token to its parent; a dirty node ($local(X) > 0$) doesn't transmit a token. When a node transmits a token, it is left without any tokens.*

A node that transmits a token moves "outside" of S and its parent is added to S if it wasn't in S previously.

Rule 1 works and is sufficient for determining if page X is garbage only on the condition that no messages were sent since the scan began.

In the presence of messages, the following situation could occur. Assume some node j that is not outside S has in its local memory a page Z, where one of its objects z points to the object x in a page X. At another node i that is outside S (and its local counter is 0) the following operation is executed: $y = z$. Node j sends page Z to node i, so that in the end y points to X and the local counter of page X at node i becomes 1 (the node becomes dirty). Thus $R0$ becomes false.

As we have mentioned, each node has an additional, remote counter. In the last example we saw that when node j sent its page Z to node i it incremented its remote counter. Now we add the possibility to paint each node in one of two colors: black or white. Let's assume all nodes to be initially white. A node turns black immediately after it increments its remote or local counter. So we change R to $R0 \cup R1$, where

$$R1 \equiv \text{at least one node not outside S is black.} \tag{3}$$

Accordingly, the rule is as follows:

Rule 2 *A node that increases local(X) or remote(X) becomes black.*

Now, assume for the last example that local(X) in node j decreases to 0, and it was the only black node. Then j finally will be in S, and because $local_j(X) = 0$, it will transmit the token to its parent. After that it will remain outside S until the new wave. Then $R0$ may be false because i may be dirty and $R1$ is false because there are no black nodes not outside S. Now we add an additional possibility: each token can be painted in black or white color. Let's consider all the tokens at leaves to be white. Then a token that is transmitted from a black node is black too. So we change R to $R0 \cup R1 \cup R2$, where

$$R2 \equiv \text{ some node in S has a black token.} \tag{4}$$

$R2$ is maintained by the following rule:

Rule 3 *A node that is black or has a black token transmits a black token, otherwise it transmits a white token.*

To prevent a node that turned black from transmitting repeatedly black tokens at successive waves, let's "paint" a black node that sends a token white. This doesn't violate R because R doesn't depend on the color of the nodes outside S. So

Rule 4 *A node that transmits a token becomes white.*

Still we cannot be sure that R holds, because in some wave of tokens the following situation could occur:
In some wave i, node A not outside S sent message Z, containing a pointer to page X, to node B, outside S. Let's assume this was the only message sent in waves i and $i + 1$. Then node A turned black, received the tokens from all of its siblings, and sent a black token to its parent when A's local(X) counter became 0. After that moment node A is white for good. Consider that message Z arrives at node B only in wave $i + 1$, when B is outside S, and all the nodes of the tree are white because none of them sent messages in waves i and $i + 1$. Then with regard to wave $i + 1$

- $R0$ doesn't hold, because there is a page with a pointer to page X encapsulated in message Z, sent by node A in wave i, received by node B outside S in wave $i + 1$, which then increases its $local(X)$ counter and becomes dirty;
- $R1$ doesn't hold, because all the nodes are white;
- $R2$ doesn't hold, because in wave $i + 1$ all the nodes were initially white and none of them turned black. Thus there aren't any black tokens in the tree.

So we change R to be $R0 \cup R1 \cup R2 \cup R3$

$R3 \equiv$ There is a message that was sent in one of the previous waves of tokens destined to some node outside S, containing a page with one of its objects pointing to page X and this message has arrived to its destination in the current wave or still hasn't arrived yet.

Now the real reason for using $remote(X)$ will become clear. First we formulate Rule 5, which maintains $R3$, and then we shall explain its meaning.

Rule 5 *Each node sums up remote counters that arrived to it via tokens from its siblings, adds to the sum its own $remote(X)$ counter and sends the result together with the token to the parent, according to Rule 1.*

In the last example, when the wave of tokens reaches node 0 it doesn't receive any black token, because all nodes were white during the wave. If we would sum $local(X)$ counters using the same method as in Rule 5, but without using $remote(X)$, then node 0 would receive the sum 0, informing it that none of the nodes points to page X. But actually, as we see, there is a pointer to that page that was sent in the "delayed" message, a fact which node 0 should notice. That is where $\sum remote(X)$ comes into use. Let's assume that the "problematic" message was sent from node A to node B (that is outside S). Clearly, A increased $remote_A(X)$ before sending this message. If B doesn't decrease its $remote_B(X)$, then at node 0 we will see that $\sum remote(X) > 0$ (this fact will be proved in Section 7). Thus we derive the following rule for node 0 to determine if page X is garbage:

$$R \cap S = \{0\}$$
$$\cap \text{ node 0 is white and } local_0(X) = 0$$
$$\cap \text{ all tokens at node 0 are white}$$
$$\cap \sum remote(X) = 0 \implies \text{ page X is garbage.}$$

Otherwise, node 0 begins the repeat wave of the algorithm and directs "repeat" signals outwards. The algorithm for the outward signals complies with the following rules:

Rule 6 *When node 0 receives a token from each of its siblings, it waits until its $local(X)$ counter becomes 0, checks if it is black or received a black token or $\sum remote(X) > 0$, and then sends a repeat signal to all its children. Otherwise it discards all the replicas of page X from the memory.*

Rule 7 *An internal node that receives a repeat signal sends the signal to all its children.*

Rule 8 *A leaf receiving a repeat signal gets a white token.*

5 Effectiveness

5.1 Saving the Network Traffic

The rules derived in Section 4 imply that by the time a wave of tokens reaches the root of the spanning tree, every node in this tree has become clean at least once since the wave began. Usually, page usage by an individual node is localized in time. Thus, if some node is clean with respect to some page, the chances are high that it won't become dirty with respect to that page again. This is the main idea by which the algorithm saves the additional network traffic the collecting process usually requires. When collecting some page X, it is most probable that there will be only one wave of tokens. Even in this case, all the repeat signals and tokens can be piggybacked, thus avoiding any additional messaging at all. Although cases exist in which a node that sent its token and was considered clean may become dirty again, the root never reclaims it. This is proven in Section 7.

Sometimes the best computation time is desired, which can be achieved by turning the garbage collection off. Most algorithms in the distributed literature do not allow to suspend the GC process, and in particular they do not allow suspension once the collection process has begun. In our algorithm, if node 0 doesn't initiate "garbage scanning", no messages are sent at all. All the work is localized in the reference counting process, which has little influence on computation time as compared to messaging process. Transmission of GC messages (tokens and signals), can be safely delayed, as we prove the safety of the algorithm in Section 7 without implying synchronous message passing. We assume that it may take an arbitrarily long time for some message to arrive. Thus, suspending the "garbage scanning" process at some point in time T is not different than delaying the arrival of GC messages for that time T.

5.2 Collecting Pages vs. Objects

In order to further reduce the messaging overhead, it was decided that the algorithm would collect pages instead of objects. Collecting objects in every GC algorithm involves local and remote operations for every object. Objects allocated by a user's program can be extremely small. Thus, collecting each one of them separately causes a significant increase in the number of messages that are sent for gathering the distributed reachability information. For example, if the algorithm that was presented in Section 4 worked with objects instead of pages, the number of signal-token waves would increase according to the average number of objects in the page.

Moreover, in the object-based GC algorithm, counters would be required for every object. In the worst case, when every allocated object has the size of only one byte, the memory overhead required for counters would be at least 50% (if each counter requires one byte as well). In practice this overhead is even greater. Let's take for example the algorithm from Section 4. For each page in every node, local and remote counters are necessary. Let's assume every one of them takes two bytes. One more byte is needed for each page to hold the information regarding node color (black or white) and from which other nodes the tokens have already been obtained. So the algorithm requests 5 bytes for each page in every node. For our page-based algorithm, if the page size is 4K, then the required memory overhead is about 0.12%. But if the algorithm is changed to work with objects instead of pages, the necessity for holding these extra 5 bytes for each object in every node that references it will not be eliminated. In this case if the average number of objects per page is M, then the memory overhead is increased exactly M times.

The extension in the size of the required data structure in every node increases not only the memory overhead, but also makes the operations on these structures more time consuming. In particular, it increases the complexity of the search operation required for changing the value of some counter. That is why the local work of the page-based solution will be much faster. In addition, utilization of a page table for keeping garbage collector information for every page avoids spending valuable time on creation and removal of GC data structure elements (counters, state), whose place can be reserved at system start-up.

As mentioned in Section 2, there is a system function for converting a virtual DSM address to the corresponding entry in a page table that contains a record on the corresponding page. This function must be used in order to access a certain local or remote counter that is stored in every page table entry. In other words, every change to the garbage collection information for some page must be preceded by the call to this converting function. One might think that an algorithm which collects objects instead of pages makes any conversion unnecessary, thus gaining a considerable amount of execution time. However, consider the case of some memory cell that doesn't reference the beginning of some object, but rather one of its fields in the middle. In that case, a function is required for converting a reference to the beginning of the corresponding object, so that the right counters in GC local data structure can be found. This function requires information concerning all the objects referenced locally. For example, a field f belonging to some object c that is created at node A, is referenced in node B. To increase the local counter of c, one must find it in the objects counters table of node B, according to its initial address. If node B contains no data about c, A must send this information (initial address, object's size). Thus, in addition to the subsequent increase in the memory overhead, one can observe an increase in the number of messages sent between the nodes. No such overhead is required in the case of the page-based GC, because the information for converting a virtual address to a page number must exist for any page-based DSM regardless of which kind of GC is used. The only case in which page-based GC will need additional data is when the size of some object is greater than the DSM page size. Then the algorithm would have to know which pages this object occupies, so that the counters corresponding to these pages could be changed simultaneously.

The memory reorganization process as an important part of every garbage collection is not considered in this paper. But it is very important to note that swapping the free memory space after page reclamation is cheaper than the reorganization of small empty "cells" left after discarding separate objects.

At the same time, page-based collection has its own drawback. Even if some object x in page X becomes unreachable, the collector will not reclaim it until all other objects in the page become garbage. But in practice every memory usage agrees with the memory locality principle, which says that memory access is localized in time and space. In other words, if some object x in page X becomes unreachable, it is very probable that all other objects in page X will soon become garbage too, and the page will finally be reclaimed.

There is another problem that is caused by collecting garbage in granularity of pages. Any collection which is based on the reference counting is not able to reclaim cycles of garbage. This problem becomes more severe when pages are collected. Let's suppose that object a in page A references object b in page B. Also, there is object c in page B that references object d in page A. Though there is no garbage cycle observed in the graph of objects references, there is a circle A-B-A in the graph of references between the pages. Thus, pages A and B won't be reclaimed by our algorithm, even if all the objects in them are unreachable from the outside. On the other hand, page-based collection solves the garbage cycles problem when some cycle is located entirely inside one page. When there are no "external" references to such a page, this cycle will be safely discarded from the memory along with the page that contains it.

6 Fault Tolerance

If the algorithm is implemented using an unreliable connection protocol (like UDP), then there is no way to avoid loss and duplication of signals and tokens. As we mentioned earlier, this sometimes causes a serious problem for a considerable number of distributed GC algorithms. In this section we show that our GC algorithm avoids unsafe reclamation of garbage pages as a result of message duplication. Message loss does not affect the safety of the collection process, but may rather delay an object's reclamation for a very long time. Our algorithm helps decrease the number of such unreclaimed objects (pages).

Let's consider collection regarding some page X and assume that some token A, sent from node i to node j, was duplicated in some wave. Every node in our algorithm can be in one of two states: either it already sent its token, and is waiting for a signal (State 1), or it already sent its signal, and is waiting for a token (State 2). If node j is in State 1, when it receives the duplicated token, it can simply ignore it. It is obvious that this decision is correct, because until node j receives the signal it is waiting for and passes it to node i, there can't be any new tokens that are sent from node i. Otherwise, node j can be in State 2 when it receives the duplicated token. In the case that this token was sent in the current wave, node j can ignore it as well. The reason to this is that node j hasn't yet received the tokens from all of its siblings, or it hasn't yet sent its own token. Thus, if j receives two or more tokens from i while its state doesn't change, then it may conclude that only one of them is valid.

The situation is more difficult when the duplicated token was sent in one of the previous waves and node j is in State 2 in the current wave. Then this message can bypass the real token from node i and be accepted by node j erroneously. To avoid this, the algorithm should be slightly altered, adding one more rule that ensures it will work correctly in the case of message duplication.

Rule 9 *Each wave in the algorithm is given its ordinal number starting from 1. Every signal and every token in the algorithm is given an integer value, containing the corresponding wave number. A token or a signal, containing a wave number that differs from the current one, is ignored by the node that receives it.*

When node 0 starts a garbage scanning process for page X, it can assign the signals that it sends to its children the number 1; then, every other node that would receive a signal with number 0 would send to all its siblings signals with the same number. After receiving a signal with number 0, a node would know that the token that it must send (receive) would contain the number of the signal that it has received. At the end of the wave 0 the root will increment this counter if an additional wave is required, and the process will go on in the same manner. As a result, when node i receives a token with a wrong wave number, it will know to neglect this token. Thus we avoid errors that could occur if some duplicated token arrives not in its wave.

There are no ideal solutions for avoiding the effects of message loss and hardware failures, because if they occur, the lost garbage collection data cannot be restored without additional overhead. We propose the minimal overhead of piggybacking control messages from time to time. For example, if node i doesn't receive a token from one of its siblings (say, node j) for a considerable period of time, then there are chances that either the token was lost or node j failed to send it because of hardware problems. In this case, node i waits for any message that goes to node j and piggybacks a control message on it. If node i doesn't receive an acknowledgment (token) for its control message, then there is some hardware problem in the network or in the node j itself. If node j receives a control message from node i, then the chances are high that the last token was lost. In that case it sends to node i a token with the appropriate wave number. Thus the tree structure of nodes used in our algorithm allows us to avoid serious problems that can arise as a consequence of using the unreliable connection protocol.

7 Safety Proof

We start the proof by showing that R is indeed invariant. Then for page X we show that if

$$S = \{0\}$$
$$\cap \quad \text{node 0 is white and } local_0(X) = 0$$
$$\cap \quad \text{all tokens at node 0 are white}$$
$$\cap \sum remote(X) = 0$$
$$\implies (R1 \cap R2 \cap R3) = \text{FALSE} \implies (R0 = \text{TRUE}) \text{ and } (R3 = \text{FALSE})$$
$$\implies \text{page X is garbage} \implies \text{page X can be safely reclaimed.}$$

Claim 1 *R is an invariant.*

Proof: Here we are going to prove that $R = R0 \cup R1 \cup R2 \cup R3$ is always true.

Let's assume by negation that it is false, and look at the wave in which R first became false. Because R became false, then $R0$ is false too, and thus some node outside S is dirty. Let's look at node i, which was the first node outside S to became dirty in the wave when R first became false. Regarding this wave we denote:

- t_{err} – time when R became false;
- t_1 – time when node i sends its token;
- t_2 – time when node i first became dirty after t_1 $(t_2 > t_1)$;
- t_3 – time when the wave of tokens first reached node 0 after t_1 $(t_3 > t_1)$ (this can be regarded as the end of the wave).

In the previous waves we know that R was always true by the way that we selected "current wave". Because i is the first node outside S that became dirty in this wave, then until t_1 all nodes outside S were clean by definition of t_1.

Node i is outside S at $t_1 < t < t_3$.

At $t_1 < t < t_2$ node i is clean by definition of t_2. So at $t_1 < t < t_2$ $R0$ =TRUE, because node i is the first node outside S that became dirty. $\Longrightarrow$ at $t_1 < t < t_2$ R =TRUE.

Now we know that $t_2 \leq t_{err} < t_3$.

The fact that node i was clean and became dirty later tells us that it has received a message containing some pointer to page X. Until this message was sent, node i was clean, so $R0$ =TRUE implies R =TRUE.

If the message was sent to node i in one of the previous waves, then $R3$ =TRUE, because it's true without dependency if it arrives at node i at $t_2 \leq t < t_3$ (node i was outside S at this time interval), or if it doesn't arrive to it until t_3. If $R3$ becomes true in the middle of some wave it stays true until the end of this wave, hence R is true from the beginning to the end of the current wave.

Therefore the message that made node i dirty must have been sent in the current wave, either by a node that is outside S, or by a node that is not outside S. But we know the message couldn't have been sent by a node outside S, because all nodes outside S were clean until t_2. In addition, the message could only have been sent by a node with a *local* counter > 0. So the only possibility is that the message was sent by a node not outside S. This node increased its *remote* counter before sending the message, and since it is not outside S, it must become black by Rule 2. So, until the time when this node sends its token (let's denote it t_5) and becomes white, $R1$ =TRUE. At $t_5 < t < t_3$, $R2$ =TRUE, because after t_5 S contains a black token. Thus, we found no place for t_{err} in the interval (t_0, t_3). Thus R is always true, and is an invariant of the algorithm.

In the proof, $\sum local$ and $\sum remote$ are denoted as global values that we would receive if we summed up the respective counters at the nodes at a certain instant of time.

Claim 2 *If there are no messages that are in the network then $\sum remote = \sum local$.*

Proof: By definition local and remote counters are decreased in the same cases. The only difference between them is that local counters are increased when the message arrives by the amount of pointers to page X contained in that message, whereas remote counters are not changed. Another, final difference in counting is that remote counters are increased when sending messages, whereas local ones are not.

The proof is by induction on the number of messages sent and received before there were no messages in the network.

Base: In the beginning local and remote counters at all nodes are initialized to 0.

Assumption: After $n-1$ messages sent and received $\sum remote = \sum local$.

Step: Let's look at node i that sends message n to another node j. Let's assume that the message contains m pointers to page X. Then, after the message was sent and before it was received, $remote_i(X)$ was increased by m and so $\sum remote$ was increased as well. After message n arrived at node j $local_j(X)$ was increased and $\sum local$ was increased as well by m. Thus, message n had no effect on the difference between $\sum local$ and $\sum remote$. So Claim 2 is correct.

Claim 3 *If there are messages in the network that were sent but have not yet arrived, then $\sum remote > \sum local$.*

Proof: Supposing that the Claim is not correct, let's check the first time T its correctness was violated, i. e., until that moment $\sum remote > \sum local$. The relation could change only if a local counter at some node was increased and the remote counter was not. This could happen only in the case of receiving a message. Let's denote T' as the time when the last message was sent or received before time T. At $T' < t < T$ $\sum remote - \sum local$ was equivalent to some positive number M that is obviously equivalent to the number of pointers to page X in all the messages that were sent but haven't yet arrived. Following the arrival at time T of some messages, $\sum local$ couldn't increase by a number greater than M. So the claim is always correct.

Claim 4 $\sum remote \geq \sum local$ *is always true.*

Proof: The correctness of Claim 4 follows directly from Claim 2 and Claim 3.

Claim 5 *If node 0 hasn't received a black token, $S = \{0\}$ and node 0 is white, then $R1 \cup R2 =$FALSE.*

Proof: $R1$ is false, because at the moment the wave reaches node 0 the only node that is not outside S is node 0 itself, and it is white.

$R2$ is false by definition because S doesn't contain black tokens.

Let's denote $\sum remote$ that was actually received by node 0, plus $remote_{node\ 0}(X)$ as $\sum remote(X)_{resv}$. We also denote by $remote_{i(t)}(X)$ the value of the remote counter of page X at node i at time t, and $local_{i(t)}(X)$ the value of the local counter of page X at node i at time t.

Claim 6 *If node 0 hasn't received a black token, $S = \{0\}$, node 0 is white and $\sum remote(X)_{resv} = 0$ then $R3 =$FALSE.*

Proof: Let's assume that the claim becomes incorrect in some wave, which will hence forth be called the *current wave*.

In the proof we denote by t_0 the time when the current wave began and by t_j the time when node j (except node 0) sent its token.

Node 0 hasn't received a black token and it is white itself. So it is obvious from Rule 3 that no node not outside S has turned black in the current wave of tokens. This means that no node j in the interval of time $[t_0, t_j]$, i.e., the interval of time when node j was not outside S, has increased its remote or local counter. Also we know that no messages were sent by node j at this time (otherwise it would increase its remote counter) and no messages were received at this time by node j (otherwise it would increase its local counter). And without sending or receiving messages at some node, remote and local counters at this node change by the same values at the same rate by definition, i.e.,

$$remote_{j(t_0)}(X) - remote_{j(t_j)}(X) = local_{j(t_0)}(X) - local_{j(t_j)}(X). \qquad (5)$$

By Rule 2 node j sends its token only if $local_j = 0$. So

$$local_{j(t_j)}(X) = 0 \qquad (6)$$

By definition, the final sum that we receive at node 0 is:

$$\sum remote(X)_{resv} = \sum_{j=1}^{n} remote_{j(t_j)}(X) \qquad (7)$$

From 5 and 6 we get:

$$remote_{j(t_j)}(X) = remote_{j(t_0)}(X) - local_{j(t_0)}(X) \qquad (8)$$

From 7 and 8 we get:

$$\sum remote(X)_{resv} = \sum remote_{j(t_j)}(X)$$
$$= \sum remote_{j(t_0)}(X) - \sum local_{j(t_0)}(X) \qquad (9)$$

If $R3 =$TRUE, then at time t_0 there was at least one message that was sent in one of the previous waves and hadn't been received yet in the current wave. So from Claim 3 we can conclude that

$$\sum remote_{j(t_0)}(X) - \sum local_{j(t_0)}(X) > 0 \qquad (10)$$

From 9 and 10 we get:

$$\sum remote(X)_{resv} > 0 \qquad (11)$$

This contradicts the condition that $\sum remote(X)_{resv} = 0$. So the assumption that $R3 =$TRUE was wrong, and the claim is correct.

Claim 7 *If node 0 hasn't received a black token, $S = \{0\}$, node 0 is white and $\sum remote(X)_{resv} = 0$ then $R1 \cup R2 \cup R3 =$FALSE.*

Proof: The correctness of Claim 7 follows directly from Claim 5 and Claim 6.

Claim 8 *If node 0 hasn't received a black token, $S = \{0\}$, node 0 is white and $\sum remote(X)_{resv} = 0$ then $R0 =$TRUE and $R3 =$FALSE.*

Proof: From Claim 1 we can conclude that if $R1 \cup R2 \cup R3 =$**FALSE**, then $R0 =$**TRUE** is also maintained. $R3 =$**FALSE** from Claim 7.

Claim 9 *If node 0 hasn't received a black token, $S = \{0\}$, node 0 is white and $\sum remote(X)_{resv} = 0$ then page X can be safely reclaimed.*

Proof: We know that the page can be safely reclaimed if there are no pointers to page X in the local memory of every node and there are no messages with pointers to page X that have not been delivered yet. Claim 8 gives us that $R0 =$**TRUE** and $R3 =$**FALSE**. From $[R0 =$**TRUE** $\cap S = 0\cap$ node 0 is white$] =$**TRUE** we conclude that there are no pointers to page X in any of the local memories of the nodes. And from $R3 =$**FALSE** we conclude that there are no messages with pointers to page X that have not been delivered yet. So page X can be safely reclaimed.

8 Conclusion

In this paper we have proposed an efficient approach to collecting garbage from the Distributed Shared Memory.

Most of the work in the algorithm is localized in the nodes. Each node of the network has two counters for indicating the number of local pointers to pages in DSM and for counting references traveling across node boundaries.

We assume that there is a static spanning tree consisting of the network workstations. The algorithm uses this spanning tree to send signals to the leaves of the tree through its nodes, and to send tokens in the direction of the root. Concerning a specific page, every node, after receiving a signal, waits for the tokens from all of its children in the tree, and sends its token only if it sees no local pointers to this page in its memory. Tokens and signals can be piggybacked on messages required for other computations. So no additional messages are sent and the GC tokens are added to messages only when it is very probable that a node no longer references the page.

The part of the algorithm that includes the exchange of tokens between nodes was actually developed as it is presented in this paper. First, the invariant components were constructed one by one. Then, for every component, a rule was formulated to maintain this component. So by the time the invariant was developed it was easy to see that the algorithm was correct and its correctness was indeed proved formally later.

Duplication of messages has no effect in the described algorithm; every node remembers all the tokens it has received after it gets a signal and before it sends its token. When a nodes sees that some token was reduplicated, it simply ignores it.

Because our algorithm knows how to manage reduplicated messages, it is safe to use idempotent messages in order to significantly reduce the effect of message lost.

The only disadvantage of the proposed algorithm is its inability to collect (page) cycles of garbage. In our future research we will work to find efficient ways for collecting garbage cycles. One approach which may prove feasible is to snapshot all the reachability information into one of the nodes, and then to detect the connected components which do not include pointers from the root set. We will elaborate on this direction in the full version.

The algorithm that is presented in this work was designed as part of a collaborative effort for distributing Java on top of the MILLIPEDE system developed at the Technion - Israel Institute of Technology [2]. MILLIPEDE is a strong virtual parallel machine for distributed computing environments. It implements its own page-based DSM, which supports several memory consistency protocols. The current version of MILLIPEDE is multithreaded, is implemented in user-space on top of the Windows-NT operating system, and supports dynamic thread migration and a variety of optimization methods [8]. We refer the reader to MILLIPEDE web site to find more about this system: `www.cs.technion.ac.il/Labs/Millipede`.

References

1. S.Owicki A. Birell, G. Nelson and T. Wobber. Network objects. Technical Report 115, Digital Equipment Corporation Systems Research center , February 1994.
2. A. Schuster A. Itzkovitz and L. Wolfovich. Supporting multiple programming paradigm on top of a single virtual parallel machine. In *Proc. of the 2nd International Workshop on High-Level Parallel Programming Models and Supportive Environments*, pages 25–34, Geneve, April 1997.
3. David I. Bevan. Distributed garbage collection using reference counting. In *Parallel Architectures and Languages Europe*, volume 258 of *Lecture Notes in Computer Science*, pages 117–187. Springer-Verlag, 1987.
4. P. Ferreira and M. Shapiro. Asynchronous distributed garbage collection in a cached store. INRIA Rocquencourt,France, May 1996.
5. Richard E. Jones and Rafael D. Lins. Cyclic weighted reference counting without delay. In *Parallel Architectures and Languages Europe*, volume 694 of *Lecture Notes in Computer Science*. Springer-Verlag, 1993.
6. D. Lester. Distributed garbage collection of cyclic structures. *4th International Workshop on the Parallel Implementation of Functional Languages*, September 1992.
7. P. Dickman M. Shapiro and D. Plainfossé. Robust, distributed references and acyclic garbage collection. In *ACM Symp. On Principles of Distributed Computing*, Vancouver, August 1992.
8. A. Schuster and L. Shalev (Wolfovich). Access histories: How to use the principle of locality in distributed shared memory systems. Technical Report CS LPCR #9701 , Technion, January 1997.
9. R.W. Topor. Termination detection for distributed computations. *Information Processing Letters*, 18:33–36, 1984.
10. P. Watson and I. Watson. An efficient garbage collection scheme for parallel computer architecture. In *Parallel Architectures and Languages Europe*, volume 258 of *Lecture Notes in Computer Science*, pages 432–443. Springer-Verlag, 1987.
11. W. Yu and A. Cox. Conservative garbage collection on distributed shared memory system. In *Proc. Of the 16th International Conf. On Distributed Computing Systems*, pages 402 – 410, 1996.

Quasi-Fully Dynamic Algorithms for Two-Connectivity, Cycle Equivalence and Related Problems*

Madhukar R. Korupolu and Vijaya Ramachandran

Department of Computer Sciences,
The University of Texas at Austin,
Austin, Texas 78712.

Abstract. In this paper we introduce a new class of dynamic graph algorithms called *quasi-fully dynamic algorithms*, which are much more general than backtracking algorithms and are much simpler than fully dynamic algorithms. These algorithms are especially suitable for applications in which a certain core connected portion of the graph remains fixed, and fully dynamic updates occur on the remaining edges in the graph.

We present very simple quasi-fully dynamic algorithms with $O(\log n)$ worst case time per operation for 2-edge connectivity and cycle equivalence. The former is deterministic while the latter is Monte-Carlo type randomized. For 2-vertex connectivity, we give a randomized Las Vegas algorithm with $O(\log^4 n)$ expected amortized time per operation. We introduce the concept of quasi-k-edge-connectivity, which is a slightly relaxed version of k-edge connectivity, and show that it can be maintained in $O(\log n)$ worst case time per operation. We also analyze the performance of a natural extension of our quasi-fully dynamic algorithms to fully dynamic algorithms.

The quasi-fully dynamic algorithm we present for cycle equivalence (which has several applications in optimizing compilers) is of special interest since the algorithm is quite simple, and no special-purpose incremental or backtracking algorithm is known for this problem.

1 Introduction

Dynamic graph algorithms have received a great deal of attention in the last few years (see, e.g., [4]). These algorithms maintain a property of a given graph under a sequence of suitably restricted updates and queries. Throughout this paper we will be concerned with edge updates (insertions/deletions) only: insertion/deletion of isolated vertices can be implemented trivially in all the known dynamic graph algorithms. The existing dynamic algorithms can be classified into three types depending on the nature of (edge) updates allowed:

- **Partially Dynamic:** Only insertions are allowed (**Incremental**) or only deletions are allowed (**Decremental**).

---

* This research was supported in part by the NSF grant CCR/GER-90-23059. E-mail:
{madhukar, vlr}@cs.utexas.edu.

- **Backtracking:** Arbitrary insertions are allowed. But only backtracking deletions (Undo operation) are allowed [18].
- **Fully Dynamic:** Arbitrary insertions and arbitrary deletions are allowed.

Fully dynamic algorithms tend to involve complicated data structures and are quite difficult to implement. The deterministic fully dynamic algorithms for 2-edge connectivity (given in [3]), 2-vertex connectivity (given in [10]) and cycle equivalence (given in [8]) are good examples of this. The randomized fully dynamic algorithms for 2-edge connectivity (given in [9, 16]) and 2-vertex connectivity (given in [12]) are pretty involved too. In fact, the 2-vertex connectivity algorithm of [12] does not work for some graphs in which the maximum degree is $\omega(polylog(n))$ ([14]). In view of this, simpler algorithms will be more useful for applications which do not require the generality of the fully dynamic algorithms.

With this motivation, we consider another class of dynamic algorithms, where both insertions and deletions are allowed but the deletions are slightly restricted. The restriction is as follows: The algorithm maintains a spanning forest F of the current graph G. Arbitrary edge insertions are allowed; Arbitrary nonforest edge deletions are allowed; But deletion of a forest edge is allowed only if it is a cut edge. An operation which attempts to delete an edge of F which is not a cut edge, will be considered *invalid*. Such a valid sequence of operations, w.r.t. F, will be referred to as a *quasi-fully dynamic sequence* of operations (w.r.t. F) and the algorithms that support such a sequence of operations will be called *quasi-fully dynamic algorithms*. The algorithm will detect if a given operation is valid or not, and if not, it would flag an error.

Why quasi-fully dynamic algorithms?

Firstly, these algorithms are more general than backtracking algorithms; i.e., a backtracking sequence of operations is a quasi-fully dynamic sequence (w.r.t. a suitable spanning forest F). By maintaining F in a natural way, we can show that during a backtracking sequence of updates, a forest edge is deleted only if it is a cut edge. This natural way of maintaining F is the following: If u and v are in different connected components when the edge (u, v) is added, then this edge is added to F. Otherwise (u, v) never enters F. In this scenario when $(u, v) \in F$ is about to be deleted, all the edges that were inserted after (u, v) would have been removed and hence (u, v) would be a cut edge.

Another useful feature of the quasi-fully dynamic algorithms is the following: these algorithms can be extended to handle the invalid deletions in a way which is more efficient than rebuilding the entire data structure. On the other hand, in the backtracking algorithm of [18], performing an invalid operation requires the rebuild of the entire data structure. As expected, these invalid deletions can be very expensive in the worst case. Section 5 discusses this feature.

Secondly, these algorithms are much simpler than the fully dynamic algorithms. For instance, the quasi-fully dynamic algorithms for 2-edge connectivity and cycle equivalence are as simple as maintaining a dynamic tree data structure [20]. For the sake of completeness, a brief review of the essential features of the dynamic tree data structure is given in the complete version of this paper ([15]).

Thirdly, these algorithms would be ideal for situations where some core structure (of the input graph) remains fixed and the updates occur only on the remaining part. The only requirement is that we should be able to extract a spanning tree from the core structure.

Throughout the paper, n denotes the number of vertices in the graph G. Unless otherwise mentioned, deletions considered in quasi-fully dynamic algorithms will be valid ones only. The current status of the dynamic algorithms for connectivity, 2-connectivity and cycle equivalence is summarized below.

Connectivity: Backtracking connectivity can be solved in $\Theta(\log n/\log\log n)$ time per operation by a straightforward application of the backtracking algorithm for the union-find problem (see [23, 22]). Currently the best deterministic fully dynamic connectivity algorithm takes $O(\sqrt{n})$ time per update and $O(1)$ per query [3]. In [9] a randomized fully dynamic algorithm, taking $O(\log^3 n)$ amortized expected time per update and $O(\log n/\log\log n)$ worst case time per query, is presented. This paper also gives a simpler deterministic fully dynamic connectivity algorithm with $O(\sqrt{n}\log n)$ time per update. An empirical study of the dynamic connectivity algorithms is presented in [1].

2-Edge Connectivity and Quasi-k-Edge Connectivity: An incremental algorithm with $O(\alpha(m, n))$ amortized time per operation was given in [24, 17]. A backtracking algorithm with $O(\log n)$ worst case time per operation is presented in [18]. The best known deterministic fully dynamic algorithm takes $O(\sqrt{n})$ time per update and $O(\log n)$ time per query [3]. A randomized fully dynamic algorithm with an $O(\log^4 n)$ expected amortized time per update and $O(\log n/\log\log n)$ worst case time per query is claimed in [9] and [11]: the details of the algorithm presented there are rather sketchy. A somewhat different randomized fully dynamic algorithm with polylog time per operation is given in [16].

In this paper, we present a simple quasi-fully dynamic algorithm with the same time bounds as the backtracking case: $O(\log n)$ worst case time per operation. We then introduce the concept of quasi-k-edge connectivity and show that the above algorithm can be extended to answer the quasi-k-edge connectivity queries within the same time bounds.

Cycle Equivalence: Two edges e_1 and e_2 of an undirected graph are *cycle equivalent* iff the set of cycles that contain e_1 is exactly the same as the set of cycles that contain e_2. Finding the cycle equivalence classes is central to several compilation problems. (See [13, 21, 6] for some applications of cycle equivalence.) As mentioned in [8], dynamic algorithms for this problem can speed up incremental compilers.

No special-purpose incremental or backtracking algorithms are known for this problem. The only dynamic algorithms known for handling an incremental or a backtracking sequence of updates are the fully dynamic algorithms. A deterministic fully dynamic algorithm with $O(\sqrt{n}\log n)$ time per update and $O(\log^2 n)$ time per query is presented in [8]. A randomized fully dynamic algorithm with $O(\log^3 n)$ amortized expected time for updates and queries is given in [9].

In this paper, we present a very simple randomized quasi-fully dynamic algorithm which takes $O(\log n)$ worst case time per operation. This algorithm is Monte-Carlo type. We also show some connection of cycle equivalence to 3-edge connectivity.

Our quasi-fully dynamic algorithm for cycle-equivalence is of special interest because of the absence of special-purpose incremental/backtracking algorithms for this problem.

2-Vertex Connectivity: Incremental algorithms with $O(\alpha(m, n))$ amortized time per operation are given in [24, 17]. A backtracking algorithm with $O(\log n)$ worst case time per operation is presented in [18]. The best known deterministic fully dynamic algorithm takes $O(\sqrt{n} \log^2 n)$ amortized time per update and $O(1)$ worst case time per query [10]. A randomized fully dynamic algorithm with an $O(\log^4 n)$ expected amortized time per update and $O(\log^2 n)$ worst case time per query is stated in [12]: however, this algorithm does not work for some graphs in which the maximum degree is $\omega(polylog(n))$ ([14]).

In this paper, we present a randomized quasi-fully dynamic algorithm that takes $O(\log^4 n)$ amortized expected time per operation. This is the largest class of dynamic operations for which a polylog time bound per operation is currently known for biconnectivity on general graphs.

Towards Fully Dynamic Algorithms: In section 5, we analyze the complexity of fully dynamic algorithms obtained from our quasi-fully dynamic algorithms by implementing the invalid deletions in a natural way. We show that these algorithms can take $\Omega(n)$ time for certain operations. We also show that if we use a uniform random spanning tree, the worst case complexity is $\Omega(\sqrt{n})$ per operation. We leave open the possibility that some other natural extension of our quasi-dynamic algorithms could give fully dynamic algorithms that run in polylog time.

2   2-Edge Connectivity

In this section, we present a straightforward quasi-fully dynamic algorithm for 2-edge connectivity. Queries ask whether a given pair of vertices, u and v are 2-edge connected (or equivalently, whether there are at least two edge-disjoint paths between u and v). The algorithm takes $O(\log n)$ worst-case time for insertions, (valid) deletions and queries.

Let $G = (V, E)$ be a graph and $F \subseteq E$ be a spanning forest of G. We use F_{uv} to denote the tree path between vertices u and v. Edges in F will be called *tree edges* and edges in $E - F$ will be called *nontree edges*. A tree edge e is said to be *covered* (with respect to F) by a nontree edge (u, v) iff e lies on F_{uv}. Equivalently, e lies on the fundamental cycle of nontree edge (u, v). For an edge $e \in E$, we define $CoverSet_F(e) = \{e' \in E - F : e$ lies on the fundamental cycle of e' w.r.t. $F\}$. Observe that for a nontree edge $e \in E - F$, $CoverSet_F(e) = \{e\}$, and for a cutedge $e \in F$, $CoverSet_F(e) = \emptyset$. We sometimes use $cover_F(e)$ to denote $|CoverSet_F(e)|$. Throughout this paper, unless otherwise mentioned, covering will be with respect to F only and the subscript F will be dropped when there is no ambiguity.

Fact 1 *[5] Two vertices u and v are 2-edge connected iff $cover_F(e) \geq 1$ for every edge $e \in F_{uv}$.*

We store F in a dynamic tree data structure ([20]) with edge costs representing the cover values. The basic idea behind our algorithm is that the cover values of the tree edges (which are sufficient to answer the queries) can be maintained easily under insertions and valid deletions. During insertion (or deletion) of non-tree edges we add $+1$ (or -1) to the cover values on the corresponding tree path. Deletion of a tree edge is allowed only if it is a cut edge. Such a deletion will not affect the cover values of other tree edges. This is because such a tree edge does not lie on the fundamental cycle of any nontree edge. Hence its removal does not change any fundamental cycles and therefore does not change the cover values. To answer a query we just need to check the minimum cover value on the corresponding tree path. The implementations of the operations are briefly described below.

Query(u, v): If u and v are in different trees, then return *no*. If they are in the same tree, then perform $evert(v)$ followed by $min\_cost(u)$. If the minimum value is zero, then return *no*. Otherwise return *yes*.

Insert(u, v): If u and v are in the same tree, then mark (u, v) as a nontree edge and perform $evert(v)$ and $add\_cost\_path(u, 1)$. If u and v are in different trees, then mark (u, v) as a tree edge and modify F as follows : perform $evert(u)$, followed by $link(u, v)$; set $cover(u, v) \leftarrow 0$.

Delete(u, v): If (u, v) is a nontree edge, then perform $evert(v)$ followed by $add\_cost\_path(u, -1)$. If (u, v) is a tree edge and if it is not a cut edge, then report *invalid deletion*. If it is a cut edge then modify F as follows: perform $evert(v)$ followed by $cut(u)$.

Complexity: To check whether a tree edge is a cut edge, we can just check whether its cover value is zero. Testing whether u and v are in the same tree can be done by performing $evert(v)$ and checking whether $root(u) = v$. Thus each of the insert, delete and query operations uses only a constant number of dynamic tree operations, and hence takes $O(\log n)$ worst case time. If amortized $O(\log n)$ time is sufficient, then a simpler implementation of dynamic trees can be used (see [20]).

Other Types of 2-Edge Connectivity Queries: We note that our algorithm can be extended to answer two other types of 2-edge connectivity queries as well.

Firstly, we can test whether a given edge e is a cut edge. For this, we just need to check whether $e \in F$ and $cover(e) = 0$. This takes $O(\log n)$ time per operation in the worst-case.

Secondly, we can also test whether the entire graph is 2-edge connected. This query is equivalent to asking whether there exists a cut edge in the entire graph. To answer this query, it suffices to find the global minimum cover value (i.e., $min_{e \in F} cover_F(e)$) and check whether it is zero. The original dynamic tree data structure (of [20]) supports only the path minimum operation and does not support the global minimum operation. We use an enhancement of the dynamic tree data structure which supports the global minimum operation in $O(\log^2 n)$ time per operation, by using a suitable auxiliary heap structure. Details are given in [15].

2.1 Quasi-k-Edge Connectivity

In this subsection, we introduce the concept of quasi-k-edge connectivity which is a restriction of the concept of k-edge connectivity to the case where only the valid edge deletions are allowed. We show that quasi-k-edge connectivity information can be maintained easily with $O(\log n)$ worst case time per operation.

Definition 1 *(Quasi-k-Edge Connectivity) Given a graph $G = (V, E)$ with a spanning forest F, two vertices $u, v \in V$ are said to be quasi-k-edge connected iff with respect to F, no quasi-fully dynamic sequence of $k - 1$ edge deletions disconnects u and v.*

Observe that u and v are quasi-2-edge connected (w.r.t. any spanning forest) iff they are 2-edge connected. For $k > 2$, it is possible that a pair of vertices u and v that are quasi-k-edge connected are not k-edge connected, since there could exist a set of $k - 1$ edges (of which at least two are tree edges), whose deletion disconnects u and v.

The relevance of this concept can be understood by considering graphs for which a core part remains fixed while updates occur on the remaining part. Assume that we can extract a (fixed) spanning tree T from the core part of the graph. Then, all valid sequences of edge deletions are quasi-fully dynamic sequences. Further, if vertices u and v are quasi-k-edge connected, then at least $k - 1$ valid deletions are needed on the current graph in order to reduce the number of paths between u and v to one, which is the minimum possible (since T forms a core part of the graph). The proof of the following Lemma is given in the complete version ([15]).

Lemma 1 *Two vertices u, v are quasi-k-edge connected, w.r.t. a spanning forest F, iff $cover_F(e) \geq k - 1$ for every edge $e \in F_{uv}$.*

The above lemma immediately implies a quasi-fully dynamic algorithm with $O(\log n)$ worst case time per operation. The update operations are implemented exactly as in the 2-edge connectivity case. A query asks whether u and v are quasi-k-edge connected (w.r.t. F). It can be answered as follows:

Query(u, v): If u and v are in different trees, then return *no*. If they are in the same tree, then perform $evert(v)$ followed by $min\_cost(u)$. If the minimum value is less than $k - 1$, then return *no*. Otherwise return *yes*.

Thus each of the insert, (valid) delete, quasi-k-edge connectivity queries can be performed in $O(\log n)$ worst-case time.

3   Cycle Equivalence

We present a simple quasi-fully dynamic randomized algorithm for the cycle equivalence problem (defined below) that takes $O(\log n)$ worst-case time for updates (insertions and valid deletions) and queries.

Definition 2 *(Cycle Equivalence) Edges* $e_1, e_2 \in E$ *are cycle equivalent iff the set of cycles that contain* e_1 *is exactly the same as the set of cycles that contain* e_2.

Note that edges e_1 and e_2 are cycle equivalent iff $CoverSet(e_1) = CoverSet(e_2)$. A pair of edges (e_1, e_2) will be called a *cut-edge pair* iff the removal of e_1 and e_2 increases the number of connected components in the graph. As observed in [8] two edges are cycle equivalent iff they are a cut-edge pair in the graph.

3.1   Isolating Lemma

In this subsection, we describe a simple probabilistic lemma which is the basis for our cycle equivalence algorithm. A *set system* (S, Q) consists of a finite set S of elements, i.e., $S = \{x_1, x_2, \ldots x_{|S|}\}$, and a family Q of subsets of S, i.e., $Q = \{S_1, S_2, \ldots S_{|Q|}\}$ with $S_j \subseteq S$, for $1 \le j \le |Q|$. We assign a weight $w(x_i)$ to each element $x_i \in S$, and define the weight of a subset S_j to be $\sum_{x_i \in S_j} w(x_i)$.

Lemma 2 *[19]* **(Unique Set Isolating Lemma)** *Let* (S, Q) *be a set system. If the elements in* S *are assigned integer weights chosen randomly and uniformly from* $[1 \ldots 2|S|]$, *then* $Pr[Minimum\ weight\ subset\ in\ Q\ is\ unique] \ge 1/2$.

The above lemma enables us to assign random polynomial weights such that a single subset in Q is isolated from the other subsets in Q. We strengthen the above lemma so that we can isolate all subsets in Q. i.e., we want each subset in Q to have a distinct weight. If exponential weights are allowed, this task becomes easy (even deterministically): just assign $w(x_i) = 2^i$, and represent each subset as a bit vector of length $|S|$. However as we restrict ourselves to polynomial sized weights only, the task is no longer trivial. The following lemma gives a randomized method for this task, using only polynomial sized weights.

Lemma 3 (All Sets Isolating Lemma) *Let* (S, Q) *be a set system. Let* $Z = |Q|^2|S|$. *If the elements in* S *are assigned integer weights chosen randomly and uniformly from* $[1 \ldots Z]$, *then* $Pr[every\ subset\ in\ Q\ gets\ a\ distinct\ weight] \ge 1/2$.

Proof: Let $W : S \to [1 \ldots Z]$ be a uniformly random weight assignment. An element $y \in S$ is said to be W-**bad** if there exist distinct subsets S_i, S_j in Q such that $y \in S_i - S_j$ and $W(S_i) = W(S_j)$.

Consider a fixed element $y \in S$. Suppose we fix $W(x)$, for all the elements, x in S, except y. $W(y)$ is still to be chosen. We are interested in knowing the number of values of $W(y)$ that would make it W-bad. For each pair (S_i, S_j) such that $y \in S_i - S_j$, there is at most one value of $W(y)$ that would make

$W(S_i) = W(S_j)$. Hence there are at most $|Q|^2/2$ values of $W(y)$ that could cause y to become W-bad. Hence $Pr_W[y$ is W-bad$] \leq |Q|^2/2Z = 1/(2|S|)$. Here, the subscript W indicates that the probability is over all choices of W. We now have,

$Pr_W[$Some two subsets in Q get the same weight with assignment $W]$
$\quad = Pr_W[\exists S_i, S_j \in Q$ such that $S_i \neq S_j$ and $W(S_i) = W(S_j)]$
$\quad = Pr_W[\exists x \in S$ such that x is W-bad$]$
$\quad \leq \sum_{x \in S} Pr_W[x$ is W-bad$]$
$\quad \leq 1/2$.

In the second step above, we used the fact that any element $x \in (S_i - S_j) \cup (S_j - S_i)$ will be W-bad. In the third step, we used the Boole's inequality.

□

Corollary: Let $Z = |Q|^2|S|$. If the elements in S are assigned integer weights uniformly at random from $[1 \ldots NZ]$, then $Pr[$some two distinct subsets in Q get the same weight$] \leq 1/(2N)$.

3.2 Algorithm for Cycle Equivalence

Let $S = E - F$ be the set of nontree edges in G and let $Q = \{CoverSet(e) : e \in E\}$. We note that (S, Q) is a set system, with $|S| \leq |E| < n^2$ and $|Q| \leq |E| < n^2$. Hence, we can choose $Z = n^6$.

For each nontree edge of G, we assign a random weight chosen uniformly from $[1 \ldots NZ]$. Let W be this random weight assignment. For an edge $e \in E$, define $cost(e) = W(CoverSet(e))$. Then it follows from the isolating lemma that with high probability, e_1 and e_2 are cycle equivalent iff $cost(e_1) = cost(e_2)$. Our quasi-fully dynamic algorithm is based on this idea, and is outlined below.

For a nontree edge $e \in E - F$, as $CoverSet(e) = \{e\}$, we have $cost(e) = w(e)$. Note that the cost of a nontree edge does not change as long as that edge is in the graph. For a tree edge $e \in F$, we maintain $cost(e)$ using a dynamic tree data structure. We store F in a dynamic tree data structure with costs on (tree) edges. The update operations (insertions and valid deletions) are very similar to those in 2-edge connectivity. The only difference is the following: when an edge $e = (u, v)$ is inserted as a nontree edge, then we pick $w(e)$ uniformly at random from $[1..NZ]$ and perform $add\_cost\_path(F_{uv}, w(e))$ (instead of $add\_cost\_path(F_{uv}, 1)$ in the 2-edge connectivity case). Correspondingly, when a nontree edge $e = (u, v)$ is deleted, we perform $add\_cost\_path(F_{uv}, -w(e))$. For a $query(e_1, e_2)$, it suffices to check whether $cost(e_1) = cost(e_2)$. For a tree edge e, the $cost(e)$ can be obtained from the dynamic tree in $O(\log n)$ time.

Complexity and Error Probability: If we choose N to be a polynomial in n, the costs will still be polynomial in n (i.e., they are only $O(\log n)$ bits long). Hence addition operations on these can still be done in $O(1)$ time, assuming that each word in memory is of size $\Theta(\log n)$. Thus updates and queries can be performed in $O(\log n)$ worst-case time.

The above algorithm is a Monte-Carlo randomized algorithm. For each query, there is a nonzero probability that it may return *yes* even when e_1 and e_2 are not cycle equivalent. The probability of error can be made as small as desired. To be precise, it can be made $O(n^{-c})$ for any constant c, by choosing $N = n^c$. When a sequence of operations is performed on the graph, the error probabilities will add up. To keep the overall error probability small, it suffices to modify the range of the weights suitably. For instance, if we know an upper bound M on the number of operations (insertions, valid deletions, queries) to be performed, then by choosing the weights from $[1..(MNZ)]$, we can show that $Pr[$Algorithm gives a wrong answer some time during the sequence$] \leq 1/(2N)$. In this case, the weights will be $O(\log M + \log n)$ bits long and each addition operation on these weights would take $O(max\{1, \log M / \log n\})$ time. Hence each operation would cost $O(\log n + \log M)$ time in the worst-case.

3.3 Relation to 3-Edge Connectivity

A 3-edge connectivity query asks whether there are 3 edge-disjoint paths between two given vertices u and v. The following lemma gives a characterization of the 3-edge connectivity property and reveals its close connection to the cycle equivalence problem; its proof is given in [15].

Lemma 4 *Two vertices u and v are not 3-edge connected iff one of the following holds:*

1. *There exists a tree edge e on F_{uv} such that $cover(e) \leq 1$; OR*
2. *There exists a tree edge e on F_{uv} and another tree edge y on $F - F_{uv}$ such that e and f are cycle equivalent.*

Condition (1) can be checked easily by storing F as a dynamic tree data structure augmented with edge costs as the cover value (as in the 2-edge connectivity algorithm, section 2). The only remaining task for obtaining a quasi-fully dynamic algorithm for 3-edge connectivity is to be able to check condition (2). In view of its close relation to the cycle equivalence problem, it would be interesting to see if this can be tested efficiently.

4   2-Vertex Connectivity

In the 2-vertex connectivity problem, a query asks whether two given vertices u and v are in the same biconnected component of G (or equivalently, whether there exist two vertex-disjoint paths between u and v). In this section, we present a quasi-fully dynamic algorithm for 2-vertex connectivity.

Let F be a spanning forest of G. For a vertex v, we use $N_F(v)$ to denote the set of neighbors of v in F. Let $x, y \in N_F(v)$. We use $(F - v)_x$ to denote the subtree of $F - v$ that contains x. We define the *neighborhood graph* of v, denoted $NG_F(v)$, as follows: the vertex set is $N_F(v)$; an edge (x, y) is present in $NG_F(v)$ iff there exists a nontree edge in G between $(F - v)_x$ and $(F - v)_y$. The following lemma is a direct consequence of the above definition.

Lemma 5 *Suppose* $x, y \in N_F(v)$. *Then* x *and* y *belong to the same biconnected component of* G *iff* x *and* y *belong to the same connected component of* $NG_F(v)$.

Our dynamic algorithm uses some additional parameters which are defined below. For $x, y \in N_F(v)$, we use $D_v(x, y)$ to denote the number of nontree edges in G from $(F - v)_x$ to $(F - v)_y$. We also define $C_v(x, y)$ to be 1 if x, y belong to the same connected component of $NG_F(v) - (x, y)$ and 0 otherwise. Here, $NG_F(v) - (x, y)$ denotes the graph obtained by deleting the edge (x, y), if it exists, from $NG_F(v)$. Observe that $C_v(x, y)$ captures the existence of an indirect path (i.e., using more than one edge) from x to y in $NG_F(v)$ while $D_v(x, y)$ captures the existence of a direct path (i.e., using exactly one edge) from x to y in $NG_F(v)$. Finally, we use $L_v(x, y)$ to denote $C_v(x, y) + D_v(x, y)$. By the above lemma, x and y belong to the same biconnected component of G iff $L_v(x, y) > 0$. Strictly speaking, the above three definitions must specify the underlying spanning forest F; but for ease of notation we drop F when there is no ambiguity.

We store F, augmented with costs at vertices, in a dynamic tree data structure. The costs on the vertices have the following interpretation: If vertex v lies on a solid path, with (v, x) and (v, y) as its solid edges, then $cost(v) = L_v(x, y)$. Otherwise (i.e., if v has at most one solid edge incident on it) $cost(v)$ is arbitrary.

Fact 2 *[7] Two vertices* u *and* v *are biconnected iff after making* F_{uv} *a solid path, no internal vertex on* F_{uv} *has cost* 0.

4.1 Overview of the Algorithm

To determine whether u and v are biconnected, we will transform the tree path F_{uv} into a solid path, and check whether the minimum cost on this solid path is zero. We will see later that even for inserting (or deleting) a nontree edge (u, v) it suffices to transform the tree path F_{uv} into a solid path and increment (or decrement) the cost of each vertex on this solid path by one.

Hence the basic requirement is to convert the tree path F_{uv} into a solid path. Using the *expose* and *evert* operations, this takes $O(\log n)$ amortized time, provided no costs are updated. However, our data structure has to update the cost of a vertex v whenever the edges incident on v change from solid to dashed or vice-versa. The basic dynamic tree operation that converts dashed edges into solid edges (and vice-versa) is the *splice* operation (see [15] for relevant details about this operation). We will show that the splice operation can be extended so that the costs at the vertices involved in the splice operation can be correctly updated in roughly $O(\log^3 n)$ amortized expected time. This would imply that each of the update/query operations takes roughly $O(\log^4 n)$ amortized expected time.

Additional Data Structures: The *modified neighborhood graph* of a vertex v, denoted by $NG'_F(v)$, is defined as follows: if vertex v has two solid edges, say (v, x) and (v, y), then $NG'_F(v) = NG_F(v) - (x, y)$; otherwise $NG'_F(v) = NG_F(v)$. Note that $NG'_F(v)$ changes if the solid edges incident on v change.

For each vertex v, we store the modified neighborhood graph, $NG'_F(v)$, in a randomized fully dynamic connectivity data structure [9].

For each vertex v, we also use a *dictionary* DDS_v to store the nonzero $D_v(x, y)$ values. To keep the space requirements small, the zero $D_v(x, y)$ values are not stored. A dictionary is a data structure that supports insert, delete and search operations on a set of items. Dictionary implementations which take $O(\log r)$ worst case time for each operation are well known (r is the number of items in the dictionary). More specifically, the DDS_v stores the set $\{D_v(x, y) : x, y \in N_T(v) \quad \text{and} \quad D_v(x, y) > 0\}$. The items in DDS_v are indexed by the (x, y) tuple.

4.2 The Update and Query Operations:

In the next subsection we present an augmented splice operation that correctly updates the costs on the vertices when their incident edges change from dashed to solid or vice-versa. In this subsection, we present simple implementations of the update and query operations assuming that the splice operation (and hence the *expose* and *evert* operations) correctly updates the costs.

Query(u, v): If u and v are in different trees, then return *no*. If u and v are in the same tree, then perform $evert(v)$ followed by $min\_cost(u)$. This gives the minimum cost on the solid path F_{uv}. If the minimum value is zero, then return *no*. Otherwise return *yes*.

Insert(u, v): If u and v are in the same tree, then mark (u, v) as a nontree edge. We convert the tree path F_{uv} into a solid path and add 1 to all the costs on this path as follows: perform $evert(v)$ followed by $add\_cost\_path(u, 1)$. If u and v are in different trees, then mark (u, v) as a tree edge and modify F (and the neighborhood graphs) as follows: perform $evert(u)$, followed by $link(u, v)$; set $cover(u, v) \leftarrow 0$; Add isolated vertices u and v in $NG'_F(v)$ and $NG'_F(u)$ respectively.

Delete(u, v): If (u, v) is a nontree edge, then we make the path F_{uv} solid and decrement the costs along the path as follows: perform $evert(v)$ followed by $add\_cost\_path(u, -1)$. If (u, v) is a tree edge and if it is not a cut edge then report *invalid deletion*. If it is a cut edge, we modify F (and the neighborhood graphs) as follows: perform $evert(v)$, followed by $cut(u)$; Remove the isolated vertices u and v from $NG'_T(v)$ and $NG'_T(u)$ respectively.

Note that each of the above operations invokes a constant number of dynamic tree operations.

Proof of Correctness: To prove that the queries give the correct answer, it suffices to argue that for every vertex v, as long as edges (v, x) and (v, y) remain solid, $cost(v)$ is always equal to $L_v(x, y)$.

Lemma 6 *Suppose that at a vertex v, edges (v, x) and (v, y) become solid at some time and that the $cost(v)$ is correctly set to $L_v(x, y)$ at that time. Then as long as both (v, x) and (v, y) remain solid, after any sequence of updates and queries, $cost(v)$ will be equal to $L_v(x, y)$.*

Proof: As $L_v(x, y) = C_v(x, y) + D_v(x, y)$, we will show that $C_v(x, y)$ will not change at all and that changes in $D_v(x, y)$ are correctly reflected in $cost(v)$.

Firstly, note that $NG'_F(v)$ does not change as long as both (v, x) and (v, y) remain solid. This is because, the only way $NG'_F(v)$ can change is if a nontree edge between $(F - v)_w$ and $(F - v)_z$ is inserted or deleted for some $w, z \in N_F(v)$ $(w, z \neq x, y)$. But when this happens edges (v, w) and (v, z) become solid and at least one of (v, x) and (v, y) becomes dashed. This implies that $C_v(x, y)$ does not change.

Secondly, insertion (deletion) of a nontree edge from $(F - v)_x$ to $(F - v)_y$ increments (decrements) both $cost(v)$ and $D_v(x, y)$.

$\square$

To complete the proof of correctness, it remains to ensure that $cost(v)$ is correctly initialized each time the solid edges at v change. We do this in the next subsection.

4.3 Modifying the Splice Operation

In this subsection, we look at the modifications to the splice operation which are needed to ensure that $cost(v)$ is correctly initialized each time the solid edges ar v change.

Let v be a parent of w and also let (v, y) and (v, x) be the solid edges incident on v. We only consider the case where both x and y exist. The other cases can be handled in a similar fashion. The $splice(w)$ operation converts the dashed edge (w, v) to solid and the solid edge (v, y) to dashed. During this conversion, we need to store and update the parameters corresponding to the outgoing solid edge (v, y). More specifically, we update the modified neighborhood graph of v (i.e., $NG'_F(v)$) and we also store the parameter $D_v(x, y)$ in DDS_v. Similarly for the incoming solid edge (w, v), we update $NG'_F(v)$ and compute the new $cost(v)$ using $DDS_v(w, x)$ and $NG'_F(v)$. The extended $splice$ operation, which performs all these updates, is outlined below in pseudocode.

Splice(w):
    { $parent(w) = v$; Edge (w, v) is solid }
    { Current solid edges at v are (v, x) and (v, y) }
begin
    Perform the steps of the original splice operation (see [15, 20]);
    {For outgoing solid edge (v, y) }
        Compute $C_v(x, y)$ by a query $Connected(x, y)$? in $NG'_F(v)$;
        $D_v(x, y) \leftarrow cost(v) - C_v(x, y)$;
        if $D_v(x, y) > 0$ **then**
            $insert\_edge(x, y)$ into $NG'_F(v)$;
            $insert\_item(x, y, D_v(x, y))$ into DDS_v;
        endif
    { For incoming solid edge (w, v) }
        $delete\_edge(w, x)$ from $NG'_F(v)$ if this edge exists;
        Compute $C_v(x, w)$ by a query $Connected(x, w)$? in $NG'_F(v)$;

Search for $D_v(x, w)$ in DDS_v;
if found **then** $delete\_item(x, w, D_v(x, w))$ from DDS_v;
else $D_v(x, w) \leftarrow 0$
endif
$cost(v) \leftarrow C_v(x, w) + D_v(x, w)$
end

Let $G = (V, E)$ be a graph with n vertices and m_0 initial edges. For the sake of simplicity, we keep all n vertices in $NG'_F(v)$, for each $v \in V$, instead of just $|N_F(v)|$. Moreover, for the sake of analysis, we build the $NG'_F(v)$ structures as follows: We first put n isolated vertices in each of them. We then insert the m_0 edges of G one at a time into our quasi-fully dynamic 2-vertex connectivity structure, which causes the appropriate modifications of $NG'_F(v)$. This ensures that each of the $NG'_F(v)$ starts with no initial edges. The proof of the following theorem is given in the complete version of the paper ([15]).

Theorem 1 *Let $G = (V, E)$ be a graph with n vertices and m_0 initial edges.*

1. **Space Bound:** *Our quasi-fully dynamic algorithm for 2-vertex connectivity uses $O(n^2 \log n)$ space.*
2. **Time Bound:** *The expected running time for a sequence of k insert, (valid) delete and biconnectivity query operations is $O(n^2 \log n + (k + m_0) \log^4 n)$.*

Thus if $k = \Omega(m_0 + n^2)$, the above lemma implies an amortized expected cost of $O(\log^4 n)$ per operation.

5 Towards Fully Dynamic Algorithms

In this section, we analyze the performance of fully dynamic algorithms obtained by a natural extension of our quasi-fully dynamic algorithms. To extend the quasi-fully dynamic algorithms to fully dynamic algorithms we need to show how the invalid tree edge deletions are to be handled. Recall that, for a tree edge $e \in F$, $CoverSet_F(e) = \{e' \in E - F : e' \text{ covers } e \text{ with respect to } F\}$. Deleting a tree edge e with $CoverSet(e) = \emptyset$ is a valid operation and it is handled by the quasi-fully dynamic algorithm. On the other hand, deletion of a tree edge e with $CoverSet(e) \neq \emptyset$ is an invalid operation in the quasi-fully dynamic case.

One natural way of implementing an invalid deletion of a tree edge e is the following: first find $CoverSet(e)$, delete the nontree edges in $CoverSet(e)$ one at a time, then delete the tree edge e (which is now a cut edge), and then reinsert the edges of $CoverSet(e)$ one at a time. Pick a random edge of $CoverSet(e)$ to replace e in F. The complexity of this implementation is analyzed below.

Finding CoverSet(e): We store G in a randomized fully dynamic connectivity data structure ([9]) with F as the spanning forest. For a tree edge e, the $CoverSet(e)$ can be found as follows: let e' be the random edge found by the algorithm in [9] to replace e in F. Remove e' and find its next replacement, remove that and so on until no more replacements exist. All these removed replacement edges form the $CoverSet(e)$.

Overall Complexity: Using the above method for finding the $CoverSet(e)$, the time taken to delete e is $O(|CoverSet(e)| \cdot polylog(n))$. However, in the worst case, $|CoverSet(e)|$ can be $\Omega(m)$ (as shown by the example in [15]). This implies that for the above natural implementation of invalid deletions, there exist sequences of operations in which the expected cost of deleting an edge is $\Omega(m)$.

Interestingly, a similar result holds even if F were chosen uniformly at random from the set of all labeled spanning trees of G (see, e.g., [2] for properties and construction of random spanning trees). The proof of the following lemma is given in [15].

Lemma 7 *Let U be the uniform distribution on the set $\mathcal{F}$ of all labeled spanning forests of $G = (V, E)$ (i.e., $U[F_i] = 1/|\mathcal{F}|$). Then, $\exists G$ and $\exists e \in E$, such that the expected cost of deleting e using the fully dynamic algorithm described above is $\Omega(\sqrt{n})$.*

We leave it as an open question to determine whether there exists some other reasonable extension of our quasi-fully dynamic algorithms that leads to fully dynamic algorithms that run in polylog time per operation.

References

1. David Alberts, Giuseppe Cattaneo, and Giuseppe F. Italiano. An empirical study of dynamic graph algorithms. In *Proceedings of the Seventh Annual ACM SIAM Symp. on Discrete Algorithms*, pages 192–201, 1996.
2. David A. Aldous. The random walk construction of uniform spanning trees and uniform labelled trees. *Siam J. Disc. Math.*, 3(4):450–465, November 1990.
3. D. Eppstein, Z. Galil, and G. Italiano. Improved sparsification. Technical Report 93-20, University of California at Irvine, Dept of Information and Computer Science, 1993.
4. J. Feigenbaum and Sampath Kannan. *Handbook of Discrete and Combinatorial Mathematics*, chapter Dynamic Graph Algorithms, pages 583–591. 1995.
5. G.N. Frederickson. Ambivalent data structures for dynamic 2-edge connectivity and k smallest spanning trees. In *Proceedings of 32nd Symp. on Foundations of Computer Science*, pages 632–641, 1991.
6. R. Gupta and M.L. Soffa. Region scheduling. In *Proc. 2nd International Conference on Supercomputing*, pages 141–148, 1987.
7. M. Rauch Henzinger. Fully dynamic biconnectivity in graphs. In *Proceedings of 33rd Symp. on Foundations of Computer Science*, pages 50–59, 1992.
8. M. Rauch Henzinger. Fully dynamic cycle equivalence in graphs. In *Proceedings of 35th Symposium on Foundations of Computer Science*, pages 744–755, 1994.
9. M. Rauch Henzinger and V. King. Randomized dynamic algorithms with polylogarithmic time per operation. In *Proceedings of 27th Annual Symp. on Theory of Computing*, pages 519–527, 1995.
10. M. Rauch Henzinger and J. A. La Poutre. Certificates and fast algorithms for biconnectivity in fully-dynamic graphs. In *Proceedings of Third Annual European Symposium on Algorithms (ESA)*, pages 171–184, 1995.

11. Monika Henzinger and Valerie King. Personal communication, July-August 1996.
12. M.R Henzinger and Valerie King. Fully dynamic biconnectivity and transitive closure. In *Proceedings 36th Symp. on Foundations of Computer Science*, pages 664–672, 1995.
13. Richard Johnson, David Pearson, and Keshav Pingali. Finding regions fast: Single entry single exit and control regions in linear time. In *Proceedings of ACM SIGPLAN '94 Conference on Programming Language Design and Implementation*, pages 171–185, 1994.
14. Valerie King. Personal communication, July 1996.
15. Madhukar R. Korupolu and Vijaya Ramachandran. Quasi-fully dynamic algorithms for two-connectivity, cycle equivalence and related problems. Technical report TR97-14, Univ. of Texas at Austin, Dept. of Computer Sciences, 1997.
16. Madhukar R. Korupolu. Randomized fully dynamic two edge connectivity: A variant of the Henzinger-King sketch. Manuscript, Univ of Texas at Austin, May 1997.
17. J.A. La Poutre. Maintenance of 2- and 3- connected components of graphs, part ii: 2- and 3- edge connected components and 2-vertex connected components. Technical Report RUU-CS-90-27, Utrecht University, 1990.
18. J.A. La Poutre and J. Westbrook. Dynamic two-connectivity with backtracking. In *Proceedings of 4th Symp. on Discrete Algorithms*, pages 204–212, 1994.
19. Ketan Mulmuley, U.V. Vazirani, and V.V. Vazirani. Matching is as easy as matrix inversion. *Combinatorica*, 7(1):105–113, 1987.
20. D.D. Sleator and R.E. Tarjan. A data structure for dynamic trees. *J. Comput. System Sci.*, 26:362–391, 1983.
21. R.E Tarjan and Valdes Jacobo. Prime subprogram parsing of a program. In *Conference record of the Seventh Annual ACM Symp. on Principles of Programming Languages*, pages 28–30, 1980.
22. J. Westbrook. *Algorithms and Data Structures for Dynamic Graph Problems*. PhD thesis, Dept of Computer Science, Princeton University, Princeton, NJ, 1989.
23. J. Westbrook and R.E. Tarjan. Amortized analysis of algorithms for set union with backtracking. *SIAM Jl. Computing*, 18:1–11, 1989.
24. J. Westbrook and R.E. Tarjan. Maintaining bridge connected and biconnected components online. *Algorithmica*, pages 433–464, 1992.

Minimum Spanning Trees in d Dimensions

Drago Krznaric[1], Christos Levcopoulos[1], and Bengt J. Nilsson[1]

Department of Computer Science, Lund University, Box 118, S-221 00 Lund, Sweden

Abstract. It is shown that a minimum spanning tree of n points in $\mathbb{R}^d$ can be computed in optimal $O(T_d(n, n))$ time under any fixed L_t-metric, where $T_d(n, m)$ denotes the time to find a bichromatic closest pair between n red points and m blue points. The previous bound was $O(T_d(n, n) \log n)$ and it was proved only for the L_2 (Euclidean) metric. Furthermore, for $d = 3$ it is shown that a minimum spanning tree can be found in optimal $O(n \log n)$ time under the L_1 and L_∞-metric. The previous bound was $O(n \log n \log \log n)$.

1 Introduction

Let S be a set of n points in d-dimensional real space where $d \geq 1$ is an integer constant. A *minimum spanning tree* (MST) of S is a tree with S as its set of nodes such that the total length of its edges is minimized, where the length of an edge equals the distance between its endpoints. For $d = 2$, Shamos and Hoey [12] showed that a Euclidean MST can be found in $O(n \log n)$ time. For $d \geq 3$, Yao [14] developed an algorithm that finds a MST in $o(n^2)$ time under the L_1, L_2 and L_∞-metric (these metrics are defined in the next section). A more efficient algorithm for computing a Euclidean MST in more than two dimensions was proposed by Agarwal et al. [1]. For $d = 3$, their algorithm runs in randomized expected time $O((n \log n)^{4/3})$, and for $d \geq 4$, in expected time $O(n^{2-2/(\lceil d/2 \rceil + 1) + \epsilon})$, where ϵ is an arbitrarily small positive constant.

The algorithm of Agarwal et al. builds on exploring the relationship between computing a Euclidean MST and finding a bichromatic closest pair between n red points and m blue points. They show that if $T_d(n, m)$ denotes the time to solve the latter problem, then a Euclidean MST can be computed in $O(T_d(n, n) \log^d n)$ time. Recently, Callahan and Kosaraju [3] improved this bound to $O(T_d(n, n) \log n)$. Both methods achieve running time $O(T_d(n, n))$ if $T_d(n, n) = \Omega(n^{1+\alpha})$ for some $\alpha > 0$. Callahan and Kosaraju also show that a spanning tree of length within a factor $1 + \epsilon$ from that of a Euclidean MST can be computed in $O(n(\log n + \epsilon^{-d/2} \log \epsilon^{-1}))$ time. Approximation algorithms with worse tradeoffs between time and quality had earlier been developed by Clarkson [4], Vaidya [13], and Salowe [11].

In this paper we show that the problem of computing a MST and that of finding a bichromatic closest pair are equivalent to within a constant factor. That is, we show that to compute a MST of S takes $\Theta(T_d(n, n))$ time in the worst case. Moreover, this is true even when distances are measured according to any given L_t-metric such that $t \geq 1$ is an integer constant or equals infinity.

In addition, our algorithm computes a bichromatic closest pair only when we have two small clusters, one red and one blue, far from each other. For $d = 3$, this enables us to find a MST of S in optimal $O(n \log n)$ time under the L_1 and L_∞-metric. The previous best bound that we are aware of is due to Gabow et al. [5], who proved that, for $d = 3$, a MST can be computed in $O(n \log n \log \log n)$ time under the L_1 and L_∞-metric.

2   Conventions

We use $d(x, y)$ to denote the distance between points $x = (x_1, x_2, \ldots, x_d)$ and $y = (y_1, y_2, \ldots, y_d)$, that is,

$$ d(x, y) = \left(\sum_{i=1}^{d} |x_i - y_i|^t \right)^{1/t} . $$

Note that the L_∞-distance is given by $\max\{|x_i - y_i| \mid 1 \leq i \leq d\}$. Throughout the paper, $\epsilon > 0$ is a sufficiently small constant. (An upper bound for ϵ is given implicitly in the proof of Lemma 4.)

3   Representing Trees of a Forest

At each step of the algorithm we have a forest. The initial forest is the set S of points, that is, each point of the input constitutes an individual edge-less tree. Then, as long as there is more than one tree in the forest, we merge two trees by producing an edge connecting two nodes, one from each tree. After this procedure, the produced edges comprise the tree that remains in the forest, and this tree constitutes the output of the algorithm. If a shortest possible edge is produced at each step then it is a well-known fact that the algorithm will output a minimum spanning tree (for example, see [10]).

For each tree T in the forest we have a nonempty subset of its nodes representing it. The nodes in such a subset will be called *leaders*. Further, every leader v of T has a set including v of so-called ϵ-*nodes*, each ϵ-node being a node of T. On the other hand, each node of T is an ϵ-node of exactly one leader. We use the name ϵ-node because each ϵ-node is within distance $< \epsilon \cdot l$ from its leader if the next edge to be produced has length $\geq l$.

All leaders of all trees in the current forest are stored in a data structure for dynamic closest pair queries. Henceforth, that data structure will be referred to as the *DCP-structure*. The DCP-structure supports the following three operations: (i) find a closest pair of points, (ii) delete a point, and (iii) insert a new point. We assume that the DCP-structure uses linear space and makes it possible to carry out each of these three operations in logarithmic worst-case time, and that it fits in the algebraic computation tree model. These requirements can be met by using the data structure of Bespamyatnikh [2].

Now, suppose that the next edge to be produced has length l. From what has been said above it follows that this edge connects an ϵ-node of one leader with an ϵ-node of another leader, where these two leaders are within distance $< l(1+2\epsilon)$ from each other and belong to different trees of the current forest. So, to determine which edge to produce next, we only need to consider such pairs of leaders. The following result from [8] states an upper bound for how fast these pairs can be found.

Fact 1 (Lemma 3.1 in [8]). *Let S be a set of n points in $\mathbb{R}^d$ where $d \geq 1$ is an integer constant, and let l be a positive real such that there are at most $O(1)$ points from S within distance less than l from any point. Then, if the points of S are stored in the DCP-structure, the set $\{(x,y) \mid d(x,y) < l$ and $x,y \in S\}$ can be found in $O(\eta \log n)$ time, where η denotes the cardinality of the set $\{(x,y) \mid d(x,y) < 3l$ and $x,y \in S\}$.*

Remark 2. The algorithm given in [8] in the proof of Lemma 3.1 fits in the algebraic computation tree model.

4 Producing Edges in a Sequence of Phases

The algorithm works in a sequence $p_1, p_2, \ldots$ of phases. The objective of a phase p_i is to produce all edges of length $< 2l_i$, where the parameter l_i is defined as follows. At phase p_1, l_1 equals the distance of a closest pair of points in S. For $i > 1$, let l be the distance of a closest pair of leaders immediately after phase p_{i-1}. Then, if $l > 2l_{i-1}$ we set l_i to l, otherwise we set l_i to $2l_{i-1}$.

We require that the following two properties hold at the beginning of each phase p_i (it is easy to verify that they hold at the beginning of phase p_1):

Property 1. Each ϵ-node is within distance $< \epsilon \cdot l_i$ from its leader.

Property 2. Each leader has at most a constant number of leaders within distance $< l_i$ from it.

A phase p_i is composed of three main parts: (A) finding candidate edges, (B) producing edges, and (C) preparing for the next phase. Below, each of these three parts is described in more detail.

A. Finding candidate edges. In this part we are going to insert into a priority queue all pairs of nodes between which we might produce an edge during phase p_i. First, compute a set L consisting of every pair of leaders such that the distance of the pair is $< 2l_i(1 + \epsilon)$. This is done according to Fact 1. Then, starting with an empty priority queue, do the following for each pair (u, v) in L:

A1. Find the trees T_u and T_v which include u and v as nodes, respectively.
A2. If T_u and T_v are different trees, then perform the following two steps:
    (a) Select "appropriate" nodes u' and v' where u' is an ϵ-node of u and v' is an ϵ-node of v.
    (b) Insert (u', v') into the priority queue.

We use a union-find algorithm for Step A1. The implementation of Step A2(a) is important, that is, how to choose an ϵ-node u' of u and an ϵ-node v' of v. Selecting (u', v') as a closest such pair does not lead to an efficient algorithm. Instead, we shall only guarantee that a closest pair of nodes which connects T_u and T_v is inserted into the priority queue at Step A2(b) (this will imply that our algorithm outputs a minimum spanning tree). Since these nodes do not have to be ϵ-nodes of the pair (u, v) currently considered at Step A2(a), we do not always select a closest pair of ϵ-nodes at Step A2(a) (however, for the time being it might help pretending that we are). In Sect. 5 we give details which lead to an efficient implementation of Step A2(a).

B. Producing edges. In this part we repeatedly merge two trees of the current forest by producing an edge connecting a pair of nodes in the priority queue. More precisely, as long as the distance of a closest pair of nodes in the priority queue is $< 2l_i$, repeat the following three steps:

B1. Remove a closest pair (u, v) from the priority queue.
B2. Find the trees T_u and T_v which include u and v as nodes, respectively.
B3. If T_u and T_v are different trees, then merge T_u and T_v into a single tree by producing the edge (u, v).

Step B2 is similar to Step A1, that is,it is implemented by using a union-find algorithm. Consequently, every merge at Step B3 is followed by performing a union of T_u's set of nodes and T_v's set of nodes.

C. Preparing for the next phase. The objective of this part is to ensure that Properties 1 and 2 hold at phase p_{i+1}. This is done by deleting leaders and assigning their ϵ-nodes to other leaders unless it might cause that Property 1 does not hold at phase p_{i+1}. First, compute a set L' consisting of each pair of leaders such that the distance of the pair is $< \epsilon \cdot l_i$. (The set L' can be obtained according to Fact 1 or by extracting relevant pairs from the set L computed in Part A.) Scanning L', we can, for each leader v that belongs to a pair in L', find all leaders that are within distance $< \epsilon \cdot l_i$ from v. Then, for each leader v that belongs to a pair in L', do the following three steps (unless v has been deleted at Step C3 below):

C1. Let $\{v_1, v_2, \ldots, v_m\}$ be the set of leaders within distance $< \epsilon \cdot l_i$ from v.
C2. Let E_v equal $E_v \cup E_{v_1} \cup E_{v_2} \cup \cdots \cup E_{v_m}$, where E_u denotes the set of ϵ-nodes of a leader u.
C3. Delete $v_1, v_2, \ldots, v_m$ from the DCP-structure.

At Step C1 we have by Property 1 that $d(v_k, w) < \epsilon \cdot l_i$ for each $w \in E_{v_k}$ and $1 \leq k \leq m$. Hence, after Step C2 it holds that $d(v, w) < 2 \cdot \epsilon \cdot l_i \leq \epsilon \cdot l_{i+1}$ for each $w \in E_v$, which implies that Property 1 holds at phase p_{i+1}. That Property 2 holds at phase p_{i+1} follows easily from the fact that ϵ is constant and by observing that the distance of each pair of leaders is $\geq \epsilon \cdot l_i$ after the above procedure (in case $l_{i+1} > 2l_i$, the distance of each pair of leaders is $\geq l_{i+1}$).

Lemma 3. *The above algorithm takes total time $O(n \log n)$ plus the time taken for Step A2(a).*

Proof. Among all pairs of leaders that exist at the beginning of phase p_i, let $\mathcal{L}_i$ be the set consisting of each pair such that the distance of the pair is $< 6l_i(1+\epsilon)$. By Fact 1, to compute the set L in Part A at phase p_i takes $O(|\mathcal{L}_i| \log n)$ time. Consider a pair (u, v) in $\mathcal{L}_i$. It is not hard to show that there exists a constant c such that $d(u, v) < \epsilon \cdot l_{i+c}$ (it holds for any $c \geq \log_2(12/\epsilon)$). Hence, not later than at phase p_{i+c}, u or v will be deleted from the DCP-structure at Step C3. Let us associate the pair (u, v) with the first one of u and v that is deleted from the DCP-structure. If we treat each other pair in $\mathcal{L}_i$ in the same manner, we associate by Property 2 at most a constant number of pairs to each leader. Moreover, doing this for every i, any leader can be associated with pairs in a set $\mathcal{L}_i$ for at most c distinct i's. As at most n leaders are deleted from the DCP-structure, it follows that $\sum_{i \geq 1} |\mathcal{L}_i| = O(n)$. We can assume that each priority queue operation and union-find operation takes $O(\log n)$ time. Thus, ignoring Step A2(a), the total time for the algorithm is $O(n \log n)$. $\square$

5 Organizing ϵ-Nodes According to Directions

In order to complete our algorithm, it remains to give a method for Step A2(a) of the previous section. Recall, at that step we are given two leaders, one of a tree T_u and the other of a tree T_v, and we want to select a pair of ϵ-nodes, one of each leader. However, for certain ϵ-nodes it is possible to detect that they are not endpoints of a shortest edge connecting T_u and T_v, and hence, it is not necessary to consider them in order to find such a shortest edge. In this section we describe how the ϵ-nodes can be organized in order to avoid considering such ϵ-nodes.

For each leader u we maintain c subsets of its ϵ-nodes, where c is a sufficiently large integer constant (a lower bound for c is given implicitly in the proof of Lemma 4). We call these subsets *direction sets of u* and denote them by $D_1^u, D_2^u, \ldots, D_c^u$. Initially, at the beginning of phase p_1, each direction set of u is just a copy of u's set of ϵ-nodes, that is, $D_k^u = \{u\}$ for each $1 \leq k \leq c$. Then, whenever a leader v receives ϵ-nodes from leaders $v_1, v_2, \ldots, v_m$ at Step C2, its direction sets are updated accordingly by letting D_k^v equal $D_k^v \cup D_k^{v_1} \cup D_k^{v_2} \cup \cdots \cup D_k^{v_m}$ for each $1 \leq k \leq c$.

Next we define what we mean by the *relevant direction labels* of a leader u with respect to a leader v. (The following definitions might be easier to comprehend by having $d = 3$.) Let H be the axis-aligned d-dimensional cube centered at u such that v lies on its boundary. Note that H has $2d$ facets, each corresponding to a $(d-1)$-dimensional cube. Partition each facet into $(d-1)$-dimensional subcubes of equal size so that the total number of subcubes equals c. (So $(c/(2d))^{1/(d-1)}$ must be an integer.) We maintain a labeling function which assigns to each subcube h a distinct integer between 1 and c. The labeling is such that the label of h uniquely specifies the position of h on the boundary of

H, that is, the label of h depends only on the direction of the ray starting at u and passing through the center of h. Now, let K' be the set consisting of each subcube whose closure contains v, and let K be the union of K' and the set consisting of each subcube that is adjacent to at least one subcube in K'. Then the relevant direction labels of u with respect to v are the labels of the subcubes in K.

At Step A2(a) we are given two leaders u and v, and we want to select an ϵ-node u' of u and an ϵ-node v' of v. This is done by performing the following six steps (where the actual selection is done within the second step):

(i) Let K_u be the set of relevant direction labels of u with respect to v, and let K_v be the set of relevant direction labels of v with respect to u.
(ii) Let (u', v') be a closest pair of ϵ-nodes such that $u' \in \bigcup_{k \in K_u} D_k^u$ and $v' \in \bigcup_{k \in K_v} D_k^v$.
(iii) For each $k \in K_u$, delete from D_k^u every marked ϵ-node.
(iv) For each $k \in K_v$, delete from D_k^v every marked ϵ-node.
(v) For each $k \in K_u$, mark every ϵ-node in D_k^u.
(vi) For each $k \in K_v$, mark every ϵ-node in D_k^v.

Lemma 4. *For a sufficiently large constant c and a sufficiently small constant ϵ, the above algorithm computes a minimum spanning tree.*

Proof. At any time during the execution of the algorithm, let (w, w') be a closest pair of nodes such that they belong to different trees of the current forest, and let T be the tree that includes w. Let p_i be the phase where T is merged with some other tree. It suffices to show that (w, w') is inserted into the priority queue at Step A2(b) during phase p_i. Consider the situation at the beginning of phase p_i. Let u be the leader of w, let u' be the leader of w', and let K_u be the set of relevant direction labels of u with respect to u'. We assume that $w \notin \bigcup_{k \in K_u} D_k^u$ and derive a contradiction. The lemma follows by observing that w' can be treated in a completely symmetrical way.

Fix an integer k in K_u. From our assumptions and Steps (i) through (vi) above, it follows that w has been marked at a phase p_j preceding p_{i-1}, and then deleted at some phase preceding p_i. More precisely, there exists a phase p_j, $j < i - 1$, at which the following three conditions hold:·

1. two leaders v and v' are considered at Step (i),
2. k is a relevant direction label of v with respect to v', and
3. w belongs to D_k^v.

Since $j < i - 1$, it follows that v' belongs to T if $\epsilon < 1$, because $d(v, v') < 2l_j(1 + \epsilon) \leq l_i(1 + \epsilon)/2$. Hence, if we prove that $d(w', v') < d(w', w)$ then we obtain a contradiction. But first we need the following notation. Denote by angle(xx', yy') the angle between the ray from x to x' and the ray from y to y'. (The angle is here given by $\arccos(\mathbf{x}^T\mathbf{y}/(\|\mathbf{x}\| \cdot \|\mathbf{y}\|))$, where $\mathbf{x} = x' - x$, $\|\mathbf{x}\| = (\mathbf{x}^T\mathbf{x})^{1/2}$, and similarly for $\mathbf{y}$.)

First, angle(wv', vv') approaches zero as ϵ approaches zero, because $d(v, v') \geq l_j$ and $d(w, v) < \epsilon \cdot l_j$. Next we observe that angle(vv', uu') approaches zero as

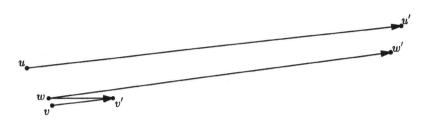

Fig. 1. An example configuration.

c grows. Also, since $d(u, u') \geq l_i$, $d(u', w') < \epsilon \cdot l_i$, and $d(u, w) < \epsilon \cdot l_i$, we realize that angle($uu', ww'$) approaches zero as ϵ approaches zero. Putting all together, we get that angle(wv', ww') approaches zero as ϵ approaches zero and c grows. It is now easy to realize that $d(w', v') < d(w', w)$ for a sufficiently small ϵ and sufficiently large c (because we are dealing with an L_t-metric), which is the contradiction we were looking for. □

It still remains to give a method for Step (ii) above. Let us first calculate how much time we can spend at that step. Say that a node w is k-*active* whenever one of the following two conditions holds at Step (ii): (1) $w \in D_k^u$ and $k \in K_u$, or (2) $w \in D_k^v$ and $k \in K_v$. For a fixed k between 1 and c, any node can be k-active at no more than two phases, by Steps (iii) to (vi). Moreover, by Property 2, any node is k-active at most a constant number of times during each phase. Combining these observations with Lemmas 3 and 4, we get

Lemma 5. *If Step (ii) takes t time per node in the union of $\cup_{k \in K_u} D_k^u$ and $\cup_{k \in K_v} D_k^v$, then the above algorithm finds a minimum spanning tree in total time $O(n \log n + nt)$.*

If the nodes in $\cup_{k \in K_u} D_k^u$ are colored red and the nodes in $\cup_{k \in K_v} D_k^v$ are colored blue, then Step (ii) corresponds to finding a bichromatic closest pair. Moreover, it is not hard to see that the above algorithm fits in the algebraic computation tree model, and in this model, finding a bichromatic closest pair takes $\Omega(n \log n)$ time, even if we restrict ourselves to 1-dimensional space (see, e.g., [10]). So, by Lemma 5, computing a minimum spanning tree is no harder than finding a bichromatic closest pair. On the other hand, it is straightforward to show that finding a bichromatic closest pair is no harder than computing a minimum spanning trees [1]. Thus we obtain the following theorem (recall that $T_d(n, m)$ denotes the time to find a bichromatic closest pair between n red points and m blue points).

Theorem 6. *Let S be a set of n points in $\mathbb{R}^d$ where $d \geq 1$ is an integer constant. Then, under any fixed L_t-metric, a minimum spanning tree of S can be computed in $O(T_d(n, n))$ time, which is optimal in the algebraic computation tree model.*

6 Finding Minimum Spanning Trees in $\mathbb{R}^3$

Consider n red points and n blue points in $\mathbb{R}^3$. Let l be the diameter of the set of red points and assume without loss of generality that the set of blue points has diameter $\leq l$. In order to compute a minimum spanning tree, we can assume that the minimum distance between any red point and any blue point is $> 10l$, say (by selecting ϵ sufficiently small, the bichromatic closest pair problem at Step (ii) of the previous section has this property). In the next two subsections we explain how a bichromatic closest pair can be found in $O(n \log n)$ time, under the L_1 and L_∞-metric.

6.1 Under the L_1-Metric

Throughout this subsection, distances are according to the L_1-metric. Let C be the axis-aligned cube centered at some red point and with edges of length $2l$. So all red points are internal to C, whereas all blue points are external to C. It suffices to show that we can build a data structure in $O(n \log n)$ time such that it supports the following operation in $O(\log n)$ time: given any point q external to C, return a red point that is closest to q. (This is a special case of the so-called post office problem.)

Let p be the point on the surface of C that is closest to the query point q, and let P be a bounding plane of C that contains p. It is easy to see that $d(q,r) = d(q,p) + d(p,r)$ for any point r internal to C. Now, suppose that we have partitioned P into regions and associated each region with a red point, in such a way that the following holds for each region: if it is associated with the red point r, then it contains only points that are at least as close to r as to any other red point. Using a standard data structure for planar point location (for example, the one developed by Kirkpatrick [6]) in this partition, we get the desired data structure. In other words, it remains only to show that such a partition of P can be computed in $O(n \log n)$ time. This can be done by, for example, using the so-called *abstract Voronoi diagrams* proposed by Klein [7]. (The details are included in [9].) This subsection is summarized by the following corollary of Theorem 6.

Corollary 7. *A minimum spanning tree of any set of n points in $\mathbb{R}^3$ can be computed in $O(n \log n)$ time under the L_1-metric, which is optimal in the algebraic computation tree model.*

6.2 Under the L_∞-Metric

Throughout this subsection, distances are according to the L_∞-metric. Consider three straight-line segments centered at some red point, each of length $2l$ and parallel with a distinct coordinate axis. Let H be the octahedron whose vertices coincide with the endpoints of these three segments. Again, all red points are internal to H and all blue points are external to H. In a manner analogous to the one in the previous subsection, we partition each bounding plane of H into regions, where each region is associated with one closest red point. We also have the following analogue: if h is the point on the surface of H that is closest to a query point q external to H, then $d(q,r) = d(q,h) + d(h,r)$ for any point r internal to H. (To see this, imagine expanding an infinitesimal cube centered at q until it hits h, and then, expanding an infinitesimal cube centered at h until it hits r.) Hence, we can find a bichromatic closest pair by doing planar point location in our partition of the surface of H. This subsection is summarized by the following corollary of Theorem 6.

Corollary 8. *A minimum spanning tree of any set of n points in $\mathbb{R}^3$ can be computed in $O(n \log n)$ time under the L_∞-metric, which is optimal in the algebraic computation tree model.*

References

1. Agarwal, P. K., Edelsbrunner, H., Schwarzkopf, O., Welzl, E.: Euclidean minimum spanning trees and bichromatic closest pairs. Discrete & Computational Geometry **6** (1991) 407–422.
2. Bespamyatnikh, S. N.: An optimal algorithm for closest pair maintenance. Proc. of the 11th Annual ACM Symposium on Computational Geometry (1995) 152–161.
3. Callahan, P. B., Kosaraju, S. R.: Faster algorithms for some geometric graph problems in higher dimensions. Proc. of the 4th Annual ACM-SIAM Symposium on Discrete Algorithms (1993) 291–300.
4. Clarkson, K.: Fast expected-time and approximate algorithms for geometric minimum spanning tree. Proc. of the 16th Annual ACM Symposium on Theory of Computing (1984) 343–348.
5. Gabow, H. N., Bentley, J. L., Tarjan, R. E.: Scaling and related techniques for geometry problems. Proc. of the 16th Annual ACM Symposium on Theory of Computing (1984) 135–143.
6. Kirkpatrick, D. G.: Optimal search in planar subdivisions. SIAM Journal on Computing **12** (1983) 28–35.
7. Klein, R.: Abstract Voronoi diagrams and their applications. Computational Geometry and its Applications (LNCS 333, Springer-Verlag, 1988) 148–157.
8. Krznaric, D., Levcopoulos, C.: Optimal algorithms for complete linkage clustering in d dimensions. Technical Report LU-CS-TR:96-180, Department of Computer Science, Lund University (1996).
9. Krznaric, D., Levcopoulos, C., Nilsson, B. J.: Minimum spanning trees in d dimensions. Technical Report LU-CS-TR:96-183, Department of Computer Science, Lund University (1996).
10. Preparata, F. P., Shamos, M. I.: Computational Geometry: An Introduction. Springer-Verlag, New York (1985).
11. Salowe, J. S.: Constructing multidimensional spanner graphs. International Journal of Computational Geometry & Applications **1** (1991) 99–107.
12. Shamos, M. I., Hoey, D. J.: Closest-point problems. Proc. of the 16th Annual IEEE Symposium on Foundations of Computer Science (1975) 151–162.
13. Vaidya, P. M.: Minimum spanning trees in k-dimensional space. SIAM Journal on Computing **17** (1988) 572–582.
14. Yao, A. C.: On constructing minimum spanning trees in k-dimensional spaces and related problems. SIAM Journal on Computing **11** (1982) 721–736.

Relaxed Balance for Search Trees
with Local Rebalancing

Kim S. Larsen[1]*, Thomas Ottmann[2], Eljas Soisalon-Soininen[3]

[1] Department of Mathematics and Computer Science, Odense University, Campusvej
55, DK-5230 Odense M, Denmark. E-mail: kslarsen@imada.ou.dk.
[2] Institute of Computer Science, Freiburg University, Am Flughafen 17, D-79110
Freiburg, Germany. E-mail: ottmann@informatik.uni-freiburg.de.
[3] Laboratory of Information Processing Science, Helsinki University of Technology,
Otakaari 1 A, FIN-02150 Espoo, Finland. E-mail: ess@cs.hut.fi.

Abstract. Search trees with relaxed balance were introduced with the
aim of facilitating fast updating on shared-memory asynchronous parallel
architectures. To obtain this, rebalancing has been uncoupled from the
updating, so extensive locking in connection with updates is avoided.
Rebalancing is taken care of by background processes, which do only a
constant amount of work at a time before they release locks. Thus, the
rebalancing and the associated locks are very localized in time as well
as in space. In particular, there is no exclusive locking of whole paths.
This means that the amount of parallelism possible is not limited by the
height of the tree.
Search trees with relaxed balance have been obtained by adapting standard sequential search trees to this new paradigm; clearly using similar
techniques in each case, but no general result has been obtained. We
show how any search tree with local bottom-up rebalancing can be used
in a relaxed variant preserving the complexity of the rebalancing from
the sequential case. Additionally, we single out the one high level locking mechanism that a parallel implementation must provide in order to
guarantee consistency.

1 Introduction

Search trees with relaxed balance are interesting for several reasons. First of all,
they give answers to a fundamental question about what is probably the most
important data structure: is it possible to uncouple rebalancing in search trees
from updating without a great loss in efficiency? The answer is yes, and there are
some immediate applications. In a sequential environment during bursts of updates, it is possible to postpone rebalancing partly or completely to speed up the

---

* Some of the work was done while this author was visiting the Department of Computer Sciences, University of Wisconsin at Madison. The work of this author was
supported in part by SNF (Denmark), in part by NSF (U.S.) grant CCR-9510244,
and in part by the ESPRIT Long Term Research Programme of the EU under project
number 20244 (ALCOM-IT).

processing of user requests. The process can then gradually catch up on delayed rebalancing operations after the burst. In a concurrent environment, the relaxed balancing allows for an efficient solution to the concurrency control problem for binary search trees, and makes a higher degree of parallelism possible.

Search trees with relaxed balance have been studied in many papers, including a survey [27]. Here, we just mention the most important, and, for this paper, relevant results. The crucial idea of uncoupling the rebalancing from the updating was first mentioned in [12]. The first partial result, dealing with insertions only, is from [16]. The first relaxed version of AVL-trees [1] was presented in [21] and proofs of complexity for the rebalancing, matching the complexity from the sequential case, was obtained in [17]. The first relaxed version of B-trees [6] is also from [21] with proofs of complexity matching the sequential ones in [18]. Since [18] really treats (a, b)-trees [15], it also provides proofs for 2-3 trees [14, 2], as well as for (a, b)-trees with other choices of a and b. A relaxed version of red-black trees [12] was introduced in [22] and complexities matching the sequential case were established in [9, 8] for a variant of the original proposal.

Both the worst-case logarithmic time results, which hold for all of the sequential structures mentioned above, as well as the amortized constant time results, which only hold for some, were eventually matched in the relaxed versions.

However, in the cases above, in order to obtain efficient rebalancing, the rebalancing operations in the relaxed version of a given tree structure ended up looking somewhat different from the original ones. Using new strategies for insertion and deletion [24], we show how to use exactly the same rebalancing operations in the relaxed version as in the standard version in such a way that the complexity analysis, with respect to the number rebalancing operations applied, carries over.

Our paper explores the limits of this idea. We discuss the question of what properties a search tree must have in order for a relaxed version to exist and for the complexity to carry over. We give a collection of requirements and prove that if a search tree fulfills these requirements, then such a relaxed version exists. The informal conclusion is that this method works for search trees with bottom-up rebalancing, where rebalancing is in the form of a sequence of local transformations removing or moving purely local conflicts. Bottom-up rebalancing means that rebalancing is performed *after* an update, moving from the location of the update towards the root. The alternative method is top-down rebalancing, where rebalancing is performed while searching for the location for the update such that when the update has been performed, no further rebalancing is required.

This means that in addition to the search trees mentioned above, we now have efficient relaxed versions of a number of other types of search trees. It should be pointed out that the present paper does not only contribute to our understanding of fundamental issues in data structures, but also gives a starting point for an efficient implementation. Before [24] and this present paper, a relaxed version of a given search tree paradigm would have to be created from scratch. Now, by applying our result, there is always a starting point, which can then be optimized for efficiency.

2 Requirements

In this section, we state sufficient conditions which a balanced search tree, along with its rebalancing operations, must fulfill in order for us to be able to equip this structure with a relaxed balancing scheme. In the following, we refer to the given search tree, which is in fact a parameter to this whole set-up, as the T-tree.

For convenience, we assume that the T-tree is binary. However, all results hold for multi-way search trees as well, with minor modifications.

Requirement 1. *The T-tree must be a binary search tree allowing for several occurrences of the same key. It must offer insertion and deletion operations as well as a finite collection of rebalancing operations.*

Explanation 1. Nodes can be leaves, or unary or binary internal nodes. The children of a node are referred to as the left and the right child. Each node contains a key, and for any node it must be the case that all keys in its left subtree are smaller than or equal to the key in the node, and all keys in its right subtree are larger than the key in the node.

Nodes may contain additional variables for balance information. These are called the *balance variables*.

Informally, rebalancing after an insertion or deletion must be bottom-up and local, i.e., a rebalancing operation transforms one local problem to another, possibly of the same type, but higher up in the tree, until the problem can be taken care of completely and removed from the tree.

Definition 1. A tuple (t_1, P, t_2, L) is called a *local transformation* if t_1 is a T-tree, P is a boolean predicate over the variables in the nodes of t_1, t_2 is a T-tree with the same number of leaves as t_1, and L is a list of assignments, where the left hand side of each assignment is a variable in a node of t_2 and the right hand side is an expression over the variables in the nodes of t_1.

Below, we give two typical examples of how local transformations are usually given. In these examples, there is only one balance variable that P and L refer to. Call this variable W. Then in the first example, the predicate P is $(u_1.W \geq 1) \wedge (u_2.W = 0) \wedge (u_3.W \geq 1) \wedge (u_4.W = 0)$, where the nodes (u_1, u_2, u_3, u_4) are numbered top-to-bottom and left-to-right. The assignment list L, also in the first example, is $v_1.W := u_1.W$; $v_2.W := 0$; $v_3.W := 0$; $v_4.W := u_3.W$, where the nodes of t_2 (v_1, v_2, v_3, v_4) are also numbered top-to-bottom and left-to-right.

Definition 2. A *rebalancing operation* on a tree T is an operation that replaces one subtree S with another S'. It is the execution of a local transformation (t_1, P, t_2, L) such that the subtree can be written as t_1 with subtrees taking the place of the leaves of t_1. S' is t_2 assigned the values specified in L, and with all the subtrees of t_1 attached to t_2 taking the place of its leaves and in the same order, from left to right, in which they appeared in S.

Also insertion and deletion must be local. This can be formulated in different ways. We choose to use the already defined locality of the rebalancing operations.

Requirement 2. *Insertion and deletion must be rebalancing operations.*

Explanation 2. Insertion must take place in two steps. The new key is inserted into the location of the T-tree, where it should go according to the search tree invariant, and then a rebalancing operation involving the new node is carried out.

For deletion, the node containing the key in question is deleted, and then a rebalancing operation involving the parent of the deleted node is carried out.

Rebalancing must be bottom-up, i.e., after an update causing problems, the problem is taken care of by applying a sequence of rebalancing operations, gradually moving the problem towards the root until it can be removed.

Requirement 3. *Rebalancing must be bottom-up.*

Explanation 3. Rebalancing must be carried out as a finite sequence of rebalancing operations, where for each two consecutive local transformations in this sequence, (t_1^a, P^a, t_2^a, L^a) and (t_1^b, P^b, t_2^b, L^b), t_1^b is rooted at the same node, or at a node above the node where t_1^a is rooted.

It is not sufficient that rebalancing is bottom-up in this simple sense. There is no reason to do the following in a sequential search tree, but if we imagine allowing new updates to be carried out *before* finishing the rebalancing of another, we can create a situation where there are a number of separate problems of imbalance in the tree at the same time. Also under these circumstances must rebalancing be bottom-up, i.e., one problem of imbalance must move towards the root (or possibly remain at the same distance from the root) even in a context where there are other problems around it in the tree. At first, one might think that this stronger requirement must follow from the requirement above, but as we discuss in section 5, this is not the case.

Definition 3. Let h be the maximal height of the t_1 trees in all of the local transformations used in the rebalancing operations including the insertion and deletion rebalancing operations.

Requirement 4. *Rebalancing must be strongly bottom-up.*

Explanation 4. Assume that we allow updates to be performed with problems of imbalance still present in the tree (the tree is still a search tree, so this is meaningful) as long as the update is performed at a distance at least h below any problem of imbalance. We also allow rebalancing operations to be carried out provided that the problem of imbalance is located at a distance at least h nodes below any other problem of imbalance. Then rebalancing must still be bottom-up, i.e., no matter which problem of imbalance we decide to apply a rebalancing operation to, the problem of imbalance must remain at the same distance from the root or move closer to the root.

Rebalancing is based on local transformations which change the tree locally (parts of the tree which are structurally identical to one of the t_1's). The fact that the t_1's from different local transformations are different (of different shape and possibly of different height), makes it difficult to formulate and prove results concerning the behavior of the rebalancing as a whole. It is also a problem in guaranteeing that two separate problems of imbalance do not interfere. For these reasons, we embed all these different shapes in some common shapes, large enough to contain each of the problems of imbalance. We introduce an *operation area*, which is as much of the tree which it can possibly be necessary for a rebalancing operation to inspect, and a *synchronization area*, which is a larger area used to avoid deadlocks when different problems of imbalance progress towards a common ancestor in the tree.

Definition 4. A *triangle* rooted at a node u in the T-tree is the part of the tree at a distance at most h edges below u. The *operation area* of a triangle rooted at u is rooted at the node v, h edges above u (or at the root of the T-tree if u is fewer than h edges below the root of the T-tree). Let $d_o = depth(u) - depth(v) + h$. The operation area is the part of the tree at a distance at most d_o edges below v, except that nothing in the subtree of u belongs to the operation area. The *synchronization area* of a triangle rooted at u is rooted at the node v, h edges above u (or at the root of the T-tree if u is fewer than h edges below the root of the T-tree). Let $d_s = depth(u) - depth(v) + 2h$. The synchronization area is the part of the tree at a distance at most d_s edges below v, except that nothing in the subtree of u belongs to the synchronization area.

Below, assume that u and its two children constitute a triangle. Then the rest of the edges and nodes shown form u's synchronization area.

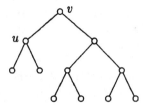

The final requirement is that all information pertinent to the rebalancing process must be kept in the tree, i.e., in the balance variables of the nodes, as opposed to, for instance, in the procedure that repeatedly applies the rebalancing operations.

Requirement 5. *Rebalancing must be oblivious.*

Explanation 5. If a rebalancing operation is carried out by applying the local transformation (t_1^a, P^a, t_2^a, L^a), where t_1^a is rooted at u, it must be possible to determine which local transformation to apply next, and where, by inspecting only nodes in the triangle rooted at u or in its operation area. In addition, the next local transformation (t_1^b, P^b, t_2^b, L^b) must have t_1^b rooted at u or at a node in its operation area.

If a local transformation can be applied somewhere, rebalancing cannot yet have terminated.

Some presentations of standard search trees do not fulfill these requirements at all. However, some implementations violate the requirements for trivial reasons, and can easily be *brought* to fulfill these. As an example, when a node is deleted from a red-black tree [12], one node may temporarily be double black. In some presentations, the procedure carrying out the rebalancing operations remembers which node is double black. So, rebalancing is not oblivious. However, it is trivial to change this to registering in the node that it is double black, thereby conforming to the requirements. In fact, any finite control that such a procedure may maintain can just as well be built into the nodes in the tree.

3 A general scheme for relaxed balancing

Now we assume that we are given a T-tree fulfilling the requirements listed in the previous section. Based on this, we define a *relaxed* T-tree. For emphasis, if we need to refer to the T-tree which is not relaxed, we use the term *standard* T-tree. Searching and updating, to be described below, is handled as in [24].

We add the following variables to the nodes of the tree: **insreq** and **delreq**, assuming that the nodes already have the variables **key**, **left**, **right**, **parent**, **external**. It is clear how these variables are intended to be used (**external** is set if the node is a leaf, and unset otherwise) and below, we only discuss the use of the new variables.

We use the tree as a *leaf-oriented* tree, i.e., all keys are kept in the leaves (also called external nodes) and the internal nodes merely contain routers that guide the search through the tree. This generally requires that the tree can contain several identical keys, which is ensured by requirement 1. Routers are of the same type as the keys (and are stored in the **key** variable), but a router is not necessarily present as a key in the tree, since we do not want to update routers when a deletion occurs. A router is greater than or equal to any key in the subtree to its left and less than any key in the subtree to its right.

The tree will at times physically contain keys which should not be there. In those nodes, the deletion request **delreq** is set. A search operation for a key k is performed as follows: First we search for the correct node as usual in a search tree. We will end at a node which is a leaf, because the tree is leaf-oriented. If the leaf has key k and `delreq` is unset, then a positive response is returned. Otherwise the search was negative.

Apart from the nodes that should be deleted, the nodes of the tree are divided into two groups: the ones that are in, what we will refer to as, the *structured part* of the tree, and the ones below the structured part. The ones below are organized in a binary tree without any balance information whatsoever. We call these parts the *unstructured parts* of the tree. The structured part may not actually be balanced (in the sense that all paths have lengths asymptotically logarithmically bounded in the size of the structured part of the tree) since rebalancing may be in progress. It merely reflects that the process of carrying out the necessary rebalancing operations caused by the insertion of the key in a node has begun. If the variable `insreq` is set in a node, then that node belongs to the unstructured part of the tree; otherwise it belongs to the structured part. When we discuss updating and rebalancing below, we also specify the values of `insreq` for the nodes involved.

Updating

Insertion of a key k is performed as follows: First the location for the insertion, a leaf, is found just as for insertion into a standard search tree. If the key in this leaf is k, then `delreq` is unset. Otherwise, the leaf l is replaced by three nodes: an internal node with two leaves as children. One leaf is the leaf l and the other leaf is new and contains the key k and `delreq` is unset. The leaf with the smaller key is made the left child, and the new internal node is given this smaller key as its router. The variable `insreq` in the internal node is set, indicating an insertion request.

Deletion of a key k is performed as follows: First the location for the deletion is found just as for a deletion of a key in a standard search tree. If the key is not found, nothing happens. Otherwise, the key will be in a leaf, since the tree is leaf-oriented. If the parent u of this leaf has `insreq` unset, then the variable `delreq` in the leaf is set. Otherwise, we are in the unstructured part of the tree and the node u is replaced with u's other child (the child which is *not* the leaf with key k).

Notice that the updating outlined above handles redundant updates correctly, i.e., an attempt to insert a key which is already present or delete a key which is not present (or has `delreq` set) will not alter any `insreq` or `delreq` values.

Rebalancing

After some insertions and deletions have been performed, the following problems might be present: there could be leaves, which should be deleted (`delreq` is set), and there could be whole subtrees of nodes that should be moved into the structured part of the tree (all the internal nodes where `insreq` is set) such that balance can be obtained. In this section, we describe how this is done. The insertion and deletion requests create triangles; each update can create one triangle. How this is done is described at the end of this section. First we explain the conditions under which triangles are moved around in the tree.

We say that a relaxed T-tree is *in balance* if no rebalancing operations can be carried out, i.e., for each local transformation (t_1, P, t_2, L), there is no connected set of nodes in the tree matching t_1 such that P holds. If the tree is in balance, then the relaxed T-tree is also a standard T-tree. The goal of the rebalancing process is to work towards obtaining this.

By requirement 5, the local part of the tree to be transformed next (one of the t_1's in a local transformation) can be associated with a triangle as defined in definition 4. When the local transformation is carried out, the next location for a local transformation (if any) can be determined (also by requirement 5) and another triangle can be associated with *this* location. It turns out that in the rules for rebalancing and the complexity proofs in the next section, it is sufficient to deal with triangles moving from one location to the next as outlined above, from when they are created by an update until they disappear.

The restrictions given below for when triangles can be moved deal with the problems of interference and deadlock. We must avoid interference between different rebalancing operations. This can be handled by forbidding that triangles move into each other. However, if this was the only requirement, then a deadlock situation could arise because neither of two triangles located side-by-side would be able to move without entering the other triangle.

Now, a triangle Δ can be moved from one location up to another, if the following two conditions are fulfilled:

- the location Δ is moving to is completely available, i.e., no other triangle occupies any part of the new location.
- if Δ is in the synchronization area of some triangle Δ', this Δ' is located at a depth greater than or equal to the depth of Δ.

If a triangle cannot be moved as a rebalancing operation requires because of the rules above, the operation cannot be performed at that point in time. Note that these rules are meaningful since by requirement 4, rebalancing is strongly bottom-up. Also observe that the second rule implies that it is not possible to move into another triangle's operation area (since triangles can move up a distance of at most h at a time), so there cannot be any interference from other operations when the location for the next triangle following a rebalancing operation has to be decided upon.

Now we comment on the creation and removal of triangles. Dealing with the latter first, when the last rebalancing operation is carried out, the triangle disappears.

Triangles are created when the top node in an unstructured part of the tree is moved into the structured part (a node with an insertion request) or when a node with a deletion request is treated. Since, by requirement 2, insertions and deletions are merely rebalancing operations, the t_1 in the first local transformation to be carried out decides the location of the first triangle. That triangle may be created when the location it wants to occupy is completely available, and does not overlap with the current operation area of any previously generated triangle.

4 Complexity

If a search tree with relaxed balance is to be used in a concurrent environment, a locking mechanism must be applied to ensure the atomicity of rebalancing operations. Even though operations may take place in parallel, for the sake of the complexity argument, we may assume that operations take place one at a time (notice that our goal is merely to count the number of rebalancing operations; not to make any direct statement about the parallel complexity). This is also the approach taken in [9, 17, 8, 18].

So, insertions, deletions, creations of triangles, removal of triangles, and moving of triangles are the basic operations in the following, and the assumption is that they can be interleaved in any way in accordance with the rules given in the previous section.

The following theorem establishes the connection between the operations in a relaxed T-tree and a standard one. In short, the implication is that the rebalancing operations carried out as a response to a number of updates in a relaxed T-tree are *exactly* the same ones that would be carried out in a standard T-tree for some ordering of the updates. The only assumption is that rebalancing in the standard T-tree after any update is finite.

In the following, a triangle rooted at u is said to be *above* another triangle rooted at v, if v is a descendent of u in the tree.

Additionally, we shall talk about nodes *having belonged* to a triangle. For the purpose of the following proof, we say that triangle movements are carried out in the following steps. Given a certain location of a triangle, at some point the next rebalancing operation for the triangle is carried out, and then the triangle is moved to its new location. All the nodes in the triangle immediately before it is moved are said to having belonged to the triangle.

Theorem 5. (Serializability) *Assume that some updates $u_1, \ldots, u_k$ and rebalancing operations $r_1, \ldots, r_p$ are carried out on a relaxed T-tree T, which is in balance. Then there exists an ordering of the updates $u_{i_1}, \ldots, u_{i_k}$ such that the multiset $\{r_{i_1}^1, \ldots, r_{i_1}^{k_{i_1}}, \ldots, r_{i_k}^1, \ldots, r_{i_k}^{k_{i_k}}\}$ contains the multiset $\{r_1, \ldots, r_p\}$, where $r_{i_j}^1, \ldots, r_{i_j}^{k_{i_j}}$ are the rebalancing operations carried out, in that order, after the update u_{i_j} on the corresponding standard T-tree, created by carrying out the operations*

$$u_{i_1}, r_{i_1}^1, \ldots, r_{i_1}^{k_{i_1}}, \ldots, u_{i_{j-1}}, r_{i_{j-1}}^1, \ldots, r_{i_{j-1}}^{k_{i_{j-1}}}$$

on T in the listed order.

Proof. We define an ordering $<$ as follows:

- if the update u_j takes place after all rebalancing operations arising from u_i have been carried out, we define $u_i < u_j$.
- if, at any point in time, a triangle arising from u_i is above a triangle from u_j, we define $u_i < u_j$.
- if a triangle arising from u_j is moved on top of a node that have belonged to a triangle arising from u_i, we define $u_i < u_j$.

— if a triangle arising from u_j is unable to move because it is in the synchronization area of another triangle arising from u_i which is located at a smaller depth, we define $u_i < u_j$.

Clearly, the graph corresponding to $<$ is acyclic, so we can choose a total ordering $u_{i_1}, \ldots, u_{i_k}$ consistent with $<$ (also known as a topological ordering of the graph).

The proof of this theorem is by induction. We prove for any $j \in \{1, \ldots, k\}$ that if only the updates $u_{i_1}, \ldots, u_{i_j}$ are performed, then the set of rebalancing operations carried out in the relaxed T-tree is contained in

$$\{r_{i_1}^1, \ldots, r_{i_1}^{k_{i_1}}, \ldots, r_{i_j}^1, \ldots, r_{i_j}^{k_{i_j}}\}.$$

The base case $j = 1$ is trivial, since with only one update, we have standard T-tree rebalancing. For the induction step, assume that up until the index j, everything holds. We now want to include the update $u_{i_{j+1}}$. Clearly rebalancing from the first j updates will progress as without update $u_{i_{j+1}}$ unless this update changes some nodes that later appear in the operation area of a triangle arising from an update u_{i_h}, $h \le j$. However, this cannot happen since then by definition, update u_{i_h} would be after update $u_{i_{j+1}}$ in the ordering.

It remains to be shown that the set of rebalancing operations due to update $u_{i_{j+1}}$ carried out in the relaxed T-tree is a subset of $\{r_{i_{j+1}}^1, \ldots, r_{i_{j+1}}^{k_{i_{j+1}}}\}$. If the triangles from $u_{i_{j+1}}$ arrive last to any node that they visit, the result is immediate, since those parts of the tree will appear exactly as if all other rebalancing had been carried out completely. The question is whether it is possible for the triangle Δ arising from $u_{i_{j+1}}$ to arrive at a location containing a node which ought to be visited by a triangle from one of the other updates, u_{i_h}, before the triangle from $u_{i_{j+1}}$ was supposed to arrive. Since the triangle from u_{i_h} will progress up the tree as long as everything looks the way it did before we took update $u_{i_{j+1}}$ into account, it will eventually reach the location where nodes that have belonged to Δ are to be found, or it will be unable to move because it is in the synchronization area of Δ. Both of these possibilities would contradict the ordering. □

With this proof, we have bounded the number of rebalancing operations which can be carried out in the relaxed version by the bound from the standard case. This could, of course, also have been obtained by defining a relaxed version in which no rebalancing operations can be applied. Thus, we also have to prove that all the triangles containing the problems of imbalance can be moved and eventually removed.

The problem is that triangles from different subtrees can meet at a node and one triangle has to move up before the other. If both triangles continue up towards the same node, a potential deadlock situation arises, where none of the triangles can move because the other is in its way. The rules involving the operation and synchronization areas have been designed to avoid this situation.

Theorem 6. (Progress) *If updating in a relaxed T-tree stops, the relaxed T-tree will eventually become a standard T-tree.*

Proof. Assume for the sake of contradiction that there are triangles in the tree, but none of them can move. Consider a triangle Δ at a smallest depth d. Since Δ is at a smallest depth, the two rules for when triangles can move imply that Δ cannot move because the location it wants to move to is partly occupied by another triangle Δ'. This Δ' must be located at a depth no larger than $d + h$; otherwise it cannot possibly occupy the next location for Δ.

Since triangles move at most h edges up at a time (requirement 5), the move leading to this situation could not have been made by Δ', since Δ' could not have moved while in the synchronization area of Δ because of its larger depth. So, assume that the last move was made by Δ. If Δ had smaller depth than Δ' before its last move, it must have moved more than once since Δ' moved. Consider the earliest point where the depth of Δ was no larger than the depth of Δ'. The move by either Δ or Δ' immediately prior to this must have been carried out while in the other triangle's synchronization area with that triangle at a smaller depth. This is impossible according to the rules, and therefore a contradiction.

Since theorem 5 bounds the number of rebalancing operations that can be carried out, the result follows. □

5 Examples

At this point, it is possible to take a concrete search tree, check the requirements from section 2, and conclude that a relaxed version exists, which is as efficient, in terms of the number of rebalancing operations necessary, as the standard version. If all requirements are not fulfilled, one can try to determine if minor changes can be made, rectifying the situation (as in the example given at the end of section 2). This inspection is easy and tedious, and documentation of the requirements being fulfilled is very space consuming, since, among other things, it includes listing all the rebalancing operations.

Here, we merely list some known search trees that fulfill the requirements and summarize the main complexity results which carry over to the relaxed version. This is also an opportunity to comment on the relative strength of worst-case and amortized results which changes in the scenario of relaxed balance.

It is no surprise, of course, that AVL-trees [1], B-trees [6] (as well as 2-3 trees [14, 2] and the more general (a, b)-trees [15]), and red-black trees [12] fulfill the requirements and have relaxed versions, since results on these data structures have already been published [21, 22] with complexity results in [9, 17, 8, 18] including worst-case logarithmic rebalancing per update, and for red-black trees and (a, b)-tree variants with $b \geq 2a$ also amortized constant rebalancing per update, just as it was obtained for the standard case in [25, 15]. For AVL-trees, we also get the amortized constant rebalancing results in insertion-only or deletion-only situations that was proven for the standard case in [19]. However, the relaxed versions that come out from applying the method from this paper are different. Another relaxed version of red-black trees has been defined in [13].

In that paper, problems that we solve in general have been worked out for the specific case of red-black trees.

Also symmetric binary B-trees [5] and half-balanced trees [23] fulfill the requirements, so worst-case logarithmic rebalancing per update is obtained. This is not too surprising since these are similar to red-black trees, in the sense that the shape of the trees arising from these two paradigms can always be colored in such a way that a red-black tree is obtained. However, the rebalancing operations and criteria are different, so they are not the same as red-black trees (rebalancing operations in response to an update would be different).

Of completely new data structures (that is new to the world of relaxed balancing), BB[α]-trees [20] fulfill the requirements and the logarithmic worst-case rebalancing per update as well as the amortized constant rebalancing per update [7] carry over.

Among search trees that were not assumed to have (efficient) relaxed versions using the methods from the earlier papers are splay trees [26]. This data structure also fulfills the requirements and we obtain amortized logarithmic rebalancing for relaxed splay trees. Notice that amortized logarithmic rebalancing is in fact no weaker than worst-case logarithmic rebalancing in this scenario because our statements only deal with the *number* of rebalancing operations carried out and not *when* they are carried out. They can be carried out significantly later than the update that caused them, since they can be interleaved with updates and rebalancing operations caused by other updates.

Note that randomized search trees [4] (also called treaps) fulfill all requirements; except that they are not strongly bottom-up. They are indeed bottom-up. However, if two problems of imbalance exist simultaneously, one above the other, then the one farthest from the root can move *down* when a rebalancing operation is applied. It is easy to construct such an example and we leave this to the reader. This means first of all that the proofs in this paper do not hold. This does not rule out, of course, that the structure could have an efficient relaxed version. However, the naive approach does not work [10] ($\Theta(k)$ rebalancing in the sequential case can become $\Theta(k^2)$ rebalancing in the relaxed version), and there is no obvious way of avoiding the problem. The problems seems to be that problems of imbalance are not *purely* local, since an incorrectly placed priority can have a conflict with not just its neighbors, but with nodes on a whole path.

Some search trees cannot possibly be brought to fulfill the requirements. For instance, search trees that apply global rebuilding, such as general balanced trees [3] and scapegoat trees [11].

6 Comments on implementation

When a search tree with relaxed balance is to be used on a parallel shared-memory architecture, a locking scheme is necessary in order to ensure consistency. Many different locking scheme for tree restructuring have been made (one can be found in [22]) and one of these can be adapted to meet our demands.

However, the important observation is that also this problem can now be solved once and for all. We need a safe mechanism for dealing with triangles: their creation, moving, and removal, as well as simple insertion and deletion into the unstructured parts of the tree. Once this has been accomplished in connection with one relaxed search tree, it can be reused in later implementations of other relaxed search trees. So, also here there is a standard starting point, which can then be optimized for a specific structure. Without any optimization, the result will probably not be satisfactory because of the size of the operation and synchronization areas.

At the algorithmic level, optimization of this standard proposal for rebalancing towards higher efficiency in a concrete implementation could involve combining updates before they are moved into the structured part of the tree, or allowing for some controlled interference between problems in triangles, as has been done in earlier proposals for relaxed search trees, the idea being that combined problems can sometimes be handled in fewer rebalancing operations total, and sometimes two or more problems of imbalance cancel out.

Acknowledgements

The first author would like to thank Rolf Fagerberg for numerous interesting discussions on this and related topics, and Joan Boyar for discussions and for comments on an earlier draft of the paper.

References

1. G. M. Adel'son-Vel'skiĭ and E. M. Landis, An Algorithm for the Organisation of Information. *Soviet Math. Doklady* **3** (1962) 1259–1263.
2. Alfred V. Aho, John E. Hopcroft and Jeffrey D. Ullman, *Data Structures and Algorithms*. Addison-Wesley, 1983.
3. Arne Andersson, Improving Partial Rebuilding by Using Simple Balance Criteria. "1st International Workshop on Algorithms and Data Structures", *Lecture Notes in Computer Science* **382** (1989) 393–402.
4. Cecilia R. Aragon and Raimund G. Seidel, Randomized Search Trees. *Proceedings of the 30th Annual IEEE Symposium on the Foundations of Computer Science* (1989) 540–545.
5. R. Bayer, Symmetric Binary B-Trees: Data Structure and Maintenance Algorithms. *Acta Informatica* **1** (1972) 290–306.
6. R. Bayer and E. McCreight, Organization and Maintenance of Large Ordered Indexes. *Acta Informatica* **1** (1972) 173–189.
7. Norbert Blum and Kurt Mehlhorn, On the Average Number of Rebalancing Operations in Weight-Balanced Trees. *Theoretical Computer Science* **11** (1980) 303–320.
8. Joan Boyar, Rolf Fagerberg and Kim S. Larsen, Amortization Results for Chromatic Search Trees, with an Application to Priority Queues. "Fourth International Workshop on Algorithms and Data Structures", *Lecture Notes in Computer Science* **955** (1994) 270–281. To appear in *Journal of Computer and System Sciences*.
9. Joan F. Boyar and Kim S. Larsen, Efficient Rebalancing of Chromatic Search Trees. *Journal of Computer and System Sciences* **49** (1994) 667–682.

10. Rolf Fagerberg, Personal communication, 1996.
11. Igal Galperin and Ronald L. Rivest, Scapegoat Trees. *Proceedings of the Fourth Annual ACM-SIAM Symposium on Discrete Algorithms* (1993) 165–174.
12. Leo J. Guibas and Robert Sedgewick, A Dichromatic Framework for Balanced Trees. *Proceedings of the 19th Annual IEEE Symposium on the Foundations of Computer Science* (1978) 8–21.
13. S. Hanke, T. Ottmann and E. Soisalon-Soininen, Relaxed Balanced Red-Black Trees. "Proceedings 3rd Italian Conference on Algorithms and Complexity", *Lecture Notes in Computer Science* **1203** (1997) 193–204.
14. J. E. Hopcroft, Title unknown, 1970. Unpublished work on 2-3 trees.
15. Scott Huddleston and Kurt Mehlhorn, A New Data Structure for Representing Sorted Lists. *Acta Informatica* **17** (1982) 157–184.
16. J. L. W. Kessels, On-the-Fly Optimization of Data Structures. *Communications of the ACM* **26** (1983) 895–901.
17. Kim S. Larsen, AVL Trees with Relaxed Balance. *Proceedings of the 8th International Parallel Processing Symposium*, IEEE Computer Society Press (1994) 888–893.
18. Kim S. Larsen and Rolf Fagerberg, Efficient Rebalancing of B-Trees with Relaxed Balance. *International Journal of Foundations of Computer Science* **7** (1996) 169–186.
19. Kurt Mehlhorn and Athanasios Tsakalidis, An Amortized Analysis of Insertions into AVL-Trees. *SIAM Journal on Computing* **15** (1986) 22–33.
20. J. Nievergelt and M. Reingold, Binary Search Trees of Bounded Balance. *SIAM Journal on Computing* **2** (1973) 33–43.
21. O. Nurmi, E. Soisalon-Soininen and D. Wood, Concurrency Control in Database Structures with Relaxed Balance. *Proceedings of the 6th ACM Symposium on Principles of Database Systems* (1987) 170–176.
22. Otto Nurmi and Eljas Soisalon-Soininen, Uncoupling Updating and Rebalancing in Chromatic Binary Search Trees. *Proceedings of the Tenth ACM SIGACT-SIGMOD-SIGART Symposium on Principles of Database Systems* (1991) 192–198.
23. H. J. Olivié, A New Class of Balanced Trees: Half-Balanced Binary Search Trees. *R.A.I.R.O. Informatique Théoretique* **16** (1982) 51–71.
24. Th. Ottmann and E. Soisalon-Soininen, Relaxed Balancing Made Simple. Institut für Informatik, Universität Freiburg, Technical Report 71, 1995.
25. Neil Sarnak and Robert E. Tarjan, Planar Point Location Using Persistent Search Trees. *Communications of the ACM* **29** (1986) 669–679.
26. Daniel Dominic Sleator and Robert Endre Tarjan, Self-Adjusting Binary Trees. *Proceedings of the 15th Annual ACM Symposium on the Theory of Computing* (1983) 235–245.
27. Eljas Soisalon-Soininen and Peter Widmayer, Relaxed Balancing in Search Trees. *Advances in Algorithms, Languages, and Complexity*, D.-Z. Du and K.-I. Ko (Eds.), Kluwer Academic Publishers (1997) 267–283.

Improved Approximations for Minimum Cardinality Quadrangulations of Finite Element Meshes

Matthias Müller–Hannemann[1] and Karsten Weihe[2]

[1] Technische Universität Berlin, Fachbereich Mathematik, Sekr. MA 6–1, Str. des 17. Juni 136, 10623 Berlin, Germany, mhannema@math.tu–berlin.de, http://www.math.tu–berlin.de/~mhannema
[2] Universität Konstanz, Fakultät für Mathematik und Informatik, Fach D188, 78457 Konstanz, Germany, karsten.weihe@uni–konstanz.de, http://www.informatik.uni–konstanz.de/~weihe

Abstract. Conformal mesh refinement has gained much attention as a necessary preprocessing step for the finite element method in the computer-aided design of machines, vehicles, and many other technical devices. For many applications, such as torsion problems and crash simulations, it is important to have mesh refinements into quadrilaterals. In this paper, we consider the problem of constructing a minimum-cardinality conformal mesh refinement into quadrilaterals. However, this problem is $\mathcal{NP}$–hard, which motivates the search for good approximations. The previously best known performance guarantee has been achieved by a linear-time algorithm with a factor of 4. We give improved approximation algorithms. In particular, for meshes without so-called folding edges, we now present a 2–approximation algorithm. This algorithm requires $\mathcal{O}(n\,m \log n)$ time, where n is the number of polygons and m the number of edges in the mesh. The asymptotic complexity of the latter algorithm is dominated by solving a T-join, or equivalently, a minimum–cost perfect b–matching problem in a certain variant of the dual graph of the mesh. If a mesh without foldings corresponds to a planar graph, the running time can be further reduced to $\mathcal{O}(n^{3/2} \log n)$ by an application of the planar separator theorem.

1 Introduction

In recent years, the conformal refinement of *finite element meshes* has gained much attention as a necessary preprocessing step for the finite element method in the computer-aided design of machines, vehicles, and many other technical devices. Much work has been done on decompositions into *triangles*; see [Ho88] and [BE95] for surveys. However, for many applications, such as torsion problems

---

[1] This author was supported by the special program "Efficient Algorithms for Discrete Problems and Their Applications" of the Deutsche Forschungsgemeinschaft (DFG) under grant Mo 446/2-2.

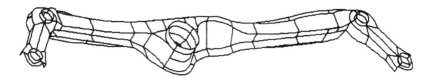

Fig. 1. A coarse mesh modeling a chassis of a car. This mesh has been constructed by a German car company.

and crash simulations, it is important to have mesh refinements into quadrilaterals [ZT89]. See also [Tou95] for a systematic survey on quadrangulations.

A *polygon* is a closed and connected region in the plane or, more generally, of a smooth surface in the three–dimensional space, bounded by a finite, closed sequence of straight line segments (*edges*). The endpoints of the line segments or curves are the *vertices*. A polygon is *convex* if the internal angle at each vertex is at most π. A *mesh* is a set of openly disjoint, convex polygons (Fig. 1). A mesh may contain *folding edges*, that is, edges incident to more than two polygons (Fig. 5). We call a mesh *homogeneous* if it does not contain folding edges.

In a *conformal refinement* of a mesh, each polygon is decomposed into strictly convex quadrilaterals, and if two quadrilaterals share more than a corner, they share exactly one edge as a whole (Fig. 2).

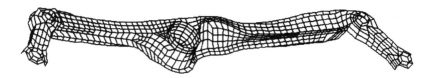

Fig. 2. The conformal refinement produced by the algorithm in [MMW95].

Workpieces are modeled interactively as meshes; see Fig. 1 for an example of an instance taken from practice. However, such meshes are usually very coarse and not conformal. To be suitable for the finite element method, the mesh has to be refined into a conformal mesh in a preprocessing step. Previous work puts emphasis on the shape of the quadrilaterals (angles should neither be too small nor too large; the *aspect ratio*, i.e. the ratio between the largest and the smallest side of a quadrilateral, should be small). This is important for the numerical accuracy in the later iterations of the cyclic design process, when the model has become mature and exact results are required for fine–tuning.

In this paper now, we focus on the early stages of this process, where the model is designed only roughly, and the numerical accuracy must only suffice to indicate the general tendency. Hence, the development time is crucial, which in turn is determined by the run time of the finite element method. This raises the following problem: Given a mesh, find a conformal refinement with a *minimum* number of quadrilaterals.

Until recently, work on this problem (cf. [MMW95] and [TA93]) has considered the number of quadrilaterals only heuristically or not at all. Usually, a *template model* is used, which restricts the possibilities of decomposing a single polygon to a few classes of *templates*. These templates are designed to achieve good angles and aspect ratios heuristically. For example, the most important template for quadrangular polygons is a $(p \times q)$–grid, where p and q are variable. However, this template uses $p \cdot q$ quadrilaterals, which is quadratic in size compared with minimal quadrangulations of size $\mathcal{O}(p + q)$ (easy to see). Therefore, algorithms often refine workpieces into too many quadrilaterals, which makes the finite element method very costly or even infeasible.

Unfortunately, it is hard to find conformal decompositions into a minimum number of quadrilaterals:

Theorem 1. [MW96] *The minimum cardinality conformal mesh refinement problem is $\mathcal{NP}$-hard even for homogeneous meshes.*

For single polygons, however, this problem is efficiently solvable. More precisely, two variants of the problem can be solved in linear time, namely the case which allows to insert additional vertices to arbitrary positions and the case which allows additional vertices only into the *interior* of the polygon, but not on its boundary, i.e. it forbids to subdivide edges.

In the mesh refinement problem, the polygons cannot be refined independently since we have to ensure that the mesh is conformal. Hence, we carefully distinguish between conformal refinements, where vertices can be inserted at arbitrary positions, and conformal decompositions (see Fig. 3 for an example). By a *conformal decomposition* of a single polygon we will always mean the variant which does not allow to subdivide edges but to place vertices into the *interior* of the polygon.

The following theorem holds:

Theorem 2. [MW96] *There is a linear–time algorithm which constructs a minimal conformal decomposition of a polygon into strictly convex quadrilaterals.*

A very similar linear–time algorithm solves the minimal conformal refinement problem for a polygon (details will be given in the complete version of this paper). There is also a well-known (see, for example [Joe95]), but important characterization of those polygons which can be decomposed into strictly convex quadrilaterals:

Lemma 3. *A simple polygon P admits a conformal decomposition if and only if the number of vertices of P is even.*

Lemma 3 and Theorem 2 give rise to the following two-stage approach: First, subdivide a couple of edges such that each polygon achieves an even number of vertices; second, refine each polygon separately according to the algorithm mentioned in Theorem 2. Clearly, the first stage determines the approximation factor. In [MW96], each edge of the mesh is subdivided exactly once, which

 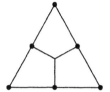 

Fig. 3. A triangular–shaped polygon with four vertices (left); an optimal *refinement*, which places two extra vertices on the boundary (middle); and an optimal *decomposition*, where no additional vertices on the boundary are allowed (right).

trivially makes all polygons even. It is also proved in [MW96] that this simple strategy already yields a 4–approximation. The analysis of this simple strategy is tight: for a conformal mesh of quadrilaterals, this algorithm obviously takes four times as many quadrilaterals as the optimum. To improve upon this performance guarantee, we apply a more sophisticated strategy.

A few related problems have found some attention. Note, for example, that it is important that polygons are by definition convex polygons, as Lubiw [Lub85] has shown that both problems, minimum refinement and minimum decomposition, are $\mathcal{NP}$–hard for single *non-convex* polygons with holes. To the best of our knowledge, the complexity status of the refinement problem for non-convex polygons *without* holes is still open. Everett et al. [ELOSU92] give lower and upper bounds on the number of quadrilaterals in a conformal refinement of simple, not necessarily convex polygons (with and without holes), but not on decompositions. Refs. [Sac82,ST81] investigate perfect decompositions of (star-shaped) rectilinear polygons into *non–strictly* convex quadrilaterals, and [Lub85] considers perfect decompositions of non-convex polygons but even allows overlapping internal edges. See [Tou95] for a systematic survey.

In Sect. 4, we will present the main results of this paper:

- There is a linear–time approximation algorithm which exceeds ratio 2 by an additional term of at most $\Delta(\mathcal{M})$. This parameter $\Delta(\mathcal{M})$ (to be defined below in Def. 4) depends on the mesh structure, but for all practical instances that we know of, $\Delta(\mathcal{M})$ is significantly smaller than the minimal number of quadrilaterals in a conformal refinement. Hence, for such instances, this yields a 3–approximation. (For general instances, this algorithm always guarantees a 4–approximation.)
- As an immediate consequence, this yields a linear-time 3–approximation for homogeneous meshes. (This is not true for the algorithm in [MW96].)
- For homogeneous meshes, we can even do better, namely, we get a 2–approximation algorithm which runs in $\mathcal{O}(n\,m\log n)$ time, where n (m) is the number of polygons (edges) in the mesh. If a homogeneous mesh corresponds to a planar graph, the running time can be further reduced to $\mathcal{O}(n^{3/2}\log n)$ by an application of the planar separator theorem.

The asymptotic complexity of the algorithm for homogeneous meshes is dominated by solving a *T–join problem*, or equivalently, a *minimum–cost perfect b–*

matching problem (see the monograph by Derigs [Der88] or the survey by Gerards [Ger95] for matching problems) in a certain variant on the dual graph of the mesh. In our application, the algorithm from [Gab83] requires $\mathcal{O}(n\,m\log n)$ time. Usually, (homogeneous) meshes are sparse, i.e. they have only $m = \mathcal{O}(n)$ edges.

All our results also carry directly over to the following, slightly more general variant on the minimum mesh refinement problem. Suppose that the given mesh is too coarse to expect reasonable results from the finite element method, but a finite element error estimation gives lower bounds on the mesh density which should be achieved. More precisely, suppose that these lower bounds on the mesh density are expressed as lower bounds on the number of vertices which have to be placed on the original edges in a feasible refinement. (There are CAD packages which pursue this strategy.) The general problem is to find a conformal refinement which respects these lower bounds, but minimizes the number of quadrilaterals.

The rest of the paper is organized as follows. In Section 2, we start with some preliminaries and introduce further terminology. Then, in Section 3 we present a combinatorial result (cf. Lemma 5). Roughly speaking, this result means that the minimum number of quadrilaterals needed for a decomposition of a polygon does not increase exorbitantly, if each edge is subdivided at most once. The proof of Lemma 5 is quite involved and omitted for lack of space.

Finally, in Section 4, we present the new approximation algorithms and prove their performance guarantees.

2 Preliminaries and Further Definitions

Let $\mathcal{P} = \{P_1, P_2, \ldots, P_n\}$ be the set of polygons forming the mesh. These polygons are convex, but not necessarily strictly convex. Two polygons are *neighbored* if they have an interval of the boundary in common which has strictly positive length. These neighborhood relationships induce an undirected graph $G = (V, E)$, which is embedded on the surface approximated by the mesh and whose faces are the polygons. More precisely, V consists of the corners of the polygons. If a corner of a polygon also belongs to the interior of a side of another polygon, it subdivides this side. Hence, we may identify common intervals of neighbored sides of polygons with each other, and E consists of these intervals after identification.

Note that the graph G of a mesh need not be planar; for example, a mesh approximating the surface of a torus has genus one. The set of all folding edges that are incident to exactly the same homogeneous components is called a *folding*.

For an edge $e_i \in E$, let E_i be the set of all those polygons which are incident to e_i. A *combinatorial description* of a mesh consists of the graph G and the hypergraph $H = (\mathcal{P}, \{E_1, \ldots, E_m\})$ with vertex set $\mathcal{P}$ and edge set $\{E_1, \ldots, E_m\}$. We will often identify a mesh with its combinatorial description.

A *non-folding path* in H is a path between two polygons $P_1, P_2 \in \mathcal{P}$ which contains only hyperedges of cardinality two, i.e. only such hyperedges which

Fig. 4. A convex polygon with 7 corners and 16 vertices and a conformal decomposition with 7 additional, internal vertices. (The decomposition is not minimal.)

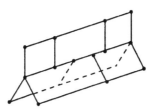

Fig. 5. A small mesh with three homogeneous components and one folding, which consists of five folding edges. The corresponding hypergraph has 13 edges of degree one (boundary edges), 4 hyperedges of degree 2, and 5 hyperedges of degree 3.

belong to exactly two polygons. Being connected by a non-folding path is an equivalence relation on the set of polygons. Its equivalence classes are exactly the *homogeneous components* of a mesh. For a mesh $G = (V, E)$ let $G_1, \ldots, G_{c(\mathcal{P})}$ denote the homogeneous components, and $c(\mathcal{P})$ the number of components. The *degree* of an edge in E is the number of incident polygons. The *boundary* of a mesh is the set of all hyperedges with degree one (the *boundary edges*).

Since all folding edges within a folding are incident to exactly the same homogeneous components, the degree of a folding is well defined. Let $D(\mathcal{M})$ denote the total sum of the degrees of all foldings that consist of an *odd* number of folding edges each. This allows us to define the parameter $\Delta(\mathcal{M})$ which appears in the performance guarantee we can achieve for meshes, in general:

Definition 4. $\Delta(\mathcal{M}) := D(\mathcal{M}) + |\mathcal{P}| - c(\mathcal{P})$.

Empirically, the mesh parameter $\Delta(\mathcal{M})$ is fairly small. In a whole bunch of real-world examples, which stem from the German car industry, the average number of odd foldings per homogeneous component is less than three, and always smaller than the minimum number of quadrilaterals needed for that component. (This means that we guarantee a 3–approximation for such instances from practice.) In fact, it seems hard to imagine a non-pathological instance, where $\Delta(\mathcal{M})$ is larger than the minimum number of quadrilaterals in an optimal mesh refinement.

A vertex of a convex polygon is a *corner* if its internal angle is strictly less than π. An *interval* of a polygon P is a path of edges on its boundary. A *segment* S is an interval between two successive corners of P.

A conformal decomposition of P is usually identified with the planar, embedded graph $G_P = (V, E)$ whose outer face is P and whose internal faces are the quadrilaterals. Let $q(G)$ denote the number of internal, quadrangular faces. Let $G^* = (V^*, E^*)$ be the variant on the *dual graph* which arises by removing the vertex corresponding to the outer face of G_P. We call a conformal decomposition of a convex polygon P *perfect* if it has no vertices other than P.

We will denote a polygon by the counterclockwise sequence of the lengths of its segments. For example, $(1, 1, 1, 1)$ denotes the strictly convex quadrilateral, $(1, 1, 2) = (1, 2, 1) = (2, 1, 1)$ the quadrilateral degenerated to a triangle, and $(4, 1, 2, 3, 2, 2, 2) = (1, 2, 3, 2, 2, 2, 4) = \cdots$ the polygon in Fig. 4. This is justified by the following observation (cf. Lemma 3.4 in [MW96]): If two polygons P_1 and P_2 have the same such sequence (up to cyclic shifts), then every graph of a conformal decomposition for P_1 is also the graph of some conformal decomposition for P_2 and vice versa.

3 Subdivisions of Polygons

In this section, we present a combinatorial result which relates the optimal conformal refinements of a polygon P to optimal refinements of those polygons, which arise if some of the edges of P are subdivided by one additional vertex. This result will be useful for conformal refinements of meshes in Sect. 4. In fact, Lemma 5 is the most difficult part of the proofs of Theorems 8 and 10.

For a convex polygon P with an even number of vertices, min(P) denotes the minimum number of quadrilaterals required by any conformal decomposition of P. For an arbitrary convex polygon P with edge set E_P, a mapping $X : E_P \to \mathbb{N}_0$ is called *feasible* if $\sum_{e \in E_P} X(e)$ has the same parity as E_P. In particular, if $|E_P|$ is even, $X \equiv 0$ is possible, too. The polygon P_X is constructed from P by subdividing each edge $e \in E_P$ exactly $X(e)$ times. Hence, X feasible means that P_X admits a conformal decomposition. Moreover, Min(P) denotes the minimum number of quadrilaterals in any conformal decomposition of any polygon P_X, Min$(P) := \min\{q(G) \mid G \text{ conformal decomposition of } P_X, X : E_P \to \mathbb{N}_0 \text{ feasible}\}$. If $X(e) \in \{0, 1\}$ for all e, we denote $|X| = |\{e \in E_P : X(e) = 1\}|$.

Lemma 5. *For a polygon P and $X : E_P \to \{0, 1\}$ we have* min$(P_X) \leq 2 \cdot$ Min$(P) + |X| - 1$, *and we even have* min$(P_X) \leq 2 \cdot$ Min$(P) + |X| - 2$ *except for the following cases:*

1. *$P = (1, 1, 1, 1)$ and $P_X = (1, 1, 1, 1)$;*
2. *$P = (1, 1, 1, 1)$ and $P_X = (2, 2, 1, 1)$;*
3. *$P = (2, 1, 1, 1)$ and $P_X = (3, 1, 1, 1)$;*
4. *$P = (2, 1, 2, 1)$ and $P_X = (4, 1, 2, 1)$;*
5. *$P = (2, 1, 1, 1, 1)$ and $P_X = (4, 1, 1, 1, 1)$;*
6. *$P = (2, 1, 1)$ and $P_X = (2, 1, 1)$.*

Proof. Omitted.

4 Approximation of Minimal Conformal Mesh Refinements

In this section, we describe the improved approximation algorithms. In the following, we will need a certain variant $G^d = (V^d, E^d)$ on the dual graph of the graph $G = (V, E)$ of a homogeneous mesh $\mathcal{M}$.

Definition 6. For a homogeneous mesh $\mathcal{M}$ and its corresponding graph $G = (V, E)$, the graph $G^d = (V^d, E^d)$ (multiple edges allowed) has a vertex for each polygon of $\mathcal{M}$. If the homogeneous mesh $\mathcal{M}$ has a non-empty boundary, exactly one more vertex is added to V^d. Two dual vertices that correspond to polygons of $\mathcal{M}$ are connected by an edge of E^d if and only if they share an edge in E. For each boundary edge (if existing) the additional vertex is connected with the dual vertex whose corresponding polygon is incident to the boundary.

We will need the next fact, Lemma 7, for all algorithmic results to follow. Let $G = (V, E)$ be an undirected graph, and let each vertex be either labeled *odd* or labeled *even*. This *odd/even* labeling is called *feasible* if the number of vertices labeled *odd* is even. A subgraph G' of a graph G is called *feasible* if the following holds: The degree of a vertex is odd in G' if and only if its label is "*odd*."

Lemma 7. *There is a linear–time algorithm that produces a feasible acyclic subgraph F of a connected graph G with respect to a feasible odd/even labeling.*

Proof. Let T be a spanning tree of G and let F be the forest comprising all edges e of T that divide $T - e$ into two subtrees, each with an odd number of vertices labeled *odd*. It is easy to see that F is feasible.

Theorem 8. *There is a linear–time algorithm that constructs a conformal refinement of a mesh G such that the number of quadrilaterals exceeds twice the optimum by at most $\Delta(\mathcal{M})$.*

Proof. First we describe the algorithm and prove that it constructs a conformal refinement. Recall from Lemma 3 that we have to ensure that every polygon becomes even. For each folding that consists of an odd number of folding edges, we select one of these edges and subdivide it once. We will see that this suffices to refine all homogeneous components separately. So let $G_i = (V_i, E_i)$ be a homogeneous component. First consider the edges in E_i that have degree one but are not folding edges of the original mesh. In other words, consider the boundary edges of G which belong to G_i. If the number of these edges is odd, we select exactly one of these edges and subdivide it once, too. After this procedure, the number of edges of degree one in E_i is even. (Note that all other edges have degree 2, because G_i is homogeneous.)

Let $G_i^d = (V_i^d, E_i^d)$ denote the variant on the dual graph of G_i as in Def. 6. Then the vertex of G_i^d added for the boundary edges has even degree in G_i^d. Therefore, the number of odd vertices in G_i^d that correspond to polygons (and hence the number of odd polygons themselves) in G_i is even. Consequently, we may apply the algorithm of Lemma 7 to construct a feasible acyclic subgraph F_i of G_i^d, where a vertex is labeled *odd/even* according to the parity of its degree. Next each edge of E_i that corresponds to an edge in F_i is subdivided exactly once. Obviously, every polygon is now even, and we apply the algorithm from [MW96] to decompose each polygon separately.

It remains to show that this construction leads to a refinement that exceeds twice the optimum by at most $\Delta(\mathcal{M})$.

For a polygon P of the homogeneous component G_i, $i = 1, \ldots, k$, let $X_P : E_P \to \{0,1\}$ be defined such that $X_P(e) = 1$ means that e corresponds to a dual edge of F_i. Analogously, let $Y_P : E_P \to \{0,1\}$ attain 1 exactly on the edges of P that are selected in the algorithm to make all foldings even. Moreover, let $Z_P : E_P \to \{0,1\}$ attain 1 on an edge if and only if this edge is selected to make the number of boundary edges of G_i even. Let $P' := P_{(X_P + Y_P + Z_P)}$. Of course, we have $X_P(e) + Y_P(e) + Z_P(e) \leq 1$ for each edge e. Therefore, Lemma 5 gives

$$\min(P') \leq 2 \cdot \operatorname{Min}(P) + |X_P| + |Y_P| + |Z_P| - 1 \ . \tag{1}$$

Let $\mathcal{P}_i$ denote the set of polygons in G_i. Since the feasible subgraph F_i of G_i^d constructed by the algorithm is acyclic, we have $\sum_{P \in \mathcal{P}_i} |X_P| \leq 2 \cdot (|\mathcal{P}_i| - 1)$, and since $\sum_{P \in \mathcal{P}_i} |Z_P| \leq 1$, Ineq. (1) sums up to

$$\sum_{P \in \mathcal{P}_i} \min(P') \leq \sum_{P \in \mathcal{P}_i} \left[2 \cdot \operatorname{Min}(P) + |Y_P| \right] + |\mathcal{P}_i| - 1 \ . \tag{2}$$

Note that $\sum_{i=1}^{k} \sum_{P \in \mathcal{P}_i} |Y_P| = D(\mathcal{M})$. Hence, Ineq. (2) sums up to

$$\sum_{P \in \mathcal{P}} \min(P') \leq 2 \cdot \sum_{P \in \mathcal{P}} \operatorname{Min}(P) + D(\mathcal{M}) + |\mathcal{P}| - k = 2 \cdot \sum_{P \in \mathcal{P}} \operatorname{Min}(P) + \Delta(\mathcal{M}) \ .$$

As an immediate consequence, we obtain for the special cases of meshes without foldings of odd degree, i.e. where $D(\mathcal{M})$ vanishes, the following corollary:

Corollary 9. *There is a linear-time algorithm that yields a 3-approximation for the minimum conformal refinement problem for the special cases where $D(\mathcal{M}) = 0$. This includes, in particular, the homogeneous meshes.*

For meshes without foldings, we can even find significantly better approximations using a nice application of matching techniques:

Theorem 10. *For homogeneous meshes, there is an $\mathcal{O}(n \, m \log n)$ algorithm that constructs a conformal refinement such that the number of quadrilaterals is at most twice the optimum.*

Like in the proof of Theorem 8, we construct a feasible acyclic subgraph F, but now we use edge weights to find subgraphs which allow for a better analysis. The idea is to choose edge weights in such a way that we get an improved lower bound if some of the expensive edges are chosen in a feasible subgraph of minimum weight.

We determine F in an auxiliary graph $G_{aux}^d = (V_{aux}^d, E_{aux}^d)$. This is necessary to cope with the exceptions 2. - 5. in Lemma 5. For these four types of polygons, we try to avoid a subdivision which leads to the exception. This is in contrast to the first and last exceptions. In particular, for the last case $P = (2,1,1)$ a solution with a subdivision of edges is favorable (see again Fig. 3). The analysis below will show that we do not have to treat polygons of type $P = (2,1,1)$ in a special way.

The graph G_{aux}^d is constructed from G^d as follows. Each polygon $v^d \in V^d$ of type $(1,1,1,1)$, $(2,1,1,1)$, $(2,1,2,1)$, or $(2,1,1,1,1)$ is replaced by a couple of vertices and edges, which respectively form subgraphs as shown in Fig. 6. Each edge $e \in E_{aux}^d$ is assigned a weight $w(e^d)$. The weights $w(e^d)$ of the edges in Fig. 6 are introduced there, too. All other edges $e^d \in E_{aux}^d$ have weight $w(e^d) = 0$. We say that a subgraph F_{aux} of G_{aux}^d is feasible if the degree of every vertex outside these four kinds of subgraphs has the same parity in G_{aux}^d and F_{aux}, and all vertices inside these subgraphs have even degree in F_{aux} except for the vertex indicated by an arrow in Fig. 6, which must have odd degree in F_{aux}.

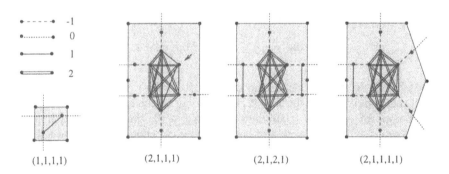

Fig. 6. The subgraphs of G_{aux}^d for polygons v^d of type $(1,1,1,1)$, $(2,1,1,1)$, $(2,1,2,1)$, and $(2,1,1,1,1)$. The weight of an edge is indicated by its line style.

Lemma 11. *Let F_{aux} be a feasible subgraph of G_{aux}^d such that the sum of all weights $w(\cdot)$ of edges in F_{aux} is minimum. Let F be a feasible acyclic subgraph of G^d such that F is constructed from F_{aux} by shrinking all subgraphs in Fig. 6 and removing all cycles from the shrunken F_{aux}. Then subdividing once all edges of G that correspond to dual edges in F yields a 2–approximation.*

Proof. The following facts are easy to see from Fig. 6:

- There are no negative cycles in G_{aux}^d with respect to the edge weights $w(\cdot)$.
- For an optimal F_{aux}, the weights $w(\cdot)$ inside a polygon in Fig. 6 sum up to zero if and only if none of the exceptional situations in Lemma 5 occurs.

- If such an exceptional situation occurs for a polygon in Fig. 6, the edges inside this polygon contribute exactly one unit of weight in total for an optimal F_{aux}.

In summary, the total weight of an optimal F_{aux} in G_{aux}^d is equal to the number r, say, of exceptional situations that occur in the corresponding conformal refinement.

Hence, r is minimum among all those conformal refinements of G wherein no edge in E is subdivided more than once, and $\sum_{P \in \mathcal{P}} \text{Min}(P) + r$ is a lower bound on the number of quadrilaterals in any such refinement.

Next we show that $\sum_{P \in \mathcal{P}} \text{Min}(P) + r$ is a lower bound on the number of quadrilaterals in *every* conformal refinement. To see this, take an arbitrary conformal refinement. For $e \in E$, let $Y(e)$ denote how often e is subdivided in this refinement. Let $Y'(e) \in \{0, 1\}$ be the remainder of $Y(e)/2$. Then subdividing each edge $e \in E$ exactly $Y'(e)$ times makes all polygons even. However, it is easy to see that a polygon P that is exceptional in the refinement induced by Y' cannot be refined with $\text{Min}(P)$ quadrilaterals in the refinement induced by Y.

This shows that $\sum_{P \in \mathcal{P}} \text{Min}(P) + r$ is a lower bound on the number of quadrilaterals in the refinement induced by Y, too.

For a polygon $P \in \mathcal{P}$, let $X_P : E_P \to \{0, 1\}$ denote which edges are subdivided in the refinement induced by F. Now it suffices to show

$$\sum_{P \in \mathcal{P}} \min(P_{X_P}) \leq 2 \cdot \sum_{P \in \mathcal{P}} \text{Min}(P) + r \ .$$

For each $i \in \mathbb{N}_0$ let n_i denote the number of vertices in F having degree i. Recall that a polygon of type $T = (2, 1, 1)$ belongs to the exceptions of Lemma 5 only if none of its edges is subdivided. If $X_T \equiv 0$, then our solution needs $\min(T) = 5 \leq 2 \cdot \text{Min}(T) = 6$, and is therefore good enough to achieve a factor of 2 for such a polygon.

Thus, Lemma 5 implies

$$\sum_{P \in \mathcal{P}} \min(P_{X_P}) \leq 2 \cdot \sum_{P \in \mathcal{P}} \text{Min}(P) - n_1 + \sum_{i \geq 3}(i - 2) \cdot n_i + r \ .$$

Hence, it suffices to show $\sum_{i > 3}(i - 2) \cdot n_i \leq n_1$. To see this, we first replace each vertex with degree $i > 3$ by a path of $i - 2$ vertices each with degree 3. In this modified graph, the claim follows from $n_3 \leq n_1$, which is well known for binary trees (and all the more true for binary forests). Obviously, the backward transformation preserves validity of the claim.

Proof of Theorem 10. Because of Lemma 11, it remains to show how to construct an optimal feasible graph F_{aux}. We solve this problem by a reduction to a *capacitated minimum–cost perfect b–matching problem* [Der88]. In order to introduce this reduction, we first state the problem we want to reduce in more general terms, also known in the literature as the *T–join problem* (see, for example [Ger95]). So let $G = (V, E)$ be an undirected graph, let $w(\cdot)$ be a weighting of E, and for $v \in V$ let *equal(v)* be a logical flag. We call a subgraph F of G feasible if the following holds: The degree of each $v \in V$ in F has the same parity as the degree in G if and only if *equal(v)* is true. The problem is to find a feasible subgraph that minimizes the sum of the edge weights $w(\cdot)$.

The reduction is as follows (and was first proposed by Edmonds and Johnson [EJ73]). For $v \in V$, let $b(v)$ equal the degree of v if *equal(v)* is true, otherwise let $b(v)$ equal the degree plus one. Let $\bar{G} = (V, \bar{E})$ denote G with all loops $\{v, v\}$, $v \in V$, added to E. The weight of such a loop is $w(\{v, v\}) := 0$. Moreover, we set $\ell(e) := 0$ for all $e \in \bar{E}$, $u(e) := 1$ for $e \in E$, and $u(\{v, v\}) := \lfloor b(v)/2 \rfloor$ for $\{v, v\} \in \bar{E} \setminus E$.

There is a straightforward one–to–one correspondence between feasible subgraphs of G and perfect b–matchings in $\bar{G}$ with lower bounds $\ell(\cdot)$ and upper bounds $u(\cdot)$. Moreover, the cost of a b–matching with respect to $w(\cdot)$ equals the sum of edge weights of the corresponding feasible subgraph.

Note that the graph of the b–matching instance is essentially as dense as G, i.e. it has $\mathcal{O}(m)$ edges. This establishes the reduction, and by an application of Gabow's [Gab83] algorithm the time bound follows.

Apart from pathological constructions, we even have $m = \mathcal{O}(n)$. In particular, this is always the case, if the number of corners of each polygon is bounded by some constant. Hence, in such cases, the b–matching instance runs on $\mathcal{O}(n)$ edges. However, the underlying graph is not planar, in general.

If the graph G of the homogeneous mesh $\mathcal{M}$ is planar, the running time of the minimum T–join algorithm can be slightly improved by an application of the famous planar separator theorem of Lipton and Tarjan.

Theorem 12 (planar separator). [LT79] *Let G be a planar graph on n vertices. Then the vertices of G can be partitioned into three sets A, B, C, such that no edge joins a vertex in A with a vertex in B, neither A nor B contains more than $\frac{2}{3}n$ vertices, and C contains no more than $2\sqrt{2n}$ vertices. Furthermore, the sets A, B, C can be found in $\mathcal{O}(n)$ time.*

Following [MNS86], we use the notion of a *good separator*.

Definition 13. A graph G on n vertices has a *good separator* if there exist two constants $c_1 < 1$ and c_2 satisfying: The vertices of G can be partitioned into three sets A, B, C such that no edge joins a vertex in A with a vertex in B, neither A nor B contains more than $c_1 n$ vertices, and C contains no more than $c_2 \sqrt{n}$ vertices.

Lemma 14. *If the graph G of the homogeneous mesh $\mathcal{M}$ is planar, then an optimal feasible subgraph can be computed in time $\mathcal{O}(n^{3/2} \log n)$.*

Proof. Barahona[Bar90] and Matsumoto et al. [MNS86] have shown how to solve the minimum T–join problem in $\mathcal{O}(n^{3/2} \log n)$ for planar graphs using the planar separator theorem.

We cannot directly use their result, as the graph G_{aux}^d on which we have to solve the T-join problem is not planar, in general. However, with a slight modification of the technique used by Matsumoto et al. [MNS86] we can show in the following that G_{aux}^d has a good separator.

Let G^d be the variant of the dual graph as in Def. 6 and $v_b \in V^d$ be the vertex corresponding to the boundary of the mesh. G^d need not be planar, if the boundary of the mesh is not connected. However, after deletion of v_b, the graph $G^d \setminus \{v_b\}$ is certainly planar, if G is planar. Hence, we can apply the planar separator theorem to $G^d \setminus \{v_b\}$. This means that we can partition the graph $G^d \setminus \{v_b\}$ into sets A, B, C, such that no edge joins a vertex in A with a vertex in B, neither A nor B contains more than $\frac{2}{3}n$ vertices and C contains no more than $2(2n)^{1/2}$ vertices. If we put v_b into the set C, we clearly also have a partition for G^d with the required properties.

The partition A, B, C naturally induces a partition of the vertices of G_{aux}^d into A', B' and C': If a vertex $v_d \in A$ ($v_d \in B$, $v_d \in C$) is replaced by a subgraph in G_{aux}^d, then all vertices of the subgraph belong to A' (B', C', respectively).

We have to show that A', B', C' yields a good separator. Let k be the number of vertices of the largest subgraph introduced for a polygon v^d. It is important that k is some constant number, namely $k = 12$ in Fig. 6. Denote by n' the number of vertices in G_{aux}^d. Hence, we have $n' \leq kn$. As $|A| \leq \frac{2}{3}n$, at least $\frac{n}{3}$ vertices of n' cannot belong to $|A'|$. This means $|A'| \leq n' - \frac{n}{3} \leq (1 - \frac{1}{3k})n' = c_1 n'$, with $c_1 := 1 - \frac{1}{3k} < 1$. By symmetry, we can also bound the number of vertices in B' by $|B'| \leq c_1 n'$. Certainly, C' contains no more than $|C'| \leq k|C| \leq 2k(2n)^{1/2} \leq 2k(2n')^{1/2}$ vertices. Hence, we can choose $c_2 := 2k\sqrt{2}$. Thus, there is a good separator for G_{aux}^d which can be found in linear time.

Very similarly, we can also show that in the whole separation tree the subgraphs partitioned by their separators all have good separators.

Now a very similar analysis as in [Bar90] yields the claimed result.

References

[Bar90] F. Barahona. Planar multicommodity flows, max cut and the Chinese postman problem. In W. Cook and P. D. Seymour, editors, *Polyhedral Combinatorics*, DIMACS Series in Discrete Mathematics and Theoretical Computer Science, vol. 1, pages 189–202. American Mathematical Society, Providence, RI, 1990.

[BE95] M. Bern and D. Eppstein. Mesh generation and optimal triangulation. In D.-Z. Du and F. Hwang, editors, *Computing in Euclidean Geometry*, 2nd Edition, pages 47–123. World Scientific, Singapore, 1995.

[Der88] U. Derigs. *Programming in networks and graphs: on the combinatorial background and near-equivalence of network flow and matching algorithms.* Springer, 1988.

[EJ73] J. Edmonds and E.L. Johnson. Matching, Euler tours and the Chinese postman. *Mathematical Programming*, 5:88-124, 1973.

[ELOSU92] H. Everett, W. Lenhart, M. Overmars, T. Shermer, and J. Urrutia. Strictly convex quadrilateralizations of polygons. In *Proceedings Fourth Canadian Conference on Computational Geometry*, pages 77–83, 1992.

[Gab83] H. N. Gabow. An efficient reduction technique for degree-constrained subgraph and bidirected network flow problems. In *Proceedings of the 15th ACM Symposium on the Theory of Computing, STOC'83*, 448–456, 1983.

[Ger95] A. M. H. Gerards. Matching. In Ball et. al, editors, Handbooks in Operations Research and Management Science, vol. 7, *"Network Models"*, chapter 3, North-Holland, 1995.

[Ho88] K. Ho-Le. Finite element mesh generation methods: a review and classification. *Computer–Aided Design*, 20:27–38, 1988.

[Joe95] B. Joe. Quadrilateral mesh generation in polygonal regions. *Computer–Aided Design*, 27:209–222, 1995.

[LT79] R.J. Lipton and R.E. Tarjan. A separator theorem for planar graphs. *SIAM J. Appl. Math.*, 36:177-189, 1979.

[Lub85] A. Lubiw. Decomposing polygonal regions into convex quadrilaterals. In *Proceedings of the ACM Symposium on Computational Geometry*, pages 97–106, 1985.

[MNS86] K. Matsumoto, T. Nishizeki, and N. Saito. Planar multicommodity flows, maximum matchings and negative cycles. *SIAM J. of Computing*, 15:495-510, 1986.

[MMW95] R. H. Möhring, M. Müller-Hannemann, and K. Weihe. Using network flows for surface modeling. In *Proceedings of the 6th Annual ACM-SIAM Symposium on Discrete Algorithms, SODA'95*, pages 350–359, 1995.

[MW96] M. Müller-Hannemann and K. Weihe. Minimum strictly convex quadrangulations of convex polygons. *Technical report No. 519/1996*, Department of Mathematics, Technische Universität Berlin, 1996, available via anonymous ftp from *ftp.math.tu-berlin.de; cd pub/Preprints/combi;* file *Report-519-1996.ps.Z* (extended abstract appears in the *Proceedings of the 13th Annual ACM Symposium on Computational Geometry, SoCG'97*, 1997).

[Sac82] J. R. Sack. An $\mathcal{O}(n \log n)$ algorithm for decomposing simple rectilinear polygons into convex quadrilaterals. In *Proceedings 20th Conference on Communications, Control, and Computing*, pages 64–74, 1982.

[ST81] J. R. Sack and G. Toussaint. A linear–time algorithm for decomposing rectilinear star–shaped polygons into convex quadrilaterals. In *Proceedings 19th Conference on Communications, Control, and Computing*, pages 21–30, 1981.

[TA93] T. Tam and C. Armstrong. Finite element mesh control by integer programming. *Int. J. Numer. Methods Eng.*, 36:2581-2605, 1993.

[Tou95] G. Toussaint. Quadrangulations of planar sets. In *Proceedings 4th International Workshop on Algorithms and Data Structures, WADS'95, Kingston, Canada, August 16-18, 1995*, pages 218–227, 1995.

[ZT89] O.C. Zienkiewicz and R.L. Taylor. *The finite element method.* McGraw Hill, London, 1989.

Dynamic Storage Allocation with Known Durations*

Joseph (Seffi) Naor[1]**, Ariel Orda[2]***, Yael Petruschka[3]

[1] Computer Science Dept., Technion, Haifa 32000, Israel (naor@cs.technion.ac.il).
[2] Electrical Engineering Dept., Technion, Haifa 32000, Israel (ariel@ee.technion.ac.il).
[3] Computer Science Dept., Technion, Haifa 32000, Israel (petrus@cs.technion.ac.il).

Abstract. This paper is concerned with a new version of on-line storage allocation in which the durations of all processes are *known* at their arrival time. This version of the problem is motivated by applications in communication networks and has not been studied previously. We provide an on-line algorithm for the problem with a competitive ratio of $O(\min\{\log \Delta, \log \tau\})$, where Δ is the ratio between the longest and shortest duration of a process, and τ is the maximum number of concurrent active processes that have different durations. For the special case where all durations are powers of two, the competitive ratio achieved is $O(\log \log \Delta)$.

1 Introduction

This paper is concerned with the classic problem of dynamic storage allocation (DSA). In a typical on-line setting of the problem, a sequence of processes, each requesting a certain amount of memory space, arrive at a storage device. Upon arrival of a process, a contiguous storage area is allocated for it. This area is liberated for reuse only when the process terminates at some later point of time, which, in general, is *unknown* at the time of allocation. As a result, "holes" of wasted space are generated over time in the storage device. The objective is to find an allocation algorithm that minimizes the wasted space. The decisions of this algorithm must be based at each point of time on the known current state of the storage device without knowledge of future events. A common way to measure the quality of an on-line algorithm is competitive analysis. Here, the costs associated with an on-line algorithm are compared with the costs expanded by an optimal off-line algorithm that knows the sequence of events in advance. The maximum ratio between their respective costs, taken over all sequences, is called the *competitive ratio*.

Broadband communication networks are an important area of potential application of dynamic storage allocation algorithms. In such networks, bandwidth

---

* This research is supported by the consortium for broadband communication administered by the chief scientist of the Israeli Ministry of Industry and Commerce.
** Supported by Technion V.P.R. Fund 120-911 - Promotion of Sponsored Research.
*** Supported by Technion V.P.R. Fund 050-862 - Promotion of Sponsored Research.

is allocated to sessions, according to their quality of service requirements. Typically, this allocation is contiguous, e.g., in a time-division multiplexed (TDM) environment a session is allocated in successive time slots. Indeed, an uncontiguous allocation would incur higher overhead in terms of bandwidth management (see [SS94] for a related discussion). Replacing the standard terms of "storage" or "memory" with "bandwidth", and "process" with "session", dynamic storage allocation concepts can be applied to the networking framework. Typically, the duration time of broadband applications is either known *a priori* or can be predicted fairly well, e.g., video on demand, reserved video conferences. Moreover, it has been recently recognized [Par94] that quality of service guarantees should often be specified on a *time interval* basis, which is facilitated when the session duration is known. Such knowledge on session duration has been utilized in the design of on-line network algorithms, e.g. [BNCK+95, AAP93, AKP+93]. However, in the context of dynamic storage allocation, this has not been done, and, in effect, is the objective of the present study.

In this paper, we investigate the dynamic storage allocation problem in the case where the duration of a process is known at the time of its arrival. That is, upon arrival of a process, a contiguous storage area is allocated for it for a prespecified period of time. Our main result is an on-line algorithm for dynamic storage allocation with known durations. Denote by τ the maximum number of *concurrent* active processes that have different durations, and denote by Δ the ratio between the longest and shortest duration of a process. The competitive ratio of our algorithm is $O(\min\{\log \Delta, \log \tau\})$. For the special case that all durations are powers of two, the competitive ratio is actually $O(\log \log \Delta)$. The main feature of our on-line algorithm is that it distinguishes between the allocation for "heavy" and "light" processes, where this distinction is based on *both* storage area requirements and durations of the processes.

The dynamic storage allocation problem has been extensively investigated since the Fifties. Basic methods are discussed in Knuth [Knu73, Section 2.5]. A survey of results in this area until the early Eighties appears in Coffman [Cof83].

Two intuitive heuristics that were developed early on are *First Fit* and *Best Fit*. First Fit finds the first free area that can fit the requirement of the current process; Best Fit finds the free area that best fits the requirement of the current process, i.e., the one that causes minimal fragmentation. In the early Seventies, Robson [Rob71, Rob74] gave an algorithm which had a competitive ratio of $O(\log(W_{max}))$, where W_{max} is the maximum memory space requested by a process. He also showed that this is the best possible bound up to constant factor. Subsequently, he showed [Rob77] that the competitive ratio of First Fit is also $\Theta(\log(W_{max}))$. Surprisingly, Robson [Rob77] also showed that the competitive ratio of Best Fit can be as bad as $\Omega(W_{max})$.

Segregated storage methods [Sta95] constitute a different category for DSA. These methods include many variations. The basic idea here is to partition the memory into blocks, such that each block only contains processes that have the same (or similar) requirement.

Recently, Luby, Naor and Orda [LNO96] showed that the competitive ratio of both First Fit and some segregated storage methods is $\Theta\left(\min\{\log k, \log W_{\max}\}\right)$, where k is the maximum number of concurrent processes. Furthermore, by proving a lower bound on the competitive ratio in the randomized case which is very close to the deterministic bound, they showed that randomization may be of little help here.

The off-line version of the storage allocation problem is NP-complete in the strong sense [GJ79, page 226]. Yet, polynomial time approximation algorithms for this problem were devised. Kierstead [Kie88] showed that if the processes are sorted by their storage requirements, then the competitive ratio of First Fit is 80. This approximation factor was later improved by Kierstead [Kie91] to 6. Recently, Gergov [Ger96] achieved an approximation factor of 5.

2 Preliminaries

The storage allocation problem is equivalent to the problem of finding a weighted coloring of a given weighted interval graph.

Definition 1. An undirected graph $G = (V, E)$ is called an *interval graph* if its set of vertices V can be put into one-to-one correspondence with a set of intervals I, such that two vertices are connected by an edge if and only if their corresponding intervals overlap. An interval graph is called *weighted* if each interval is associated with a positive weight.

Definition 2. A *weighted coloring* of an interval graph G is a coloring of each interval i with a range $[a_i, b_i]$ (where a_i, b_i are integers), such that $b_i - a_i + 1$ is equal to the weight of interval i, and there is no intersection between the ranges of two overlapping intervals.

The processes form an interval graph denoted by $G = (V, E)$, where a process corresponds to the interval defined (on a time axis) between its arrival and departure time. Each interval is associated with a weight which equals the storage area required by the corresponding process. The objective is, thus, to minimize $\max_i(b_i)$ in a weighted coloring of G.

We denote the maximum weight, taken over all intervals, by $W_{\max}$ and the weight of the heaviest clique in G by $\omega^*(G)$ (or simply by ω^* when G is clear from the context). Let τ be the maximum number of concurrent different length intervals and let Δ be the ratio between the longest and shortest interval in G.

In the following, the storage requested by a process is referred to as the weight of the process. For convenience, we refer interchangeably to the duration of a process and to the length of its corresponding interval. The storage device consists of memory cells that are addressed consecutively from zero. The space that is used by an algorithm is equal to the value of the highest address of a memory cell that was used by the algorithm.

3 The Algorithms

Herein, we present two on-line algorithms for DSA with known durations. The algorithm presented in subsection 3.2 achieves a competitive ratio of $O(\log \Delta)$. The algorithm presented in subsection 3.3, achieves a competitive ratio of $O(\log \tau)$. Since Δ and τ are incomparable in general, no algorithm is "better" than the other. In Section 3.4 we show that both algorithms may be used "concurrently" to achieve a competitive ratio of $O(\min\{\log \Delta, \log \tau\})$. The algorithms are presented under the assumption that ω^* is known in advance. Also, the algorithm in subsection 3.3 is presented under the assumption that τ is known in advance. In Section 3.4 we show how to get rid of these assumptions.

We start in subsection 3.1 by presenting algorithms for DSA with known durations in the simple case where the time intervals of the processes are not properly contained in each other. The techniques of these algorithms are used later on, in subsections 3.2 and 3.3, for handling the general case.

3.1 A Simple Case

Consider the following two special cases of interval graphs.

Definition 3. A *proper interval graph* is an interval graph induced by a set of intervals such that no interval properly contains another interval.

Definition 4. A *unit interval graph* is an interval graph induced by a set of intervals which are all of the same length.

We remark that in general, the classes of unit interval graphs and proper interval graphs coincide [Rob69].

First, we present a simple on-line algorithm for the DSA problem with known durations in the case that the processes form a proper interval graph. This algorithm achieves a competitive ratio of $O(1)$ using at most $2\omega^*$ memory space.

Let w_i denote the weight of process i. Let a_i denote the highest address of a memory cell that is occupied by process i, and let $a_0 = -1$.

The Cyclic Algorithm:
Upon arrival of process i:

- If $a_{i-1} + w_i < 2\omega^*$, then allocate w_i consecutive memory cells for process p, starting from memory address $a_{i-1} + 1$.
- Else, allocate w_i consecutive memory cells for process p, starting from memory address 0.

Lemma 5. *The Cyclic Algorithm always succeeds in allocating memory space for a process in a storage device of size $2\omega^*$.*

Proof. For ease of exposition, we consider a cyclic storage device where address 0 follows address $2\omega^* - 1$. Since the processes form a proper interval graph, it is guaranteed that for all $i \geq 2$, process i terminates after process $i - 1$. Hence, by the algorithm, at any point of time, all occupied memory cells must be contained in a single block with at most one hole of size less than $W_{\max}$. The size of this block (with the hole) is at most $\omega^* + W_{\max}$, and therefore a storage device of size $2\omega^*$ is large enough to serve all memory requests. □

We now present an alternative on-line algorithm for the case that the processes form a unit interval graph. This algorithm also achieves a competitive ratio of $O(1)$ using at most $2\omega^*$ memory space. We use it in subsection 3.3 as a procedure for the general case.

Denote by T the length of the intervals in G. Let w_i be the weight of process i and suppose it arrives at time $t = T \cdot j + k$ where $k < T$.

The Alternating Algorithm:
Upon arrival of process i:

- If j is even, allocate w_i consecutive memory cells for process i above the highest occupied address within the first area of ω^* memory cells.
- If j is odd, allocate w_i consecutive memory cells for process i above the highest occupied address within the second area of ω^* memory cells (starting from memory address ω^*).

Lemma 6. *The Alternating Algorithm always succeeds in allocating memory space for a process in a storage device of size $2\omega^*$.*

Proof. Let us consider time phases of length T, where phase j is defined to be between time $T \cdot j$ and time $T \cdot (j+1) - 1$, $(j \geq 0)$. Phase j is referred to as *even* if j is even, otherwise it is *odd*. Since the processes form a unit interval graph where the length of each interval is T, it is guaranteed that all the processes that arrive within a certain phase must terminate within the next phase and have overlapping intervals. Hence, by the algorithm, at any time, all occupied memory cells in the first (second) area of ω^* memory cells must be contained in a single block whose size is at most ω^*, and this block is vacant and ready for reuse within the next even (odd) phase. Therefore, a storage device of size $2\omega^*$ is large enough to serve all memory requests. □

3.2 An $O(\log \Delta)$ Algorithm

Based on the Alternating Algorithm, we present an on-line algorithm whose competitive ratio is $O(\log \Delta)$. We first prove the following technical lemma:

Lemma 7. *Let $G = (V, E)$ be a weighted interval graph with intervals of length T, $2^{k-1} \leq T < 2^k$, $(k \geq 1)$. Let the weight of the heaviest clique in G be $\omega^*(G) = \omega^*$. Let G' be the weighted (unit) interval graph that is generated from G by extending to the right all intervals to be of length 2^k. Then, any clique of G' weighs at most $2\omega^*$.*

Proof. Assume that G' has a clique that weighs more than $2\omega^*$, and let t_0 be a time point which is contained in all the intervals of that clique. Denote by $\omega_G(t)$

the total weight of intervals in G that contain a time point t. Then, for any time point t, $\omega_G(t) \leq \omega^*$, and especially $\omega_G(t_0) \leq \omega^*$. Correspondingly, denote by $\omega_{G'}(t)$ the total weight of intervals in G' that contain a time point t. By assumption, $\omega_{G'}(t_0) > 2\omega^*$. Now, consider the time point $t_1 = t_0 - 2^{k-1}$. Since the original intervals of G are at least of length 2^{k-1}, and they were extended by at most 2^{k-1}, it follows that $\omega_G(t_1) \geq \omega_{G'}(t_0) - \omega_G(t_0) > \omega^*$. Thus, we get a contradiction. $\qquad\Box$

In the following algorithm, the length of all intervals is rounded up to the closest power of 2. We assume that the storage area is divided into slots of size $4\omega^*$ memory cells each. The main idea of the algorithm is that, at any time, a slot which is not empty contains only processes having the same duration. A slot is referred to as a slot of *type i* if it contains processes whose duration is 2^i.

The Slots Algorithm:
Consider a process p with duration $2^{i-1} \leq T < 2^i$ that arrives at time $t = 2^i j + k$, where $k < 2^i$. Let w be the weight of process p.

1. Look for a slot of type i. (If there is no such slot, take the first empty slot.)
2. Use the Alternating Algorithm to allocate w consecutive memory cells for process p within the slot:
   (a) If j is even, allocate above the highest occupied address within the first half of the slot (i.e., the first $2\omega^*$ memory cells of the slot).
   (b) If j is odd, allocate above the highest occupied address within the second half of the slot (i.e., the last $2\omega^*$ memory cells of the slot).

Lemma 8. *The Slots Algorithm always succeeds in allocating memory space for a process in the storage device.*

Proof. It suffices to show that for each $i \geq 0$, the Alternating Algorithm in stage 2 succeeds in allocating memory space for all the processes with duration 2^i within the corresponding slot of size $4\omega^*$. For each $i \geq 0$, Lemma 7 yields that the total weight of the processes with duration 2^i is at most $2\omega^*$. Since the processes with duration 2^i form a unit interval graph, Lemma 6 yields that a slot of size $4\omega^*$ is large enough to serve all memory requests by the Alternating Algorithm. $\qquad\Box$

Performance of the algorithm: At any time there is at most one slot of each possible type, and the number of possible types is at most $\log \Delta$. Since the algorithm always chooses the first empty slot, only the first $\log \Delta$ slots of the storage area may contain processes. Thus the algorithm achieves a competitive ratio of $O(\log \Delta)$ using at most $4\omega^* \log \Delta$ memory space. $\qquad\Box$

3.3 An $O(\log \tau)$ Algorithm

Recall that τ is the maximum number of concurrent different durations. We now present an algorithm whose competitive ratio is $O(\log \tau)$. For the special case that all time intervals are powers of two, the competitive ratio is $O(\log \log \Delta)$, since $\tau \leq \log \Delta$.

The main idea of the algorithm is to separate between the allocation for "heavy" processes, i.e., processes requiring large amount of storage, and "light" processes, i.e., processes requiring small amount of storage. Both storage area requirements and durations of the processes are used to distinguish between "heavy" and "light" processes. This yields that, on one hand, the number of concurrent heavy processes is limited, while on the other hand, a relatively small size slot is required for allocating memory space for light processes with the same duration.

Let the first $6\omega^* + 2\omega^*/\tau$ memory cells of the storage area be divided into $3\tau+1$ slots of size $2\omega^*/\tau$ memory cells each. All processes contained concurrently in a slot must have the same duration, as is the case with the Slots Algorithm.

To clarify the exposition, we adopt the following notation. A slot is called *empty* if it contains no processes. Otherwise, the slot is called *active*. An active slot is referred to as a slot of *type* T if the processes it contains have duration T. An active slot of any type is called *full* if more than ω^*/τ memory cells of the slot are occupied, otherwise, it is called *absorbing* if memory address ω^*/τ of the slot is vacant, and *releasing* if the first memory address of the slot is vacant.

The Separation Algorithm:

Consider a process p with duration T and weight w.

1. If $w \geq \omega^*/\tau$ then allocate memory space for process p in the storage area above memory address $6\omega^* + 2\omega^*$ using either First-Fit or the Coloring algorithm [LNO96].
2. Else ($w < \omega^*/\tau$):
   (a) If there is no active slot of type T, look for an empty slot and allocate w consecutive memory cells for process p starting from the first memory address of the slot.
   (b) Else, look for an active slot of type T which contains the process with the latest arrival time:
      i. If it is an absorbing slot, allocate w consecutive memory cells for process p above the highest occupied address within the slot.
      ii. else, look for an empty slot and allocate w consecutive memory cells for process p starting from the first memory address of the slot.

Lemma 9. *The Separation Algorithm always succeeds in allocating memory space for a process in the storage device.*

Proof. It suffices to show that an empty slot can always be found in stages 2a and 2(b)ii of the algorithm. We show that there are at most 3τ active slots at any point of time. Since the total number of slots is $3\tau + 1$, there is always at least one empty slot.

Consider the processes with duration T; these processes form a unit interval graph. Since all occupied memory cells in a slot must be contained in a single block, we claim that an active slot of type T which is not full, must be either an absorbing slot, or a releasing slot (or both). Furthermore, at each point of time there can be at most one absorbing slot of type T, and at most one releasing slot of type T. This happens since any two slots of type T can be ordered such that the arrival time of all processes belonging to the first slot precedes the arrival time of the processes belonging to the second slot. Hence, at any point of time, there can be at most τ absorbing slots of all types, and τ releasing slots of all types. To complete the proof, we note that there can be at most τ concurrent full slots of all types, since a full slot contains at least ω^*/τ occupied memory cells. □

Performance of the algorithm: The first stage of the Separation Algorithm is First Fit with respect to the processes with weights greater than ω^*/τ. This stage requires at most $O(\omega^* \log \tau)$ memory space by the results of [LNO96], since the maximum number of concurrent processes is less than or equal to τ. The second stage of the Separation Algorithm requires $O(\omega^*)$ memory space, since it uses $3\tau + 1$ slots of size $2\omega^*/\tau$. Hence, the Separation Algorithm achieves a competitive ratio of $O(\log \tau)$. □

3.4 Implementation Issues

In all the algorithms presented so far we assumed that the value of ω^* is known in advance. Also, the Separation Algorithm is presented under the assumption that τ is known in advance. We now show how to get rid of these assumptions. We also outline how to combine the Slots Algorithm and the Separation Algorithm in order to obtain a competitive ratio of $O(\min\{\log \Delta, \log \tau\})$.

At any given point of time, the value of ω^* which should be used by the algorithm is the maximum concurrent storage requirement thus far, rounded up to the closest power of 2. This value is updated during the execution of the algorithm. Upon each such update, the algorithm "resets" the bottom of the storage area to be the highest memory address used thus far. The resets can, at most, double the storage area used in total.

Similarly, at any given point of time, the value of τ which should be used by the algorithm is the maximum number thus far of concurrent processes having different durations, rounded up to the closest power of 2. This value is updated during the execution of the algorithm. Upon each such update, we reset the area allocated for the "light" processes. As there can be at most $O(\log \tau)$ such resets, the total area allocated for the "light" processes is at most $O(\omega^* \log \tau)$. As for the "heavy" processes, it is easy to verify that if they are handled by the Coloring Algorithm of [LNO96], no resets of the area allocated for these processes is needed.

It is also easy to verify that the Slots Algorithm and the Separation Algorithm can be executed in parallel, since essentially, they both use segregated storage methods. Upon arrival of a process, memory space is allocated by the algorithm with the (current) better competitive ratio. Referring to this parallel execution of the two algorithms as to the *Combined Slots-Separation Algorithm*, we obtain the following result:

Theorem 10. *The Combined Slots-Separation Algorithm has a competitive ratio of $O(\min\{\log \Delta, \log \tau\})$.*

4 Open Problems

An intriguing open problem for dynamic storage allocation with known durations is proving non-trivial lower bounds on the competitive ratio. We note that the proofs of the lower bounds in the case of unknown durations (e.g., [Rob77, LNO96]) do not seem to carry over to the case of known durations. These proofs heavily use the fact that the adversary can dynamically determine the duration of a process based on the storage configuration chosen by the on-line algorithm.

References

[AAP93]   B. Awerbuch, Y. Azar, and S. Plotkin. Throughput-competitive of on-line routing. In *Proc. 30th IEEE Symp. on Foundations of Computer Science*, pages 32–40, October 1993.

[AKP+93]   Y. Azar, B. Kalyanasundaram, S. Plotkin, K. R. Pruhs, and O. Waarts. On-line load balancing of temporary tasks. In *Proc. Workshop on Algorithms and Data Structures*, August 1993.

[BNCK+95] A. Bar-Noy, R. Canetti, S. Kutten, Y. Mansour, and B. Schieber. Bandwidth allocation with preemption. In *Proc. 27th ACM Symp. on Theory of Computing*, 1995.

[Cof83]   E. G. Coffman. An introduction to combinatorial models of dynamic storage allocation. *SIAM Review*, 25:311–325, 1983.

[Ger96]   J. Gergov. Approximation algorithms for dynamic storage allocation. In *Proc. of the 4th Annual European Symposium on Algorithms*, pages 52–61, 1996.

[GJ79]   M. R. Garey and D. S. Johnson. *Computers and intractability - a guide to the theory of NP-completeness.* W. H. Freeman, San Francisco, 1979.

[Kie88]   H. A. Kierstead. The linearity of first-fit coloring of interval graphs. *SIAM Journal on Discrete Math*, 1:526–530, 1988.

[Kie91]   H. A. Kierstead. A polynomial time approximation algorithm for dynamic storage allocation. *Discrete Mathematics*, 88:231–237, 1991.

[Knu73]   D. E. Knuth. *Fundamental algorithms*, volume 1. Addison-Wesley, Reading, MA, second edition, 1973.

[LNO96]   M. G. Luby, J. Naor, and A. Orda. Tight bounds for dynamic storage allocation. *SIAM Journal on Discrete Math*, 9(1):155–166, 1996.

[Par94]   C. Partridge. *Gigabit Networking.* Addison-Wesley, 1994.

[Rob69]   F. S. Roberts. Indifference graphs. In F. Harary, editor, *Proof Techniques in Graph Theory*, pages 139–146. Academic Press, N.Y., 1969.

[Rob71]   J. M. Robson. An estimate of the store size necessary for dynamic storage allocation. *Journal of the ACM*, 18:416–423, 1971.

[Rob74]   J. M. Robson. Bounds for some functions concerning dynamic storage allocation. *Journal of the ACM*, 12:491–499, 1974.

[Rob77]   J. M. Robson. Worst case fragmentation of first fit and best fit storage
           allocation strategies. *Computer Journal*, 20:242–244, 1977.
[SS94]   O. Sharon and A. Segal. Schemes for slot reuse in CRMA. *IEEE/ACM
           Transactions on Networking*, 2(3):269–278, 1994.
[Sta95]   T. A. Standish. *Data structures, algorithms and software principles in C.*
           Addison-Wesley, Reading, MA, second edition, 1995.

Coloring in Sublinear Time

Andreas Nolte and Rainer Schrader

University of Cologne, Weyertal 80, 50931 Cologne, Germany

We will present an algorithm, based on SA-techniques and a sampling procedure, that colors a given random 3-colorable graph with high probability in sublinear time. This result is the first theoretical proof for the excellent experimental performance results of Simulated Annealing known from the literature when applied to graph coloring problems.

1   Introduction

JOHNSON ET AL. [10] and PETFORD and WELSH [16] report on very good practical results of Simulated Annealing coloring random graphs. Compared to deterministic algorithms such as described in [12] and [18], Simulated Annealing finds correct colorings faster on almost all random instances. Taking these practical results as a starting point, we will consider the same random model as in [16]. We assume a given set of n vertices partitioned into three color classes, each pair of vertices of different color being connected with a certain probability p. Thus, the random graphs are 3-colorable. Due to the simplicity of the representation, and due to the fact that in a sense the 3-colorability case is the most difficult, we consider only this case. But, most results seem to be extendible to the arbitrary k-colorability case.

There exist already some deterministic algorithms [3, 18, 12, 5] presented by various authors which color a given random instance correctly with three colors and with high probability. As usual, the term "with high probability" means that the probability tends to 1 as the problem size tends to infinity. All deterministic algorithms mentioned so far show the characteristic that the correct construction of the 3-coloring requires nearly equal sized color classes (up to a factor of $1+o(1)$). Moreover, this construction takes a number of steps, that is linear in the number of edges of a given random graph. In the usual sense the time complexity is of course optimal, because the size of the input is linear in the number of edges. Furthermore, a verification of the correctness of the coloring requires a number of steps that is at least as large as the number of edges.

But the extremely good performance of Simulated Annealing on random instances suggests the question, wether it is possible to construct a correct 3-coloring of a given instance even faster. This means that we try to find a correct

---
* Research supported by DFG (Schr 390/2-1)

coloring without looking at every edge. Aside from the theoretical point of view, the answer of this question is of course only of practical value for on-line applications, where the demand of a correct coloring must only be fulfilled with high probability. But we do not want to dwell further on this point.

In the following section we will answer this question in the affirmative. We will describe an algorithm that uses SA-type techniques and a sampling procedure to compute the cost function more efficiently. This algorithm will stop after a number of steps that is strictly less than the number of edges. We will show that this algorithm produces with certainty a coloring of a given random graph with equal sized color classes and that this coloring is a correct one with high probability over all graphs and random steps.

Only two results are published by now concerned with the convergence of Simulated Annealing in polynomial time. JERRUM and SORKIN [9] proved the convergence to an optimal solution on certain random instances of the Graph Bisection Problem in $O(n^3)$ steps, where n is the number of vertices. HAJEK and SASAKI [17] were engaged with the performance of Simulated Annealing applied to the Maximum Matching Problem and showed Simulated Annealing to be in a near optimal state in polynomial time. But these authors consider only the Metropolis process (Simulated Annealing at a certain fixed temperature) and thus investigate the convergence of a homogeneous Markov chain. In the following, we will show the convergence of Simulated Annealing with varying, time-dependent temperatures.

2 Random Model

We consider a quite simple random model ensuring that every graph taken from the induced probability space is 3-colorable with equal sized color classes. The model introduced here is commonly accepted and already used by several authors [18, 16, 3] to analyze the performance of various coloring algorithms.

We consider a given set V of n vertices and color $n/3$ of these vertices red, $n/3$ of these vertices blue and $n/3$ of these vertices green (we assume n to be a multiple of 3 to get equal sized color classes). Then we connect vertices of different color with constant probability p getting an edge set $E \subset V^2$ (an edge is considered as an unordered pair of vertices). In order to simplify the representation of our analysis we assume $p = 1/2$. However, the presented analysis in this chapter may be extended to some lower problem-dependent probability $p \leq 1/2$, but this is not considered further in the following.

After constructing the graph we forget the coloring. We call the induced probability space $\mathcal{G}$. The task is to color a given graph in $\mathcal{G}$ properly with 3 colors and high probability.

3 Idea of the Algorithm

The strategy to color a given $G \in \mathcal{G}$ is to recover the coloring that has been used to construct G. We take an arbitrary set A of $n/3$ vertices from the set of all vertices. Then we try to increase the imbalance in A by local exchanges. This implies that we try to increase the cardinality of the largest color class in A. After a certain small number of SA-transition steps we can guarantee a quite large imbalance. Then we use some hillclimbing steps (i.e. Simulated Annealing at temperature 0). In order to compute the cost function (i.e. the number of neighbors in A and $V \setminus A$) more efficiently we look only at a certain smaller sample of vertices in the corresponding sets. But, these samples are large enough to reflect the imbalance situation of A and $V \setminus A$. After these hillclimbing steps we can be sure that there are only vertices of two colors in A. Applying again SA-transitions followed by hillclimbing steps we will see that A will only contain vertices of one color. In order to 2-color $V \setminus A$ we apply a similar method as described above.

The main purpose of the following sections is to give a detailed description of the algorithm and some hints how to prove the following main theorem in this chapter. The detailed version of this paper containing all proofs may be found at http://www.informatik.uni-koeln.de/ls_schrader/anolte.html.

Theorem 1 *There exists a randomized algorithm using only SA-type transitions that determines a correct 3-coloring of a given $G \in \mathcal{G}$ in $O(n^{\delta})$ steps ($\delta < 2$) with high probability over all graphs and random steps, while checking the correctness of the constructed coloring requires $\Omega(n^2)$ steps.*

4 Simulated Annealing

4.1 Local Exchanges

Let a graph $G \in \mathcal{G}$ be given. We partition the set of n vertices into a set A containing $n/3$ vertices and a set containing $2/3n$ vertices. For technical reasons we consider these partition as a permutation π with the first $n/3$ vertices belonging to $A = A_{\pi}$. The set of configurations of the following Simulated Annealing algorithm will therefore be the set of permutations π. We choose an arbitrary state π_0 as a starting state. To explain the transition steps let a state π_t in step t be given.

We choose a vertex u uniformly at random from the set A_{π_t} and a vertex v uniformly at random from the set $V \setminus A_{\pi_t}$. The proposed move is the change of the positions of u and v in π_t.

The cost function Δc of a proposed move considered here is the number of neighbors of v in A_{π_t} minus the number of neighbors of u in A_{π_t}. Thus, Δc measures the change of the number of edges between the vertices in A_{π_t}. The acceptance probability of a proposed move is

$$a(\Delta c) = \begin{cases} \frac{1}{2} - \frac{\Delta c}{T_t} & \text{for} \quad \Delta c \in [-T_t/2, T_t/2], \\ 0 & \text{for} \quad \Delta c > T_t/2, \\ 1 & \text{for} \quad \Delta c < -T_t/2, \end{cases}$$

where $T_t \in \mathbb{Q}$ is the temperature in step t. (We use rational temperatures due to complexity reasons.) The sequence $(T_t)_{t \in \mathbb{N}}$ is the sequence of decreasing real numbers known as cooling schedule of SA and will be specified in the following. This implies the following: If a proposed move decreases the number of edges between vertices in A_{π_t}, we accept it with a certain larger probability than a proposed move that increases the number of edges. It is intuitively easy to see that the current state π_t will be a state with a quite low number of edges in A_{π_t} provided that the number of steps is high enough. This acceptance function is not as common as $\min\{\exp\left(\frac{-\Delta c}{T_t}\right), 1\}$, but due to its symmetry it simplifies the following analysis.

4.2 Expected Transition Probabilities

The key value that we use to analyze the performance of our local exchanges is the imbalance of a partition π. To explain this value let an arbitrary partition π be given. A_π is a set of $n/3$ vertices that are red, blue or green (known from the construction, but invisible for the algorithm). Let $r = n/9 + i_r$ be the number of red vertices, $b = n/9 + i_b$ be the number of blue vertices and $g = n/9 + i_g$ be the number of green vertices in A_π. Thus, we get $i_r + i_b + i_g = 0$. The imbalance i_π of the permutation π is the value of that i_* which corresponds to the largest color class in A_π. Therefore we get $i_\pi = \max\{i_r, i_b, i_g\}$ and $0 \le i_\pi \le 2/9n$.

We are now interested in the number of steps, that are necessary to ensure that $i_\pi \ge n^{3/4}$ with high probability over all graphs and transition steps.

The reason to consider the imbalance is that we have projected the quite complex coloring Markov chain on the natural numbers, a method analogous to that suggested by [9]. This process is not necessarily Markovian any more, and this complicates the analysis, but the process can be bounded by dominating Markov chains on the natural numbers which are easy to analyze. First of all we try to get bounds for the expected value of i_π after one transition.

Proposition 1 *Let a partition π of a graph $G \in \mathcal{G}$ and a temperature $T_t > n^{1/2}$ be given. Let $u =$ "up" be the random variable on the set of all graphs that denotes the probability over all possible transitions to increase i_π in transition step t. $d =$ "down" denotes the corresponding random variable for a decrease of*

i_π. *Then we get*
$\exists c_1, c_2 \in \mathbb{R}_+, k \geq 1/9$

$$E(u) \geq k - c_1 \frac{i_\pi}{n}$$

and

$$E(d) \leq k + c_1 \frac{i_\pi}{n} - c_2 \min\left\{\frac{i_\pi}{T_t}, 1\right\},$$

where E denotes the expectation over all graphs $G \in \mathcal{G}$.

□

Assuming $T_t \ll n$, this proposition states that the imbalance of the current partition is more likely to be increased than to be decreased, and this depends crucially on the size of the current imbalance. This is a positive hint that it is possible to reach our target getting a partition with a quite high value of the imbalance after a certain number of steps.

But so far, we know only some facts about the expected values of the transition probabilities. In the following section we are interested in the deviation of the current transition probabilities from the expected value.

4.3 Deviation

To measure the deviation from the expected values of u and d we need the following definition

Definition 2. A pair (G, π) of a graph G and a partition π is called α-deviant, if either $u = u(G, \pi)$ or $d = d(G, \pi)$ differs from its expected value (over all graphs) by more than α.

The following lemma is the first result we can get about α-deviation by a direct application of the method of bounded differences.

Lemma 3 *Let a partition π and an $\alpha > 0$ be given. The probability over all graphs $G \in \mathcal{G}$ that a pair (G, π) is α-deviant is at most $2 \exp(O(\alpha^2 T_t^2))$.*

□

But it turns out that this lemma alone is, although necessary in the following, not sufficient to get proper bounds of the deviation. These proper bounds are necessary to ensure that the current transition probabilities do not differ too much from their expected values during a certain number of subsequent steps. In the following we apply an idea of JERRUM and SORKIN [9] to improve the bound of the lemma.

The key idea is to introduce a new quantity that is a suitable sum of transition probabilities. We will see that we can get bounds for this aggregated value allowing us to derive better bounds for the whole process than looking at the individual transition probabilites of every partition π as in Lemma 3. For a given graph G let $P(\pi, \pi')$ be the generation and acceptance probability for the transition from π to π', and $EP(\pi, \pi')$ be the expected value over all graphs. The crucial value considered here is the sum over all partitions π' of the difference $|P(\pi, \pi') - EP(\pi, \pi')|$.

First of all we try to get an analogue of Lemma 3 for the new quantitiy.

Proposition 4 *Let a temperature $T_t > 0$ be given. With high probability over all graphs and for all partitions π we get*

$$\sum_{\pi'} |P(\pi, \pi') - EP(\pi, \pi')| = O\left(\frac{\sqrt{n}}{T_t}\right).$$

$\square$

With the help of the last proposition we can prove the main theorem of this section that bounds the deviation of u and d. The proof idea goes back to JERRUM and SORKIN [9], who solved randomized instances of the Graph Bisection Problem via the Metropolis process.

Theorem 2 *For any temperature $0 < T < poly(n)$ and any $\alpha > 0$, the Simulated Annealing process at constant temperature T for $t = \Theta(\alpha^2 T^3 / \sqrt{n} \log^2 n)$ steps encounters a α-deviant state with probability of at most*

$$\exp(-\Omega(\alpha^2 T^2 / \log^2 n)).$$

$\square$

4.4 Random Walks

In this section we analyze the time dependent behavior of the imbalance of the current partition during the Simulated Annealing process. We can portray this value as a projection of a Markov process on the natural numbers. This process depends on a hidden variable, namely the current partition. Thus, it is not necessarily Markovian. But, with the help of the results concerning the expected transition probability and the deviation we can construct a process that is a lower bound of the process of the imbalance values. This means that we can construct a new random walk with constant transition probabilities being quite easy to analyze. Additionally, we can be sure that our process of interest $((i_{\pi_t})_{t \in \mathbb{N}})$ has got "on the average" larger values than our random walk. The technique used to carry out this comparison is known as coupling and is made precise in the next proposition. A similar version can be found in [9].

Proposition 5 *Let $(X_t)_{t\in\mathbb{N}}$ be a Markov chain with state space Ω and $\Xi : \Omega \to \mathbb{N}$ be a projection on the natural numbers. Suppose $(\xi_t)_{t\in\mathbb{N}}$ to be a random walk on the natural numbers having $\xi_{t+1} \in \{\xi_t - 1, \xi_t, \xi_t + 1\}$ as its only allowed transitions. Suppose further that it is a probabilistic lower bound of the process $\Xi(X_t)_{t\in\mathbb{N}}$ in the following sense*

$$\xi_0 \leq \Xi(X_0)$$
$$P(\xi_{t+1} = s + 1 | \xi_t = s) \leq P(\Xi(X_{t+1}) \geq s + 1 | \Xi(X_t) = s)$$
$$P(\xi_{t+1} = s - 1 | \xi_t = s) \geq P(\Xi(X_{t+1}) = s - 1 | \Xi(X_t) = s)$$
$$P(\Xi(X_{t+1}) < s - 1 | \Xi(X_t) = s) = 0$$

for $s \in \mathbb{N}$ arbitrary.

Then there exist a coupling $(Y_t)_{t\in\mathbb{N}}$ of $(\xi_t)_{t\in\mathbb{N}}$ and a coupling $(Z_t)_{t\in\mathbb{N}}$ of $(\Xi(X_t))_{t\in\mathbb{N}}$ (i.e. Y_t (Z_t) has the same distribution as ξ_t ($\Xi(X_t)$)) with $Y_t \leq Z_t$ for all $t \in \mathbb{N}$.

$\square$

Now, we are able to analyze the stochastic process i_{π_t}. Let $0 < \gamma \leq 1/100$ be fixed. We choose a temperature $T_0 = \lceil n^{2/3-\gamma} \rceil$ and keep it fixed during the first r steps (r will be defined in the following). By defining $\alpha = n^{-1/3}$ we can conclude with the help of Proposition 2 that the imbalance process i_{π_t} does not reach a partition with a larger deviation than α with high probability during the first $\lceil n^{5/6-4\gamma} \rceil$ steps. In the following we concentrate on this case. We define a random walk $(lb_t^b)_{t\in\mathbb{N}}$ on the natural numbers, starting at 0, that plays the role of the lower bound of i_{π_t} at the beginning in the interval $[0, n^{1/3}]$

$$lb_0^b = 0,$$
$$lb_{t+1}^b \in \{lb_t^b, lb_t^b + 1, lb_t^b - 1\},$$
$$P(lb_{t+1}^b = n^{1/3} + 1 | lb_t^b = n^{1/3}) = P(lb_{t+1}^b = -1 | lb_t^b = 0) = 0$$

and

$$P(lb_{t+1}^b = s + 1 | lb_t^b = s) = k - 2\alpha \quad \text{for} \quad s \in \{0, \ldots, n^{1/3} - 1\},$$
$$P(lb_{t+1}^b = s - 1 | lb_t^b = s) = k + 2\alpha \quad \text{for} \quad s \in \{1, \ldots, n^{1/3}\}$$

with k chosen as in Proposition 1. Using Proposition 1 and the facts that $i_\pi \geq 0$ and only a deviation of α is allowed, we can see that lb_t^b and the process i_{π_t} are correlated in the same way as ξ_t and $\Xi(X_t)$ in Proposition 5. Thus, lb_t^b is in fact a probabilistic lower bound of i_{π_t} in the sense of Proposition 5.

Now, we analyze the time being necessary for the process lb_t^b to reach $n^{1/3}$. It is a random walk with one elastic barrier 0 and can be analyzed quite easily with the standard Markov chain theory.

Lemma 6 *Let $\epsilon > 0$ be given. The random walk lb_t^b will hit $n^{1/3}$ with high probability within $O(n^{2/3+\epsilon})$ steps.*

$\square$

According to Proposition 5 we obtain - with the help of the above described identification of lb_t^b with ξ_t and i_{π_t} with $\Xi(X_t)$ - that there exist couplings Y_t and Z_t of lb_t^b and i_{π_t} with $Y_t \leq Z_t$ for all $t \in \mathbb{N}$. Since Y_t has the same distribution as lb_t^b, we can derive the same result as in Lemma 6 for Y_t. Therefore, we can be sure that Z_t has hit a state $s \geq n^{1/3}$ with high probability within $O(n^{2/3+\epsilon})$ steps. Due to the fact that Z_t and i_{π_t} have got the same distribution, we get

Lemma 7 *The process i_{π_t} will hit a state $s \geq n^{1/3}$ with high probability within $O(n^{2/3+\epsilon})$ steps.*

In the following we consider the behavior of the process i_{π_t} on the interval $[n^{1/3}, n^{1/2+\gamma}]$. Again, we construct a random walk on the natural numbers playing the role of a lower bound for i_{π_t}. Strictly speaking, it is a series of random walks lb_t^j, $j \in \{0, \ldots, \lfloor \frac{1}{6\gamma} \rfloor\}$, each being a lower bound for i_{π_t} on $[\frac{n^{1/3+j\gamma}}{2}, n^{1/3+(j+1)\gamma}]$. Let

$$lb_0^j = n^{1/3+j\gamma},$$
$$lb_{t+1}^j \in \{lb_t^j, lb_t^j + 1, lb_t^j - 1\},$$

$$P(lb_{t+1}^j = n^{1/3+(j+1)\gamma} + 1 | lb_t^j = n^{1/3+(j+1)\gamma})$$
$$= P(lb_{t+1}^j = n^{1/3+j\gamma}/2 + 1 | lb_t^j = n^{1/3+j\gamma}/2)$$
$$= 0$$

and

$$P(lb_{t+1}^j = s + 1 | lb_t^j = s) = k - \Theta(n^{-1/3}) \qquad \text{for}$$
$$s \in \{n^{1/3+j\gamma}/2, \ldots, n^{1/3+(j+1)\gamma} - 1\},$$
$$P(lb_{t+1}^j = s - 1 | lb_t^j = s) = k - \Theta(n^{-1/3+(j+1)\gamma}) \qquad \text{for}$$
$$s \in \{n^{1/3+j\gamma}/2 + 1, \ldots, n^{1/3+(j+1)\gamma}\}.$$

Due to Proposition 1 and the allowed deviation of $\alpha = n^{-1/3}$ the processes lb_t^j and i_{π_t} are also in the same correlation as ξ_t and $\Xi(X_t)$. Using Proposition 5 it is again sufficient to analyze lb_t^j.

Lemma 8 *Let $j \in \{0, \ldots, \lfloor \frac{1}{6\gamma} \rfloor\}, \epsilon > 0$ be given. The process lb_t^j starting at $n^{1/3+j\gamma}$ will hit $n^{1/3+(j+1)\gamma}$ earlier than $\frac{n^{1/3+j\gamma}}{2}$ with high probability and within $O(n^{2/3+\epsilon})$ steps.*

$\square$

Since γ was fixed, and lb_t^j and i_{π_t} are correlated in a way allowing us to apply Proposition 5, we get with the same coupling argument described above the following corollary

Corollary 9. *Let $\epsilon > 0$ be given. The process i_{π_t} will hit $n^{1/2+\gamma}$ with high probability within $O(n^{2/3+\epsilon})$ steps.*

After $ln^{2/3+\epsilon}$ steps with a suitable constant $l \in \mathbb{N}$ we stop the process (a rigorous computation with exact factors instead of the O/Ω-notation yields, that $l = 9$ is sufficient). Carrying out a similar analysis to that above with a random walk on $[\frac{n^{1/2+\gamma}}{2}, n^{3/4}]$ that starts at $n^{1/2+\gamma}$, we get a random walk on the natural numbers that is a lower bound of i_{π_t}. It reaches $n^{3/4}$ earlier than $\frac{n^{1/2+\gamma}}{2}$ with high probability. Because reaching $n^{3/4}$ would take at least $n^{3/4}$ steps, we could derive that after stopping, our process i_{π_t} is in a state somewhere between $\frac{n^{1/2+\gamma}}{2}$ and $n^{3/4}$.

Now we lower the temperature T_t. The new temperature will be $T_t = \lceil n^{1/2+\gamma} \rceil$ for $t \in \{9\lceil n^{2/3+\epsilon} \rceil, \ldots\}$ and $\gamma > 0$ fixed. Setting $\alpha = n^{-1/8}$ we can derive with Proposition 2 that no α-deviant state will occur in the next $n^{3/4+3\gamma}$ steps with high probability.

To analyze the imbalance process in the interval $[n^{1/2+\gamma}/4, n^{3/4}]$ we define the lower bound process lb_t^e

$$lb_0^e = n^{1/2+\gamma}/2,$$
$$lb_{t+1}^e \in \{lb_t^e, lb_t^e + 1, lb_t^e - 1\},$$
$$P(lb_{t+1}^e = n^{3/4} + 1 | lb_t^e = n^{3/4}) = P(lb_{t+1}^e = n^{1/2+\gamma}/4 - 1 | lb_t^e = n^{1/2+\gamma}/4) = 0$$

and

$$P(lb_{t+1}^e = s + 1 | lb_t^e = s) = k_1 \quad \text{for} \quad s \in \{n^{1/2+\gamma}/4, \ldots, n^{3/4} - 1\},$$
$$P(lb_{t+1}^e = s - 1 | lb_t^e = s) = k_2 \quad \text{for} \quad s \in \{n^{1/2+\gamma}/4 + 1, \ldots, n^{3/4}\}$$

with constants $k_1 < k_2$ suitably chosen according to Proposition 1.

Carrying out an analysis in the same way as in the proof of Corollary 9 and proving as described after Corollary 9 that the process will stay above $n^{3/4}$ we get the main result of this section.

Proposition 10 *Let $\gamma, \epsilon > 0$ be sufficiently small constants. Running the Simulated Annealing process for $9n^{2/3+\epsilon}$ steps with temperature $T = n^{2/3-\gamma}$ and $16n^{3/4+\epsilon}$ steps with temperature $T = n^{1/2+\gamma}$ yields a partition π with an imbalance value i_π of at least $n^{3/4}$.*

5 Hillclimbing

Assuming a given graph G and a partition π with imbalance $i_\pi \geq n^{3/4}$, we will define in this chapter certain hillclimbing steps (i.e. Simulated Annealing at temperature 0) in order to remove one color class in A_π. Roughly speaking we will compare the number of neighbors of a randomly chosen vertex in A_π and $V \setminus A_\pi$. To save time during the computation of the cost function of one move (i.e. the number of neighbors) it is not necessary to look at the complete sets. We choose only a sample in A_π and $V \setminus A_\pi$ of size $n^{3/4-\epsilon}$ ($\epsilon > 0$ sufficiently small) uniformly at random.

In the following we will see that the imbalance situation will be reflected in the samples. Intuitively it is clear that a vertex having the same color as the minimum color class in A_π will tend to have more neighbors in the sample in A_π than in the sample in $V \setminus A_\pi$.

This idea is made precise in the following. First we consider the case of choosing a sample of size $n^{3/4-\epsilon}$ uniformly at random from A_π.

Lemma 11 Let $0 < \epsilon < 1/24$ and a partition π with imbalance $i_\pi \geq n^{3/4}$ be given. The number of vertices of a distinct color $*$ in a sample of $n^{3/4-\epsilon}$ vertices chosen uniformly at random in A_π differs only within a range of $\Delta = O(n^{1/2-2\epsilon})$ from the expected number

$$n^{3/4-\epsilon}\left(\frac{n/9 + i_*}{n/3}\right)$$

of the corresponding independent Bernoulli trials with high probability over all chosen samples.

$\square$

Using Lemma 11 we can derive that after the local exchanges in the last chapter our sampling process yields a reliable smaller copy of our partition with high probability.

Again, let a graph G and a partition with imbalance $n^{3/4}$ be given. Furthermore, we assume a given reliable sample S with error bound Δ as in Lemma 11.

Let v be a vertex of the largest color class in our sample, which we assume to be red. The expected number of neighbors over all graphs in S is

$$
\begin{aligned}
1/2(S_g + S_b) &\leq 1/2 n^{3/4-\epsilon}\left(\frac{n/9 + i_g + n/9 + i_b}{n/3}\right) + \Delta \\
&\leq 1/3 n^{3/4-\epsilon} - 3/2 n^{1/2-\epsilon} + \Delta \\
&\leq 1/3 n^{3/4-\epsilon} - 4/3 n^{1/2-\epsilon}
\end{aligned}
$$

with S_b and S_g being the number of green and blue vertices in S. Assuming green to be the least frequent color in the partition we get the following bound for the expected number of neighbors of a green vertex in S

$$1/2(S_b + S_r) \geq 1/2n^{3/4-\epsilon}\left(\frac{n/9 + i_b + n/9 + i_r}{n/3}\right) - \Delta$$
$$\geq 1/3n^{3/4-\epsilon} + 3/2n^{-1/4-\epsilon}(i_b + i_r) - \Delta$$
$$\geq 1/3n^{3/4-\epsilon} + 3/4n^{1/2-\epsilon} - \Delta$$
$$\geq 1/3n^{3/4-\epsilon} + 1/2n^{1/2-\epsilon}$$

with S_r being the number of red vertices in S.

Now we consider the case of choosing a sample of size $n^{3/4-\epsilon}$ uniformly at random in $V \setminus A_\pi$. One can show analogously:

The number of neighbors of a red vertex in this sample is at least

$$1/3n^{3/4-\epsilon} + 1/2n^{1/2-\epsilon}.$$

with high probability. As an upper bound for the number of neighbors of a green vertex we get:

$$1/3n^{3/4-\epsilon} - 1/2n^{1/2-\epsilon}.$$

In the following we will describe the hillclimbing steps:

1. Take one of the sets $\{A_\pi, V \setminus A_\pi\}$ as starting set B
2. Choose a vertex v uniformly at random in B
3. Choose a sample of size $n^{3/4-\epsilon}$ uniformly at random in A_π and $V \setminus A_\pi$
4. Compare the number of neighbors in both samples
5. If the number of neighbors in the sample in B is greater than the number of neighbors in the other sample, we move v from B to the complement, take B as this complement and go to 2.
   else go to 2.

Obviously we can be sure that our sizes of the sets remain constant (up to +/-1).

With the analysis at the beginning of this chapter we can see that a red vertex will readily move to or stay at the smaller set. In contrast to that, a green vertex will readily move to or stay at the bigger set. Moreover it is easy to see that the bias towards these directions will be reinforced during the process so that the above described behavior will continue during the hillclimbing process. After $O(n \log n)$ steps every vertex is visited at least once with high probability. This implies that there are only vertices of two colors left in the smaller set.

6 Simulated Annealing again

After the hillclimbing steps we start our initial SA-algorithm (see Chapter 4) again. To analyze this process we will introduce a new value i_π^2 similar to our initial imbalance, but this time suitable for the two color case. Let red and blue be the two colors in A_π and $r = n/6 + i_r$ and $b = n/6 + i_b$ be the corresponding cardinalities. Thus we get $i_r + i_b = 0$. The imbalance i_π^2 is the value of that i_* which corresponds to the largest color class in A_π. Therefore we get $i_\pi^2 = \max\{i_r, i_b\}$ and $0 \leq i_\pi \leq 1/6n$. We are again interested in the number of steps necessary to ensure that $i_\pi^2 \geq n^{3/4}$ with high probability over all graphs and transition steps.

We consider the case $i_\pi^2 < n^{7/8}$ first. Let green denote the color of those vertices that are not n the smaller set. With the help of Chernoff bounds it can be easily seen that the number of neighbors in the smaller set A_π is $\Omega(n)$ with high probability. Because the absolute value of the cost difference Δc will be greater than $T/2$ for our choices of temperatures $T_t \in \{n^{2/3-\gamma}, n^{1/2+\gamma}\}$, this implies the following:

A transition with a chosen green vertex in $V \smallsetminus A_\pi$ will not be accepted. Therefore we can be sure that in this case no green vertex will move to A_π.

Choosing k random pairs of vertices (u, v) with $u \in A_\pi$ and $v \in V \smallsetminus A_\pi$ we have to ensure that among these pairs there are enough pairs with no green vertices involved. The reason for this is of course that these transitions will not help us to improve the imbalance in A_π. But the number of pairs (u, v) with a green vertex v is Bernoulli distributed with probability $1/2$. Using Chernoff bounds one can easily see that there will be at least $k/4$ pairs with no green vertices involved.

Using now temperature $T = n^{2/3-\gamma}$ for the first $4n^{2/3-\gamma}$ steps and $T = n^{1/2+\gamma}$ for the next $72n^{3/4+\epsilon}$ steps one can prove with an analogous analysis to that in Chapter 4 that the process will reach an imbalance of at least $n^{3/4}$. Taking care of the case of an initial imbalance of $i_\pi^2 \geq n^{7/8}$ we can summarize as a result of our second SA-application:

Proposition 12. *Let ϵ, γ be sufficiently small constants. Running the SA- process for $4n^{2/3-\epsilon}$ with temperature $T = n^{2/3-\gamma}$ and $72n^{3/4+\epsilon}$ steps with temperature $T = n^{1/2+\gamma}$ yields a partition π with*

- $|\{red\ vertices\ in\ A_\pi\}| \geq n/6 + n^{3/4}$

- $|\{blue\ vertices\ in\ A_\pi\}| \leq n/6 - n^{3/4}$

- $|\{red\ vertices\ in\ A_\pi\}| \leq 100n^{3/4}$

With the help of the last proposition we can be sure that after our second application of the SA algorithm the largest color class in A_π dominates the other color classes heavily. Applying the hillclimbing algorithm ($O(n\log(n))$ steps) again yields a set A_π, that contains only vertices of one color class.

6.1 2-coloring

After separating one color as described in the last section we have to 2-color the remaining $2/3n$ vertices. This is not very difficult, since it is very easy to find an $O(n^2)$ algorithm that colors the remaining vertices correctly. But in order to get an overall performance that is strictly less than the quadratic bound we have to be more careful. The obvious idea to get better performance bounds is to apply the same idea as used in the last section to separate the vertices of one color from the rest.

Starting with an arbitrary bisection of the remaining vertices, we try to increase again the imbalance of an arbitrary chosen bisection with SA steps. After carrying out a number of local exchanges we can separate the two remaining color classes by applying the same hillclimbing steps as described above.

Input: arbitrary chosen bisection $\pi = (\pi_1, \pi_2)$ of the remaining $2/3n$ vertices, ϵ, δ, γ sufficiently small.

Output: 2-coloring of the remaining $2/3n$ vertices.

1. Execute $9n^{2/3+\epsilon}$ steps of the SA local exchanges with temperature $T = n^{2/3-\gamma}$.
2. Execute $16n^{3/4+\epsilon}$ steps of the SA local exchanges with temperature $T = n^{1/2+\gamma}$ and obtain a partition of the vertices in two equally sized sets V_1 and V_2.
3. Apply $2n\log(n)$ steps of the hillclimbing algorithm to the two sets of vertices V_1, V_2.

Because the analysis of this part of the coloring is analogous to the analysis carried out in the last sections we omit the detailed proofs here.

7 Concluding remarks

We have proved that our algorithm using only SA-type techniques will converge to a proper coloring in sublinear time with high probability. Although the transition steps of our algorithm are quite simple one may wonder if they can be simplified further.

PETFORD and WELSH [16] suggested the following algorithm, taking the same random model as used above. Choose an arbitrary coloring as starting state. Then choose a vertex and a color uniformly at random. The cost function is the number of wrong edges, i.e. edges between vertices of the same color, that are forbidden in a proper coloring. They show by some experimental results that the algorithm works well in practice but a theoretical ananlysis is still missing.

References

1. Alon, N.; Kahale, N.: *A spectral technique for coloring random 3-colorable graphs*, Proceedings of the 26th Symposium on Theory of Computing, 1994
2. Angluin, D.; Valiant, L.G.: *Fast probabilistic algorithms for Hamiltonian circuits and matchings*, Journal of Computer System Science 18, 1979
3. Blum, A.; Spencer, J.: *Coloring Random and Semi-Random k-Colorable graphs*, Journal of Algorithms 19, 1995
4. Chung, K.L.: *Markov chains with stationary transition probabilities*, Springer Verlag, Heidelberg, 1960.
5. Dyer, M.; Frieze, A.: *The Solution of Some Random NP-Hard Problems in Polynomial Expected Time*, Journal of Algorithms 10, 1989
6. Feller, W.: *An introduction to probability theory and its applications, Volume 1*, John Wiley & Sons, New York, 1950
7. Garey, M.R.; Johnson, S.J.: *Computers and Intractability*, W.H. Freeman and Company, 1979
8. Jensen, T.; Toft, B.: *Graph Coloring Problems*, John Wiley & Sons, 1995
9. Jerrum, J; Sorkin, G.: *Simulated Annealing for Graph Bisection*, Technical Report 1993, LFCS, University of Edinburgh
10. Johnson, D.S.; Aragon, C.R.; McGeoch, L.A.; Schevon, C.: *Optimization by Simulated Annealing: An Experimental Evaluation*, Operations Research 39, 1991
11. Kirkpatrick, S.; Gelatt, C.D.; Vecchi, M.P.: *Optimization by Simulated Annealing*, Science 220, 1983.
12. Kucera, L.: *Expected behaviour of graph coloring algorithms*, Lecture Notes in Computer Science 56, 1977
13. Laarhoven, P.J.M.; Aarts, E.H.L.: *Simulated Annealing: Theory and Applications*, Kluwer Academic Publishers, 1989.
14. Lindvall, T.: *Lectures on the Coupling method*, John Wiley & Sons, 1992
15. Metropolis, N.; Rosenbluth, M.; Rosenbluth, M.; Teller, A.; Teller, E.: *Equation of state calculations by fast computer machines*, Journal of Chemical Physics 21, 1953
16. Petford, A.D.; Welsh, D.J.A.: *A Randomized 3-Coloring Algorithm*, Discrete Mathematics 74 (1989), North Holland
17. Sasaki, G.H.; Hajek, B.: *The Time Complexity of Maximum Matching by Simulated Annealing*, Journal of the Association for Computing Machinery 35, 1988
18. Turner, J.: *Almost All k-Colorable Graphs Are Easy to Color*, Journal of Algorithms 9, 1988

Competitive Analysis of on-line Stack-Up Algorithms

J. Rethmann and E. Wanke

University of Düsseldorf, Department of Computer Science, D-40225 Düsseldorf, Germany

Abstract. Let $q = (b_1, \ldots, b_n)$ be a sequence of bins. Each bin is destined for some pallet t. For two given integers s and p, the stack-up problem is to move step by step all bins from q onto their pallets such that the position of the bin moved from q is always not greater than s and after each step there are at most p pallets for which the first bin is already stacked up but the last bin is still missing. If a bin b is moved from q then all bins to the right of b are shifted one position to the left. We determine the performance of four simple on-line algorithms called First-In, First-Done, Most-Frequently, and Greedy with respect to an optimal off-line solution for the stack-up problem.

1 Introduction

In this paper, we consider the problem of stacking up bins from a conveyer onto pallets by so-called stack-up systems that are usually located at the end of pick-to-belt order-picking systems (see [dK94, LLKS93]). Each order consists of several bins. The bins arrive the stack-up system on a conveyer. At the end of the conveyer the bins are moved by stacker cranes onto pallets, where they are stacked up. The pallets are build on certain stack-up places. Vehicles take full pallets, put them onto trucks, and bring new empty pallets to the stack-up places (see [RW97c, RW97a]).

All bins belonging to one order have to be placed onto the same pallet and bins for different pallets have to be placed onto different pallets. Unfortunately, the bins arrive the stack-up system not in a succession such that they can be placed one after the other onto pallets. Furthermore, not all bins of the sequence are known in advance. So the stack-up problem is to compute a step by step placement of bins onto pallets such that the position of the placed bin is always not greater than s (an upper bound on the storage capacity) and after each step there are at most p (an upper bound on the number of stack-up places) pallets for which the first bin is already stacked up but the last bin is still missing.

The stack-up problem is up to now not intensively investigated by other authors, although it has important practical applications. However, the following results are already known about the stack-up problem. The stack-up decision problem is in general NP-complete [GJ79] but can be solved very efficiently and even in parallel if the storage capacity s or the number of stack-up places p is bounded, see [RW97c]. In [RW97b], a polynomial time off-line approximation algorithm for the processing of sequences q with a storage capacity of $\lceil s_{\min}(q,p) \cdot \log_2(p+1) \rceil$ and $p+1$ stack-up places is introduced, where $s_{\min}(q,p)$ is the minimal storage capacity to solve the stack-up problem with p stack-up places. Off-line algorithms assume that the complete sequence of bins q is given to the input and thus known in advance.

In this paper, we analyze the worst-case behavior of simple on-line algorithms by competitive analysis [MMS88]. That means, we determine the performance of our on-line algorithms with respect to an optimal off-line solution. In an off-line processing the complete sequence is known in advance, whereas in an on-line processing in each step only the first s bins for q are known. On-line algorithms are very interesting from a practical point of view. The algorithms introduced and analyzed in this paper are called First-In (FI), First-Done (FD), Most-Frequently (MF), and Greedy. Let $s_A(q,p)$ $(p_A(q,s))$ be the minimum storage capacity (minimum number of stack-up places, respectively,) such that algorithm A processes sequence q with p stack-up places (with a storage capacity of s bins, respectively). Let q be a sequence of bins that can be processed with a storage capacity of s bins and p stack-up places. Our results can be summarized as follows.

A	FI	FD	MF & Greedy
$s_A(q,p) \leq$	$s - (p-1) + p(B_q - 1)$	$s + p(B_q - 1)$	$(p+1)s - p$
$p_A(q,s) \leq$	$s - 1 + p$	$s - 1 + p$	$\min\{s - 1 + p,\ p(\log_2(s) + 2)\}$

We give also general upper and lower bounds on $s_A(q,p)$ and $p_A(q,s)$ for arbitrary stack-up algorithms A. The results introduced in this paper show that even simple on-line algorithms like Most-Frequently and Greedy have provable good worst-case performance.

2 Preliminaries

We consider *sequences* $q = (b_1, \ldots, b_n)$ of n pairwise distinct *bins*. A sequence $q' = (b_{i_1}, \ldots, b_{i_k})$ is a *subsequence* of q if

$$1 \leq i_1 < i_2 < \cdots < i_{k-1} < i_k \leq n.$$

The *position* of a bin b_{i_j} in sequence $q' = (b_{i_1}, \ldots, b_{i_k})$ is equal to j, and denoted by $pos(q', b_{i_j})$. If X is the set of bins removed from q to obtain q' then q' is also denoted by $q - X$.

Each bin b in q is associated with a so-called *pallet symbol* $plt(b)$ which is in general some positive integer. We say bin b is destined for pallet $plt(b)$. The labeling of the pallets is arbitrary, because we only need to know whether two bins are destined for the same pallet or for different pallets. The set of all pallets of the bins in sequence q is denoted by $plts(q) = \{plt(b) \mid b \in q\}$. The set of all bins destined for some pallet t is denoted by $bins(t)$. The position of the first and the last bin in sequence q destined for some pallet $t \in plts(q)$ is denoted by $first(q,t)$ and $last(q,t)$, respectively.

Suppose we remove step by step the bins from some sequence $q = (b_1, \ldots, b_n)$. The successive removal of all bins yields n subsequences

$$q_1 = q - \{b_{j_1}\}, \quad q_2 = q - \{b_{j_1}, b_{j_2}\}, \quad \cdots, \quad q_n = q - \{b_{j_1}, \cdots, b_{j_n}\}$$

of $q = q_0$, where q_n is empty. A pallet t is called *open* in q_i, if some bin for pallet t is already removed from q but sequence q_i still contains at least one bin for pallet t. The set of all open pallets in q_i with respect to q is denoted by $open((q, q_i))$. If q_i has no bin for a pallet $t \in plts(q)$, then pallet t is called *closed* in q_i.

We consider the problem to remove step by step all bins from sequence q such that the removed bin is always on a position not greater than some integer s, i.e., $pos(q_{i-1}, b_{j_i}) \leq s$, and after each removal the number of open pallets is always not greater than some integer p, i.e., $|open((q, q_i))| \leq p$. The integers s and p correspond to the used *storage capacity* and the used number of *stack-up places* (or *places* for short), respectively. Such a step by step transformation of q into some empty sequence is called (s, p)-*processing* of q. A sequence q is called (s,p)-*sequence*, if there is an (s, p)-processing of q.

It is sometimes convenient to use in examples the pallet identifications instead of bin identifications. So we will write $q \hateq (plt(b_1), \ldots, plt(b_n))$ instead of $q = (b_1, \ldots, b_n)$.

Example 1. Consider sequence $q = (b_1, \ldots, b_8)$, where $q \hateq (1, 1, 2, 3, 2, 3, 2, 1)$. Table 1 shows a $(3, 2)$-processing of sequence q. The underlined bin is removed from q_i. It is always the first bin b in sequence q_i for pallet $plt(b)$.

i	$q_i \hateq$	$open((q, q_i)) =$	S_i	j_i
0	$(1, 1, \underline{2}, 3, 2, 3, 2, 1)$	$\emptyset$	$(1, 1, 2)$	4
1	$(1, 1, \underline{3}, 2, 3, 2, 1)$	$\{2\}$	$(1, 1, 3)$	5
2	$(1, 1, \underline{2}, 3, 2, 1)$	$\{2, 3\}$	$(1, 1, 2)$	6
3	$(1, 1, \underline{3}, 2, 1)$	$\{2, 3\}$	$(1, 1, 3)$	7
4	$(1, 1, \underline{2}, 1)$	$\{2\}$	$(1, 1, 2)$	8
5	$(\underline{1}, 1, 1)$	$\emptyset$	$(1, 1, 1)$	9
6	$(\underline{1}, 1)$	$\{1\}$	$(1, 1)$	9
7	$(\underline{1})$	$\{1\}$	(1)	9
8	$()$	$\emptyset$	$()$	9

Table 1. A $(3, 2)$-processing of q.

Let $s_{\min}(q, p)$ be the minimum storage capacity necessary to process q with p places. That is, there is an $(s_{\min}(q, p), p)$-processing of q, but no (s', p)-processing of q for $s' < s_{\min}(q, p)$. Analogously, let $p_{\min}(q, s)$ be the minimum number of places necessary to process q with storage capacity s, i.e., there is an $(s, p_{\min}(q, s))$-processing of q, but no (s, p')-processing of q for $p' < p_{\min}(q, s)$.

Let q' be a subsequence of q. Each pair (q, q') is called *configuration*. If there is a configuration $(q, q' - \{b\})$ for some bin b of q' such that $pos(q', b) \leq s$, $|open((q, q'))| \leq p$ and $|open((q, q' - \{b\}))| \leq p$, then the removal of b from q is called (s, p)-*transformation step*. A configuration (q, q') is called (s, p)-*blocking* if q' is not empty and there is no further (s, p)-transformation step.

In each (s, p)-transformation step during an (s, p)-transformation of sequence q into some sequence q' a bin on the first s positions is removed. The *active part* of q' consists of the first s bins of q', because the next (s, p)-transformation step has to remove one of these bins. Analogously, the last $|q'| - s$ bins of q' represent the so-called *passive part* of q'. For some configuration (q, q') let S denote the set of bins in the active part of q'. If bin b_j is the first bin in the passive part of q', we also write (q, S, j) instead of (q, q'), where S is called the *storage*. We write $(q, q') \cong (q, S, j)$ to symbolize that (q, q') and (q, S, j) denote the same configuration. If q' contains less than or equal to s bins then j is defined by $|q| + 1$, see table 1.

Consider an (s, p)-processing of some sequence q. Let $B = (b_{i_1}, \ldots, b_{i_n})$ be the sequence of bins in the order they are removed from q and let $P = (t_{j_1}, \ldots, t_{j_m})$ be the sequence of pallets in the order they are opened during the (s, p)-processing. B is called (s, p)-*bin solution* for q and P is called (s, p)-*pallet solution* for q. The transformation in table 1 defines $(3, 2)$-bin solution $B = (b_3, b_4, b_5, b_6, b_7, b_1, b_2, b_8)$ and $(3, 2)$-pallet solution $P = (2, 3, 1)$.

The stack-up decision problem is to determine whether there is an (s, p)-processing for a given sequence q of bins and two integers s and p. It is proved to be NP-complete in [RW97c]. A modification of the NP-completeness proof shows that the stack-up problem is NP-complete even if each pallet has the same number of bins. So there is no hope to compute $s_{\min}(q, p)$ or $p_{\min}(q, s)$ in polynomial time unless P = NP.

3 Stack-Up Strategies

There are configurations (q, q') for which it is easy to find bins b such that b can be removed from q' without eliminating an (s, p)-processing of q. This is the case, for example, if b is destined for an open pallet and $pos(q', b) \leq s$, because the positions of the bins in $q' - \{b\}$ are all less than or equal to their positions in q', and no new pallet is opened.

A similar argumentation is possible if there are less than p open pallets and all bins for some pallet t are on a position less than or equal to s in q'. Then all these bins can be removed step by step from q' by succeeding (s, p)-transformation steps without eliminating an (s, p)-processing, because after these transformation steps the same pallets are open as before and the positions of the bins in $q' - \{b\}$ are less than or equal to their positions in q'.

Moreover, bin b can be removed from q' without eliminating an (s, p)-processing of q even if not all bins for pallet $t = plt(b)$ are on a position less than or equal to s in q'. It is sufficient that each subsequence during the next transformation steps contains at least one bin for t on a position less than or equal to

s until the last bin for t is removed. Furthermore, the bins for pallet t need not to be removed one after the other. It doesn't matter if additionally some bins for other already open pallets are removed in between. The only condition that has to be satisfied is that before another pallet is opened the pallet t can be closed.

We make the following assumptions to our on-line stack-up algorithms. Let (q, S, j) be a configuration obtained during some (s, p)-transformation of q, i.e. s is the given storage capacity.

1. The bins from storage S that are destined for open pallets are removed automatically from S.
2. If storage S contains all bins for some pallet t, then pallet t is automatically opened by removing the first bin for pallet t from S.

Our on-line stack-up algorithms will only be asked for a decision if

1. each bin in the storage is not destined for some open pallet, and
2. the storage does not contain all bins for some pallet.

Such configurations are called *decision configurations* with respect to s. In a decision configuration the algorithm decides which pallet t is opened next by removing the first bin b destined for pallet t from q. In table 1 the decision configurations with respect to $s = 3$ are $(q_0, q_0) \cong (q_0, S_0, 4)$, and $(q_0, q_1) \cong (q_0, S_1, 5)$, where $S_0 = (b_1, b_2, b_3)$ and $S_1 = (b_1, b_2, b_4)$.

The discussion above allows us to consider only decision configurations. Note that for all pallets t of q the number of bins for pallet t is known from the customer order, so we are able to decide whether a configuration (q, S, j) is a decision configuration by knowing the storage S. Let $(q, q_i) \cong (q, S, j)$ be a decision configuration with respect to s. We consider the following four on-line stack-up algorithms.

1. **First-In** The FI algorithm opens pallet $t = plt(b)$, where b is the first bin in storage S.
2. **First-Done** The FD algorithm opens pallet $t \in plts(S)$ such that for all $t' \in plts(S)$ with $t' \neq t$, $last(q, t) < last(q, t')$.
3. **Most-Frequently** The MF algorithm opens some pallet $t \in plts(S)$ such that for all $t' \in plts(S)$, $|bins(t') \cap S| \leq |bins(t) \cap S|$.
4. **Greedy** Let $\#_{s,p}(q, q_i, t)$ be the number of (s, p)-transformation steps that will be executed automatically before the next decision configuration is reached. The Greedy algorithm opens some pallet $t \in plts(S)$ such that for all $t' \in plts(S)$, $\#_{s,p}(q, q_i, t') \leq \#_{s,p}(q, q_i, t)$.

If the choice of the Most-Frequently algorithm or the Greedy algorithm is not unique, then the order of the first bins of those pallets that satisfies the condition decides which pallet is opened. Strictly speaking, the First-Done algorithm and the Greedy algorithm are no on-line algorithms, because the decision which pallet has to be opened next can not be done without having certain information of the passive part. Usually there are several hundreds of bins known in advance, so this is not really a restriction.

Let $p_{FD}(q, s)$, $p_{FI}(q, s)$, $p_{MF}(q, s)$, and $p_{GR}(q, s)$ be the number of places that the FD, FI, MF, and Greedy algorithm, respectively, need to process sequence q with storage capacity s. For certain algorithms A, we can observe that $p_A(q, s)$ is sometimes greater than $p_A(q, s')$ although s is greater than s'. This property is called *stack-up anomaly*, in analogy to the Belady-Anomaly of paging algorithms; see [Tan87, BNS69]. The FD, MF and Greedy algorithms possess the stack-up anomaly as the examples in table 2 show.

$q \doteq (1, 1, 3, 2, 1, 2, 1, 4, 2, 2, 1, 4, 4, 4, 2, 3)$	$p_{FD}(q, 4) = 3$	$p_{FD}(q, 3) = 2$
$q \doteq (3, 1, 1, 1, 2, 1, 1, 3, 3, 2, 2, 2, 1, 3, 3, 3, 3, 1, 2, 2, 2, 3)$	$p_{MF}(q, 7) = 3$	$p_{MF}(q, 5) = 2$
$q \doteq (3, 1, 1, 2, 3, 2, 2, 3, 3, 1, 3, 1, 1, 1, 3, 3, 3, 1, 2, 2, 2, 3)$	$p_{GR}(q, 5) = 3$	$p_{GR}(q, 4) = 2$

Table 2. Stack-up anomalies of on-line algorithms

The FI algorithm does not possess the stack-up anomaly. If $P = (t_1, \ldots, t_m)$ is the pallet solution computed by the FI algorithm with storage capacity s, then for all pallets t_i, t_j opened by FI with $i < j$ we have $first(q, t_i) < first(q, t_j)$. A pallet opened automatically during the FI processing with storage capacity s is also opened automatically during the FI processing with storage capacity greater than s.

The assumption that certain pallets t are opened automatically, because all bins for t are in the active part, may have a negative effect for the processing of q by an on-line algorithm. Consider, for example, sequence

$$q \doteq (3, 1, 1, 1, 2, 4, 4, 1, 1, 3, 3, 2, 2, 2, 1, 3, 3, 3, 3, 1, 2, 2, 2, 2, 3)$$

and $s = 7$. The MF algorithm yields pallet solution $(4, 1, 2, 3)$ using 3 places. If pallets which bins are all in the active part are not opened automatically, then the MF algorithm would yield pallet solution $(1, 3, 2, 4)$ using only 2 places. Nevertheless, we still assume that pallets which bins are all in the active part are opened automatically.

3.1 General bounds

Let $C(q, i)$ for $1 < i \leq n$ be the set of all pallets t such that $first(q, t) < i$ and $last(q, t) \geq i$.

Theorem 1. *If there is an (s, p)-processing of some sequence $q = (b_1, \ldots, b_n)$, then $|C(q, i)| \leq s - 1 + p$ for all $1 < i \leq n$.*

Proof. Consider any configuration (q, S, i) during an (s, p)-processing of q. There are at most p open pallets in (q, S, i). The s bins in the storage are destined for at most $s - 1$ pallets not in $open((q, S, i))$, otherwise (q, S, i) is blocking with respect to s, p. Thus, $|C(q, i)| \le s - 1 + p$ for all $1 < i \le n$.

Since for each sequence q there is an $(s_{\min}(q, p), p)$-processing of q and an $(s, p_{\min}(q, s))$-processing of q by definition, theorem 1 implies that

$$|C(q, i)| \le s_{\min}(q, p) - 1 + p \quad \text{and} \quad |C(q, i)| \le s - 1 + p_{\min}(q, s).$$

Furthermore,

$$p_A(q, s) \le s - 1 + p_{\min}(q, s) \tag{1}$$

for any stack-up algorithm A, because any stack-up algorithm opens at most $\max_{1 \le i \le n} |C(q, i)|$ pallets.

Let B_q be an upper bound on the number of bins destined for the pallets in $plts(q)$.

Lemma 2. *If there is an (s, p)-processing of sequence q, then any stack-up algorithm yields an $(s', 1)$-processing of q for $s' \ge s + p(B_q - 1)$, because it does not reach a decision configuration with respect to s'.*

Proof. Let (q, S, i) be a configuration obtained during an $(s', 1)$-processing of q for some $s' \ge s + p(B_q - 1)$. Since there is an (s, p)-processing of q, the number of bins in S destined for pallets in $C(q, i)$ is less than $s + p(B_q - 1)$, cf. theorem 1. Thus, the storage contains all bins for some pallet and the processing continues automatically.

Theorem 3. *If there is an (s, p)-processing of sequence q, then any stack-up algorithm yields an (s', p)-processing of q for $s' \ge s + p(B_q - 1) - (p - 1)$.*

Proof. The algorithm can not reach a decision configuration (q, S, j) such that $|open((q, S, j))| = p - 1$. If $p - 1$ pallets are open, then by lemma 2 the storage contains either all bins for some pallet t or the last bin for an already open pallet. Here $p - 1$ places are used to extend the size of the storage by at least $p - 1$ bins.

In the following we analyze the performance of our four on-line stack-up algorithms.

3.2 The FI and FD algorithms

The worst-case behavior of the FI algorithm and the FD algorithm reaches the upper bound $p_A(q, s) \le s - 1 + p_{\min}(q, s)$ given by equation 1 for each $s \ge 2$. Consider, for example, sequence

$$q \doteq (1, 2, 3, 4, 5, 4, 5, 4, 5, 4, 5, 6, 6, 6, 6, 1, 2, 3, 4, 5),$$

and $s = 4$. Then $p_{\min}(q, 4) = 3$, $p_{FI}(q, 4) = 6$, and $p_{FD}(q, s) = 6$. This example can obviously be extended to arbitrary s.

The worst-case behavior of the FI algorithm reaches the upper bound $s_{\mathrm{FI}}(q,p) \leq s_{\min}(q,p) - (p-1) + p(B_q - 1)$ given by theorem 3 for each $p \geq 1$ and each $B_q \geq 2$. This shows the following example. Let $p = 3$ and

$$q \,\hat{=}\, (1,2,3,4,5,6,(4,5,6)^{n-2},4,5,6,4,5,6,1,2,3),$$

where the part $(4,5,6)^{n-2}$ represents an $n-2$ times repetition of three bins for the pallets $4,5,6$ for some $n \geq 2$. Then $B_q = n+1$ and $s_{\min}(q,3) = 4$. The FI algorithm will always need more than 3 stack-up places for each $s < 2 + 3 \cdot n$. This example can obviously be extended to arbitrary p.

The worst-case behavior of the FD algorithm is similar to the worst-case behavior of the FI algorithms. They differ at most by $p+1$. Consider, for example, sequence

$$q \,\hat{=}\, (1,2,3,4,5,6,(4,5,6)^{n-2},4,5,6,1,2,3,4,5,6)$$

and $p = 3$. Then $B_q = n+1$ and $s_{\min}(q,3) = 7$. The FD algorithm will always need more than 3 stack-up places for $s < 1 + 3 \cdot n$. This example can obviously be extended to arbitrary p.

3.3 The MF algorithm

Lemma 4. *If there is an (s,p)-processing of sequence q, then for any decision configuration (q,S,j) during an (s',p)-processing of q where $s' \geq (p+1)s - p$, there is a pallet t such that the storage contains at least s bins for t.*

Proof. Assume each pallet t has less than s bins in the storage S. Since at most p pallets can be open, there are at least $(p+1)s - p - p(s-1) = s$ bins in the storage for pallets that are not open. This contradicts the existence of an (s,p)-processing of q.

Theorem 5. *If there is an (s,p)-processing of sequence q, then the MF algorithm yields an (s',p)-processing of q for $s' \geq (p+1)s - p$.*

Proof. We compare simultaneously step by step the (s',p)-processing of q by the MF algorithm with the (s,p)-processing of q by some algorithm A. Let (q,S',j) be the first decision configuration obtained by the MF algorithm, and let (q,S,j) be the corresponding configuration obtained by algorithm A. The MF algorithm opens always a pallet t that has the most number of bins in the storage. By lemma 4, this is always a pallet that has at least s bins in the storage if $s' \geq (p+1)s - p$. Since the size of storage S is only s, pallet t is either open in configuration (q,S,j) or will be opened in the next (s,p)-transformation step of A. An inductive argumentation implies that for each such pair of configurations (q,S,j), (q,S',j), we have $open((q,S',j)) \subseteq open((q,S,j))$. Since A does not reach a blocking configuration with respect to s and p, the MF algorithm will not reach a blocking configuration with respect to s' and p.

To see that the bound of theorem 5 is tight, consider sequence

$$q \doteq (1,1,1,2,2,2,3,3,3,2,2,2,3,3,3,2,2,2,3,3,3,1,2)$$

and $p = 2$. Then $s_{\min}(q,p) = 4$ and $(p+1)s_{\min}(q,p) - p = 10$. The MF algorithm finds a $(10,2)$-processing of q, but finds no $(s',2)$-processing of q for each $s' < 10$. This example can obviously be extended to an arbitrary large number of stack-up places.

The worst-case behavior of the MF algorithm reaches the upper bound $p_A(q,s) \leq s - 1 + p_{\min}(q,s)$ given by equation 1 only for $s - 1 = p_{\min}(q,s)$. Consider, for example, $s = 5$ and sequence

$$q \doteq (1,2,3,4,5,6,7,8,(5,6,7,8)^n,1,2,3,4),$$

where the part $(5,6,7,8)^n$ represents an n times repetition of four bins for the pallets $5,6,7,8$ for some $n \geq 5$. Then $p_{\min}(q,5) = 4$ and $p_{MF}(q,5) = 8$. This example can not be extended to arbitrary s.

For all pallets t that have at least one bin on a position greater than or equal to j let $h(q,j,t)$ be the number of bins for pallet t on a position less than j in q. Let $\Sigma_h(q,j)$ be the sum of all bins on a position less than j destined for pallets which last bins are at least on position j.

Lemma 6. *Let (q,S,j) be a configuration during an (s,p)-processing of q, and let $T = open((q,S,j))$ be the set of open pallets in (q,S,j). Then*

$$\Sigma_h(q,j) \leq \sum_{t \in T} h(q,j,t) + s.$$

Proof. If (q,S,j) is a decision configuration, then there is no bin for an already open pallet in the storage. Furthermore, all pallets in the storage (and all open pallets) have at least one bin on a position at least j. Thus, we get

$$\Sigma_h(q,j) = \sum_{t \in T} h(q,j,t) + s.$$

If (q,S,j) is not a decision configuration, then there has to be at least one bin in the storage for an already open pallet, or the storage contains all bins for some pallet and $|open((q,S,j))| < p$. Thus, we get

$$\Sigma_h(q,j) \leq \sum_{t \in T} h(q,j,t) + s.$$

Lemma 7. *Let q be an (s,p)-sequence, and let $p' \geq p$. Let (q,S',j) be a decision configuration during an (s,p')-processing of q, and let (q,S,j) be the corresponding configuration during an (s,p)-processing of q. Let $O = open((q,S,j))$, and let $T = open((q,S',j))$. Then*

$$\frac{1}{p} \sum_{t \in T-O} h(q,j,t) \leq \max_{o \in O-T} \{ h(q,j,o) \}.$$

Proof. By lemma 6 the following holds.

$$\sum_{t \in T} h(q,j,t) + s \;=\; \Sigma_h(q,j) \;\leq\; \sum_{o \in O} h(q,j,o) + s$$

$$\Longleftrightarrow\; \sum_{t \in T} h(q,j,t) - \sum_{t \in T \cap O} h(q,j,t) \;\leq\; \sum_{o \in O} h(q,j,o) - \sum_{t \in T \cap O} h(q,j,t)$$

$$\Longleftrightarrow\; \tfrac{1}{p} \sum_{t \in T-O} h(q,j,t) \;\leq\; \tfrac{1}{p} \sum_{o \in O-T} h(q,j,o) \;\leq\; \max_{o \in O-T} \{\, h(q,j,o) \,\}$$

Theorem 8. $p_{\mathrm{MF}}(q,s) \;\leq\; \min\{s-1+p,\; p_{\min}(q,s) \cdot (\log_2(s)+2)\}$

Proof. The inequality $p_{\mathrm{MF}}(q,s) \leq s-1+p$ follows by equation 1. Consider some configuration (q, S', k) during an (s,p')-processing of q by the MF algorithm such that $|open((q,S',k))| = p'$. Let $O = open((q,S,k))$ be the set of open pallets in the corresponding configuration (q,S,k) during an (s,p)-processing by some algorithm A.

Let $(q, S_1', j_1), \ldots, (q, S_l', j_l)$ be the decision configurations during the (s,p')-processing of q by the MF algorithm such that pallet t_i is opened in configuration (q, S_i', j_i), and t_i is not in O. Let T_i be the set of open pallets in configuration (q, S_i', j_i), and let O_i denote the set of open pallets in the corresponding configuration (q, S_i, j_i) during the (s,p)-processing of q by algorithm A. Then the following holds.

1. $h(q, j_i, t_i) \geq \max\limits_{o \in O_i - T_i} \{\, h(q, j_i, o) \,\}$, otherwise MF does not open pallet t_i.
2. $\max\limits_{o \in O_i - T_i} \{\, h(q, j_i, o) \,\} \geq \tfrac{1}{p} \sum\limits_{t \in T_i - O_i} h(q, j_i, t)$, because of lemma 7.

The number of bins for some pallet t is monotone increasing up to the position of the last bin for t. Thus, for $i > k$ we get $h(q, j_i, t_k) \geq h(q, j_k, t_k)$. Furthermore, none of the pallets $t \in \{t_1, \ldots, t_{i-1}\}$ is open in any configuration (q, R, i'), $i' \leq j_i$, during the (s,p)-processing of q, because such a pallet $t \in open((q,R,i'))$ would remain open until configuration (q,S,k), and therefore pallet t has to be in O. Since these pallets are not considered, pallet t can not be in $\{t_1, \ldots, t_{i-1}\}$. Thus, we get

$$h(q, j_i, t_i) \;\geq\; \frac{1}{p} \sum_{t \in T_i - O_i} h(q, j_i, t) \;\geq\; \frac{1}{p} \sum_{k=1}^{i-1} h(q, j_k, t_k).$$

It can be shown that $h(q, j_i, t_i) \geq 2^{\lfloor \frac{i-2}{p} \rfloor}$ for each $p \geq 1$. The maximum l such that $\sum_{i=1}^{l} h(q, j_i, t_i) < s$ is an upper bound on the number of pallets the MF algorithm opens and algorithm A does not open. A simple induction yields $2^{\lfloor \frac{l-1}{p} \rfloor} \leq \sum_{i=1}^{l} h(q, j_i, t_i)$, and therefore we get

$$2^{\frac{l-1}{p} - 1} < s \quad \Longleftrightarrow \quad l \leq p \cdot (\log_2(s)+1).$$

The MF algorithm can open p additional pallets which algorithm A has opened, too. Thus, the number of open pallets during an processing of an (s,p)-sequence by the MF algorithm with storage capacity s is at most $p \cdot (\log_2(s)+2)$.

This bound seems not to be tight, but there are (s, p)-sequences q such that there is an (s, p')-processing of q by the Most-Frequently algorithm only if p' is at least $p \cdot (\log_{p+1}(s) + 1)$.

3.4 The Greedy algorithm

In some decision configuration (q, q') the greedy algorithm opens a pallet t such that the number of (s, p)-transformation steps that will be done automatically is as large as possible. This can be done only if some bins of the passive part of the current subsequence q' are known to the algorithm. Since we have to look into the passive part of q', we can also decide whether the last bin of some pallet t is automatically reached if t is opened next. By the discussion at the beginning of this section, such a pallet t can be opened without eliminating any further (s, p)-processing.

We extend the Greedy algorithm as follows. Let (q, S, j) be a decision configuration with respect to some s. If storage S contains a bin for some pallet t such that t will be completely stacked up automatically if t is opened next, then t will be opened next. If no such pallet t exists, then the decision configuration is called an *extended decision configuration* and the Greedy algorithm works as usual. With this extension the worst-case performance of the Greedy algorithm is at least as good as the worst-case performance of the MF algorithm.

If pallet t is opened in an extended decision configuration (q, q_i) during an (s, p)-processing of sequence q, then let $g_{s,p}(q, q_i, t)$ denote the number of bins for pallet t that will be removed automatically from the sequence until the next extended decision configuration is reached.

Lemma 9. *Let $(q, q_i) \cong (q, S, j)$ be an extended decision configuration during an (s', p)-processing of some (s, p)-sequence q, and let $t \in plts(S)$ be the pallet opened by the Greedy algorithm in configuration (q, S, j). Then*

$$g_{s',p}(q, q_i, t) \geq \max_{t' \in plts(S)} \{ |bins(t') \cap S| \}.$$

Proof. Let t' be a pallet such that $|bins(t') \cap S|$ is maximal. We open pallet t' and consider the configuration (q, S', j') obtained after the next $\#_{s',p}(q, q_i, t')$ automatic (s', p)-transformation steps. Since (q, S, j) is an extended decision configuration, and $\#_{s',p}(q, q_i, t) \geq \#_{s',p}(q, q_i, t')$, the last bins for t and t' are not yet in storage S'.

Consider the bins on the positions $j, \ldots, j' - 1$ in sequence q. Let λ_t and $\lambda_{t'}$ be the number of bins for pallet t and t', respectively, on the positions $j, \ldots, j' - 1$ in sequence q. Furthermore, let δ be the number of bins for pallets t'' not open in configuration (q, S, j), not in $\{t, t'\}$, and $last(q, t'') \geq j'$ on the positions $j, \ldots, j' - 1$ in sequence q. Finally, let β be the number of bins for pallets $t'' \in plts(S)$ and $last(q, t'') \leq j' - 1$ in storage S. Then we get

$$|bins(t') \cap S| + \beta = \lambda_t + \delta, \quad \text{and} \quad |bins(t) \cap S| + \beta \geq \lambda_{t'} + \delta.$$

Simple transformations yields $|bins(t) \cap S| + \lambda_t \geq |bins(t') \cap S| + \lambda_{t'}$. Since $g_{s',p}(q, q_i, t) \geq |bins(t) \cap S| + \lambda_t$ the result follows.

Lemma 10. *Let q be an (s,p)-sequence, and let $s' \geq p(s-1)+1$. Then in any extended decision configuration $(q, q_i) \cong (q, S', j)$ during an (s', p)-processing of q by the Greedy algorithm a pallet t will be opened such that $g_{s',p}(q, q_i, t) \geq s$.*

Proof by contradiction. Suppose there is an extended decision configuration $(q, q_i) \cong (q, S', j)$ during an (s', p)-processing of q by the Greedy algorithm, where a pallet t is opened, such that $g_{s',p}(q, q_i, t) = s - k$ for some $k \geq 1$. Then by lemma 9 the storage contains at most $s - k$ bins for each pallet in $plts(S')$. Therefore, during an (s,p)-processing of q there are p pallets open in the direct successor configuration $(q, R, j+1)$ of the corresponding configuration (q, S, j), because $p(s-1) + 2 - (p-1)(s-k) > s$. Let $T = open((q, R, j+1))$ be the set of open pallets in configuration $(q, R, j+1)$. We have to consider two cases.

First, let the number of bins in storage S' for each pallet in T be equal to $s - k$. Then by lemma 9 $\#_{s',p}(q, q_i, t) \geq s - k$, and the next extended decision configuration (q, S'', j') during the (s', p)-processing of q by the Greedy algorithm will be reached at some position $j' \geq j + s - k$. Since $p(s-k) + s - 1 < p(s-1) + 1 + s - k$, there has to be a pallet t', such that $g_{s',p}(q, q_i, t') > s - k$, or the last bin for pallet t' will be reached. This contradicts the assumption, that $g_{s',p}(q, q_i, t) \leq s - k$ for all $t \in plts(S')$, or it contradicts the fact, that (q, S', j) is an extended decision configuration, respectively.

Second, let the number of bins in storage S' for each pallet in T be at most $s - k$. Let Σ be the number of bins in storage S' for pallets in T, and let $\delta = p(s - k) - \Sigma$. Up to the position $\Sigma + s - 1 + \delta$ there are at most $s - 1$ bins for not open pallets. And for each open pallet that has $s - k'$ bins in storage S', there are at most $k' - k$ bins up to position $\Sigma + s - 1 + \delta$, otherwise by lemma 9, the Greedy algorithm would open some pallet t', such that $g_{s',p}(q, q_i, t') > s - k$. Since $\Sigma + s - 1 + \delta < p(s-1) + 1 + s - k$, there has to be a pallet t', such that $g_{s',p}(q, q_i, t') > s - k$, or the last bin for pallet t' will be reached. This contradicts the assumption, that $g_{s',p}(q, q_i, t) \leq s - k$ for all $t \in plts(S')$, or it contradicts the fact, that (q, S', j) is an extended decision configuration, respectively.

Theorem 11. *The Greedy algorithm yields an (s', p)-processing of an (s,p)-sequence q, if $s' \geq p(s-1)+1$.*

Proof. As in the proof of theorem 5, we compare simultaneously step by step the (s', p)-processing of q by the Greedy algorithm with the (s, p)-processing of q by some algorithm A. Let $(q, q_i) \cong (q, S', j)$ be the first extended decision configuration obtained by the Greedy algorithm, and let (q, S, j) be the corresponding configuration obtained by algorithm A. The Greedy algorithm opens always a pallet t that causes the most number of automatic movements. By lemma 10, the number of bins stacked-up for pallet t during the next automatic (s', p)-transformation steps is at least s. Since the size of storage S is s, pallet t is either open in configuration (q, S, j) or will be opened in one of the next $\#_{s',p}(q, q_i, t)$ (s, p)-transformation steps of A. As in the proof of theorem 5, an inductive argumentation implies that for each such pair of configurations (q, S, j), (q, S', j), we have $open((q, S', j)) \subseteq open((q, S, j))$. Since A does not reach a blocking configuration with respect to s and p, the Greedy algorithm will not reach a blocking configuration with respect to s' and p.

To see that the bound of theorem 11 is tight, consider sequence

$$q \hateq (1,1,1,2,2,2,3,3,3,2,2,2,3,3,3,2,2,2,3,3,3,1,2)$$

and $p = 2$. Then $s_{\min}(q,p) = 4$ and $p(s_{\min}(q,p) - 1) + 1 = 7$. The Greedy algorithm finds an $(7,2)$-processing of q, but finds no $(s',2)$-processing of q for each $s' < 7$. This example can obviously be extended to an arbitrary large number of places.

Theorem 12. *If* $p_{\min}(q,s) = 1$ *then* $p_{\mathrm{GR}}(q,s) = 1$.

Proof. Let (q, S, j) be a configuration such that $open((q,S,j)) = \emptyset$. There must be a bin b in storage S such that after b is removed all other bins for pallet $plt(b)$ are automatically removed. The extended Greedy algorithm opens first some pallet which will be closed during the following automatic transformation steps.

Theorem 13. $p_{\mathrm{GR}}(q,s) \leq \min\{s - 1 + p,\ p_{\min}(q,s) \cdot (\log_2(s) + 2)\}$.

Proof. The inequality $p_{\mathrm{MF}}(q,s) \leq s - 1 + p$ follows by equation 1. As in the proof of theorem 8 consider some configuration (q, S', k) during an (s, p')-processing of q by the Greedy algorithm such that $|open((q, S', k))| = p'$. Let $O = open((q, S, k))$ be the set of open pallets in the corresponding configuration (q, S, k) during an (s,p)-processing by some algorithm A. Let $(q, q_1) \cong (q, S'_1, j_1), \ldots, (q, q_l) \cong (q, S'_l, j_l)$ be the decision configurations during the (s, p')-processing of q by the Greedy algorithm such that pallet t_i is opened in configuration (q, S'_i, j_i), and t_i is not in O. Let T_i be the set of open pallets in configuration (q, S'_i, j_i), and let O_i denote the set of open pallets in the corresponding configuration (q, S_i, j_i) during the (s,p)-processing of q by algorithm A. Then the following holds.

1. $g_{s',p}(q, q_i, t_i) \geq \max\limits_{o \in O_i - T_i} \{ h(q, j_i, o) \}$, otherwise Greedy does not open pallet t_i, because of lemma 9.
2. $\max\limits_{o \in O_i - T_i} \{ h(q, j_i, o) \} \geq \frac{1}{p} \sum\limits_{t \in T_i - O_i} h(q, j_i, t)$, because of lemma 7.

In analogy to the proof of theorem 8, we get

$$h(q, j_{i+1}, t_i) \geq g_{s',p}(q, q_i, t_i) \geq \frac{1}{p} \sum_{t \in T_i - O_i} h(q, j_i, t) \geq \frac{1}{p} \sum_{k=1}^{i-1} h(q, j_k, t_k).$$

The maximum l such that $\sum_{i=1}^{l} h(q, j_{i+1}, t_i) < s$ is an upper bound on the number of pallets the Greedy algorithm opens and algorithm A does not open. Since $2^{\lfloor \frac{l-1}{p} \rfloor} \leq \sum_{i=1}^{l} h(q, j_{i+1}, t_i)$, we get

$$2^{\frac{l-1}{p} - 1} < s \quad \Longleftrightarrow \quad l \leq p \cdot (\log_2(s) + 1).$$

The Greedy algorithm can open p additional pallets which algorithm A has opened, too. Thus, the number of open pallets during a processing of an (s,p)-sequence by the Greedy algorithm with storage capacity s is at most $p \cdot (\log_2(s) + 2)$.

This bound seems not to be tight, but there are (s,p)-sequences q such that there is an (s,p')-processing of q by the Greedy algorithm only if p' is at least $p \cdot \log_p(s)$.

4 Conclusion

We have presented four on-line stack-up algorithms and have analyzed their worst-case behavior by competitive analysis. It is easy to see that the decision which pallet has to be opened next can be done in linear time with respect to the number of bins in the storage.

In practice, the stack-up processing is tried to be solved by fuzzy-logic approaches. Simulations with real data instances have shown that the stack-up process can considerably be improved by the MF and Greedy algorithms. We are very grateful to Bertelsmann Distribution GmbH in Gütersloh, Germany, for providing real data instances.

References

[BNS69]   L.A. Belady, R.A. Nelson, and G.S. Shelder. An Anomaly in Space-Time Characteristics of Certain Programs Running in a Paging Machine. *Communications of the ACM*, 12:349–353, 1969.

[dK94]   R. de Koster. Performance approximation of pick-to-belt orderpicking systems. *European Journal of Operational Research*, 92:558–573, 1994.

[GJ79]   M.R. Garey and D.S. Johnson. *Computers and Intractability*. W.H. Freeman and Company, San Francisco, 1979.

[LLKS93] E.L. Lawler, J.K. Lenstra, A.H.G. Rinnooy Kan, and D.B. Shmoys. Sequencing and Scheduling: Algorithms and Complexity. In S.C. Graves, A.H.G. Rinnooy Kan, and P.H. Zipkin, editors, *Handbooks in Operations Research and Management Science, vol. 4, Logistics of Production and Inventory*, pages 445–522. North-Holland, Amsterdam, 1993.

[MMS88]   M.S. Manasse, L.A. McGeoch, and D.D. Sleator. Competitive algorithms for on-line problems. In *Proceedings of the Annual ACM Symposium on Theory of Computing*, pages 322–333. ACM, 1988.

[RW97a]   J. Rethmann and E. Wanke. Storage Controlled Pile-Up Systems, Theoretical Foundations. In *Operations Research Proceedings*. Springer-Verlag, 1997.

[RW97b]   J. Rethmann and E. Wanke. An approximation algorithm for stacking up bins from a conveyer onto pallets. In *Proceedings of the Annual Workshop on Algorithms and Data Structures*, Lecture Notes in Computer Science. Springer-Verlag, to appear 1997.

[RW97c]   J. Rethmann and E. Wanke. Storage Controlled Pile-Up Systems. *European Journal of Operational Research*, to appear 1997.

[Tan87]   Andrew S. Tanenbaum. *Operating Systems*. Prentice-Hall, Englewood Cliffs, NJ, 1987.

Scheduling–LPs Bear Probabilities

Randomized Approximations for Min–Sum Criteria

Andreas S. Schulz and Martin Skutella

Technische Universität Berlin,
Fachbereich Mathematik, MA 6–1,
Straße des 17. Juni 136, 10623 Berlin, Germany,
E-mail {schulz,skutella}@math.tu–berlin.de

Abstract. In this paper, we provide a new class of randomized approximation algorithms for scheduling problems by directly interpreting solutions to so-called time-indexed LPs as probabilities. The most general model we consider is scheduling unrelated parallel machines with release dates (or even network scheduling) so as to minimize the average weighted completion time. The crucial idea for these multiple machine problems is not to use standard list scheduling but rather to assign jobs randomly to machines (with probabilities taken from an optimal LP solution) and to perform list scheduling on each of them.

For the general model, we give a $(2 + \varepsilon)$–approximation algorithm. The best previously known approximation algorithm has a performance guarantee of $16/3$ [HSW96]. Moreover, our algorithm also improves upon the best previously known approximation algorithms for the special case of identical parallel machine scheduling (performance guarantee $(2.89 + \varepsilon)$ in general [CPS+96] and 2.85 for the average completion time [CMNS97], respectively). A perhaps surprising implication for identical parallel machines is that jobs are randomly assigned to machines, in which each machine is equally likely. In addition, in this case the algorithm has running time $O(n \log n)$ and performance guarantee 2. The same algorithm also is a 2–approximation for the corresponding preemptive scheduling problem on identical parallel machines.

Finally, the results for identical parallel machine scheduling apply to both the off-line and the on-line settings with no difference in performance guarantees. In the on-line setting, we are scheduling jobs that continually arrive to be processed and, for each time t, we must construct the schedule until time t without any knowledge of the jobs that will arrive afterwards.

1 Introduction

It is by now well-known that randomization can help in the design of algorithms, cf., e. g., [MR95,MNR96]. One way of guiding randomness is the use of linear programs (LPs). In this paper, we give LP-based approximation algorithms for problems which are particularly well-known for the difficulties to obtain good lower bounds: machine (or processor) scheduling problems. Because of the random choices involved, our algorithms are rather randomized approximation algorithms. A randomized ρ–approximation algorithm is a polynomial-time algorithm that produces a feasible solution whose expected value is within a factor of ρ of the optimum; ρ is also called the expected performance guarantee of the algorithm. Actually, we always compare the output of an algorithm with a lower bound given by an optimum solution to a certain LP relaxation. Hence, at the same time we obtain an analysis of the quality of the respective LP. All our off-line algorithms can be derandomized with no difference in performance guarantee, but at the cost of increased (but still polynomial) running times.

We consider the following model. We are given a set J of n jobs (or tasks) and m unrelated parallel machines. Each job j has a positive integral processing requirement p_{ij} which depends on the machine i job j will be processed on. Each job j must be processed for the respective amount of time on one of the m machines, and may be assigned to any of them. Every machine can process at most one job at a time. In *preemptive* schedules, a job may repeatedly be interrupted and continued later on another (or the same) machine. In *nonpreemptive* schedules, a job must be processed in an uninterrupted fashion. Each job j has an integral release date $r_j \geqslant 0$ before which it cannot be processed. We denote the completion time of job j in a schedule by C_j, and for any fixed $\alpha \in (0, 1]$, the α–point $C_j(\alpha)$ of job j is the first moment in time at which an α–fraction of job j has been completed; α–points were first used in the context of approximation by Hall, Shmoys, and Wein [HSW96]. We seek to minimize the total weighted completion time: a weight $w_j \geqslant 0$ is associated with each job j and the goal is to minimize $\sum_{j \in J} w_j C_j$. In scheduling, it is quite convenient to refer to the respective problems using the standard classification scheme of Graham et al. [GLLRK79]. The nonpreemptive problem $R \mid r_j \mid \sum w_j C_j$, just described, is strongly NP-hard.

Scheduling to minimize the total weighted completion time (or, equivalently, the average weighted completion time) has recently achieved a great deal of attention, partly because of its importance as a fundamental problem in scheduling, and also because of new applications, for instance, in compiler optimization [CJM+96] or in parallel computing [CM96]. In the last two years, there has been significant progress in the design of approximation algorithms for this kind of problems which led to the development of the first constant worst-case bounds in a number of settings. This progress essentially follows on the one hand from the use of preemptive schedules to construct nonpreemptive ones [PSW95,CPS+96,CMNS97,Goe97,SS97]. On the other hand, one solves an LP relaxation and then a schedule is constructed by list scheduling in a natural order dictated by the LP solution [PSW95,HSW96,Sch96,HSSW96,MSS96,Goe97,CS97,SS97].

In this paper, we utilize a different idea: random assignments of jobs to machines. To be more precise, we exploit a new LP relaxation in time-indexed variables for the problem $R \mid r_j \mid \sum w_j C_j$, and we then show that a certain variant of randomized rounding leads to a $(2 + \varepsilon)$–approximation algorithm, for any $\varepsilon > 0$. At the same moment, the corresponding LP is a $(2 + \varepsilon)$–relaxation, i.e., the true optimum is always within a factor of $(2 + \varepsilon)$ of the optimal value of the LP relaxation; and this is tight. Our algorithm improves upon a $16/3$–approximation algorithm of Hall, Shmoys, and Wein [HSW96] that is also based on time-indexed variables which have a different meaning, however. In contrast to their approach, our algorithm does not rely on Shmoys and Tardos' rounding technique for the generalized assignment problem [ST93]. We rather exploit the LP by interpreting LP values as probabilities with which jobs are assigned to machines. For an introduction to and the application of randomized rounding to other combinatorial optimization problems, the reader is referred to [RT87,MNR96].

For the special case of identical parallel machines, i. e., for each job j and all machines i we have $p_{ij} = p_j$, Chakrabarti, Phillips, Schulz, Shmoys, Stein, and Wein [CPS+96] obtained a $(2.89 + \varepsilon)$–approximation by refining an on-line greedy framework of Hall et al. [HSW96]. The former best known LP-based algorithm, however, relies on an LP relaxation solely in completion time variables which is weaker than the one we propose. It has performance guarantee $(4 - \frac{1}{m})$ (see [HSSW96] for the details). For the LP we use here, an optimum solution can greedily be obtained by a certain preemptive schedule on a virtual single machine which is m times as fast as any of the original machines. The idea of using a preemptive relaxation on such a virtual machine was introduced before by Chekuri, Motwani, Natarajan, and Stein [CMNS97]. They show that any preemptive schedule on this machine can be converted into a nonpreemptive schedule on the m identical parallel machines such that, for each job j, its completion time in the nonpreemptive schedule is at most $(3 - \frac{1}{m})$ times its preemptive completion time. For the average completion time, they refine this to a 2.85–approximation algorithm for $P|r_j|\sum C_j$. For $P|r_j|\sum w_j C_j$, the algorithm we propose delivers in time $O(n \log n)$ a solution that is expected to be within a factor of 2 of the optimum.

Since the LP relaxation we use is also a relaxation for the corresponding preemptive problem, our algorithm is also a 2–approximation for $P|r_j, pmtn|\sum w_j C_j$. This improves upon a 3–approximation algorithm due to Hall, Schulz, Shmoys, and Wein [HSSW96].

The paper is organized as follows. In Section 2, we start with the discussion of our main result: the 2–approximation in the general context of unrelated parallel machine scheduling. In the next section, we give a combinatorial 2–approximation algorithm for $P|r_j|\sum w_j C_j$ and $P|r_j, pmtn|\sum w_j C_j$. Then, in Section 4, we discuss the derandomization of the previously given randomized algorithms. Finally, in Section 5 we give the technical details of turning the pseudo-polynomial algorithm of Section 2 into a polynomial time algorithm with performance guarantee $(2 + \varepsilon)$.

2 Unrelated Parallel Machine Scheduling with Release Dates

In this section, we consider the problem $R|r_j|\sum w_j C_j$. As in [PSW94,HSW96], we will even discuss a slightly more general problem in which the release date of job j may also depend on the machine. The release date of job j on machine i is thus denoted by r_{ij}. Machine-dependent release dates are relevant to model network scheduling in which parallel machines are connected by a network, each job is located at one given machine at time 0, and cannot be started on another machine until sufficient time elapses to allow the job to be transmitted to its new machine. This model has been introduced in [DLLX90,AKP92].

The problem $R|r_{ij}|\sum w_j C_j$ is well-known to be strongly NP-hard, even for the case of a single machine [LRKB77]. The first non-trivial approximation algorithm for this problem was given by Phillips, Stein, and Wein [PSW94]. It has performance guarantee $O(\log^2 n)$. Subsequently, Hall et al. [HSW96] gave a 16/3–approximation algorithm which relies on a time-indexed LP relaxation whose optimum value serves as a surrogate for the true optimum in their estimations. We use a somewhat similar LP relaxation, but whereas Hall et al. invoke the deterministic rounding technique of Shmoys and Tardos [ST93] to construct a feasible schedule we randomly round LP solutions to feasible schedules.

Let $T = \max_{i,j} r_{ij} + \sum_{j \in J} \max_i p_{ij} - 1$ be the time horizon, and introduce for every job $j \in J$, every machine $i = 1, \ldots, m$, and every point $t = 0, \ldots, T$ in time a variable y_{ijt} which represents the time job j is processed on machine i within the time interval $(t, t+1]$. Equivalently, one can say that a y_{ijt}/p_{ij}-fraction of job j is being processed on machine i within the time interval $(t, t+1]$. The LP, which is an extension of a single machine LP relaxation of Dyer and Wolsey [DW90], is as follows:

$$\text{minimize} \quad \sum_{j \in J} w_j C_j$$

$$\text{subject to} \quad \sum_{i=1}^{m} \sum_{t=r_{ij}}^{T} \frac{y_{ijt}}{p_{ij}} = 1 \quad \text{for all } j \in J, \tag{1}$$

$$(LP_R) \qquad \sum_{j \in J} y_{ijt} \leqslant 1 \quad \text{for } i = 1, \ldots, m \text{ and } t = 0, \ldots, T, \tag{2}$$

$$C_j = \sum_{i=1}^{m} \sum_{t=0}^{T} \left(\frac{y_{ijt}}{p_{ij}} \left(t + \tfrac{1}{2} \right) + \tfrac{1}{2} y_{ijt} \right) \quad \text{for all } j \in J, \tag{3}$$

$$y_{ijt} = 0 \quad \text{for } i = 1, \ldots, m, \, j \in J, \, t = 0, \ldots, r_{ij} - 1, \tag{4}$$

$$y_{ijt} \geqslant 0 \quad \text{for } i = 1, \ldots, m, \, j \in J, \, t = r_{ij}, \ldots, T. \tag{5}$$

Equations (1) ensure that the whole processing requirement of every job is satisfied. The machine capacity constraints (2) simply express that each machine can process at most one job at a time. Now, for (3), consider an arbitrary feasible schedule, where job j is being continuously processed between time $C_j - p_{hj}$ and C_j on machine h. Then the expression for C_j in (3) corresponds to the real completion time, if we assign the values to the LP variables y_{ijt} as defined above, i.e., $y_{ijt} = 1$ if $i = h$ and $t \in \{C_j - p_{hj}, \ldots, C_j - 1\}$, and $y_{ijt} = 0$ otherwise. Hence, (LP_R) is a relaxation of the scheduling problem under consideration.

The following algorithm takes an optimum LP solution, and then constructs a feasible schedule by using a kind of randomized rounding.

Algorithm R

1) Compute an optimum solution y to (LP_R).
2) Assign job j to a machine-time pair (i, t) at random with probability y_{ijt}/p_{ij}; draw t_j from the chosen time interval $(t, t+1]$ at random with uniform distribution.
3) Schedule on each machine i the jobs that were assigned to it nonpreemptively as early as possible in nondecreasing order of t_j.

Note that all the random assignments need to be performed independently from each other (at least pairwise).

Lemma 1. *Let y be the optimum solution to (LP_R) which is computed in Step 1 of Algorithm R. Then, for each job $j \in J$ the expected processing time of j in the schedule constructed by Algorithm R equals $\sum_{i=1}^{m} \sum_{t=0}^{T} y_{ijt}$.*

Proof. First, we fix a machine-time pair (i,t) job j has been assigned to. Then, the expected processing time of j under these conditions is p_{ij}. By adding these conditional expectations over all machines and time intervals, weighted by the corresponding probabilities y_{ijt}/p_{ij}, we get the claimed result. □

Note that the lemma remains true if we start with an arbitrary, not necessarily optimal solution y to (LP_R) in Step 1 of Algorithm R. This also holds for the following theorem. The optimality of the LP solution is only needed to get a lower bound on the value of an optimal schedule.

Theorem 2. *The expected completion time of each job j in the schedule constructed by Algorithm R is at most $2 \cdot C_j^{LP}$, where C_j^{LP} is the LP completion time (defined by (3)) of the optimum solution y we started with in Step 1.*

Proof. We consider an arbitrary, but fixed job $j \in J$. To analyze the expected completion time of job j, we first also consider a fixed assignment of j to a machine-time pair (i,t). Then, the expected starting time of job j under these conditions precisely is the conditional expected idle time plus the conditional expected amount of processing of jobs that come before j on machine i.

Observe that there is no idle time on machine i between the maximum release date of jobs on machine i which start no later than j and the starting time of job j. It follows from the ordering of jobs and constraints (4) that this maximum release date and therefore the idle time of machine i before the starting time of j is bounded from above by t.

On the other hand, we get the following bound on the conditional expected processing time on machine i before the start of j:

$$\sum_{k \neq j} p_{ik} \cdot \Pr(k \text{ on } i \text{ before } j) \leqslant \sum_{k \neq j} p_{ik} \cdot \sum_{\ell=0}^{t} \frac{y_{ik\ell}}{p_{ik}} \leqslant t+1 \ .$$

The last inequality follows from the machine capacity constraints (2). Putting the observations together we get an upper bound of $2 \cdot (t + \frac{1}{2})$ for the expected starting time of j. Unconditioning the expectation together with Lemma 1 and (3) yields the theorem. □

Theorem 2 implies a performance guarantee of 2 for Algorithm R. In the course of its proof, we also have shown that (LP_R) is a 2–relaxation. Moreover, even for the case of identical parallel machines without release dates there are instances for which this bound is asymptotically attained, see Section 3. Thus our analysis is tight.

Unfortunately, (LP_R) has exponentially many variables. Consequently, Algorithm R only is a pseudo-polynomial-time algorithm. However, we can overcome this drawback by introducing new variables which are not associated with time intervals of length 1, but rather with intervals of geometrically increasing size. The idea of using interval-indexed variables to get polynomial-time solvable LP relaxations was introduced earlier by Hall, Shmoys, and Wein [HSW96]. In order to get polynomial-time approximation algorithms in this way, we have to pay for with a slightly worse performance guarantee. For any constant $\varepsilon > 0$, we get an approximation algorithm with performance guarantee $2 + \varepsilon$ for $R \mid r_{ij} \mid \sum w_j C_j$. The rather technical details can be found in Section 5.

It is shown in [SS97] that in the absence of (non-trivial) release dates, the use of a slightly stronger LP relaxation improves the performance guarantee of Algorithm R to $3/2$. Independently, this has also been observed by Fabián A. Chudak (communicated to us by David B. Shmoys, March 1997) after reading a preliminary version of the paper on hand which only contained the 2–approximation for $R \mid r_{ij} \mid \sum w_j C_j$.

Remark 3. The reader might wonder whether the seemingly artificial random choice of the t_j's in Algorithm R is really necessary. Indeed, it is not, which also implies that we could work with a discrete probability space: From the proof of Theorem 2 one can see that we could simply set $t_j = t$ if job j is assigned to a machine-time pair (i, t) — without loosing the performance guarantee of 2. Ties are simply broken (as before) by preferring jobs with smaller indices, or even randomly. We mainly chose this presentation for the sake of giving a different, perhaps more intuitive interpretation in the special case of identical parallel machine scheduling; this will become apparent soon.

3 Identical Parallel Machine Scheduling with Release Dates

We now consider the special case of m *identical* parallel machines. Each job must be nonpreemptively processed on one of these machines, and may be assigned to any of them. The processing requirement and the release date of job j no longer depend on the machine job j is processed on and are thus denoted by p_j resp. r_j. Already the problem $P2 \mid \mid \sum w_j C_j$ is NP-hard, see [BCS74,LRKB77]. We consider $P \mid r_j \mid \sum w_j C_j$. There are several good reasons to explicitly investigate this special case:

- o There is a purely combinatorial algorithm to solve the LP relaxation. This leads to a randomized approximation algorithm with running time $O(n \log n)$.
- o The previous use of randomness obtains another interpretation in terms of scheduling by α–points (and vice versa).
- o The same algorithm also is a 2–approximation algorithm for the preemptive variant $P \mid r_j, pmtn \mid \sum w_j C_j$.
- o The algorithm can easily be turned into a randomized on-line algorithm, with no difference in performance guarantee.

We give an approximation algorithm that converts a preemptive schedule on a virtual single machine into a nonpreemptive one on m identical parallel machines. The single machine is assumed to be m times as fast as each of the original m machines, i. e., the (virtual) processing time of any job j on this (virtual) machine is p_j/m (w. l. o. g., we may assume that p_j is a multiple of m). The weight and the release date of job j remain the same. This kind of single machine relaxation has been used before in [CMNS97]. The algorithm is as follows:

Algorithm P
1) Construct a preemptive schedule on the virtual single machine by scheduling at any point in time among the available jobs the one with largest w_j/p_j ratio. Let C_j be the completion time of job j in this preemptive schedule.
2) Independently for each job $j \in J$, draw α_j randomly and uniformly distributed from $(0, 1]$ and assign j randomly (with probability $1/m$) to one of the m machines.
3) Schedule on each machine i the jobs that were assigned to it nonpreemptively as early as possible in nondecreasing order of $C_j(\alpha_j)$.

Notice that in Step 1 whenever a job is released, the job being processed (if any) will be preempted if the released job has a higher w_j/p_j ratio.

In the analysis of Algorithm P, we use a reformulation of (LP_R) for the special case of identical parallel machines. We therefore combine for each job j and each time interval $(t, t+1]$ the variables y_{ijt}, $i = 1, \ldots, m$, to a single variable y_{jt} with the interpretation $y_{jt} = y_{1jt} + \cdots + y_{mjt}$. This leads to the following simplified LP:

$$\text{minimize} \quad \sum_{j \in J} w_j C_j$$

$$\text{subject to} \quad \sum_{t=r_j}^{T} y_{jt} = p_j \quad \text{for all } j \in J, \tag{6}$$

(LP_P)
$$\sum_{j \in J} y_{jt} \leqslant m \quad \text{for } t = 0, \ldots, T, \tag{7}$$

$$C_j = \frac{p_j}{2} + \frac{1}{p_j} \sum_{t=0}^{T} y_{jt}\left(t + \tfrac{1}{2}\right) \quad \text{for all } j \in J, \tag{8}$$

$$y_{jt} = 0 \quad \text{for all } j \in J \text{ and } t = 0, \ldots, r_j - 1, \tag{9}$$

$$y_{jt} \geqslant 0 \quad \text{for all } j \in J \text{ and } t = r_j, \ldots, T. \tag{10}$$

For the special case $m = 1$, this LP was introduced by Dyer and Wolsey [DW90]. One crucial insight for the analysis of Algorithm P is that the preemptive schedule on the fast machine that is constructed in Step 1 of Algorithm P defines an optimum solution to (LP_P). If we set $y_{jt} = m$ whenever job j is being processed on the virtual machine in the period $(t, t+1]$ by the preemptive schedule, it essentially follows from the work of Goemans [Goe96] that y is an optimum solution to (LP_P).

The prior discussion especially implies that Step 1 of Algorithm P simply computes an optimum solution to the LP relaxation, as does Step 1 of Algorithm R. Moreover, the optimum solution of (LP_R) that corresponds to the preemptive schedule in Step 1 of Algorithm P is symmetric for the m machines and therefore for a job j each machine i is chosen with probability $1/m$ in Algorithm R. The symmetry also yields that for each job j the choice of t_j is not correlated with the choice of i. It can easily be seen that the probability distribution of the random variable t_j in Algorithm R exactly equals the probability distribution of $C_j(\alpha_j)$ in Algorithm P. For this, observe that the probability that $C_j(\alpha_j) \in (t, t+1]$ for some t equals the fraction y_{jt}/p_j of job j that is being processed in this time interval. Moreover, since α_j is uniformly distributed in $(0, 1]$ each point in $(t, t+1]$ is equally likely to be obtained for $C_j(\alpha_j)$. Therefore, the random choice of $C_j(\alpha_j)$ in Algorithm P is an alternative way of choosing t_j as it is done in Algorithm R. Consequently, Algorithm P is a reformulation of Algorithm R for the identical parallel machine case. In particular, the expected completion time of each job is bounded from above by twice its LP completion time and Algorithm P is a 2–approximation algorithm.

At this point, it is appropriate to briefly compare this result in the single machine case ($m = 1$) with the result of Goemans [Goe97]. In Step 2, if we only work with one α for all jobs instead of individual and independent α_j's and if we draw α uniformly from $(0, 1]$, then we precisely get Goemans' randomized 2–approximation algorithm

RANDOM$_\alpha$ [Goe97]. In particular, for arbitrary m, Algorithm P has the same running time $O(n \log n)$ as RANDOM$_\alpha$ since it is dominated by the effort to compute the preemptive schedule in Step 1. Finally, whereas Goemans' algorithm has a small sample space — the algorithm can only output $O(n)$ different schedules — Algorithm P can construct exponentially many different schedules.

Let us continue with some additional remarks on Algorithm P. Since (LP_P) is also a relaxation for $P \,|\, r_j, pmtn \,|\, \sum w_j C_j$ it follows that the (nonpreemptive) schedule constructed by Algorithm P is not worse than twice the optimum preemptive schedule. This improves upon a 3–approximation algorithm due to Hall et al. [HSSW96].

The analysis implies as well that (LP_P) is a 2–relaxation for $P \,|\, r_j \,|\, \sum w_j C_j$ and $P \,|\, r_j, pmtn \,|\, \sum w_j C_j$. In fact, this bound is best possible, for (LP_P). For this, consider an instance with m machines and one job of length m, unit weight, and release date 0. The optimum LP completion time is $\frac{m+1}{2}$, whereas the optimum completion time is m.

Furthermore, notice that Algorithm P can easily be turned into an on-line algorithm with competitive ratio 2. In particular, the preemptive schedule in Step 1 can be constructed until time t without any knowledge of jobs that are released afterwards. The random assignment of a job to a machine and the random choice of α_j can be done as soon as the job is released. Moreover, it follows from the analysis in the proof of Theorem 2 that we get the same performance guarantee if job j is not started before time t_j resp. $C_j(\alpha_j)$.

Finally, by a nonuniform choice of the α_j's one can improve the analysis for the on-line and the off-line algorithm to get a performance guarantee better than 2. This improvement, however, depends on m. For the single machine case, Goemans, Queyranne, Schulz, Skutella, and Wang elaborate on this and give a 1.6853–approximation algorithm [GQS+97].

The perhaps most appealing aspect of Algorithm P is that the random assignment of jobs to machines does not depend on job characteristics; any job is put with probability $1/m$ to any of the machines. This technique also proves useful for the problem without (non-trivial) release dates. The very same random machine assignment followed by list scheduling in order of nonincreasing ratios w_j/p_j on every machine is a randomized 3/2–approximation algorithm (see [SS97] for details). Quite interestingly, its derandomized variant precisely coincides with the WSPT-rule analyzed by Kawaguchi and Kyan [KK86]: list the jobs according to nonincreasing ratios w_j/p_j and schedule the next job whenever a machine becomes available.

4 Derandomization

Up to now we have presented randomized algorithms that compute a feasible solution the *expected* value of which can be bounded from above by twice the optimum solution to the scheduling problem under consideration. This means that our algorithms will perform well on average; however, we cannot give a firm guarantee for the performance of any single execution. From a theoretical point of view it is perhaps more desirable to have (deterministic) algorithms that obtain a certain performance in all cases rather than merely with high probability.

If we can bound the expected value of the solution computed by a randomized algorithm we know that there exists at least one particular fixed assignment of values to random variables such that the value of the corresponding solution is at least as good as the expected value and can thus be bounded by the same factor. The issue of derandomization is to find such a good assignment deterministically in reasonable time.

One of the most important techniques in this context is the method of conditional probabilities. This method is implicitly contained in a paper of Erdös and Selfridge [ES73] and has been developed in a more general context by Spencer [Spe87]. The idea is to consider the random decisions in a randomized algorithm one after another and always to choose the most promising alternative. This is done by assuming that all the remaining decisions will be made randomly. Thus, an alternative is said to be most promising if the corresponding conditional expectation for the value of the solution is as small as possible.

The purpose of this section is to derandomize Algorithm R by the method of conditional probabilities. Using Remark 3 we consider the variant of Algorithm R where we set $t_j = t$ if job j is assigned to a machine-time pair (i,t). Thus, we have to construct a deterministic assignment of jobs to machine-time pairs. Unfortunately, the analysis of Algorithm R in the proof of Theorem 2 does not give a precise expression for the expected value of the solution but only an upper bound. However, in order to apply the method of conditional probabilities we have to compute in each step exact conditional expectations. Hence, for the sake of a more accessible derandomization, we modify Algorithm R by replacing Step 3 with

3') Schedule on each machine i the jobs that were assigned to it nonpreemptively in nondecreasing order of t_j, where ties are broken by preferring jobs with smaller indices. At the starting time of job j the amount of idle time on its machine has to be exactly t_j.

Since $r_{ij} \leq t_j$ for each job j that has been assigned to machine i and $t_j \leq t_k$ if job k is scheduled after job j, Step 3' defines a feasible schedule. In the proof of Theorem 2 we have bounded the idle time before the start of job j on its machine from above by t_j. Thus, the analysis still works for the modified Algorithm R which therefore has expected performance guarantee 2. The main advantage of the modification of Step 3 is that we can now give precise expressions for the expectations resp. conditional expectations of completion times.

Let y be the optimum solution we started with in Step 1 of Algorithm R. Using the same arguments as in the proof of Theorem 2 we get the following expected completion time of job j in the schedule constructed by our modified Algorithm R

$$E(C_j) = \sum_{i=1}^{m} \sum_{t=0}^{T} \frac{y_{ijt}}{p_{ij}} \left(p_{ij} + t + \sum_{k \neq j} \sum_{\ell=0}^{t-1} y_{ik\ell} + \sum_{k<j} y_{ikt} \right).$$

Moreover, we are also interested in the conditional expectation of j's completion time if some of the jobs have already been assigned to a machine-time pair. Let $K \subseteq J$ be such a subset of jobs. For each job $k \in K$ the 0/1–variable x_{ikt} for $t \geq r_{ik}$ indicates if k has been assigned to the machine time-pair (i,t) ($x_{ikt} = 1$) or not ($x_{ikt} = 0$). This enables us to give the following expressions for the conditional expectation of j's completion time. If $j \notin K$ we get

$$E_{K,x}(C_j) = \sum_{i=1}^{m} \sum_{t=0}^{T} \frac{y_{ijt}}{p_{ij}} \left(p_{ij} + t + \sum_{k \in K} \sum_{\ell=0}^{t-1} x_{ik\ell} p_{ik} + \sum_{k \in K, k<j} x_{ikt} p_{ik} \right.$$

$$\left. + \sum_{k \in J \setminus (K \cup \{j\})} \sum_{\ell=0}^{t-1} y_{ik\ell} + \sum_{k \in J \setminus K, k<j} y_{ikt} \right) , \tag{11}$$

and, if $j \in K$, we get

$$E_{K,x}(C_j) = p_{ij} + t + \sum_{k \in K} \sum_{\ell=0}^{t-1} x_{ik\ell} p_{ik} + \sum_{k \in K, k<j} x_{ikt} p_{ik}$$

$$+ \sum_{k \in J \setminus K} \sum_{\ell=0}^{t-1} y_{ik\ell} + \sum_{k \in J \setminus K, k<j} y_{ikt} , \tag{12}$$

where (i,t) is the machine-time pair job j has been assigned to, i. e., $x_{ijt} = 1$. The following lemma is the most important part of the derandomization of Algorithm R by the method of conditional probabilities.

Lemma 4. *Let y be the optimum solution we started with in Step 1 of Algorithm R, $K \subseteq J$ and x a fixed assignment of the jobs in K to machine-time pairs. Furthermore let $j \in J \setminus K$. Then, there exists an assignment of j to a machine time pair (i,t) (i. e., $x_{ijt} = 1$) with $r_{ij} \leqslant t$ such that*

$$E_{K \cup \{j\}, x}\left(\sum_{\ell} w_\ell C_\ell\right) \leqslant E_{K,x}\left(\sum_{\ell} w_\ell C_\ell\right) . \tag{13}$$

Proof. The conditional expectation on the right hand side of (13) can be written as a convex combination of conditional expectations $E_{K \cup \{j\}, x}\left(\sum_{\ell} w_\ell C_\ell\right)$ over all possible assignments of job j to machine-time pairs (i,t) with coefficients y_{ijt} / p_{ij}. □

We therefore get a good derandomized version of Algorithm R if we replace Step 2 by

2') Set $K = \emptyset$; x:=0; for all $j \in J$ do
   i) for all possible assignments of job j to machine-time pairs (i,t) (i. e., $x_{ijt} = 1$) compute $E_{K \cup \{j\}, x}\left(\sum_{\ell} w_\ell C_\ell\right)$;
   ii) determine the machine time pair (i,t) that minimizes $E_{K \cup \{j\}, x}\left(\sum_{\ell} w_\ell C_\ell\right)$;
   iii) set $K := K \cup \{j\}$; $x_{ijt} = 1$;

Notice that we have replaced Step 3 of Algorithm R by 3' only to give a more accessible analysis of its derandomization. Since the value of the schedule constructed in Step 3 of Algorithm R is always at least as good as the one constructed in Step 3', the following theorem can be formulated for Algorithm R with the original Step 3.

Theorem 5. *If we replace Step 2 in Algorithm R with 2' we get a deterministic algorithm with performance guarantee 2. Moreover, the running time of this algorithm is polynomially bounded in the number of variables of the LP relaxation.*

Proof. Since $E(\sum_\ell w_\ell C_\ell) \le 2\sum_\ell w_\ell C_\ell^{LP}$ by Theorem 2, an inductive use of Lemma 4 yields the performance guarantee 2 for the derandomized algorithm. The computation of (11) and (12) is polynomially bounded in the number of variables. Therefore, the running time of each of the n iterations in Step 2' is polynomially bounded in this number. □

Since Algorithm P can be seen as a special case of Algorithm R, it can be derandomized in the same manner. Notice that, in contrast to the situation for the randomized algorithms, we can no longer give job-by-job bounds for the derandomized algorithms.

5 Interval-Indexed Formulation

As mentioned earlier, the Algorithm R for scheduling unrelated parallel machines suffers from the exponential number of variables in the corresponding LP relaxation. However, we can overcome this drawback by using new variables which are not associated with exponentially many time intervals of length 1, but rather with a polynomial number of intervals of geometrically increasing size. This idea was earlier introduced by Hall et al. [HSW96].

For a given $\varepsilon > 0$, L is chosen to be the smallest integer such that $(1+\varepsilon)^L \ge T+1$. Consequently, L is polynomially bounded in the input size of the considered scheduling problem. Let $I_0 = [0,1]$ and for $1 \le \ell \le L$ let $I_\ell = ((1+\varepsilon)^{\ell-1}, (1+\varepsilon)^\ell]$. We denote with $|I_\ell|$ the length of the ℓ–th interval, i. e., $|I_\ell| = \varepsilon(1+\varepsilon)^{\ell-1}$ for $1 \le \ell \le L$. To simplify notation we define $(1+\varepsilon)^{\ell-1}$ to be $\frac{1}{2}$ for $\ell = 0$. We introduce variables $y_{ij\ell}$ for $i = 1,\ldots,m$, $j \in J$, and $\ell = 0,\ldots,L$ with the following interpretation: $y_{ij\ell} \cdot |I_\ell|$ is the time job j is processed on machine i within time interval I_ℓ, or, equivalently: $(y_{ij\ell} \cdot |I_\ell|)/p_{ij}$ is the fraction of job j that is being processed on machine i within I_ℓ. Consider the following linear program in these interval-indexed variables:

$$\text{minimize} \quad \sum_{j \in J} w_j C_j$$

$$\text{subject to} \quad \sum_{\substack{i=1 \\ }}^{m} \sum_{\substack{\ell=0 \\ (1+\varepsilon)^\ell > r_{ij}}}^{L} \frac{y_{ij\ell} \cdot |I_\ell|}{p_{ij}} = 1 \quad \text{for all } j \in J, \tag{14}$$

$$(LP_R^\varepsilon) \qquad \sum_{j \in J} y_{ij\ell} \le 1 \quad \text{for } i = 1,\ldots,m \text{ and } \ell = 0,\ldots,L, \tag{15}$$

$$C_j = \sum_{i=1}^{m} \sum_{\ell=0}^{L} \left(\frac{y_{ij\ell} \cdot |I_\ell|}{p_{ij}}(1+\varepsilon)^{\ell-1} + \tfrac{1}{2} \cdot y_{ij\ell} \cdot |I_\ell| \right) \quad \text{for all } j \in J, \tag{16}$$

$$y_{ij\ell} = 0 \quad \text{for } i = 1,\ldots,m, \ j \in J, \ (1+\varepsilon)^\ell \le r_{ij}, \tag{17}$$

$$y_{ij\ell} \ge 0 \quad \text{for } i = 1,\ldots,m, \ j \in J, \ \ell = 0,\ldots,L. \tag{18}$$

Consider a feasible schedule and assign the values to the variables $y_{ij\ell}$ as defined above. This solution is clearly feasible: constraints (14) are satisfied since a job j consumes p_{ij} time units if it is processed on machine i; constraints (15) are satisfied since the total amount of processing on machine i of jobs that are processed within the interval I_ℓ cannot exceed its size. Finally, if job j is continuously being processed between $C_j - p_{hj}$ and C_j on machine h, then the right hand side of equation (16) is a lower bound on the real completion time. Thus, (LP_R^ε) is a relaxation of the scheduling problem $R \mid r_{ij} \mid \sum w_j C_j$.

Since for fixed $\varepsilon > 0$ (LP_R^ε) is of polynomial size, an optimum solution can be computed in polynomial time. The following algorithm is an adaptation of Algorithm R to the new LP:

Algorithm R^ε

1) Compute an optimum solution y to (LP_R^ε).
2) Assign job j to a machine-interval pair (i, I_ℓ) at random with probability $\frac{y_{ij\ell} \cdot |I_\ell|}{p_{ij}}$; draw t_j from the chosen time interval at random with uniform distribution.
3) Schedule on each machine i the jobs that were assigned to it nonpreemptively as early as possible in nondecreasing order of t_j.

Again, all the random assignments need to be performed independently from each other. The following lemma is a reformulation of Lemma 1 for Algorithm R^ε and can be proved analogously.

Lemma 6. *The expected processing time of each job $j \in J$ in the schedule constructed by Algorithm R^ε equals $\sum_{i=1}^m \sum_{\ell=0}^L y_{ij\ell} \cdot |I_\ell|$.*

Theorem 7. *The expected completion time of each job j in the schedule constructed by Algorithm R^ε is at most $2 \cdot (1+\varepsilon) \cdot C_j^{LP}$, where C_j^{LP} is the LP completion time (defined by (16)) of the optimum solution y we started with in Step 1 of Algorithm R^ε.*

Proof. We argue almost exactly as in the proof of Theorem 2, but rather use Lemma 6 instead of Lemma 1. We consider an arbitrary, but fixed job $j \in J$. First, we also consider a fixed assignment of j to machine i and time interval I_ℓ. Again, the conditional expectation of j's starting time equals the expected idle time plus the expected processing time on machine i before j is started.

With similar arguments as in the proof of Theorem 2, we can bound the sum of the idle time plus the processing time by $2 \cdot (1+\varepsilon) \cdot (1+\varepsilon)^{\ell-1}$. This, together with Lemma 6 and (16) yields the theorem. $\qquad \square$

Theorem 7 implies that Algorithm R^ε is a $(2+2\varepsilon)$–approximation algorithm. Furthermore, (LP_R^ε) is a $(2+2\varepsilon)$–relaxation of the problem $R \mid r_{ij} \mid \sum w_j C_j$.

Of course, as suggested by Remark 3, t_j need also not be chosen at random in Step 2 of Algorithm R^ε. Moreover, Algorithm R^ε can be derandomized with the same technique as described in Section 4. In particular, the running time of the derandomized version is again polynomially bounded in the number of variables in the corresponding LP relaxation and therefore, for fixed $\varepsilon > 0$, polynomial in the input size.

Acknowledgements: The authors are grateful to Chandra S. Chekuri, Michel X. Goemans, and David B. Shmoys for helpful comments.

References

[AKP92]   B. Awerbuch, S. Kutten, and D. Peleg. Competitive distributed job scheduling. In *Proceedings of the 24th Annual ACM Symposium on the Theory of Computing*, pages 571 – 581, 1992.

[BCS74]   J. L. Bruno, E. G. Coffman Jr., and R. Sethi. Scheduling independent tasks to reduce mean finishing time. *Communications of the Association for Computing Machinery*, 17:382 – 387, 1974.

[CJM+96]   C. S. Chekuri, R. Johnson, R. Motwani, B. K. Natarajan, B. R. Rau, and M. Schlansker. Profile–driven instruction level parallel scheduling with applications to super blocks. December 1996. Proceedings of the 29th Annual International Symposium on Microarchitecture (MICRO–29), Paris, France.

[CM96]   S. Chakrabarti and S. Muthukrishnan. Resource scheduling for parallel database and scientific applications. June 1996. Proceedings of the 8th ACM Symposium on Parallel Algorithms and Architectures.

[CMNS97]   C. S. Chekuri, R. Motwani, B. Natarajan, and C. Stein. Approximation techniques for average completion time scheduling. In *Proceedings of the 8th ACM–SIAM Symposium on Discrete Algorithms*, pages 609 – 618, 1997.

[CPS+96]   S. Chakrabarti, C. A. Phillips, A. S. Schulz, D. B. Shmoys, C. Stein, and J. Wein. Improved scheduling algorithms for minsum criteria. In F. Meyer auf der Heide and B. Monien, editors, *Automata, Languages and Programming*, number 1099 in Lecture Notes in Computer Science, pages 646 – 657. Springer, Berlin, 1996. Proceedings of the 23rd International Colloquium (ICALP'96).

[CS97]   F. A. Chudak and D. B. Shmoys. Approximation algorithms for precedence–constrained scheduling problems on parallel machines that run at different speeds. In *Proceedings of the 8th ACM–SIAM Symposium on Discrete Algorithms*, pages 581 – 590, 1997.

[DLLX90]   X. Deng, H. Liu, J. Long, and B. Xiao. Deterministic load balancing in computer networks. In *Proceedings of the 2nd Annual IEEE Symposium on Parallel and Distributed Processing*, pages 50 – 57, 1990.

[DW90]   M. E. Dyer and L. A. Wolsey. Formulating the single machine sequencing problem with release dates as a mixed integer program. *Discrete Applied Mathematics*, 26:255 – 270, 1990.

[ES73]   P. Erdös and J. L. Selfridge. On a combinatorial game. *Journal of Combinatorial Theory A*, 14:298 – 301, 1973.

[GLLRK79]   R. L. Graham, E. L. Lawler, J. K. Lenstra, and A. H. G. Rinnooy Kan. Optimization and approximation in deterministic sequencing and scheduling: A survey. *Annals of Discrete Mathematics*, 5:287 – 326, 1979.

[Goe96]   M. X. Goemans. A supermodular relaxation for scheduling with release dates. In W. H. Cunningham, S. T. McCormick, and M. Queyranne, editors, *Integer Programming and Combinatorial Optimization*, number 1084 in Lecture Notes in Computer Science, pages 288 – 300. Springer, Berlin, 1996. Proceedings of the 5th International IPCO Conference.

[Goe97]   M. X. Goemans. Improved approximation algorithms for scheduling with release dates. In *Proceedings of the 8th ACM–SIAM Symposium on Discrete Algorithms*, pages 591 – 598, 1997.

[GQS+97]   M. X. Goemans, M. Queyranne, A. S. Schulz, M. Skutella, and Y. Wang, 1997. In preparation.

[HSSW96]   L. A. Hall, A. S. Schulz, D. B. Shmoys, and J. Wein. Scheduling to minimize average completion time: Off–line and on–line approximation algorithms. Preprint

516/1996, Department of Mathematics, Technical University of Berlin, Berlin, Germany, 1996. To appear in *Mathematics of Operations Research*.

[HSW96]   L. A. Hall, D. B. Shmoys, and J. Wein. Scheduling to minimize average completion time: Off–line and on–line algorithms. In *Proceedings of the 7th ACM–SIAM Symposium on Discrete Algorithms*, pages 142 – 151, 1996.

[KK86]   T. Kawaguchi and S. Kyan. Worst case bound of an LRF schedule for the mean weighted flow–time problem. *SIAM Journal on Computing*, 15:1119 – 1129, 1986.

[LRKB77]   J. K. Lenstra, A. H. G. Rinnooy Kan, and P. Brucker. Complexity of machine scheduling problems. *Annals of Discrete Mathematics*, 1:343 – 362, 1977.

[MNR96]   R. Motwani, J. Naor, and P. Raghavan. Randomized approximation algorithms in combinatorial optimization. In D. S. Hochbaum, editor, *Approximation Algorithms for NP–Hard Problems*, chapter 11. Thomson, 1996.

[MR95]   R. Motwani and P. Raghavan. *Randomized Algorithms*. Cambridge University Press, 1995.

[MSS96]   R. H. Möhring, M. W. Schäffter, and A. S. Schulz. Scheduling jobs with communication delays: Using infeasible solutions for approximation. In J. Diaz and M. Serna, editors, *Algorithms – ESA'96*, volume 1136 of *Lecture Notes in Computer Science*, pages 76 – 90. Springer, Berlin, 1996. Proceedings of the 4th Annual European Symposium on Algorithms.

[PSW94]   C. Phillips, C. Stein, and J. Wein. Task scheduling in networks. In *Algorithm Theory — SWAT'94*, number 824 in Lecture Notes in Computer Science, pages 290 – 301. Springer, Berlin, 1994. Proceedings of the 4th Scandinavian Workshop on Algorithm Theory.

[PSW95]   C. Phillips, C. Stein, and J. Wein. Scheduling jobs that arrive over time. In *Proceedings of the Fourth Workshop on Algorithms and Data Structures*, number 955 in Lecture Notes in Computer Science, pages 86 – 97. Springer, Berlin, 1995.

[RT87]   P. Raghavan and C. D. Thompson. Randomized rounding: A technique for provably good algorithms and algorithmic proofs. *Combinatorica*, 7:365 – 374, 1987.

[Sch96]   A. S. Schulz. Scheduling to minimize total weighted completion time: Performance guarantees of LP–based heuristics and lower bounds. In W. H. Cunningham, S. T. McCormick, and M. Queyranne, editors, *Integer Programming and Combinatorial Optimization*, number 1084 in Lecture Notes in Computer Science, pages 301 – 315. Springer, Berlin, 1996. Proceedings of the 5th International IPCO Conference.

[Spe87]   J. Spencer. *Ten Lectures on the Probabilistic Method*. Number 52 in CBMS-NSF Reg. Conf. Ser. Appl. Math. SIAM, 1987.

[SS97]   A. S. Schulz and M. Skutella. Random–based scheduling: New approximations and LP lower bounds. Preprint 549/1997, Department of Mathematics, Technical University of Berlin, Berlin, Germany, February 1997. To appear in Springer Lecture Notes in Computer Science, Proceedings of the 1st International Symposium on Randomization and Approximation Techniques in Computer Science (Random'97).

[ST93]   D. B. Shmoys and É. Tardos. An approximation algorithm for the generalized assignment problem. *Mathematical Programming*, 62:461 – 474, 1993.

On Piercing Sets of Axis-Parallel Rectangles and Rings

Michael Segal

Department of Mathematics and Computer Science,
Ben-Gurion University of the Negev, Beer-Sheva 84105, Israel

Abstract. We consider the p-piercing problem for axis-parallel rectangles. We are given a collection of axis-parallel rectangles in the plane, and wish to determine whether there exists a set of p points whose union intersects all the given rectangles. We present efficient algorithms for finding a piercing set (i.e., a set of p points as above) for values of $p = 1, 2, 3, 4, 5$. The result for 4 and 5-piercing improves an existing result of $O(n \log^3 n)$ and $O(n \log^4 n)$ to $O(n \log n)$ time, and is applied to find a better rectilinear 5-center algorithm. We improve the existing algorithm for general (but fixed) p, and we also extend our algorithms to higher dimensional space. We also consider the problem of piercing a set of rectangular rings.

1   Introduction

Let $\mathcal{R}$ be a set of n axis-parallel rectangles in the plane, and let p be a positive integer. $\mathcal{R}$ is called p-pierceable if there exists a set of p piercing points which intersects every member in $\mathcal{R}$. Our problem, thus, is to determine whether $\mathcal{R}$ is p-pierceable, and, if so, to produce a set of p piercing points.

There are several papers in which the p-piercing problem for axis-parallel rectangles was investigated; let us mention only the very recent papers. The 1-piercing problem was easily solved in linear time using the observation that 1-piercing problem for rectangles is equivalent to finding whether the intersection of rectangles empty or not. In Sharir and Welzl [7] 2- and 3-piercing problems in the plane are solved in linear time, while they reach only $O(n \log^3 n)$ bound for the 4-piercing problem and $O(n \log^4 n)$ bound for the 5-piercing problem. Katz and Nielsen [2] present a linear time algorithm for d-dimensional boxes ($d \geq 2$), when $p = 2$. In this paper we present a new technique which allows to obtain simple linear time algorithms for $p = 1, 2, 3$, and obtain an $O(n \log n)$ time solution for $p = 4, 5$, thus improving the previous results of [7]. We improve the existing algorithm of [7] for general (but fixed) p, and we extend our algorithms to higher dimensional space. We also consider the problem of piercing the set of rectangular rings. The boundary of a *rectangular ring* consists of two concentric rectangles, where the inner rectangle is fully contained in the outer one, however, the vertical and horizontal widths of the ring, are not necessarily equal.

This paper is organized as follows. We first demonstrate our technique (Section 2) in the case of $p = 1$. We then describe this method (Section 3) for the case of $p = 2, 3$. In Section 4 an $O(n \log n)$ time algorithm for 4-piercing is given. In Section 5 we present an $O(n \log n)$ time algorithm for the case of $p = 5$ and describe generalizations of the problem. Section 6 deals with piercing sets of rectangular rings. We conclude in Section 7.

2 Rectilinear 1-piercing

We are given a set $\mathcal{R}$ of n axis-parallel rectangles in the plane; The goal is to decide whether their intersection is empty or not. We begin with an observation due to Samet [5].

Let P be the set of 4 dimensional points representing the parameters of $\mathcal{R}$. Let $P_x = \{p_1^x, \ldots, p_n^x\}$ be the projections of the x-intervals of $\mathcal{R}$ into the plane (c_x, d_x), and let $P_y = \{p_1^y, \ldots, p_n^y\}$ be the projections of the y-intervals of $\mathcal{R}$ into the plane (c_y, d_y).

If a shape R is described by k parameters, then this set of parameter values defines a point in a k-dimensional space assigned to the class of shapes. Such a point is termed a *representative point*. Note, that a representative point and the class to which it belongs completely define all of the topological and geometric properties of the corresponding shape.

The class of two-dimensional axis-parallel rectangles in the plane is described by a representative point in four dimensional space. One choice for the parameters is the x and y coordinates of the centroid of the rectangle, denoted by c_x, c_y, together with its horizontal and vertical extents (i.e. the horizontal and vertical distances from the centroid to the relevant sides), denoted by d_x, d_y. In this case a rectangle is represented by the four-tuple (c_x, d_x, c_y, d_y) interpreted as the Cartesian product of a horizontal and a vertical one-dimensional *interval* : (c_x, d_x) and (c_y, d_y), respectively. A query that asks which rectangles contain a given point is easy to implement (see Figure 1).

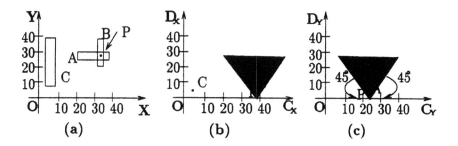

Fig. 1. (a) There are 3 rectangles, and P is a query point. All intervals containing P are in the shaded regions. Intervals appearing in the shaded regions of both (b) and (c) correspond to rectangles that contain P.

A query point P is represented by a four-tuple $(p_x, 0, p_y, 0)$. We transform the rectangles (A,B,C) in Figure 1(a) into the points in two 2-dimensional spaces $((c_x, d_x)$ and $(c_y, d_y))$ (Figure 1(b) and 1(c)). There are two points representing P in these

2-dimensional spaces. $(p_x, 0)$ in (c_x, d_x)-space, and $(p_y, 0)$ in (c_y, d_y)-space. It is easy to see that all the rectangles that contain P must be transformed into two cones in these spaces respectively (the shades cones in Figure 1). These cones have apexes on the $(p_x, 0)$ and $(p_y, 0)$ respectively and are of slope 45° and 135°. In Figure 1, A and B are in both cones and thus P is in these rectangles.

In order to find whether the set $\mathcal{R}$ is 1-pierceable, we find in each 2-dimensional space the rightmost intersection point R_x (R_y) of the 45° lines through the points of P_x (P_y) with axis c_x (c_y), and the leftmost intersection point $(L_x$ and L_y respectively) of the 135° lines with through P_x and P_y respectively, axes c_x and c_y respectively. If the intervals $[R_x, L_x]$ and $[R_y, L_y]$ exist (are not empty) then a point P whose projections are in these intervals is a piercing point.

Thus, we can conclude by the following theorem:

Theorem 1. *We can find whether a set of n axis-parallel rectangles is 1-pierceable in $O(n)$ time, and give a solution, if it exists, in the same runtime.*

3 Rectilinear 2- and 3-piercing

We begin with the 2-piercing problem. Similarly to the previous section, we have to find whether there exist four cones $C_1, C_2 \in (c_x, d_x)$ and $C_3, C_4 \in (c_y, d_y)$ such that:

1. $C_1 \cup C_2$ covers P_x.
2. $C_3 \cup C_4$ covers P_y.
3. Denote by $[C_i]$ the set of all the points of P that corresponded to the points of P_x (or P_y) covered by C_i.

At least one of the following two conditions is true:

(i) $([C_1] \cap [C_3]) \cup ([C_2] \cap [C_4])$ contains all the points of P. This will imply that the apexes of C_1, C_3 define one piercing point and apexes of C_2, C_4 define the other piercing point.

(ii) $([C_1] \cap [C_4]) \cup ([C_2] \cap [C_3])$ contains all the points of P. This will imply that the apexes of C_1, C_4 define one piercing point and apexes of C_2, C_3 define the other piercing point.

We can *constrain* the locations of the cones C_1, C_2, C_3, C_4. They are defined by minimal and maximal points of intersection of the 45° and 135° lines with the horizontal axes in the two planes (c_x, d_x) and (c_y, d_y) respectively. It is easy to see that in order for the rectangles to be 2-pierceable, we put, wlog, the apex of C_1 on R_x, C_2 on L_x, C_3 on R_y and C_4 on L_y. Clearly, if these cones cover all the points then the set $\mathcal{R}$ is 2-pierceable.

In the case of 3-piercing, we have to find six cones $C_i, 1 \leq i \leq 6$, which will define three piercing points with the following properties:

1. $C_1 \cup C_2 \cup C_3$ covers P_x.
2. $C_4 \cup C_5 \cup C_6$ covers P_y.

3. For $i, k, z \in \{1, 2, 3\}$, pairwise disjoint and $j, l, h \in \{4, 5, 6\}$, pairwise disjoint

$$|([C_i] \cap [C_j]) \cup ([C_k] \cap [C_l]) \cup ([C_z] \cap [C_h])| = n$$

for at least one combination of i, k, z (there are at most 6 combinations), where the union is taken is without repetitions.

W.l.o.g., we can find the constrained cones C_1, C_3, C_4, C_6 as in the algorithm for 2-piercing. Namely, the left boundary of C_1 (C_4) is constrained by the leftmost 135° line through the points of P_x (P_y), and the right boundary of C_3 (C_6) is constrained by the rightmost 45° line through the points of P_x (P_y).

To fulfill condition (3) we look at each combination: $[C_1] \cap [C_4]$ or $[C_1] \cap [C_6]$ or $[C_3] \cap [C_4]$ or $[C_3] \cap [C_6]$ and for these four possibilities we check in linear time, whether the rest the points is 2-pierceable. Thus we conclude:

Theorem 2. *We can check in linear time whether set of n axis-parallel rectangles is 2- or 3-pierceable and give a solution, if exists, in the same runtime.*

4 Rectilinear 4-piercing

Now we have to find eight cones $C_i, 1 \le i \le 8$ with the following properties:

1. $C_1 \cup C_2 \cup C_3 \cup C_4$ covers P_x.
2. $C_5 \cup C_6 \cup C_7 \cup C_8$ covers P_y.
3. For some pair of cones $C_i, C_j, i \in \{1, 2, 3, 4\}, j \in \{5, 6, 7, 8\}$ the set of all rectangles without those covered by $[C_i] \cap [C_j]$ is 3-pierceable.

As before, assume wlog that C_1, C_4, C_5, C_8 are constrained, so condition (3) when we choose $i \in \{1, 4\}$ and $j \in \{5, 8\}$ is easily checked in linear time, because we can find the location of C_1, C_4, C_5, C_8 in linear time and then answer the 3-piercing problem in linear time. If $i \in \{2, 3\}$ and $j \in \{6, 7\}$ then there exist $i' \in \{1, 4\}$ and $j' \in \{5, 8\}$ such that if the set of rectangles is 4-pierceable then one piercing point must be determined by the cones $C_{i'}$ and $C_{j'}$. So this case is also computed in $O(n)$ time. The worst (and the more interesting) case is when each constrained cone in one plane corresponds to a non-constrained cone in the other plane. Let us look at one such pair (there is a finite number of such pairs), wlog, C_3 in (c_x, d_x) and C_5 in (c_y, d_y). The analysis for all other such pairs is almost identical.

We sort all the 45° (135°) lines determined by P_x in (c_x, d_x) plane, and do the same to the lines determined by P_y in (c_y, d_y) plane. Clearly, the apex of C_3 is between the apexes of C_1 and C_4. So, we fix the apex of C_3 to coincide with the apex of C_1 and begin to move it rightwards towards C_4. We define an *event* when a point of P_x is inserted or deleted from C_3. Initially, we compute the set of points $A \subseteq P$ covered by $[C_3] \cap [C_5]$ (when the apex of C_3 is determined by the leftmost 135° line through the point of P_x) and apply the 3-piercing algorithm for the rest of the points $S = P - A$, allowing, only this time, C_1 to move freely. If we have a positive answer, we are done; otherwise we continue.

We move C_3 rightwards to the next event and change S accordingly (as in Figure 2). The first next event is when the leftmost point of P_x is deleted from C_3. Then

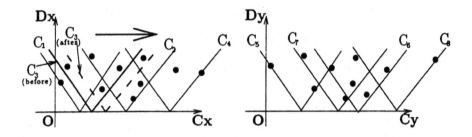

Fig. 2. Moving apex of C_3 from apex of C_1 towards apex of C_4.

we run again the 3-piercing algorithm for S. Here, too, if S is 3-pierceable then we are done. Clearly, from now on the location of the apexes of C_1, C_4 and C_8 will not change during the whole algorithm because these cones are defined by the extreme points of P_x and P_y that will never appear in both C_3 and C_5. Let C_7 be the leftmost cone covering S in (c_y, d_y). The location of C_7 will change since C_7 will move towards C_8 and back to cover points. But once C_7 moves back from C_8 it will never move towards C_8 again. This is because C_5 is constrained and C_7, the second cone from the left, moves back to cover points that got out of A. Since the leftmost point has to be covered in order to have 4-piercing, once C_7 got back to its leftmost position, it will never move to the right again. Thus, the number of changes that we perform on C_7 is $O(n)$. Our goal is to determine the location of the cones C_2 in (c_x, d_x) and C_6 in (c_y, d_y). We will check the possible combinations of pairing the cones to create piercing points. Assuming the cones C_3 and C_5 describe a piercing point, we have the following combinations for the rest of the piercing points:
(a) $(C_1, C_7), (C_2, C_6), (C_4, C_8)$, (b) $(C_1, C_7), (C_2, C_8), (C_4, C_6)$,
(c) $(C_1, C_8), (C_2, C_7), (C_4, C_6)$, (d) $(C_1, C_8), (C_2, C_6), (C_4, C_7)$,
(e) $(C_4, C_7), (C_1, C_6), (C_2, C_8)$, (f) $(C_4, C_8), (C_1, C_6), (C_2, C_7)$.

Observation 3 *The combinations of the cones at each step of the 4-piercing algorithm are* **independent,** *meaning that we check 3-pierceability for each fixed combination of the cones throughout all the steps of the 4-piercing algorithm. If we get a negative answer for a combination, we check the other combinations. If there is a solution it will be found by the algorithm in one of the steps of one combination.*

According to observation 3, because we have finite number of combinations, we can perform the 4-piercing algorithm for each one of the combinations separately. For each of the combinations, the 4-piercing algorithm is slightly different. Denote by $C_{ij} = [C_i] \cap [C_j] \cap S$. Recall that $S = P - A$, A being the points covered by (C_3, C_5).

We present a skeleton of the algorithm and then give the additional technical details.

For every combination of cones the following events happen during the 4-piercing algorithm: (exemplified by C_3 and C_5 as a piercing point and the 3-piercing combinations (a)-(f) as above):

1. Initially the left boundary point q of C_3 is getting out of C_3. If $q \in A$, we re-run the 3-piercing algorithm, otherwise no update is needed, since q did not belong to A, it was, and remains, in $S..$

2. If, when we move the apex of C_3 towards C_4, a point q' is inserted to C_3, we first check if the corresponding point to q' in (c_y, d_y) is covered by C_5. If it is not covered, then we continue moving C_3 to the next event; otherwise we have the following cases:

   2.1 If q' defines the new left boundary of the middle cone C_2 in (c_x, d_x), or q' defines the left boundary of the left cone C_7, or the left boundary of the middle cone C_6 in (c_y, d_y) for the combinations (a)-(d) (similarly, right boundary for the combinations (e)-(f)), then, for the given combination we perform the following updating scheme: we first check if q' defines the left boundary of C_7. If yes then we have to find, by binary search over S, the new left boundary for C_7 and:

      i. For combination (a). Find the new boundaries of the middle cones C_2 and C_6 in both planes and check whether they cover the rest of the points by simply computing and examining the set $S^{(1)} = S - C_{17} - C_{48}$. Note that the cones C_4 and C_8 are both constrained and do not move during the whole algorithm.

      ii. For combination (b) (similarly (e)). By computing and examining the set C_{17} (C_{14}) we found the new left boundaries of C_2 and C_6. Only thing we have to do now is to check whether the pairs (C_2, C_8) and (C_4, C_6) $((C_1, C_6))$ cover the set of all points of P not covered by (C_3, C_5) and (C_1, C_7) $((C_4, C_7))$. This could be done by computing the sets $S^{(2)} = C_{48} - C_{17}$ ($C_{18} - C_{47}$), C_4' and C_8' and updating T_1 and T_2 as described below. Using the updating scheme below, we check whether C_2 covers C_8', C_6 covers C_4', and together C_2 and C_6 cover $S^{(2)}$.

      iii. For combination (c) (similarly (f)). By computing and examining the set $S^{(3)} = S - C_{18} - C_{27}$ we find the leftmost and rightmost points of this set in both planes that should be covered in both planes by C_6 and C_4. We find the new boundaries of C_2 and C_6 and check whether C_6 and C_4 cover the leftmost and rightmost points in both plane that we just found. Note that the number of updates on C_2 in the whole algorithm is $O(n)$. This is because the left boundary of C_2 is defined by the leftmost point (of S) in (c_x, d_x) not covered by C_{18} and thus C_2 moves towards and back from C_4, but when it moves back it will never move rightwards again.

      iv. For combination (d). By computing and examining the set $S^{(4)} = S - C_{18} - C_{47}$ we find the leftmost and rightmost points of this set in both planes that should be covered in both planes by C_6 and C_2. We find the new boundaries of C_2 and C_6 and check whether C_6 and C_2 cover the leftmost and rightmost points in both plane that we just found.

   2.2 If q' does not define a left boundary of a cones as above, then for each combination we perform an identical updating scheme as in 2.1 but without computing a new left boundary of the middle cones C_2 and C_6.

   After the updates we check whether there is a 3-piercing combination for $S..$

3. If, when we move apex of C_3, a point q'' is deleted from C_3, then

3.1 If $q'' \notin C_5$ we proceed to the next event.

3.2 If in the past q'' was the left boundary of the middle cone C_2 in (c_x, d_x), or q' was the left boundary of the left cone C_7, or the left boundary of the middle cone C_6 in (c_y, d_y) for the combinations (a)-(d) (similarly, right boundary for the combinations (e)-(f)), then, for the given combination we perform the following updating scheme: If q'' defines a new left boundary C_7, then we compute a new location of C_7 and:

    i. For combination (a), find the new boundaries of the middle cones C_2 and C_6 in both planes and compute the rightmost and leftmost points of the set $S^{(1)}$ in both planes.

    ii. For combination (b), by examining the set C_{17} find the new left boundaries of C_2 and C_6, compute the sets $S^{(2)}, C_4'$ and C_8' and update T_1 and T_2.

    iii. For combination (c), find the new boundaries of C_2 and C_6. By examining the set $S^{(3)}$ find the leftmost and rightmost points of this set in both planes.

    iv. For combination (d), find the new boundaries of C_2 and C_6. By examining the set $S^{(4)}$ we find the leftmost and rightmost points of this set in both planes.

3.3 If q' does not define a left boundary of a cones as above, then for each combination we perform an identical updating scheme as in 3.2 but without computing a new left boundary of the middle cones C_2 and C_6. Notice that in this case (when q'' is deleted from C_3) the 4-piercing of P is not possible, because it wasn't possible in previous step of the algorithm.

Now we describe the technical details of the 4-piercing algorithm given above. For combination (a) we compute C_{48} at the beginning of the 4-piercing algorithm. The cones C_4 and C_8 are both constrained and do not move during the whole algorithm. For each step of the 4-piercing algorithm we maintain the set $S^{(1)} = S - C_{17} - C_{48}$. To determine whether C_2 and C_6 can cover $S^{(1)}$ we are only interested in the maxima and minima of the 45° and 135° lines through the points of this set ($S^{(1)}$) in both planes respectively. Note that the total number of updates on C_{48} and on C_7 is at most $O(n)$, thus if we maintain the points of the dynamically changing set $S^{(1)}$ sorted according to the 45° and 135° lines we can update $S^{(1)}$ and find the maxima and minima in both planes by a simple binary search. Consequently, in $O(1)$ time we check whether there exist two cones C_2 and C_6 with boundaries on these maximal and minimal values that cover these points.

Combinations (b) and (e) are similar in the sense that C_7 (that has $O(n)$ updates) is paired with a constrained cone, C_1 in (b) and C_4 in (e), and the non constrained cones C_2 and C_6 are each paired with a constrained cone. For combination (b) (similarly (e)) we compute the set C_{48} (C_{18}) at the beginning of the 4-piercing algorithm. In each step of the 4-piercing algorithm we compute the set $S^{(2)} = C_{48} - C_{17}(C_{18} - C_{47})$. Observe the set of all points of P not pierced by (C_3, C_5) and (C_1, C_7) $((C_4, C_7))$. They will have to be pierced by (C_2, C_8) and (C_4, C_6) $((C_1, C_6))$. Now the points in $S^{(2)}$ should be covered by either C_2 or C_6, whereas the points of $C_4' = [C_4] - S^{(2)}$ $(C_1' = [C_1] - S^{(2)})$ must be covered by C_6 and the points of $C_8' = [C_8] - S^{(2)}$ must be covered by C_2. C_1 and C_7 (C_4 and C_7) determine the left (right) boundary of C_2 and C_6, which are found by a binary search over the points of $S - C_{17}$ $(S - C_{47})$. As for combination (a) the number of updates on C_{48} (C_{18}), and C_7 is at most $O(n)$. The sets C'_4 (C'_1) and C'_8 are maintained sorted according to the lines throughout the whole algorithm. To check how $S^{(2)}$ is pierced, we maintain balanced binary trees T_1, T_2. The leaves of $T_1(T_2)$ contain the set $S^{(2)}$ sorted according to the 45° (135°) lines in the plane (c_x, d_x). Let T be T_1 or T_2. Initially, the leaves of T contain the sorted points of C_{48} (C_{18}) in the plane (c_x, d_x). After we compute C_{17} (C_{47}) for the first time we empty the leaves that contain the points that belong to $C_{17}(C_{47})$. Now T contains the sorted lines through the points of $S^{(2)}$. Let p be a point of $S^{(2)}$. A leaf corresponding to p contains the x value of the point of intersection of the 45° (135°) line through p with the c_x axis in (c_x, d_x). It will also contain the y value of the point of intersection of a 45° (135°) line through p with the c_y axis in (c_y, d_y). An inner node $v \in T$ will contain the maximum of the y values corresponding to 135° lines of the leaves of the subtree rooted at v, and the minimum of the 45° lines. During the algorithm we perform a sequence of updates, namely insertions and deletions, on the tree T. When a point q is add to $S^{(2)}$, then we insert it into T in a sorted x-order and update the minimum and maximum y values on the nodes of path from the leaf q to the root of T. If a point q is deleted from T, then we find the leaf of q, delete it and update the y values of the nodes on path from the leaf to the root of T. Each update of T takes $O(\log n)$ time. We can check, using the tree T, whether C_2 together with C_6 cover all the points in $S^{(2)}$. For combination (c) (similarly, (f)) at the beginning of the 4-piercing algorithm we compute C_{18}. The cones C_1 and C_8 are constrained and do not move during the whole algorithm. At the next step of the 4-piercing algorithm we work with the set $S^{(3)} = S - C_{18} - C_{27}$ and find the leftmost and rightmost points in this set that should be covered in both planes by C_6 and C_4 respectively. We maintain $S^{(3)}$ by incremental updates according to the motion of C_3.

For combination (d) we perform a scheme almost identical to that of (c), but with the difference that at each step of the 4-piercing algorithm we work with the set $S^{(4)} = S - C_{18} - C_{47}$ and find the leftmost and rightmost points that should be covered in both planes by C_6 and C_2. Again, we update $S^{(4)}$ at each motion of C_3 in logarithmic time.

From the analysis of this algorithm it follows that we have $O(n)$ updates in the whole algorithm and we can perform each update in logarithmic time. Thus,

Theorem 4. *We can determine whether set of n axis-parallel rectangles is 4-pierceable or not in $O(n \log n)$ time, and give the solution (if it exists) in the same runtime.*

5 Rectilinear 5-piercing

Now we have to find ten cones $C_i, 1 \leq i \leq 10$ with the following properties:

1. $C_1 \cup C_2 \cup C_3 \cup C_4 \cup C_5$ covers P_x.
2. $C_6 \cup C_7 \cup C_8 \cup C_9 \cup C_{10}$ covers P_y.
3. For some pair of cones $C_i, C_j, i \in \{1, 2, 3, 4, 5\}, j \in \{6, 7, 8, 9, 10\}$ the set of all rectangles without those covered by $[C_i] \cap [C_j]$ is 4-pierceable.

Due to the duality relation between our analysis and that in [7] (see Section 7) we follow the case analysis in [7]. Assume, wlog, that C_1, C_5, C_6, C_{10} are constrained and the order of the cones is from left to right. We may also assume that one of the following situations occurs:

(a) There is one pair of constrained cones C_i, C_j, $i \in \{1,5\}$ and $j \in \{6,10\}$. We try all of these possibilities, find the set of rectangles not covered by the given pair of cones, and test whether this set is 4-pierceable, using the preceding algorithm. This takes $O(n \log n)$ time.

(b) Every constrained cone is paired with a non-constrained cone. Since there are four constrained cones there are two pairs with the same constrained cones. We proceed as follows. First, we guess a *unique* constrained cone, say C_1, which is paired with a non-constrained, say C_7. Then we proceed as in 4-piercing algorithm, i.e. slide C_7 from left to right, starting at the apex of C_6 and stopping when we reach the apex of C_{10}. In each move, we check whether the set of the rest of the rectangles is 4-pierceable using the following observation by Sharir and Welzl [7]. They observe that the 4-piercing problem (that is solved at each move of C_7) has always a pair of two constrained cones in its solution. In our case they are either C_2 and C_6, or C_2 and C_{10} (C_2 becomes constrained after computing $S - C_{17}$). We process each of these cases separately. Assume, wlog, we process C_2 and C_6. Then at each move of C_7 we update C_{26} and check whether the rest of rectangles is 3-pierceable as in the update step in the 4-piercing algorithm. Omitting the easy missing details, we obtain a procedure that runs in $O(n \log n)$ time.

(c) There is some pair of the cones, where an unconstrained cone is paired with another unconstrained cone. Assume, wlog, the cones are C_4 and C_8. We also assume wlog that we have paired C_6 and C_3, C_{10} and C_2, C_1 and C_7, C_5 and C_9 (a constrained cone with an unconstrained cone). Then, as was observed in [7], either at least one of the c_x-coordinates of the apexes of C_2 and C_3 is smaller then the the c_x-coordinate of the apex of C_4 or at least one of the c_x-coordinates of the apexes of C_2 and C_3 is larger than the c_x-coordinate of the apex of C_4. Suppose one of them is smaller than C_4. Then we slide C_7 (that is paired with C_1) from left to right, starting at the apex of C_6 and stopping when we reach the apex of C_{10}. In each move, we check whether the set of the rest of rectangles is 4-pierceable. As was claimed in [7] again in each move of C_7 it has to be that C_2 is paired with either C_6 or C_{10}. Thus we have a situation identical to case (b). This can be computed as in case (b) above, implying that the computing for case (c) can also be done in $O(n \log n)$ time. Hence we obtain:

Theorem 5. *We can determine whether set of n axis-parallel rectangles is 5-pierceable or not in $O(n \log n)$ time, and give the solution (if it exists) in the same runtime.*

The result for 5-piercing improves an existing result of [7] that runs in $O(n \log^4 n)$ time, and can be applied to find a better rectilinear 5-center algorithm in time $O(n \log^2 n)$.

Higher dimensions and $p > 4$.

Our technique immediately implies a linear time algorithm for 2-pierceability of a set of axis-parallel rectangles for arbitrary (fixed) dimension $d, d \geq 2$ (there are only constrained cones) and an $O(n \log n)$ time algorithm for 3-pierceability of a set of axis-parallel rectangles for dimension $d, 3 \leq d \leq 5$ (the same result was obtained by [1] independently). In the later problem there is always a combination where $d-1$ cones are constrained and one (at most) is a non-constrained cone. At each step of the algorithm there is a finite number of the d-coupling combinations of the cones. An algorithm similar to the 4-piercing algorithm in the plane is used to solve the piercing problem. We also obtain an improved formula for general (but fixed) $p \geq 6$ (for the plane) using our approach. The general observation is that a constrained cone is always paired with a constrained or unconstrained cone. Thus for solving $p+1$-piercing problem we have to consider two cases. In the first case there is two constrained cones paired together, we can determine the rest of the (uncovered) rectangles in linear time and apply algorithm for p-piercing for the rest of rectangles. In the second case, a constrained cone is paired with a non-constrained. We move the apex of the non-constrained cone between the apexes of the constrained cones in its plane. Thus we have $O(n)$ steps (when a point is either inserted or deleted from the non-constrained cone). In each step we run the $p-4$-piercing algorithm for the rest of the points. Thus we improve the algorithm of [7] for $p \geq 6$ in the plane from $O(n^{p-4} \log^5 n)$ to $O(n^{p-4} \log n)$.

6 Piercing Sets of Rectangular Rings

A *rectangular ring* is a ring, defined by two boundaries, the outer boundary and the inner boundary. Both boundaries are axis parallel concentric rectangles, where the inner rectangle is fully contained in the outer rectangle. We do not require that the horizontal width of the ring be identical to the vertical width. We pose the piercing question on a set of rectangular rings.

1-piercing

The 1-piercing problem is equivalent to the question: Given a set R of n axis-parallel rectangular rings, is their intersection empty or not. This problem can be easily solved by decomposing the rings into $4n$ rectangles and applying the segment tree [3] to compute *the depth* of the set of rectangles in $O(n \log n)$ time. Our method, that is easily extendable to higher dimensions, also uses the *Klee measure* (the depth and the union of a set of rectangles are examples for the Klee measure, see [3, 4]).

First we use the algorithm from Section 2 to find whether the set R_r of the external rectangles defining the given rings is 1-pierceable. If R_r is not 1-pierceable then neither is the set R. Otherwise, we find the region Q (also a rectangle) where all the rectangles from R_r intersect. In our notations Q is determined as follows. Let P be the set of 4 dimensional points representing the parameters of R_r. Let $P_x = \{p_1^x, \ldots, p_n^x\}$ be the projections of the x-intervals of R_r into the plane (c_x, d_x), and let $P_y = \{p_1^y, \ldots, p_n^y\}$ be the projections of the y-intervals of R_r into the plane (c_y, d_y). We find in each plane (c_x, d_x) and (c_y, d_y) the rightmost intersection point

R_x (R_y) of the 45° lines through the points of P_x (P_y) with axis c_x (c_y), and the leftmost intersection point (L_x and L_y respectively) of the 135° lines through P_x and P_y respectively, with axes c_x and c_y respectively. The intervals $[R_x, L_x]$ and $[R_y, L_y]$ (if they exist, namely, $R_x < L_x$ and $R_y < L_y$) define Q. Now we check whether the union of the rectangles, defined by the internal boundaries of the rectangles in R, covers Q. If it does not cover Q, then R is 1-pierceable; otherwise it is not 1-pierceable.

In higher dimensional space Q is easily found as above in time $O(dn \log n)$. In order to find the union of the internal rectangles we use the algorithm of Overmars and Yap [4] who solve the Klee measure problem in higher dimensions in time $O(n^{\lfloor \frac{d}{2} \rfloor} \log n)$. Thus, this is the runtime of our 1-piercing algorithm for rings for $d \geq 2$.

2 and 3-piercing

For two and three pierceability problems a non-trivial but quadratic algorithm is as follows. We first check whether the set of the external rectangles is 2-piercable. If it is, then we continue to work with the combination of the cones that define the 2-piercing points (there might be more than one combination of cones, and we consider all of them).

Assume that the combination is (C_1, C_3) and (C_2, C_4). First we check whether there exist points of P covered by both (C_1, C_3) and (C_2, C_4). We call such points of P *joint points*. The case with no joint points is easy. We only need solve two separate 1-piercing subproblems for each pair of cones, find the rectangular regions Q' and Q'' (as was Q region in the 1-piercing algorithm) and check whether the internal rectangles corresponding to each subproblem cover Q' and Q'' respectively.

If there are joint points of P in (C_1, C_3) and (C_2, C_4) then we proceed as follows. Initially we assign the joint points to (C_1, C_3), and the points not covered by (C_1, C_3) we assign to (C_2, C_4). Similarly to the described above, we compute the intervals $[R_x, L_x]$ and $[R_y, L_y]$ for each pair of cones according to the points they cover, the joint points belonging only to (C_1, C_3). We denote these intervals by $I_{C_1}, I_{C_2}, I_{C_3}, I_{C_4}$. The intervals I_{C_1}, I_{C_3} define the rectangular region Q' where the first piercing point should be found, and the intervals I_{C_2}, I_{C_4} define the region Q'' for the second piercing point. We now check whether the internal rectangles corresponding to the points assigned to (C_1, C_3) and (C_2, C_4), respectively, cover Q' and Q'' respectively. If both Q' and Q'' are not wholly covered by the internal rectangles then we are done.

In the next steps we slide C_1 from right to left, stopping whenever a joint point q leaves C_1. The joint point q is deleted from C_1 (thus, it is now assigned to (C_2, C_4). We compute the intervals $I_{C_1}, I_{C_2}, I_{C_3}, I_{C_4}$, and check whether the corresponding internal rectangles cover Q' and Q''. We stop sliding C_1 either when it gets to the last of the joint points or when it gets to a point in (C_1, C_3) which is not a joint point. We call the latter event a *stop event*.

After we finish sliding C_1 we return it to its starting position. We now perform similar steps with C_3 moving from left to right and stopping at each joint point till the end of joint points or till a point covered by just (C_2, C_4) is met by the sliding cone C_3. There are $O(n)$ sliding steps in this algorithm. In each step we check whether the internal rectangles cover the regions Q' and Q''. This can be done in $O(n \log n)$ time using standart sweep-line algorithm. So we conclude that our algorithm runs in time $O(n^2 \log n)$.

For the 3-piercing problem we can apply a similar technique and obtain an $O(n^3 \log n)$ algorithm for solving the 3-pierceability of rings. We can improve these running time using the following observation.

Observation 6 *Throughout the motion of C_1 in the 2-piercing algorithm above, the rectangular region Q' does not shrink in any dimension, while the region Q'' does not expand in any dimension.*

More precisely, at each stop of C_1 the intervals I_{C_3} and I_{C_1} do not shrink, and if they change they can only grow, while the intervals I_{C_2} and I_{C_4} can only shrink. This is because the joint point that was deleted from C_1 has to be covered by C_2 (and also C_4) thus decreasing the freedom of movement of the apex of C_2 (C_4). Moreover, the number of internal rectangles of the rings that can cover Q' (Q'') does not increase (decrease) in each step of the algorithm.

This observation provides a monotonicity to the problem and allows us to improve the running time of our algorithm. Instead of sliding C_1 (C_3) from right to left stopping at each joint point, we move $C_1(C_3)$ in the range of joint points (between the rightmost joint point and the rightmost stop event) by a binary search, checking at each such move whether both Q' and Q'' are covered or not by the union of the corresponding inner rectangles. When we find a move where they are not covered then we done. If Q' is covered C_1 jumps to the left. If Q' is not covered but Q'' is covered, then C_1 jumps to the right. Thus we get a factor $O(\log n)$ instead of $O(n)$ for the 2-piercing algorithm, and the problem is solved in time $O(n \log^2 n)$.

The 3-piercing algorithm works just as we sketched above. Here, too, we can do the motions of C_1 and C_4 in binary skips, and then apply the just described 2-piercing algorithm, getting us to a $O(n \log^3 n)$-time algorithm.

We can solve the 2-piercing problem in higher dimensional space. We get an $O(n^{\lfloor \frac{d}{2} \rfloor} \log^2 n)$, $d \geq 3$ runtime algorithm for determining 2-pierceability of the set of input rings using as a subroutine algorithm in [4].

7 Conclusions

In this paper we present an efficient technique for solving the p-piercing problem for a set of axis-parallel rectangles. There is some duality between the analysis of [7] and that of our paper. A constrained cone in our algorithms corresponds to an edge on the boundary of the search region in [7], and two paired constrained cones in our algorithms correspond to a corner in the search region of [7]. We are looking into applying a similar technique for sets of triangles, rhombi, etc. The most intriguing question is whether we can improve the runtime of the presented algorithm for p-piercing problems where $p > 5$. We hope that our approach can help in to obtaining a better solution to these problems.

Acknowledgements: I thank Matya Katz, Klara Kedem and Yuri Rabinovich for useful discussions.

References

1. E. Assa, M. Katz, private communication.
2. M. Katz, F. Nielsen, "On piercing sets of objects", In *Proc. 12th ACM Symp. on Computational Geometry*, 1996.
3. K. Mehlhorn, *Data Structures and Algorithms 3: Multi-Dimensional Searching and Computational Geometry*, Springer-Verlag, 1984.
4. M. Overmars, C. Yap, "New upper bounds in Klee's measure problem", In *Proc. 29 Annual IEEE Symp. on the Found. of Comput. Sci.*, 1988.
5. H. Samet, "The design and analysis of spatial data structures", *Addison-Wesley*, 1990
6. M.Segal, K.Kedem, "Enclosing k points in the smallest axis parallel rectangle", *8th Canadian Conference on Computational Geometry*, 1996.
7. M. Sharir, E. Welzl, "Rectilinear and polygonal p-piercing and p-center problems", In *Proc. 12th ACM Symp. on Computational Geometry*, 1996.

Seven Problems: So Different yet Close

Sergey Sevastianov[1] *

Institute of Mathematics, Universitetskii pr. 4, 630090, Novosibirsk-90, Russia

Abstract. We show that seven discrete optimization problems from different fields of discrete mathematics (such as linear algebra, combinatorics, geometry, and functional analysis) that at first sight seem to be quite different prove to be in fact rather close to each other. This closeness enables us, given an algorithm for one problem, to construct an optimization or approximation algorithm for solving the other problems in the list. For each problem, an extremum function is defined which characterizes the performance of the optimal solution of the problem in the worst case. Relations between these extremum functions are derived.

1 Introduction

We consider seven discrete optimization problems from different fields of discrete mathematics, such as: linear algebra ("integer-making" problem), combinatorics (to find a partition of a finite set which divides each subset from a given family into parts of nearest possible sizes), geometry (to find a vertex of a given parallelepiped nearest to a given point in it; to find a vertex of n-dimensional cube which is most equidistant from each vertex of a given set of vertices), functional analysis (to find an order of summation of vectors from a given family with sum zero such that all partial sums of the series are in a ball of a minimum possible radius), and finally, a calendar-planning problem, very important for applications (to find the most even partition of an annual plan of an enterprise into 4 quarter plans, 12 monthly plans, etc.).

It will be shown that all these problems that on the face of it seem to be quite different, prove to be "close relatives" having tight connections. In some cases these connections are sufficiently evident, in other cases the connection is not trivial. In most cases this closeness between problems enables us, given an algorithm for one problem, to construct an algorithm for (optimal or approximate) solution of another problem in the list.

For each problem in the list, an extremum function is defined which characterizes the performance of the optimal solution of the problem in the worst case. Relations between these extremum functions are derived.

Now to set our problems, let us introduce some notation. We set

$IN = \{1, 2, \ldots\}$; $\quad N_n = \{1, \ldots, n\}$; $\quad B^n = \{0, 1\}^n$ is the n-dimensional cube;

* seva@math.nsc.ru. Supported by the Russian Foundation for Fundamental Research (Grant 96-01-01591).

$\mathcal{H}_{n,r} = \{H \mid H : N_n \to N_r\}$ is the set of partitions of an n-element set into r subsets;

$\mathcal{F}_{n,m} = \{F \mid F : N_m \to 2^{N_n}\}$ — each element $F \in \mathcal{F}_{n,m}$ is a family of m subsets of an n-element set;

$B_{s,m} = \{x \in I\!R^m \mid \|x\|_s \leq 1\}$ is the unit ball of a norm s in $I\!R^m$;

l_p is the norm in $I\!R^m$, defined for each vector $x = (x_1, \ldots, x_m) \in I\!R^m$ by

$$\|x\|_{l_p} = \left(\sum_{i=1}^{m} |x_i|^p \right)^{1/p} ;$$

$$\|x\|_{l_\infty} = \max_{i \in N_m} |x_i| .$$

2 Problems and Functions

2.1. First consider the following practical problem.

Problem P1 (volume-calendar planning problem [9]). *The annual plan of an enterprise consists of n items. Each item is characterized by an m-dimensional vector* $a_i = (a_1^i, \ldots, a_m^i)$, $i = 1, \ldots, n$. *The objective is to divide the annual plan into r parts as equal as possible. The following two functions were used in [9] and [12] for measuring the extent of uniformity of a division:*

$$A) \quad L_s^1(A, H) \doteq \max_{p,j} \left\| \sum_{i \in H_p} a_i - \sum_{i \in H_j} a_i \right\|_s \to \min_H ,$$

$$B) \quad L_s^2(A, H) \doteq \max_{p \in N_r} \left\| \sum_{i \in H_p} a_i - a/r \right\|_s \to \min_H ,$$

where $a = \sum_{i \in N_n} a_i$; $A = \{a_i \mid i \in N_n\}$; $H : N_n \to N_r$; $H_p = H^{-1}(p)$, $p \in N_r$; s *is a norm in* $I\!R^m$.

For each norm s, we define functions η_s^1 and η_s^2:

$$\eta_s^i(r, m) \doteq \sup_A \min_{H \in \mathcal{H}_{n,r}} L_s^i(A, H) / \max_{i \in N_n} \|a_i\|_s , \quad i = 1, 2 , \tag{1}$$

that return the worst possible values of the optimums of functionals L^i for given parameters r, m.

As it follows from results in p. 3.4, the supremum in the right hand side of (1) over all finite families $A \subset B_{s,m}$ exists for every i, r, and m, which implies that functions η^1 and η^2 are well defined. It is clear that

$$\eta_s^1(r, m) \leq 2\eta_s^2(r, m) . \tag{2}$$

Furthermore, for p that provides the maximum value to the expression in the right hand side of the functional B, we have

$$L_s^2(A, H) = \left\| \sum_{i \in H_p} a_i - a/r \right\|_s = \left\| \frac{1}{r} \left(\sum_{i \in H_p} a_i - \sum_{i \in H_1} a_i \right) + \cdots + \frac{1}{r} \left(\sum_{i \in H_p} a_i \right. \right.$$

$$\left. \left. - \sum_{i \in H_r} a_i \right) \right\|_s \leq \frac{1}{r} \sum_{j \in N_r} \left\| \sum_{i \in H_p} a_i - \sum_{i \in H_j} a_i \right\|_s \leq \frac{r-1}{r} L_s^1(A, H) \ .$$

This implies $\eta_s^2(r, m) \leq \frac{r-1}{r} \eta_s^1(r, m)$, or combining with (2),

$$\frac{r}{r-1} \eta_s^2(r, m) \leq \eta_s^1(r, m) \leq 2\eta_s^2(r, m), \quad (r \geq 2) \ . \tag{3}$$

For $r = 2$ we have the equality

$$\eta_s^1(2, m) = 2\eta_s^2(2, m) \ .$$

In the case of $\{0, 1\}$-matrix (a_j^i) the corresponding functions are denoted by η_s^{1C} and η_s^{2C}.

2.2. The following problem formulated by A. Dvoretzky in 1963 belongs to the list of unsolved problems at the 7th Symposium of AMS in Pure Mathematics [6].

Problem P2 . *To find the function* $\nu_s(m) \doteq \sup \min \left\| \sum_{i=1}^p \varepsilon_i x_i \right\|_s$, *where* sup *is taken over all finite families of vectors* $\{x_i\} \subset IR^m$, $\|x_i\|_s \leq 1$, *and* min *is taken over all p-tuples of numbers* $\{\varepsilon_i = \pm 1\}$.

2.3. Following [8], let us formulate the following "balancing" problem.

Problem P3(2) . *Given a finite set* N_n *and a family* $F = \{F_1, \ldots, F_m\}$ *of its subsets, we wish to find a partition* $N_n = X \cup \bar{X}$, $X \cap \bar{X} = \emptyset$, *of set* N_n *into two subsets such that the function*

$$q_{n,m}(F, X) \doteq \max_{i \in N_m} \left| |F_i \cap X| - |F_i \cap \bar{X}| \right|$$

takes its minimum value over all $X \subset N_n$.

In other words, we wish to find a cut $(X, \bar{X})$ which cuts every subset F_i from a given family into most equal parts. Olson and Spenser posed the question of finding the function

$$f(m) \doteq \sup_n f(n, m) \ , \tag{4}$$

where $f(n, m) \doteq \max_{F \in \mathcal{F}_{n,m}} \min_{X \subseteq N_n} q_{n,m}(F, X)$, and proved that the question is correct, since $\sup_n$ in the right hand side of (4) is bounded from above for every m.

Beck and Fiala [4] extended problem P3(2) to the case of partition of set N_n into r subsets.

Problem P3(r) . *Given a finite set N_n and a family $F = \{F_1, \ldots, F_m\}$ of its subsets, we wish to find a partition $H : N_n \to N_r$ of set N_n into r subsets which minimizes the functional*

$$q(F, H) \doteq \max_{i,j,k} \|\,|F_i \cap H_j| - |F_i \cap H_k|\,\| \;,$$

where $H_p = H^{-1}(p)$, $p \in N_r$.

Let us define the function

$$\eta^3(r, m) \doteq \sup_n \max_{F \in \mathcal{F}_{n,m}} \min_{H \in \mathcal{H}_{n,r}} q(F, H) \;.$$

Beck and Fiala [4] showed that function η^3 is well defined, since it is bounded from above by a function of m. It is clear that

$$\eta^3(2, m) = f(m) \;.$$

2.4. The following problem of finding the vertex of a parallelepiped nearest to its given point was considered in [11].

Problem P4 (nearest vertex). *For a family $A = \{a_1, \ldots, a_n\} \subset \mathbb{R}^m$ of vectors, let $P(A) = \{\sum_{i \in N_n} \alpha_i a_i \mid \alpha_i \in [0, 1], \ i \in N_n\}$ be the parallelepiped constructed on A and let $\hat{P}(A) = \{\sum_{i \in N_n} \hat{\alpha}_i a_i \mid \hat{\alpha}_i \in \{0, 1\}, \ i \in N_n\}$ be the vertex set of the parallelepiped $P(A)$. Moreover, let $x = \sum \lambda_i a_i \in P(A)$ be a given point in $P(A)$. We wish to find the vertex $y \in \hat{P}(A)$ that is nearest to the point x in norm s, i.e., the point y that minimizes $\|x - y\|_s$.*

We define the function

$$\eta_s^4(n, m) = \sup_A \ \sup_{x \in P(A)} \ \min_{y \in \hat{P}(A)} \|x - y\|_s / \max_{a_i \in A} \|a_i\|_s \;. \tag{5}$$

(For the norm $s = l_\infty$ the index of the norm will commonly be omitted.) If $s = l_\infty$ and families of vectors $\{a_i\}$ under consideration are restricted to the cases

A) $a_i \in \{-1, 0, 1\}^m$, $\forall i$,
B) $a_i \in [0, 1]^m$, $\forall i$,
C) $a_i \in \{0, 1\}^m$, $\forall i$,

then the corresponding functions will be denoted by η^{4A}, η^{4B}, and η^{4C}.

If, while calculating the value of the function (5), we consider the only point

D) $x = \sum_{i \in N_n} a_i / 2 \in P(A)$,

then every function η^{4i} will be written with "bar" (which corresponds to the distance between the center of the parallelepiped and the nearest vertex of it in the worst case). Problem P4 with a condition $X \in \{A, B, C, D\}$ will be denoted by P4X. Finally, we denote functions

$$\eta_s^4(m) \doteq \sup_n \eta_s^4(n, m), \quad \bar{\eta}_s^4(m) \doteq \sup_n \bar{\eta}_s^4(n, m) \;,$$

$$\eta^{4X}(m) \doteq \sup_n \eta^{4X}(n,m), \ \bar{\eta}^{4X}(m) \doteq \sup_n \bar{\eta}^{4X}(n,m), \ X \in \{A,B,C\} \ .$$

(It will be seen from what follows that all these functions are well-defined.)

2.5. Beck and Fiala [4] posed the following "integer-making" problem.

Problem P5 (integer-making). *Given a vector* $\alpha = (\alpha_1, \ldots, \alpha_n) \in I\!R^n$ *and a family of sets* $F = \{F_1, \ldots, F_m\}$ $(F_i \subseteq N_n)$, *we wish to find an integer-valued vector* $\bar{\alpha} = (\bar{\alpha}_1, \ldots, \bar{\alpha}_n)$, $|\bar{\alpha}_i - \alpha_i| < 1$, $(i \in N_n)$, *that minimizes the functional*

$$\chi(F, \alpha, \bar{\alpha}) \doteq \max_{F_k \in F} \left| \sum_{i \in F_k} \alpha_i - \sum_{i \in F_k} \bar{\alpha}_i \right|$$

over all vectors $\bar{\alpha}$.

In other words, given a real-valued vector $\alpha \in [0,1]^n$, we wish to find its "rounding" vector $\bar{\alpha} \in \{0,1\}^n$, "nearest" to α. As a measure of proximity of the vector $\alpha - \bar{\alpha}$ to zero, we take sums of its components over all indices from given "testing" sets $\{F_i\}$.

We define the functions

$$g(n,m) \doteq \sup_{F \in \mathcal{F}_{n,m}} \sup_{\alpha \in I\!R^n} \min_{\bar{\alpha}} \chi(F, \alpha, \bar{\alpha}) \ ;$$

$$g(m) \doteq \sup_n g(n,m) \ .$$

2.6. The following problem was introduced in [11].

Problem P6 (the best equidistance). *Given a family* $F = \{f_1, \ldots, f_m\} \subseteq B^n$ *of vertices of the n-dimensional cube, we wish to find the vertex* $x \in B^n$ *most equidistant from all vertices in* F, *i.e., that one which minimizes the functional*

$$\omega(F, x) \doteq \max_{i,j} |\rho(x, f_i) - \rho(x, f_j)| \ ,$$

where $\rho(x,y)$ *is the Hamming-distance between vectors* $x, y \in B^n$.

Denote $\mathcal{F}'_{n,m} = \{F \mid F : N_m \to B^n\}$ (there is a one-to-one correspondence between sets $\mathcal{F}'_{n,m}$ and $\mathcal{F}_{n,m}$) and define functions

$$\theta(n,m) \doteq \max_{F \in \mathcal{F}'_{n,m}} \min_{x \in B^n} \omega(F, x) \ ;$$

$$\theta(m) \doteq \sup_n \theta(n,m) \ .$$

It is clear that $\theta(n,m) \leq n$ and that $\theta(n,m)$ is a function increasing in n and m.

2.7. The setting of the following problem goes back to papers by Steinitz [15] and Gross [7].

Problem P7 (vector summation [10]). *Given a family of vectors $B = \{b_1, \ldots, b_n\} \subset I\!R^m$ satisfying*

$$\sum_{b_i \in B} b_i = 0 \; , \tag{6}$$

$$\|b_i\|_s \le 1, \; i \in N_n \; , \tag{7}$$

we wish to find a permutation $\pi = (\pi_1, \ldots, \pi_n)$ which minimizes the functional

$$d_s(B, \pi) \doteq \max_{k \in N_n} \left\| \sum_{i=1}^{k} b_{\pi_i} \right\|_s$$

In other words, given a family of m-dimensional vectors with sum zero, we wish to find an order of summation of the vectors such that all partial sums are in a ball of a minimum possible radius. Steinitz [15] showed that every family of vectors that meets (6), (7) can be summed within a ball of radius independent of the number of vectors n. This enables one to define the function

$$\varphi_s(m) \doteq \sup_n \sup_{B \in \mathcal{B}_{n,s,m}} \min_{\pi} d_s(B, \pi) \; ,$$

where $\mathcal{B}_{n,s,m} = \left\{ B \,|\, B: \, N_n \to B_{s,m}; \, \sum_{b \in B} b = 0 \right\}$ and $B_{s,m}$ is the unit ball of a norm s with the center in the origin.

Gross (1917) proved the bound $\varphi_{l_2}(2) \le \sqrt{2}$ and posed the problem of finding the functions $\varphi_s(m)$ that we now call *Steinitz functions*. Due to results by W. Banaszczyk [2], [3], we now know that $\varphi_{l_2}(2) = \sqrt{5/4}$ and $\varphi_{l_1}(2) = \varphi_{l_\infty}(2) = 3/2$.

3 Connections between Problems and Relations between Functions

3.1. (P4D $\Leftrightarrow$ P1(2) $\Leftrightarrow$ P2) An evident connection between these problems follows from relations

$$2 \left\| \sum_{i \in X} a_i - \sum_{i \in N_n} a_i / 2 \right\|_s = \left\| \sum_{i \in X} a_i - \sum_{i \in N_n \setminus X} a_i \right\|_s \doteq L_s^1(A, X) = \left\| \sum_{i \in N_n} \varepsilon_i a_i \right\|_s \; ,$$

where $X \subseteq N_n$, $\varepsilon_i = 1$ for $i \in X$ and $\varepsilon_i = -1$, otherwise. Dividing by $\max_i \|a_i\|_s$ and passing to $\min_X$, $\sup_{\{a_i\}}$ and $\sup_n$, we come to the following relations between functions:

$$2\bar{\eta}_s^4(m) = \eta_s^1(2, m) = \nu_s(m) \; . \tag{8}$$

3.2. (P3 $\Leftrightarrow$ P1) Given a family $F = \{F_1, \ldots, F_m\}$ of subsets of the set N_n in problem P3, we define a family of vectors $A = \{a_i \,|\, i \in N_n\}$ by formula

$$a_i = (a_1^i, \ldots, a_m^i); \; a_j^i = \begin{cases} 1, \text{ if } i \in F_j \; , \\ 0, \text{ otherwise} \; . \end{cases} \tag{9}$$

Then

$$q(F, H) = \max_{i,j,k} \left| |F_i \cap H_j| - |F_i \cap H_k| \right| = \max_{j,k} \max_i \left| \sum_{\nu \in H_j} a_i^\nu - \sum_{\nu \in H_k} a_i^\nu \right|$$

$$= \max_{j,k} \left\| \sum_{\nu \in H_j} a_\nu - \sum_{\nu \in H_k} a_\nu \right\|_{l_\infty} = L^1_{l_\infty}(A, H) .$$

Thus, we come to a special case of problem P1A with $\{0,1\}$ matrix (a_j^i) and norm l_∞. Consequently,

$$\eta^3(r, m) \leq \eta^{1C}_{l_\infty}(r, m) . \tag{10}$$

3.3. (P3(2) $\Leftrightarrow$ P4CD $\Leftrightarrow$ P4AD $\Leftrightarrow$ P4BD $\Leftrightarrow$ P4D) The "balancing" problem P3(2) can be easily transformed into the problem P4CD of finding the vertex of a parallelepiped nearest to its center in norm l_∞ in the case of vectors a_i with $\{0,1\}$ components:

$$q_{n,m}(F, X) = \max_{i \in N_m} \left| \sum_{j \in N_n} \varepsilon_j a_j(i) \right| = 2 \left\| y - \sum a_j / 2 \right\|_{l_\infty} ,$$

where $y = \sum_{j \in X} a_j$, $\varepsilon_j = 1$ for $j \in X$, and $\varepsilon_j = -1$ otherwise. This leads to the relations $f(n, m) = 2\bar{\eta}^{4C}(n, m)$ and

$$\eta^3(2, m) = 2\bar{\eta}^{4C}(m) .$$

To get an approximate solution of the problem of minimizing the functional $q_{n,m}$, Olson and Spenser [8] applied the following technique. Combining the vectors a_j into pairwise differences: $a'_1 = a_1 - a_2$, $a'_2 = a_3 - a_4, \ldots$, they come to a family $\{a'_j\}$ consisting of $n' = \lfloor \frac{n+1}{2} \rfloor$ vectors. And though the components of the vectors were already not in the set $\{0,1\}$, but in $\{-1,0,1\}$, this fact did not affect the algorithm they applied to solve the problem P4, since the bound $|a_j(i)| \leq 1$ was only used. For our purposes, this provides a relation between the objective functions of problem P4 with $\{0,1\}$-matrix (a_j^i) and that with $\{-1,0,1\}$-matrix (a_j^i). As a result of this, we get the relation

$$\bar{\eta}^{4C}(n, m) \leq \bar{\eta}^{4A} \left(\left\lfloor \frac{n+1}{2} \right\rfloor, m \right) . \tag{11}$$

It can be easily shown that applying the same argument, we can obtain the relation

$$\bar{\eta}^{4B}(n, m) \leq \bar{\eta}^4 \left(\left\lfloor \frac{n+1}{2} \right\rfloor, m \right) . \tag{12}$$

In addition to these relations, we have evident inequalities:

$$\eta^{4A}(n, m) \leq \eta^4(n, m), \ \eta^{4C}(n, m) \leq \eta^{4B}(n, m) , \tag{13}$$

$$\bar{\eta}^{4A}(n,m) \leq \bar{\eta}^4(n,m), \ \bar{\eta}^{4C}(n,m) \leq \bar{\eta}^{4B}(n,m) \ , \tag{14}$$

and similar relations hold for functions $\eta^{4i}(m)$ and $\bar{\eta}^{4i}(m)$. Combining (11) and (12) with (14), we obtain

$$\bar{\eta}^{4C}(n,m) \leq \left\{ \bar{\eta}^{4B}(n,m), \ \bar{\eta}^{4A}\left(\left\lfloor \frac{n+1}{2} \right\rfloor, m\right) \right\} \leq \bar{\eta}^4\left(\left\lfloor \frac{n+1}{2} \right\rfloor, m\right) \ . \tag{15}$$

Our conjecture is that the inequality

$$\bar{\eta}^{4A}(n,m) = \bar{\eta}^4(n,m)$$

holds, which would enable us to get a linear order for the four quantities above.

3.4. (P1B $\Leftrightarrow$ P4) It was already pointed out in p. 3.1 that problem P4D is a special case of problem P1 when we have to divide a given annual plan of an enterprise evenly into two parts. Now we can show how the general problem P1 can be approximately solved with a good bound on accuracy, provided we use an approximation method for solving problem P4.

Define a function $c : \ IN \to IR$ by induction:

$$c(1) = 0 \ ; \tag{16}$$

$$c(r) = \min_{k \in N_{r-1}} \max\{c(k) + 1/k, \ c(r-k) + 1/(r-k)\}, \quad r > 1 \ . \tag{17}$$

Let $k(r)$ stand for the value of $k \in N_{r-1}$ that provides the minimum to the right hand side of (17). Suppose that we have an algorithm $\mathcal{A}_4$ which solves problem P4 in time $\mathcal{T}_{\mathcal{A}_4}(n,m)$ with bound

$$\|x - y\|_s \leq \mu_s(m) \max_{i \in N_n} \|a_i\|_s \doteq C_s(m, A) \ . \tag{18}$$

Then the algorithm $\mathcal{A}_1(r)$ described below solves problem P1B(r) in time $O(r\mathcal{T}_{\mathcal{A}_4}(n,m))$ with bound

$$L_s^2(A, H) \leq c(r)C_s(m, A) \ . \tag{19}$$

Bound (19) implies relation

$$\eta_s^2(r,m) \leq c(r)\eta_s^4(m) \ . \tag{20}$$

Algorithm $\mathcal{A}_1(r)$

If $r = 1$ then the partition consists of the single set $H_1 = N_n$, and $L_s^2(A, H) = 0$. If $r > 1$ then, solving the problem P4 for the parallelepiped constructed on vectors $\{a_i \, | \, i \in N_n\}$ and for its interior point $x = \frac{k(r)}{r}\sum_{i \in N_n} a_i$, we find a vertex $y = \sum_{i \in N'} a_i$ of the parallelepiped such that $\|x - y\|_s \leq C_s(m, A)$.

Using algorithm $\mathcal{A}_1(k)$ for $k = k(r)$, we solve the problem P1B(k(r)) for the family of vectors $A' \doteq \{a_i \, | \, i \in N'\}$, i.e., we find a partition H' of the set N' into $k(r)$ subsets $H_1, \ldots, H_{k(r)}$ which meets the bound

$$L_s^2(A', H') \leq c(k(r))C_s(m, A) \ . \tag{21}$$

From (21) and (17) we obtain

$$\left\| \sum_{i \in H_\nu} a_i - \sum_{i \in N_n} a_i/r \right\|_s \leq \left\| \sum_{i \in H_\nu} a_i - \sum_{i \in N'} a_i/k(r) \right\|_s$$

$$+ \left\| \sum_{i \in N'} a_i - \frac{k(r)}{r} \sum_{i \in N_n} a_i \right\|_s /k(r) \leq c(k(r))C_s(m, A) + C_s(m, A)/k(r)$$

$$\leq c(r)C_s(m, A), \quad \nu = 1, \ldots, k(r) \ . \tag{22}$$

Similarly, solving the problem P1B$(r - k(r))$ for the family of vectors $A'' \doteq \{a_i \mid i \in N_n \setminus N'\}$, we find a partition $H'' = \{H_{k(r)+1}, \ldots, H_r\}$ of the set $N_n \setminus N'$ into $r - k(r)$ subsets which meets the bound:

$$L_s^2(A'', H'') \leq c(r - k(r))C_s(m, A) \ .$$

Using this inequality, we obtain the bound similar to (22):

$$\left\| \sum_{i \in H_\nu} a_i - \sum_{i \in N_n} a_i/r \right\|_s \leq c(r)C_s(m, A), \quad \nu = k(r) + 1, \ldots, r \ . \tag{23}$$

(22) and (23) imply (19) $\qquad\qquad\qquad\qquad\qquad\qquad\qquad\qquad \Box$

In fact, algorithm $\mathcal{A}_1(r)$ reduces to solving r copies of the problem P4. Therefore, its running time is $T_{\mathcal{A}_1}(r, n, m) = O(rT_{\mathcal{A}_4}(n, m))$. (This total includes the running time $O(r^2)$ of calculating the function $c(r)$ under natural assumption that $n \geq r$ and that function $T_{\mathcal{A}_4}(n, m)$ is at least linear in n.)

To solve problem P4, we can apply algorithm $\mathcal{A}_4$, based on the following lemma.

Lemma 1 [12, pp. 55-57]. *Suppose we are given a family of vectors $A = \{a_1, \ldots, a_n\} \subset I\!R^m$ and a vector $\alpha = (\alpha_1, \ldots, \alpha_n) \in [0, 1]^n$. Then there exists a base A' of family A and a vector $\alpha^* = (\alpha_1^*, \ldots, \alpha_n^*) \in [0, 1]^n$ such that*

$$\sum_{i \in N_n} \alpha_i^* a_i = \sum_{i \in N_n} \alpha_i a_i \ ,$$

$$\{a_i \mid \alpha_i^* \in (0, 1)\} \subseteq A' \ .$$

The base A' and the vector α^* can be found in $O(nm^2)$ time. $\qquad\qquad \Box$

(A' is called a *base* of a family A if: • $A' \subseteq A$; • A' is linear independent; • the linear hull of A' contains A.) Indeed, for any point $x = \sum_{i \in N_n} \alpha_i a_i$, $(\alpha_i \in [0, 1])$, due to Lemma 1, we can find numbers $\{\alpha_i^* \in [0, 1] \mid i \in N_n\}$ such that

- $x = \sum_{i \in N_n} \alpha_i^* a_i$, and
- almost all numbers α_i^* are equal to 0 or 1; more exactly, the set $N' \doteq \{i \in N_n \mid \alpha_i^* \notin \{0, 1\}\}$ is such that all vectors a_i, $(i \in N')$, belong to a base A', which implies $|N'| \leq m$.

Take the vertex $y = \sum_{i \in N^* \cup N''} a_i$, where $N^* = \{i \in N_n \mid \alpha_i^* = 1\}$, and N'' is a subset of set N'. Then we have $\|x - y\|_s = \left\|\sum_{i \in N'} \alpha_i^* a_i - \sum_{i \in N''} a_i\right\|_s$, which is not greater than $m \cdot \max_{i \in N_n} \|a_i\|_s$ for any subset $N'' \subseteq N'$. In the case that we take numbers $\bar{\alpha}_i \in \{0, 1\}$, $i \in N'$, obtained by the ordinary rounding of numbers α_i^*, $i \in N'$, and then choose the subset $N'' = \{i \in N' \mid \bar{\alpha}_i = 1\}$, we obtain the bound

$$\|x - y\|_s \leq m/2 \cdot \max_{i \in N_n} \|a_i\|_s, \tag{24}$$

valid for any norm s. Thus we just confirmed our ability to solve problem P4 with bound (18) for function $\mu_s(m)$ at most as much as $m/2$. (As we will see from results of Sect. 4, better upper bounds on function $\mu_s(m)$ can be suggested for certain norms, such as l_2 or l_∞.)

As far as function $c(r)$ defined by (16), (17) is concerned, a method of approximating its absolute maximum over all r was described in [14]. Implementation of this method on computer showed that this curious function attains its maximum at $r = 909$, having

$$c(r) < 2.00046, \forall r, \tag{25}$$

$c(2) = 1$, and $c(r) < 2$, for all $r \leq 251$. Using bound (25) together with (18), (19), and (24), we come to an algorithm $\mathcal{A}_1(r)$ for solving the problem P1B with bound

$$L_s^2(A, H) \leq 1.00023\, m \cdot \max_{i \in N_n} \|a_i\|_s$$

and running time $O(nrm^2)$.

Having a solution of problem P4 with bound (18), where $\mu_s(m) = \eta_s^4(m)$, and using (19) and (25), we come to the relation

$$\eta_s^2(r, m) \leq 2.00046\, \eta_s^4(m).$$

3.5. (P5 $\Leftrightarrow$ P4C) We can assume in problem P5 that $\alpha_i \in [0, 1]$, $\bar{\alpha}_i \in \{0, 1\}$, $(i \in N_n)$. Given a family $F = \{F_1, \ldots, F_m\} \in \mathcal{F}_{n,m}$, we define a family of vectors $A = \{a_1, \ldots, a_n\} \subset IR^m$ according to (9). Then we have

$$\chi(F, \alpha, \bar{\alpha}) = \max_{F_k \in F} \left| \sum_{i \in F_k} \alpha_i - \sum_{i \in F_k} \bar{\alpha}_i \right| = \max_{k \in N_m} \left| \sum_{i \in N_n} \alpha_i a_k^i - \sum_{i \in N_n} \bar{\alpha}_i a_k^i \right|$$

$$= \left\| \sum_{i \in N_n} \alpha_i a_i - \sum_{i \in N_n} \bar{\alpha}_i a_i \right\|_{l_\infty} = \|x - y\|_{l_\infty},$$

where $x = \sum_{i \in N_n} \alpha_i a_i$ is a given point in the parallelepiped constructed on the vectors $a_i \in A$, while $y \doteq \sum_{i \in N_n} \bar{\alpha}_i a_i$ is a vertex of the parallelepiped. Thus, problem P5 coincides with problem P4C, implying

$$g(n, m) = \eta^{4C}(n, m). \tag{26}$$

3.6. (P3 $\Leftrightarrow$ P5) Beck and Fiala [4] established a connection between problems P3 and P5, using the following lemma.

Lemma 2. *Given a vector* $\alpha = (\alpha_1, \ldots, \alpha_n) \in [0,1]^n$ *and a family of sets* $F \in \mathcal{F}_{n,m}$, *we can find a vector* $\alpha^* = (\alpha_1^*, \ldots, \alpha_n^*) \in [0,1]^n$ *such that* $\sum_{i \in F_j} \alpha_i^* = \sum_{i \in F_j} \alpha_i \ (\forall F_j \in F)$, *and at most m numbers* $\{\alpha_i^*\}$ *take nonintegral values (different from 0,1)* $\qquad\qquad\square$

It is clear that Lemma 2 is a special case of Lemma 1, all components of vectors $\{a_i\}$ being defined by (9). However, it should be noted that lemma 1 was proved long before by Behrend [5, p. 110] and Steinitz [15]. (Naturally, without specifying a bound on the running time of the algorithm. — It is very likely that every researcher comes to this lemma in its own time.) Getting ahead, we point out that this remarkable lemma 1 proves to be quite useful for proving the fact that, given a family of vectors $\{b_1, \ldots, b_n\} \subset I\!R^m$ in problem P7, we can find an order for their summation such that all partial sums are in the ball of radius m.

The relation between P3 and P5 seems to be of no surprise since we have already established that problem P3 is nothing but problem P1A with $\{0,1\}$-matrix (a_j^i) and norm l_∞ (see p. 3.2), whereas P5 coincides with problem P4C (again with $\{0,1\}$-vectors $\{a_i\}$, see p. 3.5), and it was shown in p. 3.4 how to solve problem P1B by means of solving problem P4 in the case of general type of matrix (a_j^i). It is natural that while solving problem P3 by using problem P5, Beck and Fiala proceeded in a way very similar to that we use in p. 3.4. The basic idea was induction and they defined a function $c'(r)$, similar to our $c(r)$, yet slightly different:

$$c'(r) \doteq \max\{c'(r_1) + 1/r_1, \ c'(r - r_1) + 1/(r - r_1)\} \ ,$$

where $r_1 = \lfloor r/2 \rfloor$. As a result, they come to the relation

$$\eta^3(r, m) < \left(12 - \frac{4}{r-1}\right) g(m) \ , \tag{27}$$

connecting problems P3 and P5. At the same time, using the "relationship" between problems just established and relations (10), (3), (20), (25), and (26), we come to the following ones:

$$\eta^3(r, m) \le \eta_{l\infty}^{1C}(r, m) \le 2\eta_{l\infty}^{2C}(r, m) \le 2c(r)\eta^{4C}(m)$$
$$< 4.001 \ \eta^{4C}(m) = 4.001 \ g(m), \ \forall r, m \ ,$$

which slightly improves relation (27).

Using $c(2) = 1$, we obtain $\eta^3(2, m) \le 2g(m)$.

3.7. (P3(2) $\Leftrightarrow$ P6) Functionals of these problems meet double-sided relations.
Let $F = \{F_1, \ldots, F_m\} \in \mathcal{F}_{n,m}$, $X \subseteq N_n$, $\bar{X} = N_n \setminus X$, $\bar{F}_i = N_n \setminus F_i$; and let $x, f_i \in B^n$ be characteristic vectors of sets X and F_i $(i \in N_m)$. Then we have

$$\left||F_i \cap X| - |F_i \cap \bar{X}|\right| = \left|(|F_i \cap X| + |\bar{F}_i \cap X|) - (|F_i \cap \bar{X}| + |\bar{F}_i \cap X|)\right|$$
$$= |\rho(x, 0) - \rho(x, f_i)| \ .$$

Thus, to solve problem P3(2), we can use a solution of problem P6 with the input family of vertices $F = \{0, f_1, \ldots, f_m\} \subseteq B^n$. Having problem P6 solved with bound $\omega(F, x) \leq \theta(n, m+1)$, we obtain a solution $(X, \bar{X})$ to problem P3(2) with bound $q_{n,m}(F, X) \leq \omega(F, x) \leq \theta(n, m+1)$ guaranteed, which implies relation

$$f(n, m) \leq \theta(n, m+1) \ .$$

From the other hand, given a family $F = \{f_1, \ldots, f_m\}$ of vertices of the unit cube B^n, one can find a vertex $x \in B^n$ (a set $X \subseteq N_n$) such that

$$|\rho(x, f_i) - \rho(x, f_j)| \leq |\rho(x, f_i) - \rho(x, 0)| + |\rho(x, f_j) - \rho(x, 0)|$$

$$= \left||F_i \cap X| - |F_i \cap \bar{X}|\right| + \left||F_j \cap X| - |F_j \cap \bar{X}|\right| \leq 2f(n, m) \ ,$$

implying $\theta(n, m) \leq 2f(n, m)$. Thus, we get

$$f(n, m - 1) \leq \theta(n, m) \leq 2f(n, m) \ . \tag{28}$$

3.8. (P1B ⇔ P7)
A connection between these two problems was established [14] which enables one to derive upper bounds on the functional η_s^2 of problem P1B and to construct "bad" instances of families of vectors for problem P7.

Let $A = \{a_1, \ldots, a_n\} \subset B_{s,m}$ be an input family of vectors in problem P1, $a = \sum a_i$; $n_1 = rt$, where $t = \lceil n/r \rceil$; $n' = n + n_1$. Complement family A to family $A_1 = \{a_i \,|\, i \in N_{n'}\}$, adding vectors $a_i = -a/n_1$ ($i \in N_{n'} \setminus N_n$). This provides equality $\sum_{i \in N_{n'}} a_i = 0$. First, solving problem P7, find a permutation $\pi = (\pi_1, \ldots, \pi_{n'})$ such that

$$\left\|\sum_{i=1}^k a_{\pi_i}\right\|_s \leq \varphi_s(m) \max_i \|a_i\|_s \ , \ k \in N_{n'} \ .$$

Next, find numbers $\{k_0, k_1, \ldots, k_r\}$ such that

- $0 = k_0 < k_1 < \cdots < k_r = n'$,
- every set $I_p \doteq \{\pi_i \,|\, i \in (k_{p-1}, k_p]\}$, $p = 1, \ldots, r$, contains exactly t "additional" indices $\pi_i > n$.

Then, setting $H_p = I_p \cap N_n$, we obtain a partition $H = \{H_1, \ldots, H_r\}$ of set N_n that meets relations

$$\left\|\sum_{i \in H_p} a_i - a/r\right\|_s \leq \begin{cases} \varphi_s(m), & p \in \{1, r\} \ , \\ 2\varphi_s(m), & p = 2, \ldots, r-1, \end{cases}$$

implying

$$\eta_s^2(r, m) \leq 2\varphi_s(m), \ \forall r \ ;$$

$$\eta_s^2(2, m) \leq \varphi_s(m) \ .$$

4 Some More Bounds for Functions

4.1. For an arbitrary norm s, it follows from Lemma 1 that

$$\eta_s^4(n, m) \leq \min\{n, m\}/2 , \tag{29}$$

and the bound is tight for the norm $s = l_1$. Furthermore, Lemma 1 implies relations

$$\eta_s^4(m) = \sup_n \eta_s^4(n, m) = \eta_s^4(m, m) . \tag{30}$$

Similar relations hold for problems P4A, P4B, and P4C, specifically,

$$\eta_s^{4C}(m) = \eta_s^{4C}(m, m) . \tag{31}$$

4.2. Beck and Fiala [4] derived the bound

$$g(m) < 2\sqrt{2m \log 2m}, \ (m \geq 5) ,$$

or applying (26) and (31),

$$\eta^{4C}(n, m) < 2\sqrt{2m \log 2m}, \ (m \geq 5) .$$

Similar arguments work the for more general problem P4 (with arbitrary matrix (a_j^i)), which leads to the bound

$$\eta^4(n, m) < 2\sqrt{2n \log 2m} ,$$

for $n > 2 \log 2m$. Using (30), we get a tighter bound

$$\eta^4(n, m) < 2\sqrt{2 \min\{n, m\} \log 2m} ,$$

for $n > 2 \log 2m$, $m \geq 5$. Finally, using (29) for $s = l_\infty$, we obtain

$$\eta^4(n, m) \leq \begin{cases} \min\{n, m\}/2, & \text{for } \min\{n, m\} \leq 32 \log 2m , \\ 2\sqrt{2 \min\{n, m\} \log 2m}, & \text{for } \min\{n, m\} \geq 32 \log 2m . \end{cases} \tag{32}$$

A vertex of the parallelepiped distant (in the norm l_∞) from its given point x by at most the value in the right hand side of (32) can be found in $O(nm)$ time.

4.3. Olson and Spenser [8] prove a lemma which implies the following bound

$$2\bar{\eta}^{4A}(n, m) < \sqrt{2n \log 2m} .$$

Applying the same argument, we can obtain a more general bound:

$$2\bar{\eta}^4(n, m) < \sqrt{2n \log 2m} , \tag{33}$$

which generalizes the following bound from [11]:

$$2\bar{\eta}^{4C}(m, m) < \sqrt{2m \log 2m} .$$

4.4. For the norm l_2, it was shown in [12] that

$$\eta_{l_2}^4(n,m) = \bar{\eta}_{l_2}^4(n,m) = \sqrt{\min\{n,m\}}/2 \ . \tag{34}$$

A vertex of the parallelepiped distant (in the norm l_2) from its given point x by at most the value $\eta_{l_2}^4(n,m)$, can be found in $O(m \cdot \min\{n^3, m^3 + mn\})$ time.

4.5. An algorithm of running time $O(n^2m^2)$ was described in [13] that finds a permutation π in problem P7 with bound

$$d_s(B,\pi) \le m$$

guaranteed for any given family of vectors $B \in \mathcal{B}_{n,s,m}$ and any norm s. This implies the bound

$$\varphi_s(m) \le m, \ \forall s \ .$$

Next we mention a few lower bounds on the functions under consideration.

4.6. Olson and Spenser [8] proved the bound

$$2\bar{\eta}^{4C}(m,m) = f(n,m) > \left(\frac{1}{2} - o(1)\right)\sqrt{n}, \ (n \le m) \ .$$

For norms l_p $(p \ge 2)$, in [14] was practically established that

$$\bar{\eta}_{l_p}^4(n,m) \ge \left(\frac{k}{m}\right)^{\frac{1}{2}-\frac{1}{p}}\sqrt{n}/2, \ (n \le m), \ (n \le m) \ ,$$

provided that there exists a Hadamard matrix H_k of order $k \le m$. This implies the bound

$$\bar{\eta}_{l_p}^4(n,m) \ge \left(\frac{1}{2} - o(1)\right)\sqrt{n}, \ (p \ge 2, \ n \le m) \ , \tag{35}$$

due to the fact (see [1, p. 194]) that the orders of Hadamard matrices are "dense" on the lattice of integers.

For norms l_p $(p \in [1,2])$, the following bound can be obtained, using the family $A = \{a_1, \ldots, a_n\}$ of $n \le m$ basis vectors:

$$\bar{\eta}_{l_p}^4(n,m) \ge n^{1/p}/2, \ (p \in [1,2], \ n \le m) \ . \tag{36}$$

Since for the norm l_∞ bound (35) is derived for families of vectors $\{a_i\}$ with components $a_j^i \in \{-1,1\}$, we come to the bound

$$\bar{\eta}^{4A}(n,m) \ge \left(\frac{1}{2} - o(1)\right)\sqrt{n}, \ (n \le m) \ .$$

4.7. We finally formulate two lower bounds for functions $\varphi_s(m)$ from [14]. If a Hadamard matrix H_n exists for $n \le m$, then

$$\varphi_{l_p}(m) \ge \sqrt{n+3}/2, \ (p \ge 2) \ .$$

This implies the lower bound

$$\varphi_{l_p}(m) \geq \left(\frac{1}{2} - o(1)\right) \sqrt{m}, \ p \geq 2 \ .$$

The other bound

$$\varphi_{l_p}(m) \geq (1 + (m-1)2^{-p})^{1/p}, \ (p \geq 1) \ ,$$

implies already unconditional lower bound

$$\varphi_{l_p}(m) \geq \sqrt{m+3}/2, \ (p \leq 2) \ ,$$

which increases as $p \to 1$ and becomes linear for $p = 1$:

$$\varphi_{l_1}(m) \geq (m+1)/2 \ .$$

5 Concluding Remarks

From the previous consideration, we can come to the conclusion that
• the seven problems formulated have numerous connections;
• the problem P4 (of finding the vertex of a parallelepiped nearest to its given point) occupies the central place in this family of problems, and therefore, it requires additional attention and research.

The following tree conjectures are aimed to stimulate further research of this problem.

$$\eta_s^4(m, m) = \bar{\eta}_s^4(m, m), \ \forall s, m \ , \tag{37}$$

i.e., the center of the "worst" parallelepiped is its "worst" point, most distant from the nearest vertex. (It is true for $s = l_1$ and $s = l_2$, as it follows from (29), (34), and (36).)

$$\eta_s^4(m, m) \geq \bar{\eta}_{l_2}^4(m, m), \ \forall s, m \ , \tag{38}$$

i.e., the Euclidean norm is the best norm for problem P4. (It is true for the case $m = 2$, see [14].)

$$\bar{\eta}^4(n, m) = \bar{\eta}^{4A}(n, m), \ \forall n, m \ , \tag{39}$$

i.e., the worst family of vectors $\{a_i\} \subset B_{l_\infty, m}$, at which the value of the function $\bar{\eta}^4(n, m)$ is attained, has components $\{a_i\} \subset \{-1, 0, 1\}^m$.

In particular, if conjectures (37) and (38) are true, we can derive a nontrivial lower bound on the Dvoretzky function for arbitrary norm s:

$$\nu_s(m) \geq \sqrt{m} \ .$$

References

1. Alon N., Spencer, J., Erdös, P.: The probabilistic method. A Willey-Interscience Publication (1992), 254 p.
2. Banaszczyk, W.: The Steinitz constant of the plane. J.Reine und Angew.Math.**373** (1987) 218–220
3. Banaszczyk, W.: A note on the Steinitz constant of the Euclidean plane. C.R. Math. Rep. Acad. Sci. Canada **12** (1990), no. 4, 97–102.
4. Beck, J., Fiala, T.: "Integer-making" theorems. Discrete Appl. Math. **3** (1981) 1–8
5. Behrend, F. A.: The Steinitz-Gross theorem on sums of vectors. Can. J. Math. **6** (1954) 108–124
6. Dvoretzky, A.: Problem. In: Proceedings of Symposia in Pure Mathematics, **Vol.7** Convexity, Amer. Math. Soc., Providence, RI, (1963) p. 496
7. Gross, W.: Bedingt Konvergente Reihen. Monatsh. Math. und Physik. **28** (1917) 221–237
8. Olson, J., Spencer, J.: Balancing families of sets. J. Comb. Theory (Ser. A) **25** (1978) 29–37
9. Sevastianov, S. V.: Asymptotical approach to some scheduling problems. (In Russian) Upravlyaemye Sistemy **14** (1975) 40–51
10. Sevastianov, S. V.: Approximate solution of some problems of scheduling theory. (In Russian) Metody Diskret. Analiz. **32** (1978) 66–75
11. Sevastianov, S. V.: On a connection between calendar-planning problem and one problem on the unit cube. (In Russian) Metody Diskret. Analiz. **35** (1980) 93-103
12. Sevastianov, S. V.: Approximate solution to a calendar-planning problem. (In Russian) Upravlyaemye Sistemy **20** (1980) 49–63
13. Sevastianov, S. V.: Approximation algorithms for Johnson's and vector summation problems. (In Russian) Upravlyaemye Sistemy **20** (1980) 64–73
14. Sevastianov, S. V.: Geometry in scheduling theory. (In Russian) In: Models and Methods of Optimization, Trudy Inst. Mat. **10** (Novosibirsk, 1988) 226–261
15. Steinitz, E.: Bedingt Konvergente Reihen und Convexe Systeme. J.Reine und Angew. Math. **143** (1913) 128–175

Linear-Time Reconstruction of Delaunay Triangulations with Applications

Jack Snoeyink*
Dept. of Computer Science
University of British Columbia

Marc van Kreveld†
Dept. of Computer Science
Utrecht University

Abstract

Many of the computational geometers' favorite data structures are planar graphs, canonically determined by a set of geometric data, that take $\Theta(n \log n)$ time to compute. Examples include 2-d Delaunay triangulation, trapezoidations of segments, and constrained Voronoi diagrams, and 3-d convex hulls. Given such a structure, one can determine a permutation of the data in $O(n)$ time such that the data structure can be reconstructed from the permuted data in $O(n)$ time by a simple incremental algorithm.

As a consequence, one can permute a data file to "hide" a geometric structure, such as a terrian model based on the Delaunay triangulation of a set of sampled points, without disrupting other applications. One can even include "importance" in the ordering so the incremental reconstruction produces approximate terrain models as the data is read or received. For the Delaunay triangulation, we can also handle input in degenerate position, even though the data structures may no longer be canonically defined.

1   Introduction

Many of the favorite data structures in computational geometry can be represented as planar graphs that are canonically determined by n input data elements. Well-known examples include:

1. The Voronoi diagram of n point or line segment *sites* in the plane. This oft-reinvented data structure records proximity [1, 16]; it is the decomposition of the plane into maximally-connected regions that have the same set of sites as closest neighbors.

2. The Delaunay triangulation of n points. This dual of the Voronoi diagram joins two points with an edge if some circle touches those two sites and no other. The Delaunay is important in GIS and meshing [2, 18].

---

*Supported in part by grants from NSERC and Facet Decision Systems
†Supported in part by ESPRIT IV LTR project 21957 (CGAL).

3. The trapezoidation (vertical visibility map) of n line segments. Decomposing the plane into trapezoids—by extending vertical segments from each given segment endpoint to the segments above and below—imposes structure on the plane that can be exploited by intersection and other algorithms [17, 15].

4. The convex hull of n points in 3-space: The convex hull is the smallest convex set enclosing the points [17, 19]. Its surface can be represented as a planar graph.

Each of these have construction algorithms that run in optimal $\Theta(n \log n)$ time; the lower bounds are proved by reduction from sorting [14, 19]. Presorting the data in natural orders does not necessarily reduce the construction time. Seidel [22] has shown that after sorting by x-coordinate, computing a 3-d convex hull or 2-d Delaunay triangulation still requires $\Omega(n \log n)$ time. Djijev and Lingas [6] have shown the same for the Voronoi diagram.

We use ideas from Kirkpatrick's point location scheme [13] to find a permutation of the n data elements in linear time so that a simple incremental algorithm can reconstruct the desired planar graph from the permuted data in $O(n)$ time and space. We need only the following properties:

1. The graph must be canonical, depending on the data and not the order of the data.

2. There must always be a constant fraction of the data that can be deleted independently. (Kirkpatrick's "constant-degree" restriction is not necessary.)

3. It must be possible to reintroduce the deleted data by independent insertions.

This result is a natural combination of work in hierarchical representation of planar subdivisions [13] and incremental construction algorithms [10, 21]. It can be seen as a form of geometry compression and progressive expansion, which is of interest in computer graphics [5, 12]. We illustrate the idea with one application: com-

Fig. 1: TIN in Java implementation

pressing and transmitting terrain models. TIN (triangulated irregular network) terrain models [18] in geographic information systems (GIS) often fit a surface to points with elevations by taking the Delaunay triangulation of their 2-d projections and raising the triangles into 3-d.

Suppose that you wanted to distribute results of some GIS analysis on a terrain. You could encode and transmit maps or images, but geometric models may have more utility. Not only does rotating a terrain model gives a better 3-dimensional "feel," but a model can provide enough information to answer queries locally, such as steepest descent paths for fluid flow. However, sending all the triangles and their interconnections (say with a winged-edge data structure) requires five times more bytes than sending the point coordinates alone. Our result allows the structure to be encoded in a permutation of the points, giving three advantages:

1. The reconstruction algorithm is simple. The asymptotic constants are small for the construction of the permutation and reconstruction of the triangulation.

2. Reconstruction is incremental; after reading k data elements and spending $O(k)$ time and space, the algorithm can present partial results while the data is still arriving.

3. Storage and transmission overhead is reduced. A flat file of permuted data implicitly stores the planar graph structure, but is still readable by many other applications that are not concerned with this structure.

4. "Importance" can be incorporated into the permutation to support progressive transmission and reconstruction of geometric models by moving data that characterizes the final output early in the permutation. This can often be done without changing the asymptotic running time.

The next section outlines linear-time deconstruction and reconstruction algorithms for the Delaunay triangulation. These algorithms can choose methods for canonical traversal and computation of independent sets; Sections 3 and 4 describe our choices. Section 5 shows how to handle degenerate point sets, for which the Delaunay triangulation is not canonically defined. Section 6 compares reconstruction time to standard construction algorithms.

2 Permuting data for incremental reconstruction

The structures listed in the previous section have randomized incremental construction algorithms that run in expected $O(n \log n)$ time, where the expectation is taken over random choices in the algorithm for a worst-case input distribution. In each case, the expected cost of all changes to the data structure is linear, but the expected time to locate where the changes must occur is $\Theta(n \log n)$. Our permutation must ensure that the cost of change is linear not just in expectation, and that location is also linear in n. Fortunately, this is easily done.

In this section, we use the computation of the Delaunay triangulation of n points P in the plane to give an example of the steps of our approach. Our first algorithm produces a permutation of the data (in reverse order); our second takes this permutation and reconstructs the graph.

Permutation Construction Algorithm

P1. **Bound the problem.** Our terrain models come with bounding rectangles, so we begin with a triangulation of four points. One could instead extend the triangulation to the entire plane by adding a vertex at infinity and joining it to the convex hull vertices.

P2. **Independent set.** Find a constant fraction of vertices that form an independent set—no two vertices in the set are adjacent. Low degree is useful, but not mandatory. A simple, greedy algorithm in Section 4 guarantees an independent set of size at least 1/6 of all vertices, using vertices of degree at most nine.

P3. **Delete and locate data.** Delete the vertices of the independent set and repair the remaining triangulation, noting which new triangles now contain the deleted vertices. There will be at most one deleted vertex per triangle.

P4. **Traverse graph & output permutation.** Enumerate triangles by any canonical method: depth-first, breadth-first, etc. (Section 3 describes our favorite traversal, which requires no changes to the data structure.) Output the deleted vertices in the reverse of the order that their triangles are visited by the enumeration.

P5. **Output end of phase signal.** Our current applet inserts a special marker to signal the end of a phase. Section 3 describes other methods that don't use special markers, but use the points themselves.

P6. **Repeat from P2.** Continue until only the bounding box remains.

Lemma 1 *Given the Delaunay triangulation of n points, the Permutation Construction Algorithm can be implemented to run in $O(n)$ time.*

Proof: Step P1 executes once; either approach mentioned can be implemented to run in linear time.

Let us say that a single execution of steps P2–P5 constitutes a phase. Each phase executes in time proportional to the number of vertices involved: this is true for steps P2 and P4 by Euler's relation and simple graph traversal techniques. If we restrict the independent set to low-degree vertices, then Step P3 is linear by independent retriangulation of polygons of constant size. Otherwise, we can use the linear-time constrained-Delaunay algorithm of Wang and Chin [23].

Step P2, which is standard in hierarchical triangulations, implies that the number of vertices decreases by a constant in each phase. Using the 1/6 guaranteed in Section 4, the total time for all phases satisfies $T(n) \leq T(5n/6) + cn + O(1) \leq 6cn + O(1)$, where cn is the time for a phase with n vertices. ∎

If we now reverse the ordered list of points and reconstruct the Delaunay triangulation, our traversal can accomplish a linear amount of point location in linear time.

Reconstruction Algorithm

R1. **Initialize.** Start with the bounding box (or with an infinite vertex).

R2. **For each phase do R3–R4.** Use the signal from P5 to recognize where one phase ends and another begins.

R3. **Traverse and locate.** Visit the triangles by the same traversal as in P4 and link vertices to the triangles that contain them as they are found.

R4. **Insert entire phase.** Using Guibas and Stolfi's incremental insertion, update the Delaunay to incorporate the new vertices.

Lemma 2 *The Reconstruction Algorithm for the Delaunay triangulation can be implemented to take $O(k)$ time after reading k points.*

Proof: We first observe, by induction, that the phases of Reconstruction R2–R3 produce the triangulations of the phases of P2–P5 in reverse order.

In the base case, the initialization step R1 recreates the triangulation of the bounding box with which the Permutation Construction algorithm terminated.

Now assume, as an induction hypothesis, that R2–R3 begins its phase $i - 1$ with the correct triangulation and list of vertices. The enumerations in P4 and R3 visit the triangles (and vertices) of phase $i - 1$ in the same order. Thus, R3 will link all the vertices of phase $i - 1$; since the algorithm identifies the ends of phases it ensures that it does not link any vertices of phase i. Insertion of all linked vertices then computes the triangulation to begin phase i. Therefore, by induction, the phases of Reconstruction compute the correct triangles.

Analyzing the running time is easy. Step R1 takes constant time. In a single phase, step R3 runs in time proportional to the number of triangles in the triangulation. Step R3 can use the incremental procedure of Guibas and Stolfi to perform each insertion in time proportional to the degree of the vertex, which is bounded by the number of edges of the triangulation in the next phase. Euler's relation ties both these quantities to the number of vertices, which increases according to the geometric series of Lemma 1. Stopping after k vertices results in a truncated series that sums to $O(k)$. ∎

3 End-of-phase signals and canonical traversal

It is important for the running time of the reconstruction algorithm to be able to detect the end of a phase. If one iteration of the loop R2 takes points from two phases, then subsequent iterations may not even complete a phase—they will be traversing different triangulations than the ones used to determine the permutation order.

On the other hand, if we avoid storing special end-of-phase markers then we may be able to "hide" terrain models without affecting other applications by simply permuting data points in a file.

Observation 3 *There are protocols to signal end-of-phase without adding markers to the permutation.*

One protocol would be to use fixed-length phases. Because the minimum phase size depends on the size of the smallest guaranteed independent set, we should use multiples of 6/5 if we use independent sets of size at least 1/6 of the total number of vertices. This leads to more phases and a higher constant factor in the reconstruction time.

Ideally, we could arrange for the traversal to run out of triangles at the end of a phase—that is, to ensure that the first point of the next phase in R2 does not lie in one of the triangles visited after the last point of the current phase. A simple protocol based on this idea is to move the first point of a phase to the end of the phase. Then that point will not be found during the traversal, but can be inserted separately after the traversal completes. After its insertion, the next phase can begin.

In the rest of this section, we specify our favorite canonical traversal for steps P4 and R3. Remove from a planar graph the edges of any spanning tree of vertices; the duals of the edges that remain form a spanning tree of the faces. This observation can be used to give a simple traversal of any planar graph with convex faces [4, 9]. As a bonus, we can use this traversal to do hidden surface elimination by the painter's algorithm.

Observation 4 *Any planar graph with convex faces has a canonical traversal that requires no data structure modifications and can list faces from back to front.*

Choose a direction, which we draw in Figure 2 as downward. Each vertex except the lowest rightmost selects the edge immediately below or to the right of the vertex. We can prove that this is a spanning tree: No edge is selected by both endpoints, so we have selected $v - 1$ edges. The resulting graph is connected because we can start at any vertex and repeatedly follow the edge selected by the current vertex; we always move right or down, so we stop at the lowest rightmost point.

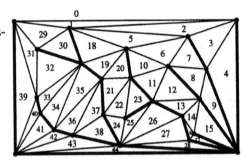

Fig. 2: Traversal from spanning tree

Now, treat spanning tree edges as the walls of a maze, and traverse the maze, keeping your left hand on a wall. Because the code is short, we include it in Figure 3. The function **wall()** recognizes selected edges by examining the coordinates of three consecutive vertices in counter-clockwise order. **drawTraverse()**

implements the traversal without making changes to the data structure. These are based on a quadedge-like data structure [11] that uses directed edges: Sym(e) reverses edge e, and Lnx(e) and Lpr(e) give the ccw and cw edges around the face to the left of e.

Figure 2 shows the faces numbered in the order that the traversal enters them. Along any vertical line, the numbers increase. Because the spanning tree of vertices consists of paths to the right, ending at the rightmost vertex, this traversal will draw the faces from back to front.

```
boolean wall(int e) {
  return p(e).greatereq(p(Lnx(e))) && p(Lnx(e)).greater(p(Lnx(Lnx(e))));
}

boolean entered(int e) {
  return p(e).greatereq(p(Lpr(e))) && p(Lnx(e)).greater(p(Lpr(e)));
}

void drawTraverse(Graphics g, int startedge) {
  int e = startedge, guard = 2*numedges;
  do {
    if (entered(e)) drawFace(g, e);
    if (wall(e) || wall(Sym(e))) e = Lpr(e);
    else e = Lpr(Sym(e));
  } while ((e != startedge) && (guard-- >= 0));
  if (guard < 0) throw new triangulationCorruptException();
}
```

Fig. 3: Traversing the faces of a convex subdivision

4 Bounds on maximal independent sets

Kirkpatrick [13] used a simple greedy algorithm to compute a maximal independent set (MIS) of low-degree vertices: Start by marking vertices of degree 12 or more. While there is an unmarked vertex, choose one for the MIS and mark it and its neighbors. His analysis gave an MIS with at least 1/24 of the vertices, which can be improved to 1/18 by initially marking vertices of degree 9 or more. Edelsbrunner [7, Problem 9.9] notes that choosing vertices of degree less than 13 in order of increasing degree gives an MIS of 1/7 of the vertices. We show that if vertices are deleted, not just marked, then choosing vertices of degree less than 10 in order of increasing degree gives an MIS of 1/6 of the vertices.

The algorithm is simple: Mark *high-degree* vertices, of degree 10 or more, then "delete" all marked vertices by decrementing the degrees of their neighbors. While vertices remain, choose a vertex of lowest degree for the MIS and delete it and its neighbors by marking them and decrementing the degrees of their neighbors. This can be implemented to run in linear time by storing a degree field for each vertex and maintaining ten lists (doubly-linked to support $O(1)$-time insertion and deletion of nodes) for vertices with degrees 0 through 9.

Let n_i, for $0 \leq i \leq 9$, be the number of vertices in the MIS that were chosen when their degree field was i. Thus, the size of the MIS is

$$\sum_{0 \leq i \leq 9} n_i. \tag{1}$$

Let n_h be the number of high-degree vertices. Because each vertex is deleted once,

$$n_h + \sum_{0 \leq i \leq 9} (i+1)n_i = n. \tag{2}$$

The deletion of high-degree vertices removes, at a minimum, $10n_h$ edges minus the number of edges in the graph induced by the high-degree vertices: zero edges when $n_h \leq 1$, at most one edge when $n_h = 2$, and at most $3n_h - 6$ when $n_h > 2$. Thus, the first deletion removes at least $\min\{10n_h, 9n_h + 1, 7n_h + 6\}$ edges. Later, when a vertex of degree i is deleted, its i neighbors all have degree at least i, too. At a minimum, deletion eliminates one edge for each n_1, and $(i+1)i - (3(i+1) - 6) = i(i-2) + 3$ edges for each n_i with $2 \leq i \leq 9$. Since there are initially at most $3n - 6$ edges,

$$\min\{10n_h, 9n_h + 1, 7n_h + 6\} + n_1 + \sum_{2 \leq i \leq 9} (i(i-2) + 3)n_i \leq 3n - 6. \tag{3}$$

This gives us three linear programming problems to find the minimum-size MIS according to (1), subject to (2) & (3). One minimum feasible solution has $n_4 = 3$, $n_5 = (n - 15)/6$, and the rest zero, for an independent set of more than 1/6 of the vertices. Patrice Belleville (private communication) has shown that this analysis is still pessimistic, and that the worst-case denominator is in [5.2, 5.5]. In our tests with Delaunay triangulations, the average MIS found by this algorithm contains 30% of the vertices.

Since our algorithms do not require the MIS to have low-degree vertices, 4- or 5-colorings could be used to obtain in-

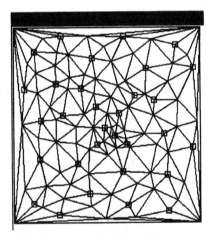

Fig. 4: MIS of 31/100 vertices

dependent sets guaranteeing 1/4 or 1/5 of the vertices. The best algorithm known for 4-coloring a planar graph takes quadratic time and must handle 633 cases [20], but 5-coloring in linear time is not especially difficult [3, 20]. The greedy algorithm above does avoid high degree vertices; it also gives us more freedom to select "important points" for our terrain model application.

5 Dealing with degenerate point sets

Thus far we have assumed that the Delaunay triangulation is determined canonically from the data, which need not be true for degenerate input: in a quadrilateral formed by four co-circular points, either of the two diagonals can be chosen to complete the triangulation. Strict adherence to the empty circle condition will choose both or neither, depending on the definition of "empty." Unfortunately, if the reconstruction chooses a different triangulation than the one used to construct the permutation, then the canonical traversal may not locate all of the points in a phase and the reconstruction will slow down.

One natural way to define a canonical triangulation in the presence of degeneracies is to enforce general position by an infinitesimal perturbation of the points. This succeeds with trapezoidations of line segments, where degeneracies are endpoints having the same x coordinate—if ties are broken by y-coordinates, then this is equivalent to perturbing the endpoints to general position by an infinitesimal skew transformation. Most perturbation methods described for the Delaunay triangulation, however, use a global index or memory address of a point. Since we permute the points, this is not an option. ("Simulation of Simplicity" [8] can be implemented by using local indices determined by lexicographic ordering.)

The *Delaunay graph* of a set S of points is a graph with vertex set S in which there is an edge between $p, q \in S$ if and only if there is a closed disc D in the plane with $D \cap S = \{p, q\}$. The Delanunay graph is canonically determined by the set S and has convex faces whose vertices are co-circular; when S is in general position, the Delaunay graph is the Delaunay triangulation.

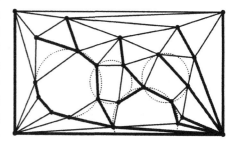

Fig. 5: Spanning tree in Delaunay graph of degenerate point set

One could obtain a canonical triangulation by triangulating each face of the Delaunay graph canonically; e.g., by connecting vertices to the highest vertex in each face. This solution works, but it does increase the degree of the highest vertex on a face, making such vertices less likely to be chosen in an independent set even though they may otherwise be unimportant.

Notice, however, that reconstruction depends on a canonical *traversal* of the faces of the Delaunay graph, and not a canonical *triangulation*. It is sufficient to triangulate each convex face such that the traversal visits all triangles within a face before it exits that face. Doing so uses fewer cases than Simulation of Simplicity.

Figure 5 illustrates the spanning tree chosen in a Delaunay graph by the selection rule of Section 3. A triangulation of this graph having the same spanning tree is called *traversable*, since our traversal visits all triangles within any face before it exits the face. We can build traversable triangulations with a standard incremental construction [11] if we carefully consider diagonal swaps for quadrilaterals with co-circular vertices.

Lemma 5 *An incremental algorithm builds a traversable triangulation for a set of points S. After point location, insertion of a point p takes time proportional to its degree.*

Proof: For this abstract, we simply sketch the changes to the standard incremental algorithm [11] for inserting a point p.

To insert p, we locate the triangle containing p and perform a standard insertion [11], creating edges joining p to the triangle vertices, and swapping diagonals to eliminate adjacent triangles whose circumcircles contain p. This forms all edges of the Delaunay graph that are incident on p. We now check that the incident faces of the Delaunay graph have traversable triangulations. In other words, that the vertices that are now neighbors of p do not select new spanning tree edges using the rule of Section 3.

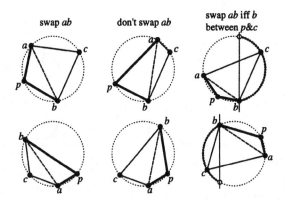

Fig. 6: Cases for swapping ab when new point p is on circumcircle of $\triangle abc$

Consider a quadrilateral $pbca$, formed from adjacent triangles, with all four points co-circular. As illustrated in Figure 6, we decide whether to swap diagonal $\overline{ab}$ for $\overline{pc}$ based on the x-coordinate order of a, p, b, and maybe c. If p is right or left of both a and b then swap $\overline{ab}$ iff a is between p and b. Otherwise, swap $\overline{ab}$ iff b is between p and c.

In the first case of figure 6, one must check whether the next edge out of a is in the Delaunay graph, and perform swaps as long as it is not. In all other cases, at most one "degenerate swap" is necessary to make the triangulation of the incident face traversable again, since both $\overline{pa}$ and $\overline{pb}$ are edges of the Delaunay graph. ∎

Theorem 6 *The Permutation Computation and Reconstruction algorithms work correctly in $O(n)$ time if degeneracies are handled according to Lemma 5.*

Proof: We sketch the main difficulty, which is to show that the independent set that was deleted from one triangulation will be an independent set when it is reinserted in another triangulation.

Consider replacing the vertices of the independent set. If the interior of any circle associated with a face of the Delaunay graph contained two vertices, then they would be neighbors and the set would not be independent. Thus, each face has at most one vertex associated with one of its triangles. Inserting the vertices into the faces first ensures that the Delaunay graph is unchanged during traversal and initial insertion; we then restoring the traversability of the triangulation by swaps.

Lemma 5 then insures that the total change is proportional to vertex degrees, which sum to linear. ∎

6 Experiments

With Bernd Juenger, we have been performing experiments to compare reconstruction and construction times using various simple point location schemes. Figure 7 shows timing of Java code (using the Sun beta JIT compiler on a 170Mhz ultraSparc) for the incremental construction in Guibas and Stolfi [11] in which points are located by walking through triangles from edge zero, a central edge, or the most recent edge, versus the reconstruction using the traversal and using the walk through triangles in the order given by the permutation (with explicit signals for end-of-phase.)

The good performance of the reconstruction that walks the permutation order leads to two questions related to multiple point location: In a triangulation of n points in the plane, show that there exists a path on m given vertices that has $O(n)$ crossings with triangulation edges. Then, find such a path in $O(n + m)$ time.

7 Conclusion

We have shown that simple ideas give a method to hide geometric structure in a permutation of the data so that, for example, a Delaunay triangulation can be reconstructed in $O(n)$ time from the permutation. These ideas also apply to trapezoidations of segments and constrained Voronoi diagrams in the plane, and to convex hulls in 3-d.

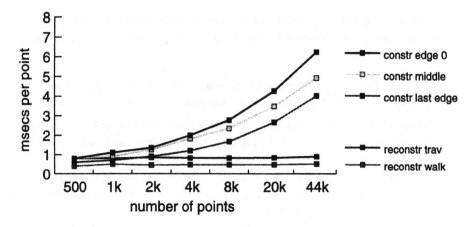

Fig. 7: Timing results for construction vs reconstruction

Acknowlegement

The data for figure 1 was provided by Facet Decision Systems. We thank them for their support of this research. Bernd Juenger performed the timing comparisons in figure 7.

References

[1] Franz Aurenhammer. Voronoi diagrams—A survey of a fundamental geometric data structure. *ACM Computing Surveys*, 23(3):345–405, 1991.

[2] Marshall Bern and David Eppstein. Mesh generation and optimal triangulation. In F. K. Hwang and D.-Z. Du, editors, *Computing in Euclidean Geometry*. World Scientific, March 1992.

[3] Norishige Chiba, Takao Nishizeki, and Nobuji Saito. A linear 5-coloring algorithm of planar graphs. *Journal of Algorithms*, 2:317–327, 1981.

[4] M. de Berg, M. van Kreveld, R. van Oostrum, and M. Overmars. Simple traversal of a subdivision without extra storage. *Int. J. of GIS*, 11, 1997.

[5] M. Deering. Geometry compression. *Computer Graphics*, pages 13–20, 1995. Proceedings of SIGGRAPH '95.

[6] H. Djidjev and A. Lingas. On computing the Voronoi diagram for restricted planar figures. In *WADS '91: Second Workshop on Data Structures and Algorithms*, number 519 in Lecture Notes in Computer Science, pages 54–64. Springer-Verlag, 1991.

[7] Herbert Edelsbrunner. *Algorithms in Combinatorial Geometry*. Springer-Verlag, Berlin, 1987.

[8] Herbert Edelsbrunner and Ernst Peter Mücke. Simulation of simplicity: A technique to cope with degenerate cases in geometric algorithms. *ACM Transactions on Graphics*, 9(1):66–104, 1990.

[9] Michael T. Goodrich and Roberto Tamassia. Dynamic trees and dynamic point location. In *Proceedings of the 23rd Annual ACM Symposium on Theory of Computing*, pages 523–533, 1991. To appear in *SIAM Journal on Computing*, 1997.

[10] L. J. Guibas, D. E. Knuth, and M. Sharir. Randomized incremental construction of Delaunay and Voronoi diagrams. *Algorithmica*, 7:381–413, 1992.

[11] Leonidas Guibas and Jorge Stolfi. Primitives for the manipulation of general subdivisions and the computation of Voronoi diagrams. *ACM Transactions on Graphics*, 4(2):74–123, 1985.

[12] H. Hoppe. Progressive meshes. *Computer Graphics*, pages 99–108, 1996. Proceedings of SIGGRAPH '96.

[13] D. Kirkpatrick. Optimal search in planar subdivisions. *SIAM Journal on Computing*, 12:28–35, 1983.

[14] D. T. Lee and F. P. Preparata. Computational geometry: a survey. *IEEE Trans. Comput.*, C-33:1072–1101, 1984.

[15] Ketan Mulmuley. *Computational Geometry: An Introduction Through Randomized Algorithms*. Prentice-Hall, Englewood Cliffs, N.J., 1993.

[16] Atsuyuki Okabe, Barry Boots, and Kokichi Sugihara. *Spatial Tessellations: Concepts and Applications of Voronoi Diagrams*. John Wiley & Sons, 1992.

[17] Joseph O'Rourke. *Computational Geometry in C*. Cambridge University Press, 1994.

[18] T. K. Peucker, R. J. Fowler, J. J. Little, and D. M. Mark. The triangulated irregular network. In *Amer. Soc. Photogrammetry Proc. Digital Terrain Models Symposium*, pages 516–532, 1978.

[19] Franco P. Preparata and Michael I. Shamos. *Computational Geometry—An Introduction*. Springer-Verlag, New York, 1985.

[20] Neil Robertson, Daniel P. Sanders, Paul Seymour, and Robin Thomas. Efficiently four-coloring planar graphs. In *Proceedings of the 28th Annual ACM Symposium on Theory of Computing*, pages 571–575, 1996.

[21] R. Seidel. Backwards analysis of randomized geometric algorithms. In J. Pach, editor, *New Trends in Discrete and Computational Geometry*, volume 10 of *Algorithms and Combinatorics*, pages 37–68. Springer-Verlag, 1993.

[22] Raimund Seidel. A method for proving lower bounds for certain geometric problems. In Godfried T. Toussaint, editor, *Computational Geometry*, pages 319–334. North Holland, Amsterdam, 1985.

[23] Cao An Wang and Francis Chin. Finding the constrained Delaunay triangulation and constrained Voronoi diagram of a simple polygon in linear time. In Paul Spirakis, editor, *Algorithms–ESA '95*, number 979 in Lecture Notes in Computer Science, pages 280–294. Springer-Verlag, 1995.

Approximating Satisfiable
Satisfiability Problems
[Extended Abstract]

Luca Trevisan*

Abstract. We study the approximability of the Maximum Satisfiability Problem (Max SAT) and of the boolean k-ary Constraint Satisfaction Problem (Max kCSP) restricted to satisfiable instances. For both problems we improve on the performance ratios of known algorithms for the unrestricted case.

Our approximation for satisfiable MAX 3CSP instances is better than any possible approximation for the unrestricted version of the problem (unless P= NP). This result implies that the requirements of perfect completeness and non-adaptiveness weaken the acceptance power of PCP verifiers.

We also present the first non-trivial results about PCP classes defined in terms of free bits that collapse to P.

1 Introduction

In the Max SAT problem we are given a boolean formula in conjunctive normal form (CNF) and we are asked to find an assignment of values to the variables that satisfies the maximum number of clauses. More generally, we can assume that each clause has a non-negative weight and that we want to maximize the total weight of satisfied clauses.

Max SAT is a standard NP-hard problem and a considerable research effort has been devoted in the last two decades to the development of approximation algorithms for it. An r-approximate algorithm for Max SAT (where $0 \le r \le 1$) is a polynomial-time algorithm that given a formula finds an assignment that satisfies clauses of total weight at least r times the optimum.

Max SAT is also the prototypical element of a large family of optimization problems in which we are given a set of weighted *constraints* over (not necessarily boolean) variables, and we want to find an assignment of values to such variables that maximizes the total weight of satisfied constraints. Problems of this kind, called constraint satisfaction problems, are of central interest in Artificial Intelligence. Their approximability properties are of interest in Theory of Computing since they can express the class MAX SNP [23, 19] and the computation of PCP verifiers [2, 25]; complete classifications of their approximability properties, for the case of boolean variables, appear in [9, 20]. We call Max kCSP

---

* trevisan@cui.unige.ch. Centre Universitaire d'Informatique, Université de Genève, Rue Général-Dufour 24, CH-1211, Genève, Switzerland.

the constraint satisfaction problem where every constraint involves at most k variables.

In this paper we consider the following restriction of the problem of r-approximating MAX SAT and MAX kCSP: given a satisfiable instance of MAX SAT (resp. MAX kCSP), find in polynomial time an assignment that satisfies at a fraction r of the total weight of clauses (resp. constraints). The problem of approximating constraint satisfaction problems restricted to satisfiable instances has been considered by Petrank [24], and called *approximation problem at gap location one*. Petrank observed that MAX SAT remains MAX SNP-complete when restricted to satisfiable instances, and proved that the same is true for other problems, such as MAX 3-COLORABLE SUBGRAPH and MAX 3-DIMENSIONAL MATCHING. More recenlty, Khanna, Sudan and Williamson [20] proved that for any MAX SNP-complete constraint satisfaction problem for which deciding satisfiability is NP-hard, the restriction to satisfiable instances remains MAX SNP-complete.

In partial constrast with the results of Petrank and of Khanna et al. we prove that restricting MAX SAT and MAX kCSP to satisfiable instances makes the problems somewhat easier, since we can exploit satisfiability to develop new algorithms with improved approximation guarantees. Our result for MAX 3CSP is particularly strong, since we will present a .514-approximate algorithm for satisfiable MAX 3CSP, while .501-approximation is NP-hard for the unrestricted MAX 3CSP problem [17]. Thus, the satisfiability restriction is not sufficient to turn a MAX SNP-complete problem into a PTAS problem, but *can change the approximation threshold*.[2] Our result for GL1-MAX 3CSP can also be reworded in the PCP terminology, and yields the interesting fact that *verifiers with perfect completeness are strictly weaker than verifier with completeness* $1 - \epsilon$.

In the rest of this section we describe in more details our results, partly clarifying the obscure terminology of the previous paragraph.

The Maximum Satisfiability Problem. The MAX SAT problem appears in a paper of Johnson [18] which is the first paper where the term "approximation algorithm" was introduced. Johnson proved that his algorithm was 1/2-approximate. It has been recently showed that Johnson's algorithm is indeed 2/3-approximate [8]. In the last five years, several improved approximation algorithms for MAX SAT and its restricted versions MAX 2SAT and MAX 3SAT have been developed; we summarize such previous results in Table 1. There is a corresponding history of continuous improvements in the non-approximability; we do not mention it here (the interested reader can find it in [5]), and we only recall that the best known hardness is $7/8+\epsilon$ due to Håstad [17], and it still holds when restricting to satisfiable instances with exactly three literals per clause.

---

[2] The approximation threshold r_A of an optimization problem A is defined as

$$r_A = \sup\{r : A \text{ admits an } r\text{-approximate algorithm }\}$$

Max SAT	Max 3SAT	Due to
.75	.75	[27]
.75	.75	[14]
.758	.765*	[15] (using [14])
.762*	.77*	[11] (using [14, 15])
.765	.769	[22] (using [27, 14, 15])
	.801	[26] (using [11])
.768		[1] (using [14, 15, 11, 22, 26])
.8	.826	This paper for satisfiable instances

Table 1. Evolution of the approximation factors for Max SAT and Max 3SAT. The factors depicted with a * do not appear explicitly in the referenced papers [15, 11].

OUR RESULTS. We present a polynomial-time algorithm that, given a satisfiable Max SAT instance, satisfies a fraction .8 of the total weight of clauses, and an algorithm that, given a satisfiable Max 3SAT instance, satisfies a fraction .826 of the total weight of clauses.

SOURCE OF OUR IMPROVEMENT. In both cases, we show how to reduce the given instance to an instance without unit clauses. The reduction sequentially applies a series of substitutions of values to variables. The .826 approximation for Max 3SAT then follows by adapting the analysis of [26] to the case of no unit clauses. The .8 approximation for Max SAT involves the use of known algorithms, with a couple of small changes.

Maximum k-ary Constraint Satisfaction Problem. The approximability of the Max kCSP problem is an algorithmic rephrasing of the accepting power of PCP verifiers that non-adaptively read k bits of the proof. The restriction to satisfiable instances of Max kCSP corresponds to the restriction to non-adaptive PCP verifiers with *perfect completeness*.[3] The requirement of perfect completeness and non-adaptiveness appeared in the first defitions of PCP and in several improved proofs of it [3, 2, 6, 7]. Recently, adaptiveness (with perfect completeness) was used in [5], and a verifier without perfect completeness (but non-adaptive) appears in [17]. The latter result was of particular interest, because it formerly appeared that "current techniques" could only yield PCP constructions with perfect completeness. The study of which PCP classes lie in P was initiated in [5]. The best known approximation for Max kCSP, for general k, is 2^{1-k} [25].

OUR RESULTS. We improve the approximation to $(k+1)2^{-k}$ for satisfiable instances.

---

[3] A verifier has perfect completeness if it accepts a correct proof with probability 1.

Source of our improvement. We use again substitutions (but of a more general kind) as a preprocessing step. The substitutions reduce the problem to an instance where any k-ary constraint has at least $k+1$ satisfying assignments, and any such assignment is consistent with the set of linear constraints. We then take a random feasible solution for the set of linear constraints, and this satisfies each constraint with probability at least $(k+1)2^{-k}$.

Maximum 3-ary Constraints Satisfaction Problem (and 3-query PCP)
The PCP Theorem states that membership proofs for any NP language can be probabilistically checked by a verifier that uses logarithmic randomness, has *perfect completeness, soundness*[4] $1/2$ and *non-adaptively* reads a constant number of bits from the proof. Since its appearence, there was interest in understanding the tightest possible formulation of the PCP Theorem, especially in terms of how low the number of query bits could be made.

It is easy to see that, with two queries, it is impossible to get perfect completeness, while with 3 it is possible (see e.g. [5]). The challenging question arises of determining which is the best soundness achievable with three bits and perfect completeness. The state of the art for this question is that NP can be checked with soundness $.75 + \epsilon$ [17], while this is impossible with soundness $.367$ [26], unless $P = NP$. Furthermore, it is possible to check NP with three queries, soundness $.5 + \epsilon$ and completeness $1 - \epsilon$ for any $\epsilon > 0$ [17]. The latter result implies that MAX 3SAT is hard to approximate within $7/8 + \epsilon$, but not when restricted to satisfiable instances. A different and more complicated proof was needed to prove the $7/8 + \epsilon$ hardness result also for satisfiable instances [17]. It was an open question whether soundness $.5 + \epsilon$ is achievable with three queries and perfect completeness.

Satisfiable instances	Arbitrary instances	Due to
.125	.125	[23]
.299		[5]
	.25	[25]
.367	.367	[26]
.514		**This paper**

Table 2. Evolution of the approximation factors for MAX 3CSP with and without the satisfiability promise.

Our result. We show that for PCP verifiers of NP languages with three non-adaptive queries and perfect completeness, the soundness is bounded away from .5, and has to be at least .514 (unless $P = NP$).

---

[4] Roughly speaking, the *soundness* is the probability of accepting a wrong proof (see Definition 6).

SOURCE OF OUR IMPROVEMENT. We give a .514-approximate algorithm for satisfiable instances of MAX 3CSP. A preprocessing step, which is a simplification of the one used for our MAX kCSP result, reduces the instance to an instance where any constraint has at least 3 satisfying assignments and each satisfying assignment is consistent with the set of linear constraints. We then apply two algorithms and take the best solution. In one algorithm, we reduce all the constraints to 2SAT using gadgets, extending an idea of [26]. In the other algorithm we take a random solution for the set of linear constraints.

Free bits. Besides the number of query bits, there is another very important parameter of the verifier that is studied in the field of probabilistic proof-checking: the number of *free bits*. It is a *relaxation* of the notion of query bit: if a verifier queries q bits on the proof, than it uses at most f free bits, but a verifier using f free bits can red arbitrarily many bits. The interest in this parameter (implicit in [13] and explicitly introduced in [7]) lies in the fact that the "efficiency" of the reduction from PCP to MAX CLIQUE [12] depends only on the number of free bits of the verifier (indeed, it depends only on the *amortized* number of free bits, but we will not exploit the latter notion here). Since the same reduction is used to derive the best known hardness result for MIN VERTEX COVER, further improvements in the hardness of approximating MIN VERTEX COVER could be obtained by improved PCP constructions with low free bits complexity. Roughly speaking, a verifier uses f free bits if, after making its queries to the proof, there at most 2^f possible answers that make him accept (this is why f cannot be larger than the number of query bits.) This definition has been used almost always, including in Håstad's papers on MAX CLIQUE (where he used the free bit-efficient *complete test.*) One exception is [5], where an *adaptive* version of the definition of free bits is used. We also mention that the free bit parameter has almost always been used for verifiers with perfect completeness (Bellare et al. [5] also show that one can always reduce the free bit complexity by reducing the completeness.) However, the currently best hardness result for MIN VERTEX COVER is due to Håstad [17] and uses a verifier with low free bit complexity and completeness $1 - \epsilon$, for any $\epsilon > 0$.

Even in the simple case of the *non-adaptive* definition and of *perfect completeness* there were basically no result about PCP classes with low free bit complexity collapsing to P. The only result was that, with perfect completeness, it is impossible to characterize NP with only 1 free bit, while $\log 3$ free bits are sufficient [5]. It has been conjectured that with $\log 3$ free bits and perfect completeness it is possible to achieve any soundness.

OUR RESULT. Under the weak (non-adaptive) definition of free bits, we prove that a verifier with perfect completeness, that uses f free bits, and whose soundness is less than $2^f/2^{2^f-1}$ can only capture P.

SOURCE OF OUR IMPROVEMENT. We adapt the previously described reductions and algorithms.

Organization of the Paper. Basic definitions on constraint satisfaction problems, PCP, and gadgets are given in Section 2. We prove a simple combinatorial result in Section 3. We present the MAX SAT approximation algorithms in Section 4 and the MAX kCSP approximation algorithms (as well as the implications with PCP classes) in Sections 5 and 6. The free bit parameter is discussed in Section 7. Several proofs are omitted or sketched in this extended abstract. The reader is referred to the full version of this paper for more details.

2 Definitions

For an integer n, we denote by $[n]$ the set $\{1, \ldots, n\}$. We begin with a definition of constraint satisfaction problem, that unifies the definitions of all the problems we are interested in.

Definition 1. A (k-ary) *constraint function* is a boolean function $f : \{0,1\}^k \to \{0,1\}$.

When it is applied to variables $x_1, \ldots, x_k$ (see the following definitions) the function f is thought of as imposing the constraint $f(x_1, \ldots, x_k) = 1$.

Definition 2. A *constraint family* $\mathcal{F}$ is a finite collection of constraint functions. The *arity* of $\mathcal{F}$ is the maximum number of arguments of the functions in $\mathcal{F}$. A *constraint* C over a variable set $x_1, \ldots, x_n$ is a pair $C = (f, (i_1, \ldots, i_k))$ where $f : \{0,1\}^k \to \{0,1\}$ is a constraint function and $i_j \in [n]$ for $j \in [k]$. The constraint C is said to be *satisfied* by an assignment $\mathbf{a} = a_1, \ldots, a_n$ to $x_1, \ldots, x_n$ if $C(a_1, \ldots, a_n) \overset{\text{def}}{=} f(a_{i_1}, \ldots, a_{i_k}) = 1$. We say that constraint C is *from* $\mathcal{F}$ if $f \in \mathcal{F}$.

We will sometimes write a constraint $(f, (i_1, \ldots, i_k))$ as $(f(x_{i_1}, \ldots, x_{i_k}) = 1)$.

Definition 3 (Constraint famillies). A *literal* is either a variable or the negation of a variable. We define the following constraint families:

kCSP: the set of all h-ary functions, $h \leq k$.
kCSP^i: the set of all k-ary functions with i satisfying assignments.
kSAT: the set of all functions expressible as the **or** of at most k literals.
SAT: the set of all functions expressible as the **or** of literals.

A constraint function $f(x_1, \ldots, x_k)$ is *linear* if either $f(x_1, \ldots, x_k) = x_1 \oplus \ldots \oplus x_k$ or $f(x_1, \ldots, x_k) = 1 \oplus x_1 \oplus \ldots \oplus x_k$, where $\oplus$ is the **xor** operator.

Definition 4 (Constraint satisfaction problems). For a function family $\mathcal{F}$, MAX $\mathcal{F}$ is the optimization problem whose instances consist of m weighted constraints from $\mathcal{F}$, on n variables, and whose objective is to find an assignment to the variables which maximizes the total weight of satisfied constraints.

Note that Definitions 3 and 4 give rise to the problems MAX SAT, MAX 3SAT, and MAX kCSP, that are defined in the standard way.

Given an instance φ of a constraint satisfaction problem, we denote by $LIN(\varphi)$ the set of linear constraints of φ.

GL1-MAX $\mathcal{F}^5$ is the restriction of MAX $\mathcal{F}$ to instances where all the constraints are simultaneously satisfiable.

We say that a maximization problem is r-approximable $r < 1$ if there exists a polynomial-time algorithm that, for any instance, finds a solution whose cost is at least r times the optimum (such a solution is said to be r-approximate).

We also need the definition of *gadgets*.

Definition 5 (Gadget [5]). For $\alpha \in \mathcal{R}$, a function $f : \{0,1\}^k \to \{0,1\}$, and a constraint family $\mathcal{F}$: an α-*gadget* reducing f to $\mathcal{F}$ is a finite collection of constraints C_j from $\mathcal{F}$ over *primary variables* $x_1, \ldots, x_k$ and *auxiliary variables* $y_1, \ldots, y_n$, and associated real weights $w_j \geq 0$, with the property that, for boolean assignments $\mathbf{a}$ to $x_1, \ldots, x_k$ and $\mathbf{b}$ to $y_1, \ldots, y_n$, the following are satisfied:

$$(\forall \mathbf{a} : f(\mathbf{a}) = 1) \ (\forall \mathbf{b}) : \sum_j w_j C_j(\mathbf{a}, \mathbf{b}) \leq \alpha, \tag{1}$$

$$(\forall \mathbf{a} : f(\mathbf{a}) = 1) \ (\exists \mathbf{b}) : \sum_j w_j C_j(\mathbf{a}, \mathbf{b}) = \alpha, \tag{2}$$

$$(\forall \mathbf{a} : f(\mathbf{a}) = 0) \ (\forall \mathbf{b}) : \sum_j w_j C_j(\mathbf{a}, \mathbf{b}) \leq \alpha - 1. \tag{3}$$

Gadgets can be used in approximation algorithms in the following way [26]. Assume we have a satisfiable instance of a constraint satisfaction problem, with constraints of total weight m, and there is α-gadget reducing each such constraint to 2SAT. Then we can build a 2SAT instance ψ whose optimum is αm and such that any solution of cost c for ψ has cost at least $c - (\alpha - 1)m$ for the old instance.

In a more general setting, assume that, for $i = 1, \ldots, k$, we have type-i constraints of total weight w_i, and that there exists an α_i-gadget reducing type-i constraints to 2SAT. Assume also that the whole CSP instance be satisfiable. Then the optimum of the instance is $\sum_i w_i$; applying all the gadgets we have a 2SAT instance ψ whose optimum is $\sum_i \alpha_i w_i$.

Applying a β-approximate algorithm to ψ, we obtain a solution for the original instance whose cost is at least

$$\sum_i \beta \alpha_i w_i - \sum_i (\alpha_i - 1)w_i = \sum_i (\beta - (1 - \beta)(\alpha_i - 1))w_i .$$

In the following, we will refer to such kind of reductions as the *TSSW method*. The FGW [15, 11] algorithm for MAX 2SAT is .931-approximate.

We conclude this section with the definition of PCP classes and their relation with the approximability of MAX kCSP.

---

[5] GL1 stands for "Gap Location 1", which is the terminology of Petrank [24].

Definition 6 (Restricted verifier). A *verifier* V for a language L is a probabilistic polynomial-time Turing machine that during its computations has oracle
access to a string called *proof*. We denote by $\mathbf{ACC}[V^{\pi}(x)]$ the probability over
its random tosses that V accepts x when accessing proof π. We also denote by
$\mathbf{ACC}[V(x)]$ the maximum of $\mathbf{ACC}[V^{\pi}(x)]$ over all proofs π. We say that

- V *has query complexity* q (where q is an integer) if for any input x, any proof
  π, and any outcome of its random bits, V reads at most q bits from π;
- V *has soundness* s if, for any $x \notin L$, $\mathbf{ACC}[V(x)] \leq s$;
- V *has completeness* c if, for any $x \in L$, $\mathbf{ACC}[V(x)] \geq c$. V has *perfect
  completeness* if it has completeness 1.

Definition 7 (PCP classes). $L \in \mathrm{PCP}_{c,s}[\log, q]$ if L admits a verifier V with
completeness c, soundness s, query complexity q, and that uses $O(\log n)$ random
bits, where n is the size of the input. We say that $L \in \mathbf{naPCP}_{c,s}[\log, q]$ if V, in
addition, queries the q bits *non-adaptively*.

Theorem 8 [2]. *If* GL1-MAX kCSP *is r-approximable, then* $\mathbf{naPCP}_{1,s}[\log, k] \subseteq$
P *for any $s < r$.*

3 Some Applications of the Linear Algebra Method

The linear algebra method in combinatorics [4] is a collection of techniques that
prove combinatorial results making use of the following well-known fact: if we
have a set of n-dimensional vectors that are linearly independent, then the size
of the set is at most n. In this section we will provide some definitions and prove
easy bounds using linear algebra. Despite the triviality of the results, they will
have powerful applications in Sections 5 and 6.

In the following, we consider vectors in $\{0,1\}^n$ and denote by $\oplus$ the bitwise
exclusive-or operation between vectors.

Definition 9. A *satisfying* table for a constraint function $f : \{0,1\}^k \to \{0,1\}$
with s satisfying assignments is a $s \times k$ boolean matrix whose rows are the
satisfying assignments of f.

The satisfying table is not unique since the matrix representation imposes an
order to the assignments. Even if it would be more natural to represent the
satisfying assignments as a set of vectors rather than a matrix, the latter representation is more suitable for combinatorial arguments, especially because we
can sometimes *see it as a set of k vectors of length s*.

Definition 10. A collection $\mathbf{x}_1, \ldots, \mathbf{x}_m$ of elements of $\{0,1\}^n$ is *k-dependent* if
there are values $a_0, \ldots, a_m \in \{0,1\}$ such that $1 \leq |\{i = 1, \ldots, m : a_i = 1\}| \leq k$
and $a_1 \mathbf{x}_1 \oplus \ldots \oplus a_m \mathbf{x}_m = a_0 \mathbf{1}$. A collection is *dependent* if it is k-dependent for
some k. A collection is *(k-)independent* if it is not *(k-)dependent*.

More intuitively, the vectors $x_1, \ldots, x_m$ are k-independent if any **xor** of at most k of them is different from 0 and from 1.

Lemma 11. *If* $x_1, \ldots, x_m \in \{0,1\}^n$ *are 2-independent, then* $m \le 2^{n-1} - 1$. *The bound is tight.*

Proof. All the $2m+2$ vectors $0, x_1, \ldots, x_m, 1, (1 \oplus x_1), \ldots, (1 \oplus x_m)$ are distinct. Therefore $2m + 2 \le 2^n$. We omit the proof of tightness. ☐

Lemma 12. *If* $x_1, \ldots, x_m \in \{0,1\}^n$ *are independent, then* $m \le n-1$. *The bound is tight.*

Proof. The $m + 1$ vectors $1, x_1, \ldots, x_m$ are distinct and linearly independent in the ordinary sense. Therefore $m + 1 \le n$. We omit the proof of tightness. ☐

Let now f be a k-ary constraint function with s satisfying assignments, and M be a satisfying table for f. If the columns of M are 2-independent, then $k \le 2^{s-1} - 1$, that is $s \ge 1 + \lceil \log(k+1) \rceil$, which implies $s = 2$ if $k = 1$ and $s \ge 3$ if $k \ge 2$. If the columns of M are independent, then we can draw the stronger statement $s \ge k + 1$.

4 The MAX SAT Algorithms

Lemma 13. *If* GL1-MAX SAT *(resp.* GL1-MAX 3SAT*) restricted to instances without unit clauses is r-approximable, then it is r-approximable for arbitrary instances.*

Proof (Sketch). Let φ be a generic instance of GL1-MAX SAT. We will show how to produce an instance ψ of GL1-MAX SAT with no unit clauses such that given an assignment satisfying a fraction r of the clauses of ψ we are able to find an assignment satisfying a fraction at least r of the clauses of φ. If φ has no unit clauses then we are done. Otherwise we apply the following transformation:

1. For any unit clause $(x) \in \varphi$, we substitue 1 in any occurrence of x in φ.
2. For any unit clause $(\bar{x}) \in \varphi$, we substitue 0 in any occurrence of x in φ.

The transformation preserves satisfiability, does not contradict any clause, satisfies a certain number $s \ge 0$ of clauses. An assignment that satisfies a fraction r of the clauses in the new instance (i.e. $r(m - s)$ clauses) can be extended to an assignment to the old instance that satisfies $r(m - s) + s \ge rm$ clauses. After the transformation, there can still be unit clauses (produced from the shrinking of formerly longer clauses); in this case we recurse until we are left with a formula without unit clause (the process must eventually terminate after a linear number of transformations, since each transformation step reduces the size of the input.) ☐

Lemma 14. *There exists a polynomial-time .826-approximate algorithm for* GL1-MAX 3SAT *without unit clauses.*

Proof (Sketch). We adapt the analysis of [26]. □

Lemma 15. *There exists a polynomial-time .8-approximate algorithm for* GL1-MAX SAT *without unit clauses.*

Proof (Sketch). We use: (i) Johnson's algorithm [18]; (ii) the FGW algorithm, extended to length-3 and lenght-4 clauses with the TSSW method, and to longer clauses with a method of [15]; (iii) we solve the 2SAT sub-instance and then we apply a method of [10]. □

The gadget for lenght-4 clauses is new, as well as the idea of combining the reduction technique of [10] with a 2SAT algorithm.

Theorem 16. *There exists a .8-approximate algorithm for* GL1-MAX SAT *and a .826-approximate algorithm for* GL1-MAX 3SAT.

5 The MAX kCSP Algorithm

Lemma 17. *There exists a polynomial-time algorithm that, given an instance of* MAX kCSP φ *and a set of linear constraints S, such that $(\varphi \cup S)$ is satisfiable, produces an assignment that satisfies all the constraints of S and a fraction $(k+1)/2^k$ of the constraints of φ.*

Proof. We say that the instance (φ, S) is *simplified* if, for any constraint C of φ, C is not linear, the columns of the satisfying table of C are independent, and any satisfying assignment of C is consistent with S. Observe that if $h \leq k$ is the arity of a constraint C in a simplified instance, then, by Lemma 12 C has at least $h+1$ satisfying assignments, and a random assignment satisfies it with probability at least $(h+1)/2^h \geq (k+1)/2^k$. If the instance is simplified, then we take a random feasible solution for S; it satisfies all constraints of S and, on the average, a fraction at least $(k+1)/2^k$ of the total weight of the constraints of φ. Derandomization is possible with the method of conditional expectation. If the instance is not simplified then we repeatedly apply the following procedure until we are left with a simplified instance:

1. If $\exists C \in \varphi$ that is linear, then $\varphi := \varphi - \{C\}$ and $S := S \cup \{C\}$;
2. If $\exists C \equiv (f(x_{i_1}, \ldots, x_{i_k}) = 1) \in \varphi$ the columns whose satisfying table are not independent, then C enforces a linear relation $x_{i_j} = a_0 \oplus \bigoplus_{h \neq j} a_h x_{i_h}$. Then we replace C by $(f(x_{i_1}, \ldots, x_{i_{j-1}}, a_0 \oplus \bigoplus_{h \neq j} a_h x_{i_h}, x_{i_{j+1}}, \ldots, x_{i_k}) = 1$ and we add the equation $x_{i_j} = a_0 \oplus \bigoplus_{h \neq j} a_h x_{i_h}$ to S.
3. If $\exists C \in \varphi$ one whose satisfying assignment is inconsistent with S, then we remove such satisfying assignment from the satisfying table of C.

Note that all the actions above reduce the size of φ, so we can only perform a linear number of actions. After an action is performed that transforms (φ, S) into (φ', S'), the following invariants are preserved:

1. There exists an assignment satisfying $(\varphi' \cup S')$.
2. A solution satisfying S' and a fraction r of the constraints of φ' also satisfies S and a fraction r of the constraints of φ.

□

Theorem 18. *There exists a polynomial-time $(k+1)/2^k$-approximate algorithm for GL1-MAX kCSP.*

Proof. Let φ be a satisfiable instance of MAX kCSP. Apply the algorithm of Lemma 17 to the instance $(\varphi, \emptyset)$.

□

Theorem 19. *For any $q \geq 3$, for any $s < (q+1)/2^q$, $naPCP_{1,s}[\log, q] \subseteq P$.*

The bound of Lemma 18 above is $1/2$ for MAX 3CSP. We will do better with semidefinite programming.

6 The MAX 3CSP Algorithm

Lemma 20. *Assume that GL1-MAX 3CSP is r-approximable in instances φ such that all constraints C of φ satisfy the following conditions*

1. *the columns of a satisfying table of C are 2-independent;*
2. *either C is linear or all its satisfying assignments are consistent with $LIN(\varphi)$.*

Then GL1-MAX 3CSP is r-approximable.

Proof (Sketch). Given a general instance φ of GL1-MAX 3CSP, we reduce it to an instance ψ satisfying properties 1 and 2. As usual, we run a series of modification steps until the required instance is generated. Each step is as follows

1. If a constraint $C \equiv (f(x_{i_1}, \ldots, x_{i_k}) = 1)$ has a 2-dependent satisfying table, then there are indices $j, h \in [k]$ and values $a_0, a_h \in \{0, 1\}$ such that $x_{i_j} = a_0 \oplus a_h x_{i_h}$ Then, we replace each occurrence of x_{i_j} by $a_0 \oplus a_h x_{i_h}$.
2. If a non-linear constraint C has an assignment that is inconsistent with $LIN(\varphi)$, then we remove such assignment from the satisfying table of C.

□

Lemma 21. *GL1-MAX 3CSP restricted to the instances of Lemma 20 is .5145-approximable.*

Proof. From Lemma 11, φ has no unit constraint, the 2-ary constraints can only be from 2SAT, the 3-ary constraints must have at least three satisfying assignments. Let m_2 be the total weight of 2SAT constraints, $m^{(3)}$, $m^{(4)}$, $m^{(5)}$, $m^{(6)}$, $m^{(7)}$ be the total weight of 3-ary constraints that have, respectively, 3, 4, 5, 6, and 7 satisfying assignments. We also let $m^{(4L)}$ be the total weight of 3-ary linear constraints and $m^{(4O)} = m^{(4)} - m^{(4L)}$.

We use to algorithms and take the best solution.

In the first algorithm, we simply consider a random feasible solution for $LIN(\varphi)$. On the average, the total weight of satisfied constraints is at least

$$\frac{3}{4}\, m_2 + \frac{3}{8}\, m^{(3)} + \frac{4}{8}\, m^{(4O)} + m^{(4L)} + \frac{5}{8}\, m^{(5)} + \frac{6}{8}\, m^{(6)} + \frac{7}{8}\, m^{(7)} \qquad (4)$$

Derandomization is possible using the method of conditional expectation.

The other algorithm uses the TSSW method and the FGW algorithm. We have to find gadgets reducing the various possible 3-ary constraints to 2SAT constraints. The new constructions (and the old ones that we use) are listed in Table 3. All the gadgets are computer-constructed using the linear programming method of [26] and are the best possible. Using the FGW algorithm with the

Source Constraint	Target Constraint	α	Due to
3SAT	2SAT	3.5	[26]
4SAT	2SAT	6	This paper
3CSP^3	2SAT	5.5	This paper
3CSP^4 not linear	2SAT	5.5	This paper
3CSP^4 linear	2SAT	11	[5]
3CSP^5	2SAT	8.25	This paper
3CSP^6	2SAT	5.5	This paper

Table 3. Gadgets used.

TSSW method and the gadgets of Table 3, we have an algorithm that satisfies constraints of total weight at least

$$.931\, m_2 + .6205\, m^{(3)} + .241\, m^{(4L)} + .6205\, m^{(4O)} + .43075\, m^{(5)} + .6205\, m^{(6)} + .7585\, m^{(7)} \qquad (5)$$

If we take the maximum of Equation (4) and (5), we have that the total weight of satisfied constraints is at least $.5145m$, where $m = m_2 + m^{(3)} + m^{(4L)} + m^{(4O)} + m^{(5)} + m^{(6)} + m^{(7)}$. $\qquad\square$

Theorem 22. *There exists a polynomial-time .5145-approximate algorithm for* GL1-MAX 3CSP.

Theorem 23. naPCP$_{1,.514}$[log, 3] $\subseteq$ P.

7 Free Bits

We define free bits as a property of boolean functions. There are two possible definitions.

Definition 24. A function $f : \{0,1\}^q \to \{0,1\}$ uses f *non-adaptive free bits* if it has at most 2^f satisfying assignments. It uses f *adaptive free bits* if it can be expressed by a DNF with at most 2^f terms such that any two terms are inconsistent. A PCP verifier uses f adaptive (resp. non-adaptive) free bits if for any input, and any fixed random string, its computation (which is a function of the proof) can be expressed as a boolean function that uses f adaptive (resp. non-adaptive) free bits. $\text{FPCP}_{c,s}[\log, f]$ is the class of languages admitting a PCP verifier with logarithmic randomness, completeness c, soundness s, that uses f adaptive free bits. The class $\textbf{naFPCP}_{c,s}[\log, f]$ is defines analogously by using the non-adaptive free bit parameter.

Regarding recent constructions of verifiers optimized for the free bit parameter, the verifiers that use the *Complete Test* [16] are non-adaptive, while the verifier that uses the *Extended Monomial Basis Test* [5] is adaptive.

We now state some results (the first ones with $f > 1$) about **naFPCP** classes that collapse to P.

Theorem 25. *The following statements hold:*

1. $\textbf{naFPCP}_{1,s}[\log, f] \subseteq \textbf{naPCP}_{1,s}[\log, 2^{2^f - 1} - 1]$.

2. $\textbf{naFPCP}_{1,s}[\log, f] \subseteq P$ *for all* $s > (2^f)/2^{2^f-1}$ *and* $f \geq \log 3$.

Acknowledgements

I thank Greg Sorkin and Madhu Sudan for having checked some of the gadget constructions of this paper. I am grateful to Pierluigi Crescenzi and, again, to Madhu for helpful discussions on free bits.

References

1. G. Andersson and L. Engebretsen. Better approximation algorithms and tighter analysis for SET SPLITTING and NOT-ALL-EQUAL SAT. Manuscript, 1997.
2. S. Arora, C. Lund, R. Motwani, M. Sudan, and M. Szegedy. Proof verification and hardness of approximation problems. In *Proc. of FOCS'92*, pages 14–23.
3. S. Arora and S. Safra. Probabilistic checking of proofs; a new characterization of NP. In *Proc. of FOCS'92*, pages 2–13, 1992.
4. L. Babai and P. Frankl. *Linear Algebraic Methods in Combinatorics (2nd Preliminary version)*. Monograph in preparation, 1992.
5. M. Bellare, O. Goldreich, and M. Sudan. Free bits, PCP's and non-approximability – towards tight results (4th version). Technical Report TR95-24, ECCC, 1996. Preliminary version in *Proc. of FOCS'95*.

6. M Bellare, S. Goldwasser, C. Lund, and A. Russell. Efficient probabilistically checkable proofs and applications to approximation. In *Proc. of STOC'94*, pages 294–304.

7. M. Bellare and M. Sudan. Improved non-approximability results. In *Proc. of STOC'94*, pages 184–193.

8. J. Chen, D. Friesen, and H. Zheng. Tight bound on Johnson's algorithm for MaxSAT. In *Proc. of CCC'97*. To appear.

9. N. Creignou. A dichotomy theorem for maximum generalized satisfiability problems. *JCSS*, 51(3):511–522, 1995.

10. P. Crescenzi and L. Trevisan. MAX NP-completeness made easy. Manuscript, 1996.

11. U. Feige and M. Goemans. Approximating the value of two provers proof systems, with applications to MAX 2SAT and MAX DICUT. In *Proc. of ISTCS'95*, pages 182–189.

12. U. Feige, S. Goldwasser, L. Lovász, S. Safra, and M. Szegedy. Interactive proofs and the hardness of approximating cliques. *J. ACM*, 43(2):268–292, 1996. Also *Proc. of FOCS91*.

13. U. Feige and J. Kilian. Two prover protocols - low error at affordable rates. In *Proceedings of STOC'94*, pages 172–183.

14. M. Goemans and D. Williamson. New 3/4-approximation algorithms for the maximum satisfiability problem. *SIAM J. Disc. Math.*, 7(4):656–666, 1994. Also *Proc. of IPCO'93*.

15. M.X. Goemans and D.P. Williamson. Improved approximation algorithms for maximum cut and satisfiability problems using semidefinite programming. *J. ACM*, 42(6):1115–1145, 1995. Also *Proc. of STOC'94*.

16. J. Håstad. Testing of the long code and hardness for clique. In *Proc. STOC'96*, pages 11–19.

17. J. Håstad. Some optimal inapproximability results. In *Proc. STOC'97*, pages 1–10.

18. D.S. Johnson. Approximation algorithms for combinatorial problems. *JCSS*, 9:256–278, 1974.

19. S. Khanna, R. Motwani, M. Sudan, and U. Vazirani. On syntactic versus computational views of approximability. In *Proc. FOCS'94*, pages 819–830.

20. S. Khanna, M. Sudan, and D.P. Williamson. A complete classification of the approximability of maximization problems derived from boolean constraint satisfaction. In *Proc. STOC'97*, pages 11–20.

21. H.C. Lau and O. Watanabe. Randomized approximation of the constraint satisfaction problem. In *Proc. of SWAT'96*, pages 76–87.

22. T. Ono, T. Hirata, and T. Asano. Approximation algorithms for the maximum satisfiability problem. In *Proc. of SWAT'96*.

23. C. H. Papadimitriou and M. Yannakakis. Optimization, approximation, and complexity classes. *JCSS*, 43:425–440, 1991. Also *Proc. of STOC'88*.

24. E. Petrank. The hardness of approximations : Gap location. *Computational Complexity*, 4:133–157, 1994. Also *Proc. of ISTCS'93*.

25. L. Trevisan. Positive linear programming, parallel approximation, and PCP's. In *Proc. of ESA'96*, pages 62–75.

26. L. Trevisan, G.B. Sorkin, M. Sudan, and D.P. Williamson. Gadgets, approximation, and linear programming. In *Proc. of FOCS'96*, pages 617–626.

27. M. Yannakakis. On the approximation of maximum satisfiability. *J. of Algorithms*, 17:475–502, 1994. Also *Proc. of SODA'92*.

Algorithms for Computing Signs of 2 × 2 Determinants: Dynamics and Average–Case Analysis

Brigitte Vallée

GREYC-URA 1526, Département d'Informatique,
Université de Caen, 14032 Caen Cedex, France
(Brigitte.Vallee@info.unicaen.fr)

Abstract. An algorithm for computing signs of 2 × 2 determinants is analysed. Equivalently, this algorithm compares two rationals by using their continued fraction expansions. It is shown that the algorithm has similarities with the usual algorithm that operates with expansions in base b. The worst-case number of iterations is linear in the size of the entries. In contrast, the average number of iterations is found to be asymptotically constant and thus essentially independent of the size of data. The distribution of the number of iterations decays geometrically. The constants that intervene in the analysis are related to the spectral properties of transfer operators arising in dynamical systems theory.

Introduction

Most decisions in computational geometry are based on determining signs of determinants of small order, and the evaluation of this sign plays the same rôle as the comparison between numbers in one-dimensional problems. The decision must be both exact and fast to compute. One direction consists in using exact integer arithmetic. The computation of a 2 × 2 determinant with ℓ-bits integer entries

$$\det \begin{pmatrix} x_1 & y_1 \\ x_2 & y_2 \end{pmatrix}, \text{ with } x_1, x_2, y_1, y_2 \text{ nonzero,}$$

then needs two products $x_1 y_2$ et $x_2 y_1$ and requires integers of length 2ℓ bits, that is to say, double precision. *When only the sign of a determinant is needed, is it possible to operate efficiently with exact integer arithmetics and single precision?* The problem is homogeneous: when the entries are nonzero, divisions can be performed on the columns,

$$\det \begin{pmatrix} 1 & 1 \\ x & y \end{pmatrix} = y - x, \text{ with } y = \frac{y_2}{y_1} \text{ and } x = \frac{x_2}{x_1},$$

and the problem reduces to evaluating the sign of the difference between two rationals, or equivalently comparing two rationals x and y. The main idea is to expand each rational x and y in a common *number representation system*, and proceed in a "lazy" manner. As long as the expansions coincide, continue to expand, and stop as soon as a discrepancy is detected. The observation of the last digits determined is sufficient to compare the rationals themselves. In a way, this principle is similar to the one that underlies digital searching, where binary expansions are used.

Here, we study algorithms based on *continued fraction expansions*. We follow an idea first proposed in the HAKMEM memo [Ha] and later used by Knuth [Kn] in his METAFONT system. This method bas been also adapted by Avnaim *et al.* within the framework of computational geometry [ABDPY]. We evaluate precisely the number of steps to be performed in order to compare two rationals. In the worst-case, the number of steps is linear in the size of the entries, an old result due to Lamé [La] in the 1850's. In the average case, the sign algorithm based on continued fractions (CF–Sign) has features that resemble the sign algorithm based on expansions in base b (Base–Sign). We establish here that, asymptotically, the average number of iterations does *not* depend on the size of the entries, and the distribution of the number of iterations is closely approximated by a *geometric law*. However, the constants (the expectation and the limit ratio of the geometric law) that intervene in the analyses of the two algorithms, CF-Sign and Base-Sign, are of quite a different nature. A contribution of this paper is to relate the structural constants of the CF-Sign algorithm to spectral properties of the functional *transfer operator* of Ruelle–Mayer type that captures the main properties of the continued fraction dynamics.

We observe that the CF–Sign algorithm can be wieved as a two-dimensional generalization of the continued fraction algorithm. There exists another two-dimensional-generalization of continued fraction algorithms, namely the Gaussian algorithm for reducing integer lattices in two dimensions. The analysis conducted here thus closely parallels the analysis of the Gaussian algorithm [DFV], with squares appearing in lieu of disks.

In the sequel, we consider two rational numbers with numerators and denominators both having ℓ bits and belonging to the interval $\mathcal{I} =]0, 1[$. Then, the entry (x, y) belongs to the unit square $\mathcal{C} := \mathcal{I} \times \mathcal{I}$; we denote by $\mathcal{K}$ the diagonal of this square. The basic probabilistic model considered is a continuous one: the point (x, y) has real coordinates and is distributed inside the unit square $\mathcal{C}$ according to a density $F(x, y)$. This density may be of course uniform. One of the characteristics of the operator approach developed here is to allow for non-uniform densities. This appears to be well-suited to practical situations occurring in computational geometry where, quite often, determinants to be evaluated have small values, and thus their sign is much harder to compute. In particular, it is natural to consider densities that give a heavier weight to the "difficult" input configurations. This is consistent with the approach of Avnaim *et al.* [ABDPY] who have already provided empirical simulation data for the CF-Sign algorithm under such conditions. In a way, the present paper is also an attempt at elucidating analytically the way more degenerate data entail higher computational costs.

The densities considered here are taken to be functions of the absolute value of the result, *i.e.*, the distance $|x - y|$ between the two numbers x and y and also the distance of point (x, y) to diagonal $\mathcal{K}$. To formalize this model, we introduce a family of densities that are proportional to $|x - y|^r$; they will be said of valuation r. In this case, the measure of a square "built on the diagonal", with sides parallel to axes of length c, is equal to c^{r+2},

$$\mu([a, b] \times [a, b]) = |b - a|^{r+2} \qquad (1).$$

In order to have a proper probabilistic model, one must take $r > -1$. The case $r = 0$ corresponds to the uniform density model; when r approaches -1, the model gives predominantly higher weights to nearly degenerate configurations, a case thus worthy of special attention. In fact, more general classes of distributions with qualitatively similar features can be treated and the essential characteristics of the analysis are preserved. This can be proven by adapting the theory of generalized transfer operators developed in [Va] in the context of lattice reduction.

A perhaps surprising outcome of the analysis of the CF-Sign algorithm is that only a small number of iterations are needed on average, for instance about 1.35 for the uniform density model, and, essentially, this holds irrespective of the size of the inputs. Higher-dimensional determinant sign algorithms are proposed, for instance Avnaim $et\ al.$, Clarkson, Yvinec discuss algorithms that are based on principles similar to lattice reduction. Since lattice reduction in higher dimensions is also known [DaVa] to involve an expected number of iterations that is a constant (for each dimension), this suggests that the results of the present paper might hold true for some higher-dimensional determinant-sign algorithms, like those of [ABDPY], [Cl], [Yv].

Plan of the paper. We first describe the behaviour of the sign algorithm that uses expansions in base b (the Base–Sign algorithm). Next, we introduce the CF-Sign algorithm based on continued fractions representations and we describe its worst–case complexity. In the sequel of the paper, we focus on the average–case complexity of the CF–Sign algorithm: We first analyse the continuous model, and evaluate the average–number of iterations. By using spectral properties of Ruelle–Mayer operators, we show that the distribution of the number of iterations is closely approximated by a geometric law, whose ratio is equal to the dominant eigenvalue of the operator. Section 5 is devoted to the analysis in the discrete model, and we conclude by a precise comparison between the performances of these two Sign–Algorithms (with base 2 and with continued fraction representations).

1   The Sign algorithm using expansions in base b

The shift operator B associated to expansions in base b is defined for a real number x of $\mathcal{I} =]0, 1[$ by

$$B(x) = bx - [bx] \quad \text{where } [x] \text{ denotes the integer part of } x.$$

The integer $n(x) := [bx]$ is the first digit of the expansion of x in base b. The iteration of the shift B produces the full expansion of x in base b:

Expansion–Base(x)

$k := 1;$
While $x \neq 0$ **do**
$\quad \{n_i := n(x); x := B(x); k := k+1; \}$

When applied to a rational number, this computation has two well-known major drawbacks. First, the expansion may be infinite, second, this expansion cannot be computed in single precision, since each step consists in a multiplication by base b, and uses integers with $\ell + \log_2 b$ bits.

The following algorithm expands both rationals x et y in base b by performing two iterations of B in parallel, and compares x and y by comparing their expansions.

Base–Sign(x, y)

Input .- $(x, y) \in \mathcal{C}$
Output.- The sign σ of $y - x$
If $n(x) < n(y)$ **then** $\sigma := +1$
$\quad$ **Else if** $n(x) > n(y)$ **then** $\sigma := -1$
$\quad\quad\quad$ **Else** **The expansions are the same, one iterates B** **
$\quad\quad\quad\quad\quad\quad \sigma(x, y) := \sigma(B(x), B(y)).$

In this simplified form, it may be that the algorithm does not terminate: this is the case when x and y are equal, even if they are rational. If they have both at most ℓ-bits, one can add a counter, and conclude to equality in the case when the expansions remain the same for a number of steps equal to $\lceil 2\ell/\log_2 b \rceil$.

We now study the average number of steps in a continuous model, where the entry (x, y) is real and distributed inside square $\mathcal{C}$ according to density F of valuation r. .
The event $[L \geq k+1]$ is formed with real pairs (x, y) of $\mathcal{C}$ which have the same expansion in base b till rank k; the real numbers x and y belong to the same interval of length $1/b^k$ defined from their common expansion $(n_1, n_2, \ldots n_k)$ in base b,

$$\mathcal{N} := [\frac{n_1}{b} + \frac{n_2}{b^2} + \ldots + \frac{n_k}{b^k}, \ \frac{n_1}{b} + \frac{n_2}{b^2} + \ldots + \frac{n_k+1}{b^k}[. \tag{2}$$

Thus, the event $[L \geq k+1]$ is equal to the union of b^k squares of type $\mathcal{N} \times \mathcal{N}$: All these squares are "built on the diagonal" and have a side equal to $1/b^k$. Figure 1 shows the events $[L = k+1]$ alternatively in black and white.
In our continuous model, we can easily describe the expectation and the distribution of the variable L "number of iterations" for the family of densities of valuation r.

Proposition 1. *Assume that the unit square $\mathcal{C}$ is endowed with a density of valuation r, i.e. proportional to $|x - y|^r$. Then, the probability that the Base–Sign algorithm performs at least $k + 1$ iterations is equal to*

$$\beta_k^{(r)} := \Pr[L \geq k+1] = (\frac{1}{b^{r+1}})^k,$$

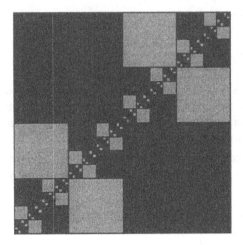

Fig. 1. The events $[L = k + 1]$ for rank $k \leq 6$ and base 2.

and the average number of iterations is

$$\beta^{(r)} := \frac{1}{1 - \frac{1}{b^{r+1}}}.$$

In the case of the uniform density, one has:

$$\beta_k = (\frac{1}{b})^k \quad \text{and} \quad \beta := \frac{1}{1 - \frac{1}{b}}.$$

When the parameter r tends to -1, the expectation $\beta^{(r)}$ admits the equivalent form

$$\beta^{(r)} \simeq \frac{1}{\log b} \frac{1}{r+1}.$$

2 The Sign algorithm using continued fractions

The shift operator U for continued fraction is defined in $\mathcal{I}$ by

$$U(x) = (\frac{1}{x}) - [\frac{1}{x}] \quad \text{where } [x] \text{ denotes the integer part of } x.$$

The integer $m(x) := [1/x]$ is the first digit of the continued fraction expansion of x. The iteration of the U operator produces the full continued fraction expansion

CF–Expansion(x)

$k := 1;$
While $x \neq 0$ **do**
  $\{m_i := m(x); x := U(x); k := k + 1; \}$

This expansion presents two advantages (as opposed to expansions in base b): First, the expansion of a rational number is always finite, second, the computation of the expansion can be performed in single precision, by using ℓ-bits integers.

To a real number x, the CF algorithm associates the sequence of iterates $x_0 = x, x_1, x_2, \ldots, x_k$, and, provided the k-th iterate exists,

$$x_0 = \cfrac{1}{m_1 + \cfrac{1}{m_2 + \cfrac{1}{\ddots \atop m_k + x_k}}},$$

where by construction $m_j \geq 1$. The relation between x_0 and x_k is described by a homography of depth k, which is associated to the k-uple $(m_1, m_2, \ldots, m_k)$,

$$h_{m_1, m_2, \ldots, m_k}(z) = \cfrac{1}{m_1 + \cfrac{1}{m_2 + \cfrac{1}{\ddots \atop m_k + z}}}.$$

The depth of h is denoted by $|h|$. The collection of all the homographies of depth k gives all the inverse branches of U^k. As is well known, an homography is expressible in terms of continuant polynomials [RS],

$$h(z) = \frac{P_k + z P_{k-1}}{Q_k + z Q_{k-1}},$$

where

$$\begin{array}{llll}
Q_k &=& Q_k(m_1, \ldots, m_k), & Q_{k-1} &=& Q_{k-1}(m_1, \ldots, m_{k-1}), \\
P_k &=& Q_{k-1}(m_2, \cdots, m_k), & P_{k-1} &=& Q_{k-2}(m_2, \ldots, m_{k-1}).
\end{array}$$

The continuant polynomial are determined by the recurrence

$$Q_k(m_1, m_2, \ldots, m_k) = m_k Q_{k-1}(m_1, \ldots, m_{k-1}) + Q_{k-2}(m_1, \ldots, m_{k-2}), \quad (3)$$

where $Q_0 = 1$, $Q_1(m_1) = m_1$. It is also well-known that the continuant polynomial $Q_k(m_1, \ldots, m_k)$ is obtained by taking the sum of monomials obtained from the product $m_1 m_2 \cdots m_k$ by crossing out in all possible ways pairs of adjacent variables $m_i m_{i+1}$. Hence, continuants satisfy the symmetry property

$$Q_k(m_1, \ldots, m_k) = Q_k(m_k, \ldots, m_1), \quad (4)$$

as well as the determinant identity

$$Q_k P_{k-1} - Q_{k-1} P_k = (-1)^k. \quad (5)$$

For a homography of depth k, the transform of segment $\mathcal{I}$ by h is called a fundamental interval of rank k: it contains all the real numbers whose expansion starts with $m_1, \ldots, m_k$. It plays the same rôle as the interval $\mathcal{N}$ in (2). The fundamental intervals $h(\mathcal{I})$ are expressible in terms of continuants

$$h(\mathcal{I}) =]\frac{P_k}{Q_k}, \frac{P_{k-1} + P_k}{Q_{k-1} + Q_k}[, \qquad (6)$$

(the bounds are ordered or not, depending on the parity of k) and are of length equal to

$$|h(\mathcal{I})| = \frac{1}{Q_k(Q_k + Q_{k-1})}. \qquad (7)$$

The Sign–Algorithm that uses continued fraction (CF–Sign) compares two real numbers x and y by running "in parallel" two iterations of U. It halts as soon as a discrepancy between the digits $m(x)$ and $m(y)$ is detected. Comparison between digits permits to compare the numbers themselves. Special rational cases are easily tested and subjected to an appropriate treatment.

CF–Sign(x, y)

Input.- $(x, y) \in \mathcal{C}$
Output.-The sign σ of $y - x$

Trivial cases
If $x = 0$ and $y = 0$ then $\sigma := 0$
   Else if $x \neq 0$ and $y = 0$ then $\sigma := -1$
      Else if $x = 0$ and $y \neq 0$ then $\sigma := +1$
         Else ****x and y are nonzero****
            If $m(x) < m(y)$ then $\sigma := -1$
               Else if $m(x) > m(y)$ then $\sigma := +1$
                  Else ** the expansions coincide **
$$\sigma(x, y) := -\sigma(U(x), U(y)).$$

The worst-case analysis of CF–Sign algorithm can be deduced from an old result due to Lamé [La], namely: the depth of the continued fraction expansion of a rational number of given size is maximal when all the digits m_i are minimal and equal to 1. Such rational numbers are the quotients of two successive Fibonacci numbers, and the infinite expansion associated is linked to the golden ratio $\phi := (1 + \sqrt{5})/2$.

Proposition 2. *For entries (x, y) of the unit square $\mathcal{C}$ formed of two rational numbers with at most ℓ bits, the CF– Sign algorithm performs at most L iterations, with*

$$L = \lceil \frac{\ell}{\log_2 \phi} \rceil.$$

In the following, we focus on the average–case complexity of CF–Sign algorithm, and we show that it has a behaviour similar to Base–Sign; in particular, the distribution of iterations decays geometrically.

3 The average–case complexity of CF–Sign

The event $[L \geq k + 1]$ is formed with all real pairs (x, y) of $\mathcal{C}$ which have the same continued fraction expansion till depth k. Then the two real numbers x and y belong to the same fundamental interval $h(\mathcal{I})$ of rank k, and the pair (x, y) belongs to the square $h(\mathcal{I}) \times h(\mathcal{I})$. Such a square is called a fundamental square of rank k. The event $[L \geq k + 1]$ is thus equal to the union of all the fundamental squares of rank k,

$$[L \geq k + 1] = \bigcup_{|h|=k} h(\mathcal{I}) \times h(\mathcal{I}).$$

Figure 2 shows the events $[L = k + 1] \cap [0, 1/2]^2$ alternatively in black and white, for rank $k \leq 4$.

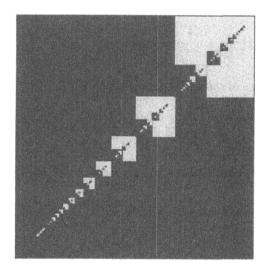

Fig. 2. Fundamental squares for the CF algorithm

Since fundamental squares are disjoint up to boundary sets of measure 0, the probability that the algorithm performs at least $k + 1$ iterations is

$$\Pr([L \geq k + 1]) = \sum_{|h|=k} \mu\left(h(\mathcal{I}) \times h(\mathcal{I})\right), \tag{8}$$

(here μ is the measure on the unit square $\mathcal{C}$), and the average number of iterations is expressible as

$$E[L] = \sum_{h} \mu\left(h(\mathcal{I}) \times h(\mathcal{I})\right). \tag{9}$$

If now the measure μ is associated to a density of valuation r, the measure of each fundamental square $h(\mathcal{I}) \times h(\mathcal{I})$ can be easily computed with (1) and (7) in terms of continuants:

$$\mu(h(\mathcal{I}) \times h(\mathcal{I})) = \frac{1}{Q_k^{r+2}(Q_k + Q_{k-1})^{r+2}} \tag{10}$$

and then, the probability that CF–Sign performs at least $k+1$ iterations is equal to

$$\Pr([L \geq k+1]) = \sum_{|h|=k} \frac{1}{Q_k^{r+2}(Q_k + Q_{k-1})^{r+2}}. \tag{11}$$

By (9) and (10), the expectation of L involves a sum over all possible continuants, and then admits a form that no longer involve continuants.

Theorem 1. *Endow the unit square $\mathcal{C}$ with density of valuation r, i.e. proportional to $|x - y|^r$. Then the average number of iterations of CF–Sign is given by the double sum*

$$\rho^{(r)} := E[L] = 1 + \frac{1}{2^{r+2}} + \frac{2}{\zeta(2r+4)} \sum_{d=2}^{\infty} \frac{1}{d^{r+2}} \sum_{d<c<2d} \frac{1}{c^{r+2}}.$$

In the particular case of the uniform density $(r = 0)$,

$$\rho = \frac{5}{4} + \frac{2}{\zeta(4)} \sum_{d=2}^{\infty} \frac{1}{d^2} \sum_{d<c<2d} \frac{1}{c^2} = 1.35113\ldots\ldots$$

When the parameter r tends to -1, the expectation $\rho^{(r)}$ admits the asymptotic form

$$\rho^{(r)} \simeq \frac{\log 2}{\zeta(2)} \frac{1}{r+1} = \frac{6\log 2}{\pi^2} \frac{1}{r+1}.$$

Proof. Each rational number c/d of $]0,1[$ admits two continued fraction expansions: the proper one that terminates with $m_k \geq 2$, and the improper one that terminates with $m_{k+1} = 1$. Thus, any integer pair (c, d) satisfying $\gcd(c, d) = 1, d \geq 2$ and $0 < c < d$ can be represented in two different ways as a pair of continuants P_k/Q_k, or equivalently as a pair of continuants (Q_{k_1}, Q_k) (by using the symmetry between Q_k and P_{k-1}). Then, taking into account boundary cases, namely numbers 0 and 1, we find

$$E[L] = 1 + \frac{1}{2^{r+2}} + \sum \frac{1}{c^{r+2}(c+d)^{r+2}},$$

where the sum is taken over all the pairs (c, d) satisfying $\gcd(c, d) = 1, d \geq 2$ et $0 < c < d$. The general term in the last sum is homogeneous of degree $2r + 4$, so that the gcd condition is eliminated provided one divides the sum by $\zeta(2r + 4)$. The asymptotic form is obtained by comparing the internal sum to an integral (by the Euler–Maclaurin formula) ∎

4 The distribution of the number of iterations and Ruelle-Mayer operators

The expression (11) that gives the distribution of the number of iterations is more difficult to deal with, because it is necessary to classify homographies with respect to their depth. Ruelle–Mayer operators are well suited to this problem. The Ruelle operators are associated to dynamical systems [Ru], and the Ruelle-Mayer operators are their specialization to continued fractions. They have been studied in an extensive way by Mayer [Ma]. In [DFV] and [FV], it is shown that these operators play a central rôle in the analyses of algorithms that are based on continued fraction expansions.

The operator $\mathcal{G}_s$ is defined for complex numbers s satisfying $\Re(s) > 1$ by

$$\mathcal{G}_s[f](z) = \sum_{m \geq 1} \frac{1}{(m+z)^s} \, f(\frac{1}{m+z}).$$

The iterate of order k is expressible in terms of continuants of rank k

$$\mathcal{G}_s^k[f](z) = \sum_{m_1 \ldots m_k} \frac{1}{(Q_{k-1}z + Q_k)^s} \, f(\frac{P_{k-1}z + P_k}{Q_{k-1}z + Q_k});$$

in particular, at $x = 0$,

$$\mathcal{G}_s^k[f](0) = \sum_{m_1 \ldots m_k} \frac{1}{Q_k^s} \, f(\frac{P_k}{Q_k}) = \sum_{m_1 \ldots m_k} \frac{1}{Q_k^s} \, f(\frac{Q_{k-1}}{Q_k}),$$

(the second equality results of the symmetry property). Remark that the distribution of the number of iterations of CF–Sign algorithm is expressible in these terms,

$$\rho_k^{(r)} := \Pr([L \geq k+1]) = \sum_{|h|=k} \frac{1}{Q_k^{r+2} (Q_k + Q_{k-1})^{r+2}} = \mathcal{G}_{2r+4}^k \left[\frac{1}{(1+x)^{r+2}}\right](0).$$

Also the expectation $E[L]$ admits the expression

$$E[L] = (I - \mathcal{G}_{2r+4})^{-1} \left[\frac{1}{(1+x)^{r+2}}\right](0).$$

Here, asymptotic properties are needed, and they are closely related to dominant spectral properties of the operator $\mathcal{G}_s$. First, one makes precise the functional space to which the operators are applied. Let $\mathcal{V}$ and $\mathcal{J}$ denote the disk and the interval of center 1 and radius $3/2$. The disks are mapped strictly within themselves by all the homographies. The Ruelle-Mayer operators $\mathcal{G}_s$ act on the space $A_\infty(\mathcal{V})$ formed with all functions f that are holomorphic in the disk $\mathcal{V}$ and continuous on the closed disk $\bar{\mathcal{V}}$. Endowed with the sup-norm, $A_\infty(\mathcal{V})$ is a Banach space and operator $\mathcal{G}_s$ is bounded, and therefore compact (by classical properties of bounded sequences of analytic functions). Its spectrum is thus discrete, with only an accumulation point at 0.

When the parameter s is real, the operator $\mathcal{G}_s$ satisfies a strong positivity property related to Perron–Frobenius theory: it has a unique dominant eigenvalue (of largest modulus) $\lambda(s)$ that is positive and has multiplicity 1. The spectral ratio $\nu(s)$, which is equal by definition to the modulus of the ratio between $\lambda(s)$ and a subdominant eigenvalue, satisfies $\nu(s) < 1$. Furthermore, the dominant eigenvalue satisfies the inequality

$$\lambda(s+t) \leq \frac{1}{\phi^t}\lambda(s) \quad \text{for any } s > 1 \text{ and } t \geq 0.$$

Such dominant spectral properties provide an asymptotic form for the iterates of Ruelle-Mayer operators:

Theorem. [Mayer] *For any real number $s > 1$, and any function f of $A_\infty(\mathcal{V})$ strictly positive on $\mathcal{J}$, there exists a constant C, strictly positive, that depends on s and f for which*

$$\mathcal{G}_s^k\,[f](0) = C\lambda(s)^k\,[1 + O(\nu(s)^k)],$$

for all $k \geq 1$. The constant in the O-error term depends also on s and f.

The value $s = 2$ is very specific and closely linked to properties of CF–algorithm: The dominant eigenvalue $\lambda(2)$ is equal to 1, and the associated eigenfunction corresponds to the limit–density for CF-algorithm, namely $1/(1+x)$. The absolute value of the subdominant eigenvalue is known as the Gauss-Kusmin-Wirsing constant. It is equal to 0.30366. The derivative of the function $s \to \lambda(s)$ at $s = 2$ is equal to the Lévy constant $\lambda'(2) = -\pi^2/(12\log 2)$ that intervenes in the average–case analysis of the Euclidean algorithm. Here, we obtain, at $s = 2r+4$,

$$\lambda(2r+4) \leq (\frac{1}{\phi^2})^{r+1}, \quad \text{and} \quad \lambda(2r+4) - 1 \simeq \frac{\pi^2}{6\log 2}(r+1), \quad (r \to -1)$$

We come back now to CF–Sign Algorithm and we apply the previous result. This shows that the random variable expressing the number of iterations has an asymptotically geometric behaviour, with a limit-ratio equal to the dominant eigenvalue $\lambda(2r+4)$ of the Ruelle-Mayer operator $\mathcal{G}_{2r+4}$.

Theorem 2. *Assume that the unit square $\mathcal{C}$ is endowed with the density of valuation r, i.e. proportional to $|x - y|^r$. Then, the probability $\rho_k^{(r)}$ that the CF-Sign Algorithm performs at least $k+1$ iterations satisfies*

$$\rho_k^{(r)} = C_r\,\lambda(2r+4)^k\,[1 + O(\nu(2r+4)^k)].$$

Here, $\lambda(2r+4)$ denotes the dominant value of the Ruelle-Mayer operator $\mathcal{G}_{2r+4}$, $\nu(2r+4)$ is the spectral ratio; the constant C_r and the constant in the O-error term depend on r.

In particular, for the uniform density ($r = 0$), one has

$$\lambda(4) = 0.19945\,88183\,43767 \pm 10^{-15}, \quad \nu(4) \approx 0.37972, \quad C_0 \approx 1.3.$$

When the parameter r tends to -1, the dominant eigenvalue tends to $\lambda(2) = 1$ (convergence becomes slow), the spectral ratio tends to the Gauss-Kusmin-Wirsing constant $\nu(2) \approx 0.30366$, and the constant C_r tends to 1.

5 Average-case analysis in the discrete model

We come back now to the practical case where the entries a, b, c, d of the determinant are integral: we consider integral entries a, b, c, d satisfying

$$1 \le a < c, \quad 1 \le b < d, \quad N < c, d \le 2N. \quad ad - bc \ne 0,$$

and, more precisely, we work with the set

$$\mathcal{C}_N := \bigcup_{N < c, d \le 2N} \mathcal{C}_{(c,d)} \quad \text{with} \quad \mathcal{C}_{(c,d)} = \{(\frac{a}{c}, \frac{b}{d}) \in \mathcal{C}, \ \frac{a}{c} \ne \frac{b}{d}\}.$$

Each set $\mathcal{C}_{(c,d)}$ is endowed with the discrete density of valuation r, which is associated to a measure $\mu_{(c,d)}$ satisfying

$$\mu_{(c,d)}(A) := \Big(\sum_{(x,y) \in \mathcal{C}_{(c,d)}} |x - y|^r \Big)^{-1} \sum_{(x,y) \in A \cap \mathcal{C}_{(c,d)}} |x - y|^r,$$

for any subset A of $\mathcal{C}$. The union $\mathcal{C}_N$ of sets $\mathcal{C}_{(c,d)}$ is endowed with measure μ_N

$$\mu_N(A) := \frac{1}{N^2} \sum_{N < c, d \le N} \mu_{(c,d)}(A).$$

The two analyses –in the continuous model and in the discrete model– can be compared when the parameter N tends to ∞: one uses the Gauss principle (that links the number of integral points in a domain and the area of the domain) and the relation between an integral and its Riemann sums. One obtains:

Theorem 3. *Let L_N be the number of iterations of CF–Sign Algorithm when applied to entries $\mathcal{C}_N$ distributed according to the discrete density of valuation r for a real number $r > -1/2$. The random variable L_N is always less than*

$$L_N(x, y) \le k_N \quad \text{with} \quad k_N := \lceil \frac{\log N}{2 \log \phi} \rceil.$$

The random variable L_N converges in distribution towards the continuous variable L associated to the same valuation r. In particular, the distribution $\rho_{k,N}^{(r)}$ of variable L_N relative to the density of valuation r uniformly converges to the distribution $\rho_k^{(r)}$ of the variable L:

$$\rho_{k,N}^{(r)} - \rho_k^{(r)} = O(\frac{1}{N^{2r+1}}), \quad \text{for } -1/2 < r \le 0; \quad \rho_{k,N}^{(r)} - \rho_k^{(r)} = O(\frac{1}{N}), \quad \text{for } r \ge 0,$$

$$\rho_{k,N}^{(r)} = 0 \text{ for } k > k_N.$$

The expectation $\rho_N^{(r)} := E[L_N]$ satisfies

$$\rho_N^{(r)} - \rho^{(r)} = O(\frac{\log N}{N^{2r+1}}) \text{ for } -1/2 < r \leq 0; \quad \rho_N^{(r)} - \rho^{(r)} = O(\frac{\log N}{N}) \text{ for } r \geq 0$$

6 Comparison between the two Sign–algorithms

Table 3 below displays the average–number of iterations for different valuations, starting with the uniform density $r = 0$ and approaching the limit–case $r = -1$. The values are given by slowly convergent series and an acceleration process based on Euler-Maclaurin summation has been employed. These values appear to obey the asymptotic law described in Proposition 1 and Theorem 1. The CF–Sign algorithm always performs a smaller number of iterations than the Base–Sign algorithm with base 2. Naturally, the arithmetic operations needed at each iterations in the CF– Sign are more expensive: these are divisions, whereas Base–Sign only performs shifts and subtractions. For 64-bits integer, we estimate that the cost of an internal loop of CF-Sign is about twice more expensive than in Base-Sign (with base 2). One concludes that, especially for "difficult" entries, the CF-Sign algorithm provides a completely robust and reasonably fast alternative method in order to decide the sign of integral determinants.

Density valuation, r	Expectation in base 2	Expectation with CF	Density valuation, r	Expectation in base 2	Expectation with CF
0	2.000000000	1.351137367	−0.91	16.53514327	5.592689232
−0.1	2.154646435	1.397773700	−0.92	18.53830862	6.177362124
−0.2	2.349343515	1.455857259	−0.93	21.11397220	6.929237083
−0.3	2.601268735	1.530332126	−0.94	24.54838290	7.931921591
−0.4	2.939049594	1.629480413	−0.95	29.35678875	9.335909309
−0.5	3.414213565	1.768258619	−0.96	36.56968605	11.44218780
−0.6	4.129812958	1.976655795	−0.97	48.59156689	14.95306175
−0.7	5.326299680	2.324772684	−0.98	72.63590621	21.97544566
−0.8	7.725023959	3.023147098	−0.99	144.7700829	43.04391318
−0.9	14.93272608	5.125054012			

Acknowledgements. Many thanks to Philippe Flajolet for the numerical computations of Table 3. and the drawings of Figures 1 and 2.

References

[ABDPY] Avnaim, F., Boissonnat, J.-D., Devillers, O., Preparata, F., and Yvinec, M., Evaluation of a new method to compute signs of determinants, In *Eleventh Annual ACM Symposium on Computational Geometry* (1995), pp. C16–C17. Full paper to appear in *Algorithmica*, 1997.

[Cl] Clarkson, K. L. Safe and effective determinant evaluation, In *Proc. 33rd Annu. IEEE Sympos. Found. Comput. Sci.*, 387-395, 1992.

[DFV] Daudé, H., Flajolet, P., and Vallée, B., An average-case analysis of the Gaussian algorithm for lattice reduction, Oct. 1995, 30 pages, To appear in *Combinatorics, Probability and Computing*,

[DV] Daudé, H., and Vallée, B., An upper bound on the average number of iterations of the LLL algorithm, *Theoretical Computer Science 123*, 1 (1994), 95–115.

[FV] Flajolet, P., and Vallée, B., Continued fraction algorithms, functional operators, and structure constants, (Invited lecture at the 7th Fibonacci Conference, Graz, July 1996), Les cahiers du GREYC, 1996, Université de Caen, to appear in *Theoretical Computer Science*

[Ha] Beeler, M., Gosper, R. W., and Schroeppel, R., HAKMEM, Memorandum 239, M.I.T., Artificial Intelligence Laboratory, Feb. 1972.

[Kn] Knuth, D.E., Volume D of Computers and Typesetting, MF: the program, Addison Wesley, Reading, Massachussetts, 1986.

[La] Lamé, G. Note sur la limite du nombre de divisions dans la recherche du plus grand commun diviseur entre deux nombres entiers, Comptes-rendus de l'Académie des Sciences XIX (1845), 867-870.

[Ma] Mayer, D. H. Continued fractions and related transformations. In Ergodic Theory, Symbolic Dynamics and Hyperbolic Spaces, T. Bedford, M. Keane, and C. Series, Eds. Oxford University Press, 1991, pp. 175–222.

[RS] Rockett, A., and Szüsz, P. Continued Fractions. World Scientific, Singapore, 1992.

[Ru] Ruelle, D. Dynamical Zeta Functions for Piecewise Monotone Maps of the Interval, vol. 4 of *CRM Monograph Series*. American Mathematical Society, Providence, 1994.

[Va] Vallée, B. Opérateurs de Ruelle-Mayer généralisés et analyse des algorithmes d'Euclide et de Gauss, Rapport de Recherche de l'Université de Caen, Les Cahiers du GREYC # 4, 1995, To appear in *Acta Arithmetica*.

[Yv] Yvinec, M. Evaluating signs of $d \times d$ determinants using single precisions arithmetic, preprint.

Reconstructing the Topology of a CAD Model
– A Discrete Approach –

Karsten Weihe and Thomas Willhalm

Universität Konstanz, Fakultät für Mathematik und Informatik
Fach D188, 78457 Konstanz, Germany
{karsten.weihe,thomas.willhalm}@uni-konstanz.de
http://www.informatik.uni-konstanz.de/~{weihe,willhalm}

Abstract. We consider a problem which arises in the computer-aided design of machines, motor vehicles, and other technical devices: convert the data describing a surface model of a three-dimensional workpiece from a low-information format to a high-information format. In the worst case, the input format only contains a list of independent descriptions of the elementary pieces, whereas the output format also contains the neighborhoods of these pieces—the *topology* of the model. Reconstructing the topology is a crucial step in this process.

This problem is ill-posed and hence cannot be solved fully automatically. In fact, manual operations and previous format conversions make the data so "dirty" that human judgement is still required. Therefore, the aim is to approximate the topology as closely as possible to reduce the effort needed to correct the result afterwards. In this paper, we present a discrete approach to this problem and report the results of a computational study.

1 Introduction

In this paper, we consider the following problem: given a list of *mesh elements*, which together approximate a surface, reconstruct the information how these mesh elements are neighbored (incident) on this surface (see Fig. 1). This information is usually called the *topology* of the model. Figure 1 suggests that two neighbored mesh elements always meet on a common segment of a side. However, in general, two mesh elements that are intended to be neighbored are only located (more or less) close to each other in the three-dimensional space. Figures 3, 4, 6(b), and 7 demonstrate this fact. It is this subtlety which makes the problem non-trivial.

In the introduction, we will describe the input and the output of this problem in detail and give a survey of the remaining sections. Although the topic of the paper touches several disciplines (discrete and analytical computational geometry, numerical analysis, engineering), the core of the paper is self-contained and belongs entirely to the field of discrete computational geometry. For clarity of exposition, we will essentially ignore the technical problems caused by numerical calculations and formulate the algorithm as if infinite precision were

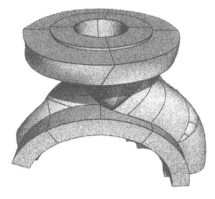 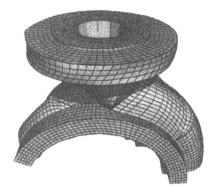

Fig. 1. a CAD model of a pump. The black curves are the edges of the wire model, and the faces of the wire model are the mesh elements.

Fig. 2. a *conformant* model of the same workpiece (generated from **Fig.** 1 by the algorithm in [4]).

available and if geometric constructions such as the perpendicular onto a curve were always well defined and unique.

1.1 Input

An instance of the problem (called a *mesh*) consists of a list of geometric descriptions of *mesh elements*, see Fig. 1. In our specific application, these mesh elements are triangles and quadrilaterals in the three-dimensional space. Each mesh element is given by a set of points: the corners of the polygon, one additional point in the open interior of each side, and possibly an auxiliary point in the open internal area of the mesh element. Each side of a mesh element is a certain parabola and interpolates the incident corners and the additional point on this side. In turn, the open interior of a mesh element is a certain quadratic interpolation of its sides and (if existing) of the additional internal point.

For notational convenience, we assume that each side of each mesh element is given an orientation.

Remark. The geometric details of the input definition arise from a concrete practical scenario, which motivated our work. Our implementation is tailored to this specific scenario. However, these details are not central to the work presented in this paper. Hence, we do not discuss them further. The crucial point is that the mesh elements cannot be assumed to be plane.

1.2 Output

A *feasible output* is a triple (S, E^+, E^-). In that, S is a set of curves. More specifically, each side of each mesh element is covered by a set of closed, openly disjoint segments, and S is the union of these interval partition sets over all sides of all mesh elements. On the other hand, E^+ and E^- are symmetric binary relations on S with $(s, s) \in E^+$ for all $s \in S$. These two relations are mutually exclusive, that is no pair of segments belongs to both relations. Moreover, $E = E^+ \dot\cup E^-$ is an equivalence relation on S. In that, for $(s_1, s_2), (s_2, s_3), (s_1, s_3) \in E$, we have $(s_1, s_3) \in E^-$ if and only if either $(s_1, s_2) \in E^+$ and $(s_2, s_3) \in E^-$ or $(s_1, s_2) \in E^-$ and $(s_2, s_3) \in E^+$. For convenience, we will often write (S, E) instead of (S, E^+, E^-). For $s_1, s_2 \in S$, $s_1 \neq s_2$, and $(s_1, s_2) \in E$, we will say that s_1 and s_2 are *neighbored*.

Each segment $s \in S$ is regarded as being oriented: the orientation of s is inherited from the mesh element side to which s belongs. Hence, we may define the point $t(s) \in \mathbb{R}^3$ from which s starts (the *tail*), and the point $h(s)$ where s ends (the *head*).

The problem is to find a specific feasible output $(S_0, E_0) = (S_0, E_0^+, E_0^-)$, which describes the topology intended by the designer of the mesh. Intuitively, (S_0, E_0) determines how the sides of the mesh elements must be identified with each other to yield a *wire model* of the surface (see Figs. 1 and 5):

Definition 1 (wire model). Let (S, E^+, E^-) be a feasible output, and let $P = \bigcup_{s \in S}\{t(s), h(s)\}$. The symmetric binary relation V on P is defined as follows: for $(s_1, s_2) \in E^+$, we have $(t(s_1), t(s_2)), (h(s_1), h(s_2)) \in V$, and for $(s_1, s_2) \in E^-$, we have $(t(s_1), h(s_2)), (h(s_1), t(s_2)) \in V$. Let V' be the equivalence relation generated by V (i.e. the smallest equivalence relation on P containing V). The *wire model* induced by (S, E) is an undirected graph $\tilde{G} = (\tilde{V}, \tilde{E})$, which consists of the equivalence classes of V' and E, that is $\tilde{V} = P/V'$ and $\tilde{E} = S/E$. In that, $e \in \tilde{E}$ is incident to the (unique) nodes $v, w \in \tilde{V}$ such that there is $s \in e$ fulfilling $t(s) \in v$ and $h(s) \in w$ (or vice versa).

To avoid vertices of degree 2 in the wire model, we will restrict our attention to feasible outputs (S, E) which are irreducible in the following sense.

Definition 2 (irreducible). A feasible output (S, E) is called *reducible* if there are two equivalence classes, $e_1, e_2 \in S/E$, $e_1 \neq e_2$, related to each other as follows: for every $s_1 \in e_1$ there is $s_2 \in e_2$ (and vice versa) such that s_1 and s_2 belong to the same side of the same mesh element and either $h(s_1) = t(s_2)$ or $t(s_1) = h(s_2)$. Otherwise, (S, E) is *irreducible*. Without loss of generality, (S_0, E_0) is assumed to be irreducible.

The following points are important to note:

- More than two mesh elements may be neighbored to the same segment of a side, see Fig. 5. This situation occurs quite frequently in instances that model hollow bodies such as tubes, containers, and bodyworks. Moreover, it occurs in surface models of solid bodies, namely when such a model is composed by connecting models of the body's parts and these models are glued together at common mesh elements [5].

– Two mesh elements may be neighbored on more than one segment. Figure 6 shows two examples, which occurred in real-world instances.

Although a look at one of the Figs. 1-7 might suffice for a human eye to correctly determine almost all details of the topology and the wire model, no sufficient formal definition is available. It seems that the data is too "dirty" to allow a fully formal description. For example, Fig. 4 shows the connection of two tubes in the interior of the model in Fig. 1. The gap shown in Fig. 4 is so large that it seems to be intended by the designer. However, from an engineering point of view, this gap does not make any sense and is likely to have resulted from a sloppy design or a conversion of the data into a non-isomorphic format. Therefore, the mesh elements around this gap should be regarded as neighbored, which closes the gap from a purely structural point of view (afterwards, the gap can be closed *geometrically* using CAD model repair techniques [2,5]).

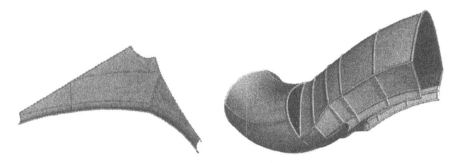

Fig. 3. an example of a significant gap. **Fig. 4.** a highly pathological example, extracted from the interior of Fig. 1. The black, semicircular hole does not make any sense from an engineering point of view and might be an extreme case of a sloppy design.

Since the problem is so ill-posed, there is no hope for an exact automatic solver of this problem (actually, our solver does not close the gap in Fig. 4 either). All one can hope for is a good approximation of the intended result (S_0, E_0): the better the approximation is, the less effort must be spent afterwards on manual corrections. However, the effort for manual corrections is acceptable only if it is easy and convenient for the designer to find the errors in the result. This might be best achieved by a graphical visualization of the output which allows a fast check of all details by eye.

1.3 Result

In this paper, we introduce a new, discrete approach to this problem, and we report the results of a computational study. Besides the encouraging experimental results, our algorithm is more convenient for the user than algorithms based

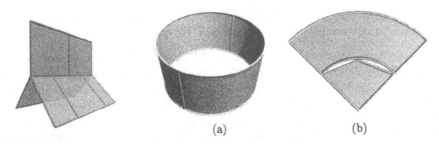

(a)   (b)

Fig. 5. a mesh with 6 mesh
elements. The wire model
consists of 14 vertices and 19
edges.

Fig. 6. two examples of how two mesh elements may be
neighbored on more than one segment.

on traditional approaches (cf. Sect. 2): the user need not choose an appropriate
threshold value, nor any other kind of numerical parameter, to achieve acceptable
results.

The asymptotic complexity of the algorithm is $\mathcal{O}(n \log n + N \log N + \sum_{i=1}^{K} M_i^2)$,
where n denotes the number of mesh elements. The values N, K, and $M_1, \ldots, M_K$
depend on certain intermediate results of the algorithm. In general, we have
$N \in \mathcal{O}(n^2)$ and $M_1, \ldots, M_K \in \mathcal{O}(n)$. However, in our computational studies
on real-world data, we constantly observed $N \in \mathcal{O}(n)$ and $M_1, \ldots, M_K \in \mathcal{O}(1)$,
which reduces the asymptotic complexity to $\mathcal{O}(n \log n)$.

We will develop a formal framework to analyze the accuracy of the algorithm
more rigorously. Using this framework, we will conclude that a certain coloring
of the output suffices to allow a fast correctness check by eye. To our knowledge,
this is the first attempt to formalize a problem like this and to give guarantees
for an algorithm based on such a formal framework.

Of course, we cannot ignore all numerical and geometric issues and entirely
concentrate on the discrete aspects of the problem. However, in our approach,
these issues only come into play in three situations, in which (to our experience)
the dirtiness of the data has little impact:

- in *very* conservative tests for wrong neighborhoods (Sect. 3.4).
- in direct comparisons of two conflicting candidates for neighborhoods, that is,
  the assumption that both neighborhoods belong to the correct output leads
  to a contradiction, and hence the "less plausible" one must be eliminated
  (Sects. 3.3 and 3.6).
- in a final *clean-up* (Sect. 3.7).

1.4 Overview

In Sect. 2, we will briefly introduce the practical background and review the
related literature. The algorithm is introduced in Sect. 3. Then, in Sect. 4, we

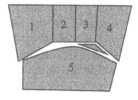 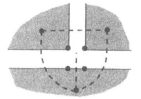

Fig. 7. mesh element 5 is neighbored with mesh elements 1 and 2 and (partially) with 4, but not with 3 because of the small triangle in between.

Fig. 8. dropping the perpendiculars of two equivalent points onto the same segment induces an unintended segment, which is not neighbored to any other segment.

will analyse the asymptotic complexity. In Sect. 5, we will introduce the formal framework and prove guarantees for the accuracy of the algorithm. Finally, in Sect. 6, we will report the results of a computational study on real-world instances. In practice, the run time is dominated by the further steps of the CAD process, and the accuracy of the output is much more important. Therefore, we will focus on the accuracy in Sect. 6.

Because of lack of space, all proofs were omitted.

2 Background

This problem arises in the computer-aided design of various kinds of technical devices. The concrete background of our work is the design of motor vehicles. Our research resulted from a cooperative effort with the *Dr. Krause Software GmbH*, Berlin, which sells CAD packages for this application domain. In particular, the finite-element preprocessor ISAGEN [3] contains a solver for the problem considered in this paper. The plan is to integrate our solver in ISAGEN in the next future.

The topology of the model is needed for mathematical techniques such as finite-element methods, which determine the physical properties of the workpiece modeled by the mesh. Moreover, it is already needed in the preprocessing phase: for example, for validity checking, model repair, and slicing. See [5] for a detailed discussion. Furthermore, the topology is needed in another step of the preprocessing phase: the refinement of the mesh into conformant quadrilaterals. In that, *conformant* means that two mesh elements are neighbored either along a whole side or at a single corner, or not at all. Figure 2 shows a conformant mesh. This refinement step is the topic of [4], and Fig. 2 shows the result of the algorithm in [4] for the model in Fig. 1.

However, many instances are only available in formats which do not contain the neighborhoods. This is the reason why the problem considered in this paper is of practical interest. As pointed out above, the problem is highly ill-posed. Basically, there are three reasons for that.

- Many instances are interactively constructed step by step, that is, element by element. This design process is difficult and tedious and may take several weeks. Due to the restricted freedom in defining the geometric details of a mesh element, it is not always possible to exactly fit two neighbored mesh elements into each other, if the neighborhood between these two mesh elements is not conformant. Restricting the freedom even further to conformant meshes would cause major redesign steps during the interactive design process and significantly increase the development time. Consequently, CAD systems allow to design non-conformant meshes, although this may result in poor, sloppy designs. It is the CAD system's task to guess the designers' intentions.

- Many instances have been designed by automatic tools, for example, from the output of a tomographer. Errors in the processing of this data are unavoidable.
- Many instances have not been created in the format in which they are stored, but were converted from other formats. Typically, these formats are not isomorphic. Hence, the conversion cannot map geometric details in a one-to-one fashion, but must simulate the features of the original format using the features of the target format. This may result in poorer quality. In particular, this occurs quite frequently when the original format is more flexible than the target format.

See [2] for a more detailed discussion of these issues.

Existing approaches. To the best of the authors' knowledge, the literature on CAD model repair focuses entirely on simpler data formats. For example, the *StereoLithography Interface Specification* [6], which is a common data exchange standard for solid freeform fabrication, only allows conformant, plane triangles. [2] summarizes and discusses the errors caused by converting CAD data to the STL format. Apparently, these errors are rather harmless and can be corrected by simple rounding techniques.

Nonetheless, many commercial CAD packages are faced with the general problem and must have implemented solvers for this problem. To our knowledge, all of these solvers attack the problem as follows:[1] a certain *distance function* is defined, which measures the distance of two sides of two different mesh elements, and two sides of two different mesh elements are regarded as neighbored if and only if the distance is smaller than a certain threshold value.

This approach is very problematic: no threshold value accurately separates the correct neighborhoods from the incorrect ones. Typically, the threshold value may be influenced by the user of a CAD package. This is necessary because, apparently, no threshold value is appropriate for all instances. However, in general, even this opportunity for manual interventions does not suffice, because even single instances do not allow globally valid threshold values. Figure 7 demonstrates this: except for the small triangle between mesh elements 3, 4, and 5, the figure is perfectly mirror-symmetric.

---

[1] This knowledge results from personal correspondence with our associates in the co-operative effort mentioned above and with other engineers.

Barequet and Sharir consider a similar problem, where large clusters of mesh elements and the topology inside the clusters are given and only the neighborhoods between these clusters must be filled in [1]. In the authors' own words, "there was a large enough range of valid ε settings" (that is, of threshold values for their distance function, which only measures the distances of pairs of points). Therefore, the data underlying the experimental results in [1] are definitely less dirty than the data with which we are faced.

3 The Algorithm

Our algorithm consists of several stages, which we consider one after the other in the following subsections. The algorithm relies on the following assumption.

Assumption 1. *No two segments of the same mesh element are neighbored.*

The rationale behind Assumption 1 is as follows. Suppose that two segments, s_1 and s_2, of the same mesh element x are neighbored. Let $p_1 \in s_1$ and $p_2 \in s_2$ be two points which correspond to each other. Then the geodetic line from p_1 to p_2 on x is a parabola with maximum width at p_1 and p_2. Hence, if p_1 and p_2 are to be identified, the two branches of the geodetic line should also be identified, which means that the mesh element is helplessly degenerate. In such a case, it is preferable that the algorithm produces a wrong output, because then a well-prepared graphical visualization (cf. Theorem 7) gives the designer a chance to identify and correct this degenerate situation.

3.1 Overestimating the Incident Mesh Elements

In the first stage, we compute a set of pairs of mesh elements, which are our "candidates" and are intended to overestimate the actual pairs of neighbored mesh elements. The general approach is as follows: we define a set $\ell_1, \ldots, \ell_\kappa$ of straight lines in $\mathbb{R}^3$, a set $f_1, \ldots, f_\mu : \mathbb{R}^3 \to \mathbb{R}^3$ of congruential rotation functions, and a discrete threshold value $\theta \in \mathbb{N}$. For $i \in \{1, \ldots, \kappa\}$ and $j \in \{1, \ldots, \mu\}$, we compute the orthogonal projection of each mesh element onto each straight line $f_j(\ell_i)$. A pair of mesh elements is a candidate, if for all $j \in \{1, \ldots, \mu\}$, the projections of these two mesh elements overlap on at least θ of the lines $f_j(\ell_1), \ldots, f_j(\ell_\kappa)$.

For our computational experiments, we used $\kappa = 15$, $\mu = 5$, and $\theta = 10$. This choice is justified by experience: a significantly smaller or larger value of κ does not seem to allow an appropriate threshold value θ; the threshold value seems to be best chosen significantly larger than $\kappa/2$ and significantly less than κ; finally, a larger value of μ does not seem to make a difference and hence results in a waste of time.

We applied evolution strategies to choose $\ell_1, \ldots, \ell_{15}$ and $f_1, \ldots, f_5$. The straight lines $\ell_1, \ldots, \ell_{15}$ were determined using the following objective function: for each triple, ℓ_i, ℓ_j, and ℓ_k, we determine the angle α_{ijk} between ℓ_k and the plane spanned by ℓ_i and ℓ_j. Let β_{ij} denote the fourth smallest value α_{ijk} over all $k = 1, \ldots, 15$, $k \neq i, j$. Then the value of the objective function is defined as $\min\{\beta_{ij} : i, j = 1, \ldots, 15, i \neq j\}$. This evolution strategy yielded the following result, on which Theorem 9 in Sect. 5 relies:

Fact 2. *For every pair ℓ_{i_1}, ℓ_{i_2}, there are at most three straight lines ℓ_{i_3}, $i_3 \neq i_1, i_2$, such that the angle between ℓ_{i_3} and the plane spanned by ℓ_{i_1} and ℓ_{i_2} is 18 degrees or smaller.*

To determine $f_1, \ldots, f_5$, we took the minimum of all $\binom{15 \cdot 5}{2}$ angles between pairs of straight lines $f_j(\ell_i)$ as the objective function of the evolution strategy.

3.2 Overestimating the Incident Sides and Segments of Mesh Elements

First we compute a list of pairs of sides, which are intended to overestimate the real neighborhoods of sides of mesh elements. Let (x_1, x_2) be one of the pairs of mesh elements determined in the first stage, let $s_1^1, \ldots, s_k^1$ be the sides of x_1, and let $s_1^2, \ldots, s_l^2$ be the sides of x_2. The list of candidate sides consists of all (s_i^1, s_j^2) for all such pairs (x_1, x_2). Let $(s_1, s_2), (s_3, s_4), \ldots, (s_{2k-1}, s_{2k})$ be an enumeration of this list. For $i \in \{1, \ldots, k\}$, we approximately compute the pair (s'_{2i-1}, s'_{2i}) of segments on which s_{2i-1} and s_{2i} are really neighbored. This is simply done by projecting s_{2i-1} onto s_{2i} and vice versa (in the final clean-up, Sect. 3.7, the segments are re-computed more accurately).

3.3 Resolving Conflicts between Candidate Neighborhoods

Again let (x_1, x_2) be a pair of mesh elements determined in the first stage. If we knew that at most one pair of sides of x_1 and x_2 is really neighbored, then we could choose the pair which is geometrically "most plausible." However, Fig. 6 demonstrates that more than one pair of sides may be incident. Therefore, at this stage, we are much more conservative in removing pairs of sides from the initial list of candidates.

More specifically, assume that (s'_{2i-1}, s'_{2i}) and (s'_{2j-1}, s'_{2j}) are two of the pairs of segments constructed in Stage 3.2, that s'_{2i-1} and s'_{2j-1} belong to the same mesh element x_1, s'_{2i} and s'_{2j} belong to the same mesh element x_2, and $s'_{2i-1} \cap s'_{2j-1} \neq \emptyset$. Then we remove the "less plausible" pair from the list of candidates. The plausibility of a pair of segments is measured according to Definition 3 below.

In the following, $\|\cdot\|$ denotes the Euclidean norm in $\mathbb{R}^3$. Moreover, for $p \in \mathbb{R}^3$ and for a finite, closed segment s of a parabola in $\mathbb{R}^3$, $d(p, s)$ denotes the minimum Euclidean distance from p to any point on s. Moreover, let $p(s)$ be the internal point of s "in the middle" (more precisely, in the middle of the parameter values of the endpoints with respect to the parameterization of the parabola). Roughly speaking, Definition 3 is intended to approximately relate the distances of endpoints of two segments, the distance between the "middle points" of the segments, and the lengths of the segments to each other. In Definition 3, corresponding endpoints are grouped by taking the minimum of the Euclidean distances.

Definition 3 (weighted distance and plausibility). Let s and s' be closed segments of parabolas in $\mathbb{R}^3$. The *weighted distance* $w(s, s')$ of (s, s') is defined as

$$\min\{\|h(s) - h(s')\|, \|h(s) - t(s')\|\} + \min\{\|t(s) - h(s')\|, \|t(s) - t(s')\|\}$$

$$+ \frac{1}{3}d(p(s), s') + \frac{1}{3}d(p(s'), s)$$

and the plausibility as
$$\frac{w(s, s')}{\|h(s) - t(s)\| + \|h(s') - t(s')\|} \ .$$

3.4 Conservative Error Corrections

The list of pairs of segments constructed in the previous stage is further reduced by applying a few checkers. All checkers are parameterized, and we chose all parameters such that the respective check is *very* conservative. In fact, we have determined all of these parameters such that all correct incident pairs of segments in all real-world instances available to us pass each of these checks, and none of these pairs comes close to the limits. A candidate pair of segments is dropped if at least one of these tests fails. We omit the details of the checkers because of lack of space.

3.5 Computing the First Intermediate Result

Let $(s''_1, s''_2), \ldots, (s''_{2m-1}, s''_{2m})$ be the pairs of segments in $(s'_1, s'_2), \ldots, (s'_{2k-1}, s'_{2k})$ that have passed all tests in Stage 3.4. Clearly, for $i \in \{1, \ldots, m\}$, s''_{2i-1} and s''_{2i} belong to different mesh elements. However, in general, the segments $s''_1, \ldots, s''_{2m}$ need not even be pairwise disjoint. We can partition each side of each mesh element into closed, openly disjoint segments such that all points in the open interior of such a segment are shared exactly by the same subset of $\{s''_1, \ldots, s''_{2m}\}$. In particular, we consider the coarsest partition, which is unique. Let S_1 denote the union of these partition sets over all sides of all mesh elements. The set S_1 is our first guess at S_0. Our first guess E_1 at E_0 is defined as follows: a pair (s, s') of segments in S_1 belongs to E_1 if and only if there is $i \in \{1, \ldots, m\}$ such that $s \subseteq s''_{2i-1}$ and $s' \subseteq s''_{2i}$ (or vice versa).

3.6 Fixing the Transitivity

The binary relation E_1 is symmetric (and trivially made reflexive), but not necessarily transitive. In this stage, we construct a new intermediate result

(S_2, E_2) from (S_1, E_1), where $S_2 = S_1$ and $E_2 \subseteq E_1$. The set E_2 is an inclusion-maximal transitive subset of E_1, that is, for (s_1, s_2), $(s_1, s_3) \in E_2$, we also have $(s_2, s_3) \in E_2$. So consider s_1, s_2, $s_3 \in E_1$ such that (s_1, s_2), $(s_1, s_3) \in E_1$, but $(s_2, s_3) \notin E_1$. To resolve this conflict, either (s_1, s_2) or (s_1, s_3) must be dropped. Here is our rule for deciding which one to drop:

1. if there is $s_4 \in S_1$ such that $(s_3, s_4) \in E_1$, s_4 belongs to the same mesh element as s_2, and $s_4 \neq s_2$, then we drop (s_1, s_3);
2. otherwise, we drop the less plausible pair (cf. Definition 3).

Remark. The first rule applies quite often. We have found constellations in real-world instances where the first rule applies and drops the more plausible pair (unfortunately, the crucial properties of these constellations cannot be shown in a two-dimensional figure). However, in all of these cases, the first rule was right. This is not surprising, because the first rule makes use of discrete, structural information, whereas the second rule relies on mere numerical computations.

Figure 7 demonstrates the rationale behind the first rule: let T denote the triangle between elements 3, 4, and 5. If (S_1, E_1) actually overestimates the real neighborhoods as intended, elements 1, 2, and 5 are pairwise neighbored in (S_1, E_1). Since the situation is perfectly mirror-symmetric except for T, elements 3, 4, 5, and T are also pairwise neighbored in (S_1, E_1). More precisely, there are segments $s_1, \ldots, s_4 \in S_1$ such that s_1 belongs to element 3, s_3 to element 5, s_1 and s_3 face T, and s_2 and s_4 are the sides of T facing element 3 and 5, respectively. Then we have (s_1, s_2), (s_1, s_3), $(s_3, s_4) \in E_1$. Hence, the first rule applies and removes the wrong neighborhood (s_1, s_3), even if, for whatever geometric reasons, (s_1, s_3) is more plausible than (s_1, s_2).

3.7 Clean-Up

In the previous stages (notably Sect. 3.2), we determined the endpoints of segments in S_1 and S_2 only approximately, because we did not care much about the numerical problems inherent in operations such as dropping a perpendicular. In fact, our experience suggests that determining the endpoints at an early stage easily results in numerical inconsistencies, which are hard to resolve later on. Therefore, we postpone this task to the last stage and use the structural information in (S_2, E_2) for that.

Let P, V, and V' be defined according to Definition 1 for $(S, E) = (S_2, E_2)$. The numerical inaccuracies have a strong impact on S_2: dropping perpendiculars from two V'-equivalent points onto the same side induces a small segment between the endpoints of the two perpendiculars (see Fig. 8). Therefore, the clean-up essentially amounts to finding and eliminating these segments.

After computing V', it is very easy to find all candidates for elimination: $s \in S_2$ has to be eliminated if and only if $(t(s), h(s)) \in V'$. Eventually, we compute the final endpoints of perpendiculars: for each equivalence class of endpoints in V', we compute the center of gravity of all its elements. From this point, we drop a perpendicular to every side whose open interior contains an element of this equivalence class. This operation defines the final endpoints of segments and hence the final solution.

4 Complexity

Remember that n denotes the number of mesh elements in the input. In the first stage, all mesh elements are projected onto $15 \cdot 5$ straight lines. For $i \in \{1, \ldots, 15\}$, $j \in \{1, \ldots, 5\}$, let $n_{i,j}$ denote the number of pairs of mesh elements whose projections overlap on $f_j(\ell_i)$. For $j \in \{1, \ldots, 5\}$, N_j denotes the tenth smallest value out of $n_{1,j}, \ldots, n_{15,j}$, and $N = \max\{|E_1|, \max\{N_j : j = 1, \ldots, 5\}\}$. For $s \in S_1$, M_s denotes the *degree* of s in E_1, which is defined as $|\{s' \in S_1 : (s, s') \in E_1\}|$. Let $K = |S_1|$. For convenience, we will often write $\{M_1, \ldots, M_K\}$ instead of $\{M_s : s \in S_1\}$.

Theorem 4. *The asymptotic complexity of the algorithm is* $\mathcal{O}(n \log n + N \log N + \sum_{i=1}^{K} M_i^2)$.

As mentioned in the introduction, N and n are strongly linearly correlated in the real-world instances available to us. In fact, although a particular value $n_{i,j}$ may be $\Theta(n^2)$ (for example, when all mesh elements are contained in the plane orthogonal to $f_j(\ell_i)$), $n_{i,j} \notin \mathcal{O}(n)$ seems next to impossible for the tenth smallest value out of $n_{1,j}, \ldots, n_{15,j}$. Therefore, it is reasonable to assume $N \in \mathcal{O}(n)$. Moreover, it might be evident that $M_1, \ldots, M_K \in \mathcal{O}(1)$ is a reasonable practical assumption.

Corollary 5. *If* $N \in \mathcal{O}(n)$ *and* $M_1, \ldots, M_K \in \mathcal{O}(1)$, *the asymptotic complexity is* $\mathcal{O}(n \log n)$.

5 Accuracy

As mentioned in Sect. 1.3, we aim at an output (S, E) in which the errors can be easily checked by eye. Our correctness check is simple: we use different colors, $c_1, c_2, c_3, \ldots$, to visualize the segments in S, and for each equivalence class $e \in S/E$, the segments in e are assigned color $c_{|e|}$. The check by eye succeeds if and only if each $s \in S$ is assigned the color that is expected by the human observer.

We can show that any *systematical overestimation* of (S_0, E_0) fulfills this condition. Formally, this means the following.

Definition 6 (overestimation). Let (S, E) and (S', E') be two feasible outputs. We say that (S, E) is an *overestimation* of (S', E'), if there is a surjective mapping $f : S \to S'$ such that $s \subseteq f(s)$ for $s \in S$ and the image $f(e) = \bigcup_{s \in e} f(s)$ of an equivalence class $e \in S/E$ is the union of a selection of equivalence classes in S'/E'.

Theorem 7. *If (S, E) is an overestimation of (S_0, E_0), then the segments on which the colorings of (S, E) and (S_0, E_0) differ are exactly the segments on which (S, E) and (S_0, E_0) themselves differ.*

Corollary 8. *Since the algorithm produced an (irreducible) overestimation of (S_0, E_0) for every real-world instance in Table 1, the errors in the solution to any of them can be found by eye.*

From a practical point of view, Corollary 8 seems to be the best one can hope for.

Next we formulate sufficient conditions for our algorithm to deliver an overestimation of (S_0, E_0). In Theorem 9 below, we will give a sufficient condition for the first stage to deliver an overestimation of the set of neighbored mesh elements. The second stage is correct if no two segments of the same mesh element are neighbored (cf. Assumption 1). The third stage is correct if, in case of a conflict as described in Sect. 3.3, the plausibility distinguishes the correct neighborhood from the wrong one. Sufficient conditions for the third stage are obvious. The fourth stage is a mere "format transformation" and does not require prerequisites. The conditions for the fifth and the sixth stage are also obvious.

In the following, we distinguish between points and directions in $\mathbb{R}^3$, and directions are indicated by arrows. Moreover, $\alpha(\vec{u_1}, \vec{u_2})$ denotes the angle between directions $\vec{u_1}$ and $\vec{u_2}$.

Theorem 9. *For $(s_1, s_2) \in E_0^+$, let $\vec{v_1} = h(s_2) - t(s_1)$ and $\vec{v_2} = t(s_2) - h(s_1)$; on the other hand, for $(s_1, s_2) \in E_0^-$, let $\vec{v_1} = h(s_2) - h(s_1)$ and $\vec{v_2} = t(s_2) - t(s_1)$. The first stage delivers all pairs of neighbored mesh elements (and potentially many more pairs), if for $(s_1, s_2) \in E_0$, the angle $\alpha(\vec{v_1}, \vec{v_2})$ is at least 162 degrees.*

Remark. The real-world instances listed in Table 1 contain 33 correct neighborhoods in total for which $\alpha(\vec{v_1}, \vec{v_2})$ is less than 162 degrees.

6 Computational Results

We have implemented a variant of the algorithm in C++ on Sun workstations using the GNU compiler of the Free Software Foundation, Inc. For technical reasons, the implementation differs from the algorithm described in Sect. 3. For example, the intermediate results are not really constructed as data structures, but evaluated in a pipelining process to reduce the space requirements.

Table 1 shows the results of a computational study. The study is based on a reference set of neighborhoods, which we regard as correct. This set has been compiled in a long-term process and was originally intended to fix errors in conformant refinements produced in [4] that were not caused by the algorithm, but by missing neighborhoods.

The first 15 instances stem from our concrete application and consist mainly of quadrilaterals. The other instances have been collected from various public sources. Instances nos. 16-20 consist solely of plain mesh elements, which are bounded by straight sides. These instances are conformant (nonetheless, instance 16 is helplessly dirty). Moreover, instances nos. 17-20 are entirely composed of triangles. For each instance, this table gives the number of mesh elements in the input (#elements), the number of correct neighbored pairs of segments in the reference data (#pairs), and the number of pairs falsely deliverered by our algorithm (#wrong); no correct pair was missing.

The average run time of the implementation of the algorithm taken over all of these instances is less than one minute on a Sparc Ultra 4000/5000 (incl. input/output). As mentioned in the introduction, these run times are negligible, because the finite-element method takes orders of magnitude more time.

References

1. Gill Barequet and Micha Sharir. Filling gaps in the boundary of a polyhedron. *Computer Aided Geometric Design*, 12:207–229, 1995.
2. Georges M. Fadel and Chuck Kirschman. Accuracy issues in CAD to RP translations. In *Internet Conference on Rapid Product Development*. MCB University Press, 1995.
   http://www.mcb.co.uk/services/conferen/dec95/rapidpd/fadel/fadel.html.
3. Günter Krause. Interactive finite element preprocessing with ISAGEN. *Finite Element News*, 15, 1991.
4. Rolf H. Möhring, Matthias Müller-Hannemann, and Karsten Weihe. Using network flows for surface modelling. In *Proceedings of the 6th Annual ACM-SIAM Symposium on Discrete Algorithms, SODA'95*, pages 350–359, 1995.
5. Stephen J. Rock and Michael J. Wozny. Generating topological information from a 'bucket of facets'. In H.L. Marcus and et al., editors, *Solid Freeform Fabrication Symposium Proceedings*, pages 251–259, 1992. http://cat.rpi.edu/~rocks/PUBS/.
6. StereoLithography interface specification (STL), 1989. 3D Systems Publication, Inc., Valencia, CA.

Table 1. computational results for real-world instances.

	#elements	#pairs	#wrong		#elements	#pairs	#wrong
1.	34	55	0	11.	24	40	0
2.	103	210	3	12.	179	409	2
3.	295	676	44	13.	154	321	43
4.	68	128	2	14.	237	446	0
5.	251	573	1	15.	158	341	14
6.	156	321	0	16.	156	261	73
7.	342	823	15	17.	1242	1844	0
8.	131	222	0	18.	1243	1786	0
9.	608	1280	4	19.	1025	1529	10
10.	530	1205	2	20.	1417	2126	0
				Total:	8353	14596	213

Author Index

Springer
and the
environment

At Springer we firmly believe that an international science publisher has a special obligation to the environment, and our corporate policies consistently reflect this conviction.

We also expect our business partners – paper mills, printers, packaging manufacturers, etc. – to commit themselves to using materials and production processes that do not harm the environment. The paper in this book is made from low- or no-chlorine pulp and is acid free, in conformance with international standards for paper permanency.

Springer

Lecture Notes in Computer Science

For information about Vols. 1–1211

please contact your bookseller or Springer-Verlag